Joseph Rahi

quantum physics

Stephen Gasiorowicz
University of Minnesota

quantum physics

JOHN WILEY & SONS, New York • Chichester • Brisbane • Toronto

Library of Congress Cataloging in Publication Data:

Gasiorowicz, Stephen.
  Quantum physics.

  Bibliography: p.
  1. Quantum theory.   I. Title

QC174.12.G37   530.1'2   73-22376
ISBN 0-471-29280-X

Printed in the United States of America

10 9 8

To Hilde

preface

This book is intended to serve as an introduction to quantum physics. In writing it, I have kept several guidelines in mind.

1. First, it is helpful for the development of intuition in any new field of study to start with a base of detailed knowledge about simple systems. I have therefore worked out a number of problems in great detail, so that the insight thus obtained can be used for more complex systems.

2. Every aspect of quantum mechanics has been helpful in understanding some physical phenomenon. I have therefore laid great stress on applications at every stage of the development of the subject. Although no area of quantum physics is totally developed, my intention is to bridge the gap between a modern physics course and the more formal development of quantum mechanics. Thus, many applications are discussed, and I have stressed order-of-magnitude estimates and the importance of numbers.

3. In keeping with the level of the book, the mathematical structure has been kept as simple as possible. New concepts, such as operators, and new mathematical tools necessarily make their appearance. I have dealt with the former more by analogy than by precise definition, and I have minimized the use of new tools insofar as possible.

In approaching quantum theory, I chose to start with wave mechanics and the Schrödinger equation. Although the state-vector approach gets at the essential structure of quantum mechanics more rapidly, experience has shown that the use of more familiar tools, such as differential equations, makes the theory more accessible and the correspondence with classical physics more transparent.

The book probably contains a little more material than can comfortably be covered in one year. The basic material can be covered in one academic quarter.

It consists of Chapters 1 to 6, 8, and 9, in which the motivation for a quantum theory, the Schrödinger equation, and the general framework of wave mechanics are covered. A number of simple problems are solved in Chapter 5, and their relevance to physical phenomena is discussed. The generalization to many particles and to three dimensions is developed. The second-quarter material deals directly with atomic physics problems and uses somewhat more sophisticated tools. Here we discuss operator methods (Chapter 7), angular momentum (Chapter 11), the hydrogen atom (Chapter 12), operators, matrices, and spin (Chapter 14), the addition of angular momenta (Chapter 15), time-independent perturbation theory (Chapter 16), and the real hydrogen atom (Chapter 17). This material prepares the student to cope with a large variety of problems that are discussed during the third and last quarter. These problems include the inter-action of charged particles with a magnetic field (Chapter 13), the helium atom (Chapter 18), problems in the radiation of atoms and related topics (Chapters 22 and 23), collision theory (Chapter 24), and the absorption of radiation in matter (Chapter 25). This material is supplemented by a more qualitative discussion of the structure of atoms and molecules (Chapters 19 to 21). The last chapter on elementary particles and their symmetries serves the dual purpose of describing some of the recent advances on that frontier of physics and of showing how the basic ideas of quantum theory have found applicability in the domain of very short distances.

Several topics arise naturally as digressions in the development of the subject matter. Instead of lengthening some long chapters, I have placed this material in a separate "Special Topics" section. There, relativistic kinematics, the equivalence principle, the WKB approximation, a detailed treatment of lifetimes, line widths and scattering resonances, and the Yukawa theory of nuclear forces are discussed. For the same reason, a brief introduction to the Fourier integral, the Dirac delta function, and some formal material dealing with operators have been placed in mathematical appendices at the end of the book.

I am indebted to my colleagues at the University of Minnesota, especially Benjamin Bayman and Donald Geffen, for many discussions on the subject of quantum mechanics. I am grateful to Eugen Merzbacher, who read the manu-script and made many helpful suggestions for improvements. I also thank my students in the introductory quantum mechanics course that I taught for several years. Their evident interest in the subject led me to the writing of the supple-mental notes that later became this book.

Stephen Gasiorowicz

contents

chapter 1

The Limits of Classical Physics

The end of the nineteenth century and the beginning of the twentieth witnessed a crisis in physics. A series of experimental results required concepts totally incompatible with classical physics. The development of these concepts, in a fascinating interplay of radical conjectures and brilliant experiments, led finally to the *quantum theory*.[1] Our objective in this chapter is to describe the background of this crisis and, armed with hindsight, to expose the new concepts in a manner that, while not historically correct, will make the transition to quantum theory less mysterious for the reader. The new concepts, *the particle properties of radiation, the wave properties of matter*, and *the quantization of physical quantities* will emerge in the phenomena discussed below.

A. Black Body Radiation

When a body is heated, it is seen to radiate. In equilibrium the light emitted ranges over the whole spectrum of frequencies v, with a spectral distribution that depends both on the frequency or, equivalently, on the wavelength of the light λ, and on the temperature. One may define a quantity $E(\lambda, T)$, the emissive power, as the energy emitted at wavelength λ per unit area, per unit time. Theoretical research in the field of thermal radiation began in 1859 with the work of Kirchhoff, who showed that for a given λ, the ratio of the emissive power E to the absorptivity A, defined as the fraction of incident radiation of wavelength λ that is absorbed by the body, is the same for all bodies. Kirchhoff considered two emitting and absorbing parallel plates and showed from the equilibrium condition that the energy emitted was equal to the energy absorbed (for each λ), that the ratios E/A must be the same for the two plates. Soon

---

[1] An interesting account of the development of quantum theory may be found in M. Jammer, *The Conceptual Development of Quantum Mechanics*, McGraw-Hill, New York, 1966.

thereafter, he observed that for a *black body*, defined as a surface that totally absorbs all radiation that falls on it, so that $A = 1$, the function $E(\lambda, T)$ is a universal function.

In order to study this function it is necessary to obtain the best possible source of black body radiation. A practical solution to this problem is to consider the radiation emerging from a small hole in an enclosure heated to a temperature T. Given the imperfections in the surface of the inside of the cavity, it is clear that any radiation falling on the hole will have no chance of emerging again. Thus the surface presented by the hole is very nearly "totally absorbing," and consequently the radiation coming from it is indeed "black body radiation." Provided the hole is small enough, this radiation will be the same as that which falls on the walls of the cavity. It is therefore necessary to understand the distribution of radiation inside a cavity whose walls are at a temperature T.

Kirchhoff showed that the second law of thermodynamics requires that the radiation in the cavity be isotropic, that is, that the flux be independent of direction; that it be homogeneous, that is, the same at all points; and that it be the same in all cavities at the same temperature—all of this for each wavelength.[2] The emissive power may, by simple geometric arguments, be shown to be connected with the energy density $u(\lambda, T)$ inside the cavity. The relation is

$$u(\lambda, T) = \frac{4E(\lambda, T)}{c} \qquad (1\text{-}1)$$

The energy density is the quantity of theoretical interest, and further understanding of it came in 1894 from the work of Wien, who, again using very general arguments,[3] showed that the energy density had to be of the form

$$u(\lambda, T) = \lambda^{-5} f(\lambda T) \qquad (1\text{-}2)$$

with f still an unknown function of a single variable. If, as is convenient, one deals instead with the energy density as a function of frequency, $u(\nu, T)$, then it follows from the fact that

$$u(\nu, T) = u(\lambda, T) \left| \frac{d\lambda}{d\nu} \right|$$

$$= \frac{c}{\nu^2} u(\lambda, T) \qquad (1\text{-}3)$$

---

[2] These matters are discussed in many textbooks on modern physics and statistical physics. References can be found at the end of this chapter.

[3] Wien considered a perfectly reflecting spherical cavity contracting adiabatically. The redistribution of the energy as a function of λ has to be caused by the Doppler shift on reflection. See Chapter V in F. K. Richtmyer, E. H. Kennard, and J. N. Cooper *Introduction to Modern Physics*, McGraw-Hill, New York, 1969.

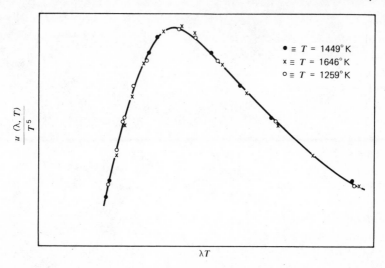

Fig. 1-1. Experimental verification of Eq. 1-2 in the form $u(\lambda, T)/T^5 = $ a universal function of λT.

that the Wien law reads

$$u(\nu, T) = \nu^3 g\left(\frac{\nu}{T}\right) \tag{1-4}$$

The implications of this law, which was confirmed experimentally (Fig. 1.1), are twofold:

1. Given the spectral distribution of black body radiation at one temperature, the distribution at any other temperature can be found with the help of the expressions given above.

2. If the function $f(x)$—or, equivalently, the function $g(x)$—has a maximum for some value of $x > 0$, then the wavelength λ_{max} at which the energy density, and hence the emissive power, has its maximum value, has the form

$$\lambda_{max} = \frac{b}{T} \tag{1-5}$$

where b is a universal constant.

Wien used a model (of no interest, except to the historian) to predict a form for $g(\nu/T)$. The form was

$$g(\nu/T) = Ce^{-\beta\nu/T} \tag{1-6}$$

and, remarkably enough, this form, containing two adjustable parameters, fit the high frequency (low wavelength) data very well. The formula is not, how-

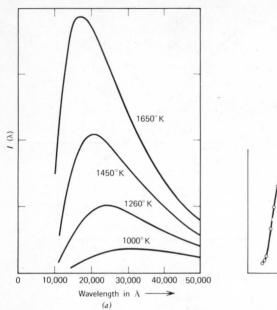

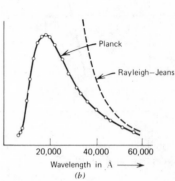

Fig. 1-2. (*a*) Distribution of power radiated by a black body at various temperatures. (*b*) Comparison of data at 1600°K with Planck formula and Rayleigh-Jeans formula.

ever, in accord with some very general notions of classical physics. Rayleigh, in 1900, derived the result

$$u(\nu, T) = \frac{8\pi\nu^2}{c^3} kT \tag{1-7}$$

where k is Boltzmann's constant, $k = 1.38 \times 10^{-16}$ erg/deg and c is the velocity of light, $c = 3.00 \times 10^{10}$ cm/sec. The ingredients that went into the derivation were (1) the classical law of equipartition of energy, according to which the average energy per degree of freedom for a dynamical system in equilibrium is, in this context,[4] kT, and (2) the calculation of the number of modes (i.e., degrees of freedom) for electromagnetic radiation with frequency in the interval $(\nu, \nu + d\nu)$, confined in a cavity.[5]

---

[4] The equipartition law predicts that the energy per degree of freedom is $kT/2$. For an oscillator—and the modes of the electromagnetic field are simple harmonic oscillators—a contribution of $kT/2$ from the kinetic energy is matched by a like contribution from the potential energy, giving kT.

[5] We will need this result again, and derive it in Chapter 23. The number of modes is $4\pi\nu^2/c^3$, further multiplied by a factor of 2 because transverse electromagnetic waves correspond to two-dimensional harmonic oscillators.

The Rayleigh-Jeans law (1-7) (Jeans made a minor contribution to its derivation) does not agree with experiment at high frequencies, where the Wien formula works, though it does fit the experimental curve at low frequencies (Fig. 1.2). The Rayleigh-Jeans law cannot, on general grounds, be correct, since the total energy density (integrated over all frequencies) is predicted to be infinite!

In 1900, Max Planck found a formula by an ingenious interpolation between the high-frequency Wien formula and the low-frequency Rayleigh-Jeans law. The formula is

$$u(\nu,\ T) = \frac{8\pi h}{c^3} \frac{\nu^3}{e^{h\nu/kT} - 1} \tag{1-8}$$

where h, *Planck's constant*, is an adjustable parameter whose numerical value was found to be $h = 6.63 \times 10^{-27}$ erg sec. This law approaches the Rayleigh-Jeans form when $\nu \to 0$, and reduces to

$$u(\nu,\ T) = \frac{8\pi h}{c^3}\, \nu^3\, e^{-h\nu/kT} \left(1 - e^{-h\nu/kT}\right)^{-1}$$

$$\cong \frac{8\pi h}{c^3}\, \nu^3\, e^{-h\nu/kT} \tag{1-9}$$

when the frequency is large, or, more accurately, when $h\nu \gg kT$. If we rewrite the formula as a product of the number of modes [we obtain this from (1-7) by dividing the energy density by kT] and another factor that can be interpreted as the average energy per degree of freedom

$$u(\nu,\ T) = \frac{8\pi\nu^2}{c^3} \frac{h\nu}{e^{h\nu/kT} - 1}$$

$$= \frac{8\pi\nu^2}{c^3}\, kT\, \frac{h\nu/kT}{e^{h\nu/kT} - 1} \tag{1-10}$$

we see that the classical equipartition law is altered whenever the frequencies are not small compared with kT/h. This alteration in the equipartition law shows that the modes have an average energy that depends on their frequency, and that the high frequency modes have a very small average energy. This effective cut-off removes the difficulty of the Rayleigh-Jeans density formula: the total energy in a cavity of unit volume is no longer infinite. We have

$$U(T) = \frac{8\pi h}{c^3} \int_0^\infty d\nu\, \frac{\nu^3}{e^{h\nu/kT} - 1}$$

$$= \frac{8\pi h}{c^3} \left(\frac{kT}{h}\right)^4 \int_0^\infty \frac{(h\nu/kT)^3\, d(h\nu/kT)}{e^{h\nu/kT} - 1}$$

$$= \frac{8\pi k^4}{h^3 c^3}\, T^4 \int_0^\infty dx\, \frac{x^3}{e^x - 1} \tag{1-11}$$

The integral can be evaluated,[6] and the result is the Stefan-Boltzmann expression for the total radiation energy per unit volume

$$U(T) = aT^4 \tag{1-12}$$

with $a = 7.56 \times 10^{-15}$ erg/cm^3 deg^4, derived much earlier, except for the numerical constant in front, on the basis of thermodynamical reasoning. A departure from the pure equipartition law was not entirely unexpected: one consequence of it was the Dulong-Petit law of specific heats, according to which the product of the atomic (or molecular) weight and the specific heat is a constant for all solids; yet departures from the Dulong-Petit predictions were observed as early as 1872.[7] These departures indicated that the specific heat decreased at lower temperatures.[8]

The unqualified success of his formula drove Planck to search for its origin, and within two months he found that he could derive it by assuming that the energy associated with each mode of the electromagnetic field did not vary continuously (with average value kT) but was an integral multiple of some minimum quantum of energy ε. Under these circumstances a calculation of the average energy associated with each mode, using the Boltzmann probability distribution in a system of equilibrium at temperature T,

$$P(E) = \frac{e^{-E/kT}}{\sum_E e^{-E/kT}} \tag{1-13}$$

led to

$$\bar{E} = \sum_E EP(E)$$

$$= \frac{\sum_{n=0}^{\infty} n\varepsilon\, e^{-n\varepsilon/kT}}{\sum_{n=0}^{\infty} e^{-n\varepsilon/kT}}$$

[6] $\int_0^\infty dx\, x^3 (e^x - 1)^{-1} = \int_0^\infty dx\, x^3 e^{-x} \sum_{n=0}^{\infty} e^{-nx} = \sum_{n=0}^{\infty} \frac{1}{(n+1)^4} \int_0^\infty dy\, y^3 e^{-y} = 6 \sum_{n=1}^{\infty} \frac{1}{n^4} = \frac{\pi^4}{15}$

[7] According to the equipartition law an assembly of N oscillators (and a lattice of atoms with elastic forces between them may be so viewed) will have energy $3NkT$, the factor 3 coming from the fact that the oscillators in a solid are three-dimensional, rather than two-dimensional as for the radiation field in an enclosure. The specific heat for a mole is obtained by differentiating with respect to T and setting $N = N_0$, Avogadro's number, so that $C_v = 3N_0k = 3R$ where $R = 8.28 \times 10^7$ erg/deg.

[8] Specific heats will be discussed very briefly in Chapter 20.

$$= \frac{-\varepsilon \dfrac{d}{dx} \displaystyle\sum_{n=0}^{\infty} e^{-nx}}{\displaystyle\sum_{n=0}^{\infty} e^{-nx}}\Bigg|_{x=\varepsilon/kT}$$

$$= \varepsilon \frac{e^{-x}}{1 - e^{-x}}\Bigg|_{x=\varepsilon/kT}$$

$$= \frac{\varepsilon}{e^{\varepsilon/kT} - 1} \tag{1-14}$$

This agrees with (1-10) provided we make the identification

$$\varepsilon = h\nu \tag{1-15}$$

and do not change the number of modes.

Planck argued that for some unknown reason the atoms in the walls of the cavity emitted radiation in "quanta" with energy $nh\nu$ ($n = 1,2,3, \ldots$), but consistency demanded, as established by Einstein a few years later, that *electromagnetic radiation behaved as if it consisted of a collection of energy quanta with energy $h\nu$.*[9]

The energy carried per quantum is extremely small. For light in the optical range, with, say, $\lambda = 6000$ Å,

$$h\nu = h\,\frac{c}{\lambda} = \frac{6.63 \times 10^{-27} \times 3.00 \times 10^{10}}{6 \times 10^{-5}} \simeq 3.3 \times 10^{-12}\ \text{erg}$$

so that the number of light quanta of this wavelength, emitted by a 100-watt source, say, is

$$N = \frac{100 \times 10^{7}}{3.3 \times 10^{-12}} \cong 3 \times 10^{20}\ \text{quanta/sec}$$

With so many quanta present, it is perhaps not surprising that we do not experience the particle nature of light directly; we shall see that on a macroscopic scale no deviations from classical optics are expected. Nevertheless, Planck's interpretation of his formula radically changes our picture of radiation.

B. The Photoelectric Effect

As successful as the Planck formula was, the conclusion from it of the quantum nature of radiation is hardly compelling. An important contribution to its acceptance came from the work of Albert Einstein, who in 1905 used the

---

[9] For a given frequency ν there may be any integral number of quanta present, and hence the energy can take on the values $nh\nu$, with $n = 0,1,2,3, \ldots$.

concept of the quantum nature of light to explain some peculiar properties of metals, when these are irradiated with visible and ultraviolet light.

In 1887, the photoelectric effect was discovered by Hertz, who, while engaged in his famous experiments on electromagnetic waves, found that the length of the spark induced in the secondary circuit was reduced when the terminals of the spark gap were shielded from the ultraviolet light coming from the spark in the primary circuit. His observations attracted much interest and the following facts were established by further experiments:

1. When polished metal plates are irradiated, they may emit electrons;[10] they do not emit positive ions.

2. Whether the plates emit electrons depends on the wavelength of the light. In general there will be a threshold that varies from metal to metal: only light with a frequency greater than a given threshold frequency will produce a photoelectric current.

3. The magnitude of the current, when it exists, is proportional to the intensity of the light source.

4. The energy of the photoelectrons is independent of the intensity of the light source but varies linearly with the frequency of the incident light.

Although the existence of the photoelectric effect can be understood within the framework of classical electromagnetic theory, since it was known that there were electrons in metals, and one could imagine them to be accelerated by absorption of radiation, the frequency-dependence of the effect is not comprehensible within that framework. The energy carried by an electromagnetic wave is proportional to the intensity of the source, and frequency has nothing to do with it. Furthermore, a classical explanation of the effect, which would have to involve the concentration of the energy deposited on single photoelectrons, would carry with it an implied time delay between the arrival of the radiation and the departure of the electron, the delay being longer when the intensity is decreased. In fact, no such time delays were ever observed, at least none longer than 10^{-9} sec, even with incident radiation of very low intensity.

Einstein considered the radiation to consist of a collection of quanta of energy $h\nu$, where ν is the frequency of the light. The absorption of a single quantum by an electron—a process that may take less time than the upper limit quoted above—increases the electron energy by an amount $h\nu$. Some of the energy must be expended to separate the electron from the metal. This amount, W (called the *work function*), might be expected to vary from metal to metal, but should not depend on the electron energy. The rest is available for the electron kinetic energy, so that on the basis of this picture one expects that

---

· [10] This was established by an e/m measurement.

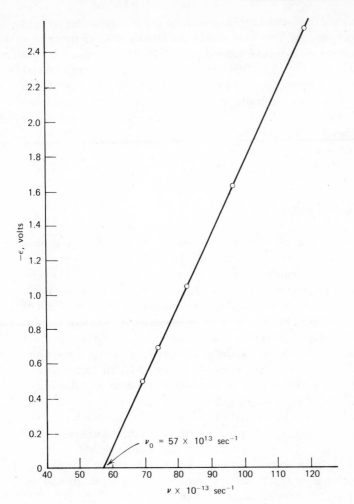

Fig. 1-3. Photoelectric effect data showing a plot of retarding potential necessary to stop electron flow from a metal (lithium), or equivalently, electron kinetic energy, as a function of frequency of the incident light. The slope of the line is h/e.

the following relation between electron velocity v and light frequency ν

$$\tfrac{1}{2}mv^2 = h\nu - W \qquad (1\text{-}16)$$

should hold. The threshold effect and the linear relation between electron kinetic energy and the frequency are contained in this formula. The proportionality of the current and the source intensity can also be understood in terms of these light quanta, or *photons*, as they came to be called: a more intense light source emits more photons, and these in turn can liberate more electrons.

Millikan carried out extensive experiments and established the correctness of the Einstein formula (Fig. 1.3). What Millikan's and the earlier experiments proved was that sometimes light behaves like a collection of particles, and that these "particles" can act individually, so that it is possible to contemplate the existence of a single photon and ask what its properties are. A by-product of these experiments was information about metals. It was found that W was of the order of several electron volts (1 eV = 1.6×10^{-12} erg), and this could be correlated with other properties of the metals.

C. The Compton Effect

The experiment that provided the most direct evidence for the particle nature of radiation is the so-called Compton effect. Compton discovered that radiation of a given wavelength (in the X-ray region) sent through a metallic foil was scattered in a manner not consistent with classical radiation theory. According to classical theory, the mechanism for the effect is the re-radiation of light by electrons set into forced oscillations by the incident radiation, and this leads to the prediction of intensity observed at an angle θ that varies as $(1 + \cos^2 \theta)$, and does not depend on the wavelength of the incident radiation. Compton found that the radiation scattered through a given angle actually consists of two components: one whose wavelength is the same as that of the incident radiation, the other of wavelength shifted relative to the incident wavelength by an amount that depends on the angle (Fig. 1.4). Compton was able to explain the "modified" component by treating the incoming radiation as a beam of photons of energy $h\nu$, with individual photons scattering elastically off individual electrons. In an elastic collision, momentum as well as energy must be conserved, and we must first assign a momentum to the photon. By analogy with relativistic particle kinematics we argue that

$$p = h\nu/c \tag{1-17}$$

The argument is that it follows from the relativistic relation between energy and momentum

$$E = [(m_0 c^2)^2 + (pc)^2]^{1/2} \tag{1-18}$$

where m_0 is the rest mass of the particle, that the velocity at this momentum is

$$v = \frac{dE}{dp} = \frac{pc^2}{E} = \frac{pc^2}{(m_0^2 c^4 + p^2 c^2)^{1/2}} \tag{1-19}$$

For a photon this is always c, and hence the *photon rest mass must be zero*. Thus the relation (1-18) becomes

$$E = pc \tag{1-20}$$

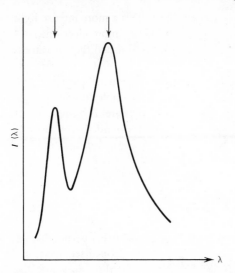

Fig. 1-4. The spectrum of radiation scattered by carbon, showing the unmodified line at 0.7078 Å on the left and the shifted line at 0.7314 Å on the right. The former is the wavelength of the primary radiation.

which yields (1-17) when we substitute $E = h\nu$. One may also derive (1-20) from consideration of the energy and momentum of an electromagnetic wave, but the analogy argument is simpler.

Consider, now, a photon with initial momentum $\mathbf{p}$, incident upon an electron at rest. After the collision, the photon momentum is $\mathbf{p'}$, and the electron recoils with momentum $\mathbf{P}$. Conservation of momentum yields (Fig. 1.5)

$$\mathbf{p} = \mathbf{p'} + \mathbf{P} \tag{1-21}$$

from which it follows that

$$\mathbf{P}^2 = (\mathbf{p} - \mathbf{p'})^2 = \mathbf{p}^2 + \mathbf{p'}^2 - 2\mathbf{p} \cdot \mathbf{p'} \tag{1-22}$$

Energy conservation reads

$$h\nu + mc^2 = h\nu' + (m^2c^4 + P^2c^2)^{1/2} \tag{1-23}$$

where m is the electron rest mass. Hence

$$m^2c^4 + P^2c^2 = (h\nu - h\nu' + mc^2)^2$$
$$= (h\nu - h\nu')^2 + 2mc^2(h\nu - h\nu') + m^2c^4$$

On the other hand (1-22) may be rewritten in the form

$$P^2 = \left(\frac{h\nu}{c}\right)^2 + \left(\frac{h\nu'}{c}\right)^2 - 2\frac{h\nu}{c} \cdot \frac{h\nu'}{c} \cos\theta$$

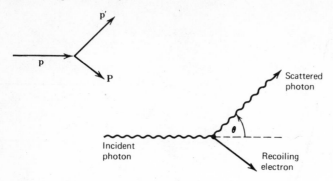

Fig. 1-5. Kinematics for Compton effect.

that is,

$$P^2c^2 = (h\nu - h\nu')^2 + 2(h\nu)(h\nu')(1 - \cos\theta) \tag{-124}$$

where θ is the photon scattering angle. Thus

$$h\nu\nu'(1 - \cos\theta) = mc^2(\nu - \nu')$$

or equivalently

$$\lambda' - \lambda = \frac{h}{mc}(1 - \cos\theta) \tag{1-25}$$

The measurements of the modified component agree very well with the above prediction. The unmodified line is presumably due to the scattering by the whole atom; if m is replaced by the mass of the atom, the shift in the wavelength is very small, since an atom is many thousands times more massive than an electron. The quantity h/mc has the dimensions of a length. It is called the Compton wavelength of the electron, and its magnitude is

$$\frac{h}{mc} \cong 2.4 \times 10^{-10} \text{ cm} \tag{1-26}$$

Measurements of the electron recoil were also made, and these are in agreement with the theory. It was furthermore determined by good time resolution coincidence experiments, that the outgoing photon and the recoil electron appear simultaneously. There is no question of the correctness of the interpretation of the collision as an ordinary "billiard ball" type of collision, that is, of the particlelike behavior of the photon. Since radiation also has wave properties and exhibits interference and diffraction, we might expect some conceptual difficulties. These exist, and we shall discuss them at the end of the chapter.

D. Electron Diffraction

In 1923 De Broglie, guided by the analogy of Fermat's principle in optics, and the least-action principle in mechanics, was led to suggest that the dual wave-particle nature of radiation should have its counterpart in a dual particle-wave nature of matter. Thus particles should have wave properties under certain circumstances, and De Broglie suggested an expression for the wavelength associated with the particle.[11] This is given by

$$\lambda = \frac{h}{p} \qquad (1\text{-}27)$$

where h is Planck's constant, and p is the momentum of the particle. De Broglie's work attracted much attention, and many people suggested that verification could be obtained by observing electron diffraction.[12] The experimental observation of this effect occurred in experiments of Davisson and Germer, who found that in the scattering of electrons by a crystal surface, there was preferential scattering in certain directions.

Figure 1-6 is a simplified picture of what happens. In the scattering of waves by a periodic structure, there will be a phase difference between waves coming from adjacent scattering "planes," whose magnitude is given by $(2\pi/\lambda)\, 2a \sin \theta$. There will be constructive interference whenever this phase difference is equal to $2\pi n$, where n is an integer, that is, when

$$\lambda = \frac{2a \sin \theta}{n} \qquad (1\text{-}28)$$

The interference pattern observed in electron scattering by Davisson and Germer could be correlated with the above formula, provided the association (1-27) was made. This verification constituted a major step in the development of wave mechanics.

The particle diffraction experiments have since been carried out with molecular beams of hydrogen and helium, and with slow neutrons. Neutron diffraction is particularly useful in the study of crystal structure. To get a rough idea of the kind of energies needed for the diffraction experiments, we note that the crystal spacings are of the order of Angstroms. The grating constant in the Davisson-Germer experiment, in which nickel was used, was $a = 2.15$ Å. Hence λ is of the order of 10^{-8} cm, so that $p = h/\lambda \cong 6.6 \times 10^{-19}$ gm cm/sec. Thus for electrons the kinetic energy is $p^2/2m_e = (6.6 \times 10^{-19})^2/$

---

[11] Chapter 2 contains a discussion of wave packets in which the De Broglie relation emerges as a very plausible result.

[12] The history of the verification of De Broglie's conjecture can be found in Jammer, *The Conceptual Development of Quantum Mechanics.*

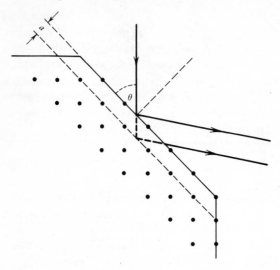

Fig. 1-6. Schematic drawing of electron scattering geometry.

$(2 \times 0.9 \times 10^{-27}) \cong 2.5 \times 10^{-10}$ ergs, and for neutrons the kinetic energy is $p^2/2m_n = (m_e/m_n) \times$ (electron energy) $\cong (1/1840) \times 2.5 \times 10^{-10}$ ergs $\cong 1.3 \times 10^{-13}$ ergs. In terms of the more convenient electron volt, these energies are approximately 160 eV and 0.08 eV, respectively.

On a macroscopic scale, the wave aspects of particles are beyond our ability to observe them. A droplet 0.1 mm in size, moving at 10 cm/sec will have a *De Broglie wavelength* of $\lambda = 6.6 \times 10^{-27}/4 \times 10^{-5} \cong 1.6 \times 10^{-22}$ cm. Since the "size" of a proton is about 10^{-14} cm, clearly there is no way in which the wave properties of an object of dimensions significantly larger than 10^{-4} cm can be observed. As for the particle properties of radiation, it is the smallness of h that determines the classical properties, in the sense that the dual aspects become apparent only when the product of momentum and dimension is of the order of h. We shall see that the formalism of quantum mechanics describes the situation very well.

E. The Bohr Atom

Experiments carried out in 1908 by Geiger and Marsden on the scattering of α particles by thin foils showed significant large angle scattering, totally inconsistent with expectations based on the Thomson model of the atom, according to which electrons were embedded in a continuous distribution of positive charge. Rutherford proposed a new model that accounted for the data:

all of the positive charge and essentially all of the mass of the atom were concentrated in a region that is small compared with the dimensions of the atom, that is, in the *nucleus* of the atom. The electrons, attracted to the nucleus with a $1/r^2$ force, traveled in planetary orbits about it. Although the model explained α particle scattering quantitatively, it faced two insuperable difficulties. Since it implied a periodic motion for the electrons, it could not account for the spectra of radiation from atoms, which did not have the expected harmonic structure (cf. a vibrating string), but instead had the structure

$$\frac{1}{\lambda} = \text{const.}\left(\frac{1}{n_1{}^2} - \frac{1}{n_2{}^2}\right) \tag{1-29}$$

where n_1 and n_2 were integers. It also lacked a mechanism for stabilizing atoms: an electron in a circular or elliptic orbit is constantly accelerating, and according to electromagnetic theory, should be radiating. The constant loss of energy would, within a very short time, (of the order of 10^{-10} sec) lead to the collapse of the atom, with the electrons plunging into the nucleus.

Just two years after this model was proposed, Niels Bohr in 1913 advanced a series of postulates, which, while sharply breaking with classical theory, explained the spectral structure and bypassed the stability problem. Bohr proposed that:

1. The electrons move in orbits restricted by the requirement that the angular momentum be an integral multiple of $h/2\pi$, that is, for circular orbits of radius r, the electron velocity v is restricted by

$$mvr = \frac{nh}{2\pi} \tag{1-30}$$

and furthermore the electrons in these orbits do not radiate in spite of their acceleration. They were said to be in stationary states.

2. Electrons can make discontinuous transitions from one allowed orbit to another, and the change in energy, $E - E'$ will appear as radiation with frequency

$$\nu = \frac{E - E'}{h} \tag{1-31}$$

An atom may absorb radiation by having its electons make a transition to a higher energy orbit.

The consequences of these postulates are very simply deduced for one-electron atoms such as hydrogen, singly ionized helium, and so on, if we deal with the circular orbits.[13] If the nuclear charge is Ze and that of the electron is

[13] When elliptical orbits are allowed, a much richer structure emerges. This will be discussed in Chapter 12.

$-e$, and if the radius of the orbit is r, then, taking the nuclear mass to be infinite, we balance the Coulomb force against the centrifugal force

$$\frac{Ze^2}{r^2} = \frac{mv^2}{r} \tag{1-32}$$

This, when combined with (1-30) leads to

$$v = \frac{2\pi e^2 Z}{hn} \tag{1-33}$$

and

$$r = \frac{1}{4\pi^2} \frac{n^2 h^2}{Ze^2 m} \tag{1-34}$$

The energy is

$$E = \tfrac{1}{2}mv^2 - \frac{Ze^2}{r} = -\frac{2\pi^2 e^4 Z^2 m}{h^2 n^2} \tag{1-35}$$

which, by postulate (2) immediately leads to the general form (1-29) (Fig. 1.7).

Before evaluating these quantities to obtain an idea of their magnitude, we will introduce some notations that will be very useful. First of all, it is $h/2\pi$ rather than h that appears in most formulas in quantum mechanics. We therefore define

$$\hbar = \frac{h}{2\pi} = 1.0545 \times 10^{-27} \text{ erg sec} \tag{1-36}$$

To keep the expressions for the energy simple, we shall deal with the angular frequency ω, rather than ν, where

$$\omega = 2\pi\nu \tag{1-37}$$

Thus (1-31) reads

$$\omega = \frac{E - E'}{\hbar} \tag{1-38}$$

Similarly, the quantum of radiation carries energy

$$E = \hbar\omega \tag{1-39}$$

It is convenient to introduce the "reduced wavelength"

$$\lambdabar = \frac{\lambda}{2\pi} = \frac{c}{\omega} \tag{1-40}$$

so that the De Broglie relation reads

$$p = \frac{\hbar}{\lambdabar} \tag{1-41}$$

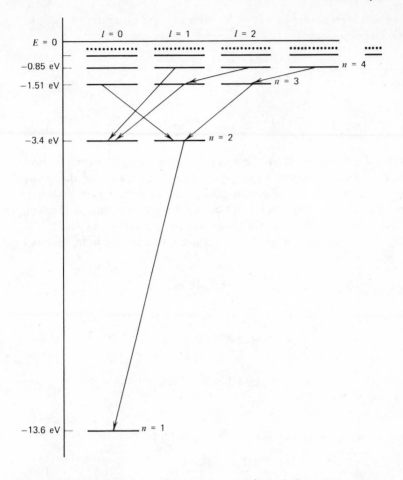

Fig. 1-7. Spectrum for hydrogen atom as derived from Bohr atomic model. The existence of the quantum numbers l emerges from a discussion of elliptical orbits. The lines connecting energy levels represent the dominant atomic transitions.

The Bohr angular momentum quantization condition reads

$$mvr = n\hbar \; (n = 1, 2, 3, \ldots) \tag{1-42}$$

It is also very convenient to introduce the dimensionless "fine structure constant"

$$\alpha = \frac{e^2}{\hbar c} = \frac{1}{137.0388} \tag{1-43}$$

which we will approximate by $1/137$. In terms of these quantities we find the much simpler expressions

$$\frac{v}{c} = \frac{Z\alpha}{n} \qquad r = \frac{n^2}{Z\alpha} \frac{\hbar}{mc} \tag{1-44}$$

and

$$E = -\tfrac{1}{2} mc^2 \frac{(Z\alpha)^2}{n^2} \tag{1-45}$$

Notice that the radius, which has the dimensions of a length, is written in terms of $\hbar/mc$, the reduced Compton wavelength of the electron, and that the energy is written in terms of mc^2. In all atomic calculations we shall express our results in terms of mc^2, $\hbar/mc$, $\hbar/mc^2$, and mc for energy, length, time, and momentum, respectively. Angular momenta will always appear as multiples of $\hbar$.

Let us now calculate some of the quantities that emerge from the Bohr theory. We calculate

$$mc^2 \cong 0.51 \times 10^6 \text{ eV}$$

$$\cong 0.51 \text{ MeV}$$

$$\frac{\hbar}{mc} \cong 3.9 \times 10^{-11} \text{ cm}$$

$$\frac{\hbar}{mc^2} \cong 1.3 \times 10^{-21} \text{ sec} \tag{1-46}$$

and thus obtain

(a) the radius of the lowest ($n = 1$) Bohr orbit is

$$a_0 = \frac{137}{Z} \frac{\hbar}{mc} = \frac{0.53}{Z} \overset{\circ}{\text{A}} \tag{1-47}$$

(b) the binding energy of the electron in the lowest Bohr orbit, that is, the energy required to put it in a state with $E = 0$ (corresponding to $n = \infty$) is

$$E = \tfrac{1}{2}mc^2 (Z\alpha)^2 = 13.6Z^2 \text{ eV} \tag{1-48}$$

Thus, for example, a transition from the $n = 1$ state to the $n = 2$ state in hydrogen ($Z = 1$) corresponds to a change in energy of $13.6 (1 - \tfrac{1}{4})$ eV $= 10.2$ eV. The frequency of the emitted radiation can be calculated by converting this into ergs, but it is more convenient to work this out in the form

$$\omega = \frac{mc^2\alpha^2(1 - \tfrac{1}{4})}{2\hbar} = \frac{3\alpha^2}{8} \frac{1}{1.3 \times 10^{-21}} \text{ rad/sec}$$

$$\cong 1.5 \times 10^{16} \text{ rad/sec}$$

Equivalently

$$\lambda = 2\pi \frac{c}{\omega} = \frac{16\pi}{3\alpha^2} \frac{\hbar}{mc}$$

$$\cong 1200 \text{ Å}$$

which lies in the ultraviolet.

The success of the Bohr theory with hydrogenlike atoms gave great impetus to further research on the "Bohr atom." In spite of some extraordinary achievements by Bohr[14] and others, it was clear that the theory was provisional. It said nothing about when the electrons would make their jumps; also the quantization rule was restricted to periodic systems; a more general statement, by Sommerfeld and Wilson,

$$\int_{\substack{\text{closed} \\ \text{path}}} p \, dq = n\hbar \qquad (1\text{-}49)$$

where p is the momentum associated with the coordinate q, was of no help in treating problems other than those associated with atomic levels of hydrogen. From the Bohr theory emerged:

1. The *correspondence principle*, which, in essence states that classical physics results should be contained as limiting cases of quantum mechanical results. The limit should be reached when the "quantum numbers" are large, for example, for large n in the Bohr atom. Once a consistent theory of quantum phenomena was constructed, it automatically contained classical physics as a limit, but the principle was very helpful in guiding theoretical guesses, and led Heisenberg to the point from which he could make his giant leap to quantum mechanics. To illustrate how the correspondence principle is satisfied by the Bohr atomic model, consider the frequency of the radiation emitted when an electron makes a "jump" from the orbit with quantum number $n + 1$ to the orbit with quantum number n, when n is very large. This is a good domain to ask for the classical limit, since the angular momentum $n\hbar$ is indeed much larger than $\hbar$. Classically an electron moving in a circular orbit with velocity v would be expected to radiate with the frequency of its motion, that is,

$$\nu_{cl} = \frac{v}{2\pi r} = \frac{Z\alpha c}{n} \frac{Z\alpha mc}{2\pi n^2 \hbar} = \frac{(Z\alpha)^2 mc^2}{2\pi\hbar} \frac{1}{n^3} \qquad (1\text{-}50)$$

On the other hand, the frequency of the radiation associated with the transition is, according to (1-31),

$$\nu = \frac{\omega}{2\pi} = \frac{1}{2\pi\hbar} \frac{mc^2}{2} (Z\alpha)^2 \left[\frac{1}{n^2} - \frac{1}{(n+1)^2} \right] \qquad (1\text{-}51)$$

[14] See S. Rozental (ed.), *Niels Bohr*, North Holland Publishing, Amsterdam, 1967.

which approaches ν_{cl} for $n \gg 1$. Note that this is a significant result, since it is only the frequency associated with an $n + 1 \rightarrow n$ transition that corresponds to the fundamental classical frequency. The radiation associated with the jump $n + 2 \rightarrow n$ has no classical counterpart even in the large n limit. We shall see in Chapter 22 that there are no $n + 2 \rightarrow n$ transitions for "circular orbits" in quantum mechanics.[15]

2. The quantization of angular momentum held in other situations as well. Its application to elliptic orbits gave a more complete picture of the specturm of hydrogenlike atoms, and it was directly observed in the experiments of Stern and Gerlach[16] in 1922.

F. The Wave-Particle Problem

The fact that radiation exhibits both wave and particle properties raises a deep conceptual difficulty, as can be seen from the following considerations:

1. Our discussion of the photoelectric effect, in particular the correlation of the number of electrons emitted with the intensity of the radiation, strongly suggests that the intensity of electromagnetic radiation is proportional to the number of photons emitted by the source. Let us now consider a *Gedanken-experiment*[17] in which radiation is diffracted by a two-slit system. Imagine that the intensity of the source is reduced to the point where, on the average, one photon per hour arrives at the screen. Note that we have to deal with entire photons: as the Compton effect as well as the photoelectric effect show, it is not possible to split a photon into parts with frequency ω but energy less than $\hbar\omega$. The decrease of intensity in the incident radiation should not affect the classical diffraction pattern, since, in effect, we are only stretching out the time scale on which the transmission from the source to the photographic plate of a large number of photons takes place. Photons that come to the plate an hour apart clearly cannot be correlated, and we may therefore think about this process one photon at a time. A photon, as a particle, will presumably go through one slit or the other. If we add to our Gedankenexperiment apparatus a small monitor that

---

[15] Such transitions can occur for elliptical orbits (not considered here), and this is consistent with the correspondence principle.

[16] These matters are discussed in any textbook in modern physics (see references at the end of this chapter).

[17] A *Gedankenexperiment* (thought experiment) is one that may be imagined, that is, one that is consistent with the known laws of physics, even though it may not be technically feasible. Thus, measuring the acceleration due to gravity on the surface of the sun is a Gedankenexperiment, whereas measuring the Doppler shift of sunlight as seen from a space ship moving with twice the velocity of light is nonsense. In Chapter 2 we shall see how careful one must be to insist on consistency with the laws of physics in setting up a Gedankenexperiment.

tells us whether the photon went through slit "1" or slit "2," we can divide the photons into two classes, associated with the two slits. For the first class, we could have closed down slit 2, since the photon did not go through it; for the second class we could have closed down slit 1. We might thus expect that the pattern on the photographic plate should be the same if we repeated the experiment with one slit closed for half the time, and the other slit closed for the other half of the time. This, however, cannot be, since the second experiment does not give an interference pattern. Thus there is an inconsistency that will be traced to the assumption that the presence of the monitor that tells us which slit the photon went through does not affect the experiment. When we discuss the *Heisenberg uncertainty principle*, we shall see that the action of the monitor destroys the interference pattern, so that there is no inconsistency. At this stage it is sufficient to point out that when there is no monitor, each photon acts as a wave, and it does not make sense to ask which slit the photon went through. Presumably, we can still speak of an average intensity of radiation at each slit: this must mean that for individual photons we can only speak of a probability of going through one slit or another.

2. The notion of probability must again be invoked in understanding the passage of polarized radiation through an analyzer. As is well known, a beam of radiation of intensity I_0 will be attenuated to $I_0 \cos^2 \alpha$, where α is the angle between the axis of the polarizer and that of the analyzer. In terms of single photons that are *indivisible*, such an attenuation is only explainable if we state that a given photon will either go through or be blocked by the system, with a probability of transmission governed by the construction of the apparatus, that is, by the angle α.

3. In the same way, consider radiation from a distant star. The star is the source of a spherical wave of electromagnetic field excitation, spreading with velocity c. In terms of individual photons it is not sensible to think of the photon as spread thinly over a sphere of radius ct (where t is the time since the photon was emitted), since the collapse of that photon to a single point on a photographic plate, or on the retina of the eye, would violate common sense, if it were "really" happening. We may however interpret the spherical distribution as giving us the probability of finding a photon at a given solid angle.

4. Sometimes it is possible to interpret a given experiment both in particle and in wave language, but then a nonclassical aspect creeps in elsewhere. Dicke and Wittke[18] have proposed the following Gedankenexperiment (Fig. 1.8). Consider a cylindrical bird cage with the bars spaced regularly, and spacing

$$a = 2\pi \frac{R}{N}$$

[18] R. H. Dicke and J. P. Wittke, *Introduction to Quantum Mechanics*, Addison-Wesley, Reading, Mass., 1960.

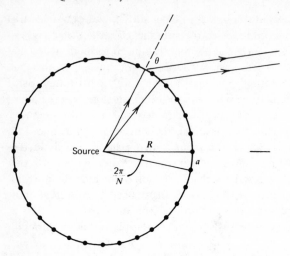

Fig. 1-8. End view of Dicke-Wittke "cage" showing equally spaced bars and geometrical quantities connected with it.

where R is the radius of the cylinder and N is the number of bars. Consider radiation emitted from a source placed on the axis of the cylinder. The bars act as a diffraction grating. If the beam emerges at an angle θ with the original direction, we have maximum intensity if the angle and the wavelength are related by

$$a \sin \theta = n\lambda \qquad (n = 1, 2, 3, \ldots)$$

that is,

$$\lambda = \frac{2\pi R \sin \theta}{Nn} \qquad (1\text{-}52)$$

We could also interpret the intensity peak by assuming that the particles scattered through an angle θ off the bars of the bird cage. The momentum transferred to the cage is $p \sin \theta$ and hence the angular momentum transferred to the cage is

$$L = pR \sin \theta \qquad (1\text{-}53)$$

If we now make the De Broglie association, $p = 2\pi\hbar/\lambda$ we obtain

$$L = \frac{2\pi\hbar Nn}{2\pi R \sin \theta} \cdot R \sin \theta = Nn\hbar \qquad (1\text{-}54)$$

that is, angular momentum is quantized! The factor N is associated with the fact that the bird cage looks the same when it is rotated through an angle $2\pi/N$, as will become clear later.

In 1925 the modern theory of quantum mechanics started with the work of Heisenberg, Born, Jordan, Schrödinger and Dirac. This theory provides a way of reconciling all of the conflicting concepts at the cost of making us abandon a certain amount of classical thinking. It is one of the joys of being a student of physics to be able to appreciate this beautiful theory and the monumental advances in our understanding the properties of matter that the theory enabed us to make.

Problems

1. Prove the relation (1-1) between the energy density in a cavity and the emissive power. [*Hint*. To do so, look at the figure. The shaded volume element

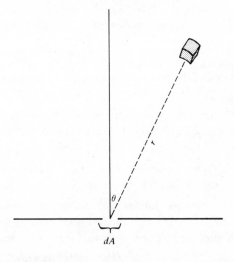

is of magnitude $r^2 \, dr \sin\theta \, d\theta \, d\phi = dV$ where r is the distance to the origin (at the aperture of area dA), θ is the angle with the vertical and ϕ is the azimuthal angle about the perpendicular axis through the opening. The energy contained in the volume element is dV multiplied by the energy density. The radiation is isotropic, so that what emerges is given by the solid angle $dA \cos\theta / 4\pi r^2$ multiplied by the energy. This is to be integrated over the angles ϕ and θ and, if the flow of radiation in time Δt is wanted, over dr from 0 to $c\Delta t$—the distance from which the radiation will escape in the given time interval.]

2. Use (1-1) and (1-12) to obtain a formula for the total rate of radiation per unit area of a black body. Assume that the sun radiates as a black body.

You are given the radius of the sun $R_\odot = 7 \times 10^{10}$ cm, the average distance of the sun to the earth $d_\odot = 1.5 \times 10^{13}$ cm, and the solar constant, the amount of energy falling on the earth when the sun is overhead 1.4×10^6 ergs/cm^2 sec. Use this information to estimate the surface temperature of the sun.

3. Given (1-9), calculate the energy density in a wavelength interval $\Delta\lambda$. Use your expression to calculate the value of $\lambda = \lambda_{max}$, for which this density is maximal. Show that λ_{max} is of the form b/T, calculate b, and use your estimate of the sun's surface temperature to calculate λ_{max} for solar radiation. [*Hint.* In calculating b you will need the solution x of the equation $(5 - x) = 5e^{-x}$. Solve this graphically or by a successive approximation method, in which you first write $x = 5 - \epsilon$, with $\epsilon \ll 1$.]

4. How much of the sun's energy is radiated in the range of wavelengths 4000 Å–7000 Å? Use the T estimated in problem 1. Plot the energy density on graph paper to obtain the numerical result.

5. There is some experimental evidence that the universe contains black body radiation corresponding to an equilibrium temperature of 3°K. Calculate the energy of a photon whose wavelength is λ_{max} corresponding to this temperature.

6. Ultraviolet light of wavelength 3500 Å falls on a potassium surface. The maximum energy of the photoelectrons is 1.6 eV. What is the work function of potassium?

7. The maximum energy of photoelectrons from aluminum is 2.3 eV for radiation of 2000Å and 0.90 eV for radiation of 3130 Å. Use this data to calculate Planck's constant and the work function of aluminum.

8. A 100 MeV photon collides with a proton that is at rest. What is the maximum possible energy loss for the photon?

9. A 100 keV photon collides with an electron at rest. It is scattered through 90°. What is its energy after the collision? What is the kinetic energy in eV of the electron after the collision, and what is the direction of its recoil?

10. An electron of energy 100 MeV collides with a photon of wavelength 3×10^7 Å (corresponding to the universal background of black body radiation). What is the maximum energy loss suffered by the electron?

11. A beam of X rays is scattered by electrons at rest. What is the energy of the X rays if the wavelength of the X rays scattered at 60° to the beam axis is 0.035 Å?

12. A nitrogen nucleus (mass $\cong 14 \times$ proton mass) emits a photon of energy 6.2 MeV. If the nucleus is initially at rest, what is the recoil energy of the nucleus in eV?

13. What is the DeBroglie wavelength of (a) a 1 eV electron, (b) a

10 MeV proton, (c) a 100 MeV electron? (*caution!* use the relativistic energy formula), (d) a thermal neutron? (defined as a neutron whose kinetic energy is $3kT/2$ with $T = 300°K$).

14. Consider a crystal with planar spacing 3.2 Å. What order of magnitude of energies would one need for (a) electrons, (b) helium nuclei (mass $\cong 4 \times$ proton mass) to observe up to 3 interference maxima?

15. The smallest separation resolvable by a microscope is of the order of magnitude of the wavelength used. What energy electrons would one need in an electron microscope to resolve separations of (a) 150 Å, (b) 5 Å?

16. If one assumes that in a stationary state of the hydrogen atom the electron fits into a circular orbit with an integral number of wavelengths, one can reproduce the results of the Bohr theory. Work this out.

17. The distance between adjacent planes in a crystal are to be measured. If X rays of wavelength 0.5 Å are detected at an angle of 5°, what is the spacing? At what angle will the second maximum occur?

18. Use the Bohr quantization rules to calculate the energy levels for a harmonic oscillator, for which the energy is $p^2/2m + m\omega^2 r^2/2$, that is, the force is $m\omega^2 r$. Restrict yourself to circular orbits. What is the analog of the Rydberg formula? Show that the correspondence principle is satisfied for all values of the quantum number n used in quantizing the angular momentum.

19. Use the Bohr quantization rules to calculate the energy states for a potential given by

$$V(r) = V_0 \left(\frac{r}{a}\right)^k$$

with k very large. Sketch the form of the potential and show that the energy values approach $E_n \simeq Cn^2$.

20. The power, that is, the energy radiated by an accelerated charge e is classically given by the formula

$$P = \frac{2}{3} \frac{e^2}{c^3} a^2 \text{ erg/sec}$$

where a is the acceleration. In a circular orbit $a = v^2/r$. Calculate the power radiated by an electron in a Bohr orbit characterized by the quantum number n. When n is very large, this should agree with a proper quantum mechanical result according to the correspondence principle.

21. The decay rate for an electron in an orbit may be defined to be the power radiated, P, divided by the energy emitted in the decay. Use the Bohr theory expression for the energy radiated, and the expression for P from problem 20 to calculate the "correspondence" value of the decay rate when the electron makes a transition from orbit n to orbit $n - 1$. What is the value of this decay rate when $n = 2$? (This will not agree exactly with the true quantum theory

result, since the correspondence principle will not hold for such small values of the quantum number.) What is the decay rate when the transition is from an orbit n to an orbit $n - m$? What is the lifetime = (decay rate)$^{-1}$?

22. The classical energy of a plane rotator is given by

$$E = L^2/2I$$

where L is the angular momentum and I is the moment of inertia. Apply the Bohr quantization rules to obtain the energy levels of the rotator. If the Bohr frequency condition is assumed for the radiation in transitions from states labeled by n_1 to states labeled by n_2, show that (a) the correspondence principle holds, and (b) that it implies that only transitions $\Delta n = \pm 1$ should occur.

23. Molecules sometimes behave like rotators. If rotational spectra are characterized by radiation of wavelength of order 10^7 Å, and this is used to estimate interatomic distances in a molecule like H_2, what kind of separations (in Å) are obtained?

References

F. K. Richtmyer, E. H. Kennard, and J. N. Cooper, *Introduction to Modern Physics*, McGraw-Hill, New York, 1969.

Robert Martin Eisberg, *Fundamentals of Modern Physics*, Wiley, New York, 1961.

Arthur Beiser, *Perspectives of Modern Physics*, McGraw-Hill, New York, 1969.

John D. McGervey, *Introduction to Modern Physics*, Academic Press, New York, 1971.

Robert B. Leighton, *Principles of Modern Physics*, McGraw-Hill, New York, 1959.

Martin Karplus and Richard N. Porter, *Atoms and Molecules*, W. A. Benjamin, New York, 1970.

Eyvind H. Wichmann, *Quantum Physics*, McGraw-Hill, New York, 1969.

Richard P. Feynman, Robert B. Leighton, and Matthew Sands, *The Feynman Lectures on Physics*, Addison-Wesley, Reading, Mass., 1963.

The first five books on this list cover the main topics of a standard modern physics course, with variations in level and emphasis, so that they should all be consulted for a not-too theoretical treatment of the subject. Wichmann's book provides an unconventional introduction to quantum theory. It stresses all the important points, ranges over a very wide field of qualitative applications, and provides a new perspective to the reader who already has some background in the subject. The *Feynman Lectures* cannot be characterized in any simple way. They are brilliant and should be read by every student, undergraduate and graduate. Instructors already know that, and read them a great deal.

chapter 2

Wave Packets and the Uncertainty Relations

Quantum mechanics provides us with an understanding of all of the phenomena discussed in Chapter 1. It is indispensable to the understanding of atoms, molecules, atomic nuclei, and aggregates of these. We will approach the study of quantum mechanics through the Schrödinger equation and the appropriate interpretation of its solutions.[1] There is no way of deriving this equation from classical physics, since it lies outside the realm of classical physics. It can only be guessed, which is what Schrödinger did, following the earlier insights of De Broglie. We will motivate the guess somewhat differently, by seeing how one might try to reconcile the wave and particle properties of electrons.

It is difficult to think of configurations of particles that somehow simulate wave behavior. This is why the diffraction experiments of Fresnel and Young led to the unanimous acceptance of the wave theory of light. On the other hand, it is possible to imagine configurations of waves that are very localized. (A clap of thunder is an example of a superposition of waves leading to an effect localized in time.) Such localized "wave packets" can be achieved by superposing waves with different frequencies in a special way, so that they interfere with each other almost completely outside of a given spatial region. The technical tools for doing this involve Fourier integrals, and Appendix A summarizes them for the reader who is familiar with Fourier series and who does not insist on mathematical rigor.

As an example, consider the function defined by

$$f(x) = \int_{-\infty}^{\infty} dk\, g(k)\, e^{ikx} \tag{2-1}$$

---

[1] A different approach can be found in R. P. Feynman, R. B. Leighton, and M. Sands, *The Feynman Lectures on Physics*, Vol. III, Addison-Wesley, Reading, Mass., 1964.

The real part of $f(x)$ is given by $\int_{-\infty}^{\infty} dk\ g(k)\cos kx$, and this is a linear super-position of waves of wavelength $\lambda = 2\pi/k$, since for a given k the wave reproduces itself when x changes to $x + 2\pi/k$.
Choose

$$g(k) = e^{-\alpha(k-k_0)^2} \tag{2-2}$$

The integral can be done: with $k' = k - k_0$ we have

$$
\begin{aligned}
f(x) &= \int_{-\infty}^{\infty} dk\ g(k)\ e^{i(k-k_0)x}\ e^{ik_0 x} \\
&= e^{ik_0 x} \int_{-\infty}^{\infty} dk'\ e^{ik'x}\ e^{-\alpha k'^2} \\
&= e^{ik_0 x} \int_{-\infty}^{\infty} dk'\ e^{-\alpha[k'-(ix/2\alpha)]^2}\ e^{-(x^2/4\alpha)}
\end{aligned}
$$

where in the last step we have completed squares. It is justified to let $k' - (ix/2\alpha) = q$ and still keep the integral along the real axis.[2] Making use of

$$\int_{-\infty}^{\infty} dk\ e^{-\alpha k^2} = \sqrt{\frac{\pi}{\alpha}} \tag{2-3}$$

we obtain

$$f(x) = \sqrt{\frac{\pi}{\alpha}}\ e^{ik_0 x}\ e^{-(x^2/4\alpha)} \tag{2-4}$$

The factor $e^{ik_0 x}$ is known as a "phase factor," since $|e^{ik_0 x}|^2 = 1$. Thus the absolute square of $f(x)$ is

$$|f(x)|^2 = \frac{\pi}{\alpha}\ e^{-x^2/2\alpha} \tag{2-5}$$

This function shows a peaking that can be very pronounced when α is chosen to be small. It represents a function localized about $x = 0$, with a width of the order $2\sqrt{2\alpha}$, since when $x = \pm\sqrt{2\alpha}$, the function drops off to $1/e$ of its peak value. The width in x-space is correlated with that in k-space. The square of $g(k)$ is a function peaked about k_0 with width $2/\sqrt{2\alpha}$. There is a reciprocity here: a function strongly localized in x is broad in k and vice versa. The product of the two "widths" is

$$\Delta k\ \Delta x \sim \frac{2}{\sqrt{2\alpha}} \cdot 2\sqrt{2\alpha} = 4 \tag{2-6}$$

---

[2] The reader familiar with the theory of complex variables will have no trouble convincing himself of this.

The actual value of the numerical constant is not important; what matters is that it is independent of α. This is a general property of functions that are Fourier transforms of each other (Fig. 2.1). We represent it by the formula

$$\Delta x \, \Delta k \gtrsim O(1) \tag{2-7}$$

where Δx and Δk are the "widths" of the two distributions, and we imply by $O(1)$ that this is a number that may depend on the functions that we are dealing with, but is not significantly smaller than 1. *It is impossible to make both Δx and Δk small.* This is a general feature of wave packets, but we shall soon see that it has some very deep implications for quantum mechanics.

In Eq. 2-1 we considered a function $f(x)$ that is made up of a continuous superposition of simple waves e^{ikx}. How will such a wave packet propagate in

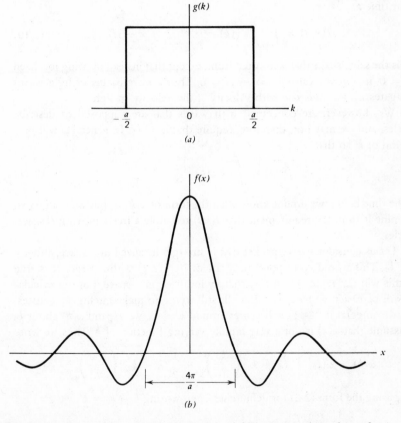

Fig. 2-1. Relation between wave packet and its Fourier transform for a square-shaped wave packet.

time? The answer to that depends on how the individual waves propagate. In general we shall write for the simple *plane wave* (so called because it only has a spatial variation in x, but not in y or z) the form

$$e^{ikx - i\omega t} \tag{2-8}$$

Here $\omega = 2\pi\nu$ is the angular frequency. The quantity k is related to the wavelength by $k = 2\pi/\lambda$ so that we may write for the simple wave another form

$$e^{2\pi i[(x/\lambda) - \nu t]} \tag{2-9}$$

If we are considering the propagation of a light wave in a vacuum, then there is a simple relation between ν and $1/\lambda$, namely, $\nu = c/\lambda$, so that the simple wave becomes

$$e^{2\pi i(x - ct)/\lambda} = e^{ik(x - ct)}$$

If we now take the superposition, with amplitude $g(k)$ of these simple waves, we get, at time t,

$$f(x, t) = \int_{-\infty}^{\infty} dk \, g(k) \, e^{ik(x - ct)} = f(x - ct) \tag{2-10}$$

This is the same shape that we started from, except that instead of being localized at $x = 0$, it is now localized at $x - ct = 0$. Thus a wave packet of light waves propagates *without distortion* with velocity c, the velocity of light.

We, however, are concerned with waves that are supposed to describe particles, and we may not, therefore, require that $\omega = kc$. In general ω will be a function of k, so that

$$f(x,t) = \int dk \, g(k) \, e^{ikx - i\omega(k)t} \tag{2-11}$$

For the time being, we do not know what the form of $\omega(k)$ is, but we shall try to determine it from the requirement that $f(x,t)$ resemble a freely moving classical particle.

Let us consider a wave packet that is strongly localized in k-space, about a value k_0. This would correspond to a choice like (2-2) with α large. It is true that this will not represent an $f(x)$ sharply localized in x-space, but our calculation will be easier, and we are, after all, still trying to make intelligent guesses. Since the integral in (2-11) will center around $k = k_0$, we expand $\omega(k)$ about k_0 and assume that $\omega(k)$ is not a very rapidly varying function of k. Thus we write

$$\omega(k) \approx \omega(k_0) + (k - k_0)\left(\frac{d\omega}{dk}\right)_{k_0} + \frac{1}{2}(k - k_0)^2 \left(\frac{d^2\omega}{dk^2}\right)_{k_0} \tag{2-12}$$

Then, using the form (2-2) for definiteness, and writing $k - k_0 = k'$, we get

$$f(x,t) = e^{ik_0 x} e^{-i\omega(k_0)t} \int dk' \, e^{-\alpha k'^2} e^{-i(k'^2/2)[(d^2\omega/dk^2)]_0 t} e^{ik'\{x - [(d\omega/dk)]_0 t\}} \tag{2-13}$$

Aside from the phase factor in front, the x and t coordinates appear in a form that strongly suggests that the velocity of propagation of the packet, the *group velocity*, is[3]

$$v_g = \left(\frac{d\omega}{dk}\right)_{k_0} \tag{2-14}$$

Thus, defining

$$\frac{1}{2}\left(\frac{d^2\omega}{dk^2}\right)_{k_0} = \beta \tag{2-15}$$

we have

$$f(x,t) = e^{i[k_0 x - \omega(k_0)t]} \int dk' \, e^{ik'(x-v_g t)} \, e^{-(\alpha+i\beta t)k'^2} \tag{2-16}$$

This is just the integral that led to (2-4), so that replacing x by $x - v_g t$ and α by $\alpha + i\beta t$, we get

$$f(x,t) = e^{i[k_0 x - \omega(k_0)t]} \left(\frac{\pi}{\alpha+i\beta t}\right)^{1/2} e^{-[(x-v_g t)^2/4(\alpha+i\beta t)]}$$

and the absolute square of this function is

$$|f(x,t)|^2 = \left(\frac{\pi^2}{\alpha^2+\beta^2 t^2}\right)^{1/2} e^{-[\alpha(x-v_g t)^2/2(\alpha^2+\beta^2 t^2)]} \tag{2-17}$$

This represents a wave packet whose peak is traveling with velocity v_g, but it does not have a definite width: the quantity that was α at $t = 0$ now becomes $\alpha + (\beta^2 t^2/\alpha)$, that is, *the packet is spreading.* Since the width is proportional to

$$\left(\alpha + \frac{\beta^2 t^2}{\alpha}\right)^{1/2} = \sqrt{\alpha}\left(1 + \frac{\beta^2 t^2}{\alpha^2}\right)^{1/2}$$

the rate of spreading will be small if α is large, that is, if the packet is spatially large to begin with.

The most important result is that if (2-11) is to represent a particle with momentum p and kinetic energy $p^2/2m$, then we must require that

$$v_g = \frac{d\omega}{dk} = \frac{p}{m} \tag{2-18}$$

[3] This is certainly in agreement with what we found in the special case of light propagation, where $\omega = kc$. The argument used more generally depends on the fact that the peak of a packet will tend to be where the phase $kx - \omega t$ has a minimum as a function of k, that is, where $x - \left(\dfrac{d\omega}{dk}\right)t = 0$.

If we further make the association that

$$E = \hbar\omega \qquad (2\text{-}19)$$

suggested by the quantum relation for radiation, so that

$$\omega = \frac{p^2}{2m\hbar} \qquad (2\text{-}20)$$

then consistency demands that we make the association

$$k = \frac{2\pi}{\lambda} = \frac{p}{\hbar} \qquad (2\text{-}21)$$

first derived in a somewhat similar way by De Broglie.

In terms of p, the expression (2-11) can be rewritten in the form

$$\psi(x,t) = \frac{1}{\sqrt{2\pi\hbar}} \int dp\; \phi(p)\; e^{i(px-Et)/\hbar} \qquad (2\text{-}22)$$

The wave packet $\psi(x,t)$ is a general solution of the partial differential equation

$$i\hbar\frac{\partial \psi(x,t)}{\partial t} = \frac{1}{\sqrt{2\pi\hbar}} \int dp\; \phi(p)\; E\; e^{i(px-Et)/\hbar}$$

$$= \frac{1}{\sqrt{2\pi\hbar}} \int dp\; \phi(p)\; \frac{p^2}{2m}\; e^{i(px-Et)/\hbar}$$

$$= -\frac{\hbar^2}{2m}\frac{\partial^2\psi(x,t)}{\partial x^2} \qquad (2\text{-}23)$$

provided, as we have done above, we describe the motion of the "particle" in a potential-free region, where $E = p^2/2m$. It is this equation, and its generalization to the case of a particle moving in a potential, that represents the important abstraction from the arguments outlined above. It should be stressed that the equation represents a guess: there was no justification on the basis of classical physics for the replacement of ω by $E/\hbar$, nor for the replacement of the wave number k by $p/\hbar$.

We must still face the difficulty of the spreading of the wave packets. If we consider a *Gaussian* packet (2-17), we see that no matter how large α is, there will be a time when the spreading will become noticeable. This is in contradiction with experience, which shows very clearly that nuclei, for example, that are very tiny, have not changed during a period of 3×10^9 years (10^{20} sec). We shall see in Chaper 3 that the notions of probability, hinted at in Chapter 1, play a role here, and the spreading really refers to a growing probability that the particle is far from where it was localized at $t = 0$.

One of the most important qualitative observations that we made in our wave packet discussion is the reciprocity relation between the widths in x- and k-space

$$\Delta k \, \Delta x \gtrsim 1 \qquad (2\text{-}24)$$

If we multiply this by $\hbar$ and use $\hbar k = p$, we obtain the Heisenberg uncertainty relations

$$\Delta p \, \Delta x \gtrsim \hbar \qquad (2\text{-}25)$$

Since the width represents a region in which a particle is likely to be in x-space or in momentum space, (2-25) states that if we try to construct a highly localized wave packet in x-space, then it is impossible to associate a well-defined momentum with it, in contrast with what is taken for granted in classical physics. By the same token, a wave packet characterized by a momentum defined within narrow limits must be spatially very broad. This limitation is one that is imposed on the classical description, which insists on being able to specify both position and momentum. In quantum physics position and momentum, just like particle behavior and wave aspects of a system, are complementary properties of the system, and the theory does not admit the possibility of an experiment in which both could be established simultaneously. The smallness of $\hbar$ guarantees that only for microscopic systems will the usual notions of classical physics fail. For example, for a dust particle of mass 10^{-4} gm moving with a velocity of 10^4 cm/sec with an uncertainty in the product of one part in a million implies $\Delta p \sim 10^{-6}$ and thus $\Delta x \sim 10^{-21}$ cm, which is 10^{-7} times smaller than the radius of a proton! This is not so for an electron in a Bohr orbit. It we take $\Delta p \sim p \sim mc\alpha/n$, then $\Delta x \sim \hbar n/mc\alpha$, of the order of magnitude of the radii of the orbits.

In what follows we will discuss a number of Gedankenexperiments in which we will show in detail how the wave-particle duality acts to conspire to prohibit a violation of the relation (2-25).

(a) *Measurement of position of an electron.* Consider the experimental set up in Fig. 2.2, whose purpose is to measure the position of an electron. The electrons are in a beam having well-defined momentum p_x and moving in the positive x direction. The microscope (lens + screen) is to be used to see where the electron is located by observing the light that is scattered off the electron. We shine light along the negative x-axis; a particular electron will scatter a particular photon, and the latter recoils through the microscope. The resolution of the microscope, that is, the precision with which the electron can be localized is known from optics. It is

$$\Delta x \sim \frac{\lambda}{\sin \phi} \qquad (2\text{-}26)$$

where λ is the wavelength of the light.
It would appear that by making λ small enough, and by making $\sin \phi$ large, Δx can be made as small as desired. This, we will now show, can only be done at the

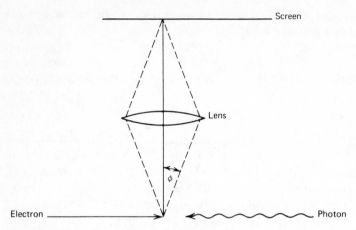

Fig. 2-2. Schematic drawing of the Heisenberg microsocpe for the measurement of electron position.

expense of losing information about the x-component of the electron momentum. Quantum theory tells us that what registers on the screen behind the lens are really individual photons that got there because they scattered off the electrons. The direction of the photon after scattering is undetermined within the angle subtended by the aperture. Hence the magnitude of the recoil momentum of the electron is uncertain by

$$\Delta p_x \sim 2 \frac{h\nu}{c} \sin \phi \qquad (2\text{-}27)$$

Hence

$$\Delta p_x \, \Delta x \sim 2 \frac{h\nu}{c} \sin \phi \, \frac{\lambda}{\sin \phi} \sim 4\pi\hbar \qquad (2\text{-}28)$$

Can we get around this difficulty? After all, the direction of the photon is correlated with its momentum, and if we could somehow measure the recoil of the screen, we could specify the photon (and hence electron) momentum better. True, but once we include the microscope as part of the "observed" system, we must worry about its location, since its momentum is to be specified. But the microscope, too, must obey the uncertainty relation, and if its momentum is to be specified, its position will be less determined. The final "classical" observation apparatus will always be faced with the indeterminacy.

(b) *The two-slit experiment.* In Chapter 1 we suggested that the interference pattern observed in the passage of an electron[4] through two slits was logically

---

[4] We actually discussed photons, but the difficulty is the same for electrons, which are also diffracted.

incompatible with our being able to know which slit the electron went through, as such knowledge would imply that the pattern is a superposition of electrons coming from one slit or the other. This, however, cannot give an interference pattern. We may use the uncertainty relation to show that a "monitor" that identifies the slits of passage will destroy the interference pattern. Let the slits be separated by a distance a, and let the distance from the slits to the screen be d. The condition for constructive interference is

$$\sin \theta = n \frac{\lambda}{a} \tag{2-29}$$

so that the distance between adjacent maxima on the screen is $d \sin \theta_{n+1} - d \sin \theta_n = d\lambda/a$. Consider, now a monitor that determines the position of an electron just behind the screen to an accuracy $\Delta y < a/2$, that is, it tells us which slit the electron went through (Fig. 2.3). In doing so, it must impart to the electron a momentum in the y direction whose amount is imprecise, with

$$\Delta p_y > \frac{2h}{a} \tag{2-30}$$

Hence

$$\frac{\Delta p_y}{p} > \frac{2}{a} \frac{h}{p} = \frac{2\lambda}{a} \tag{2-31}$$

Such an uncertainty introduces an indeterminacy in the position of the electron at the screen, whose magnitude is $2\lambda d/a$, at the very least. This, however, is larger

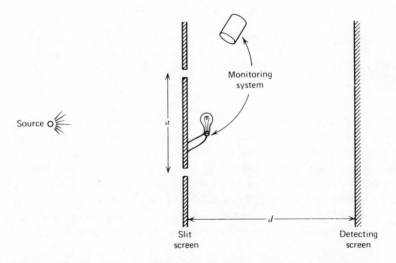

Source O

Monitoring system

Slit screen

Detecting screen

Fig. 2-3. The two-slit experiment with monitor.

than the spacing between maxima, so we conclude that a working monitor will wipe out the interference pattern, and there is no logical contradiction. Conversely, we could, of course, argue that logical consistency demanded that

$$\Delta p_y \, \Delta y > h \tag{2-32}$$

(c) *The "reality" of orbits in the Bohr atom.* As noted in Chapter 1, the Bohr atomic model deals with orbits whose radii are given by $R_n = \hbar n^2/\alpha mc$. Thus an experiment designed to measure the outlines of a given orbit must be such that a position measurement of the electron in the atom is done with an accuracy

$$\Delta x \ll R_n - R_{n-1} \cong \frac{2\hbar n}{\alpha mc} \tag{2-33}$$

This implies an uncontrollable momentum transfer to the electron that is of magnitude $\Delta p \gg mc\alpha/2n$. This implies an uncertainty in the energy of the electron of magnitude

$$\Delta E \simeq \frac{p\Delta p}{m} \gg \frac{mc\alpha}{n} \cdot \frac{\alpha c}{2n} = \frac{1}{2} \frac{mc^2\alpha^2}{n^2} \tag{2-34}$$

that is, much larger than the binding of the electron in the orbit. Thus such a measurement, as likely as not, will kick the electron out of the orbit, so that no such mapping of the orbit is possible.

(d) *The energy-time uncertainty relation.* If we take the relation (2-25) and write it in the form

$$\frac{p\Delta p}{m} \cdot \frac{\Delta x m}{p} \gtrsim \hbar$$

we may interpret the first factor as a measure of the uncertainty in the energy of the system, and the second factor, $\Delta x/v$, as a measure of Δt, an uncertainty in its localizability in time. This suggests the energy-time uncertainty relation

$$\Delta E \, \Delta t \gtrsim \hbar \tag{2-35}$$

Such a relation might also be deduced from the form of the wave packet (2-22) since E and t appear in the same reciprocal relation as p and x, and it is also suggested by the theory of relativity, since space and time, like momentum and energy are intimately connected with each other.[5] Actually, in nonrelativistic quantum mechanics, space and time play a somewhat different role, and whereas we shall be able to derive (2-25) from the formalism of quantum mechanics, this is not true of (2-35). Nevertheless the energy-time uncertainty relation is as much a part of the qualitative structure of quantum mechanics as (2-25).

---

[5] Both $(ct, \mathbf{r})$ and $(E/c, \mathbf{p})$ are *four-vectors* that transform among themselves under Lorentz transformations.

In spite of his fundamental contributions to the development of quantum mechanics, Einstein always felt uneasy about its implications, and at the Solvay Congress of 1930[6] he suggested a Gedankenexperiment that apparently avoided the limitations suggested by (2-35). Einstein suggested that a box containing radiation have a shutter controlled by a clock within the box. The shutter mechanism could be arranged such that a hole is opened for an arbitrarily short time Δt. The energy of the photon escaping from the box could be determined very accurately by weighing the box before and after the opening of the shutter.

Bohr's rebuttal of the argument is a beautiful illustration of the requirement that a Gedankenexperiment must conform to the laws of physics. Taking into consideration the apparatus shown in Fig. 2-4, Bohr made the following points:

1. A weighing implies the reading of a scale pointer with an accuracy Δx. This implies an uncertainty in the momentum of the box given by $\Delta p \gtrsim \hbar/\Delta x$.

2. If a change of mass Δm is to be detected, the weighing must take a time T, that is long enough so that the impulse due to the change in mass, that is, $gT \Delta m$ (g = acceleration due to gravity) is much larger than Δp, that is,

$$gT \Delta m \gg \hbar/\Delta x \qquad (2\text{-}36)$$

3. The well-established *equivalence principle*[7] implies that a change in the vertical position Δx in a gravitational field implies a change in the rate of the clock, given by

$$\frac{\Delta T}{T} = \frac{g \Delta x}{c^2} \qquad (2\text{-}37)$$

This yields

$$\frac{\Delta T}{T} \gg \frac{g}{c^2} \frac{\hbar}{gT \Delta m}$$

that is,

$$\Delta mc^2 \, \Delta T = \Delta E \, \Delta T \gg \hbar \qquad (2\text{-}38)$$

This shows that the energy-time uncertainty relation is maintained.

The uncertainty relations may be used to make rough numerical estimates in microscopic physics. Let us illustrate this with several examples, the first of

---

[6] See the beautiful essay by Niels Bohr, "Discussion with Einstein on Epistemological Problems in Atomic Physics," which appeared in *Atomic Physics and Human Knowledge*, John Wiley & Sons (1958).

[7] The equivalence principle is discussed in the Special Topics section 2 at the end of this book. It is amusing in this context that the principle was formulated by Einstein!

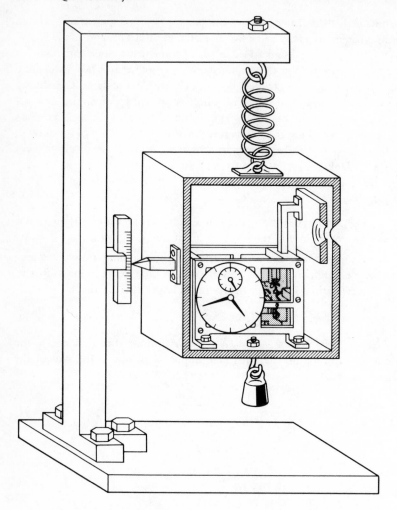

Fig. 2-4. Quasi-realistic drawing of Einstein experiment designed to show violation of $\Delta E \, \Delta t > h$ relation. Reprinted from Niels Bohr, *Atomic Physics and Human Knowledge*, John Wiley (1958), by permission of North Holland Publishing Company, Amsterdam.

which is the hydrogen atom. If we say that the electron's position inside the atom is unknown, then, if r is its radial coordinate

$$pr \sim \hbar \tag{2-39}$$

This allows us to express the energy in terms of r:

$$E = \frac{p^2}{2m} - \frac{e^2}{r}$$

$$= \frac{\hbar^2}{2mr^2} - \frac{e^2}{r} \tag{2-40}$$

The minimum value of the energy is obtained from

$$\frac{\partial E}{\partial r} = - \frac{\hbar^2}{mr^3} + \frac{e^2}{r^2} = 0$$

that is,

$$r = \frac{\hbar^2}{me^2} = \frac{\hbar}{mc\alpha} \tag{2-41}$$

and the corresponding value of E is

$$E = - \frac{1}{2} mc^2\alpha^2 \tag{2-42}$$

The fact that we obtained the exact value of the energy is, of course, a swindle, since we could equally well have written $pr \sim h$ instead of (2-39), and we would then have obtained a different result. The value of E would, however, have differed from the correct value only by a numerical constant, and the general order of magnitude would still have been the same. The main point is that in contrast to classical theory, the energy is bounded from below because of the uncertainty principle: an increase in the (negative) potential energy, obtained by decreasing r, that is, localizing the electron closer to the nucleus carries with it the necessity for increasing the kinetic energy.

As another example, consider the problem of nuclear forces. These have the range of the order of one fermi, that is, 10^{-13} cm. This implies that $p \sim \hbar/r \sim 10^{-14}$ gm cm/sec. The kinetic energy corresponding to this momentum is

$$\frac{p^2}{2M} \sim \frac{10^{-28}}{3.2 \times 10^{-24}} \sim 3 \times 10^{-5} \text{ ergs} \tag{2-43}$$

where M is the nucleon (proton or neutron) mass, which is 1.6×10^{-24} gm. Since the potential that gives rise to the binding must more than compensate for this we require that

$$|V| \sim 3 \times 10^{-5} \text{ ergs} \sim 20 \text{ MeV} \tag{2-44}$$

Again, this is only a rough order of magnitude, but it does indicate that the potential energy is to be measured in MeV rather than in eV, as in atoms.

Yet another illustration comes from the Yukawa meson theory of nuclear forces. In 1935 Yukawa proposed that the nuclear force arises through the emission of a new quantum, the pi-meson (also called pion), by one of the nucleons, and its absorption by the other.[8] If the mass of the quantum is denoted

[8] This is discussed briefly in the Special Topics section 5 on the Yukawa theory.

by μ, then its emission introduces an energy imbalance $\Delta E \sim \mu c^2$, which can only take place for a time $\Delta t \sim \hbar/E \sim \hbar/\mu c^2$. The range corresponding to a particle traveling for this time is of the order of $c\Delta T \sim \hbar/\mu c$. If we take for the range $r_0 = 1.4 \times 10^{-13}$ cm, then we find that

$$\mu c^2 \cong \frac{\hbar c}{r_0} = \frac{10^{-27} \times 3 \times 10^{10}}{1.4 \times 10^{-13}} \text{ ergs}$$

$$\cong 130 \text{ MeV} \tag{2-45}$$

When the pion was finally discovered, it was found that this estimate was remarkably accurate, since for the pion $\mu c^2 \cong 140$ MeV.

In summary, our tentative attempt to wed wave and particle properties consistent with experiment in the most naive way, has led us to an uncertainty in the description of atomic phenomena at the classical level, and this uncertainty is both necessary for a consistent description of (Gedanken) experiments, and in accord with what is observed.

Problems

1. Consider a wave packet defined by (2-1) with $g(k)$ given by
$$g(k) = 0 \qquad k < -K$$
$$= N \qquad -K < k \quad < K$$
$$= 0 \qquad K < k$$

(a) Find the form $f(x)$.

(b) Find the value of N for which
$$\int_{-\infty}^{\infty} dx |f(x)|^2 = 1$$

(c) How is this related to the choice of N for which
$$\int_{-\infty}^{\infty} dk |g(k)|^2 = \frac{1}{2\pi}$$

(d) Show that a reasonable definition of Δx for your answer to (a) yields
$$\Delta k \, \Delta x > 1$$
independent of the value of K.

2. Given that
$$g(k) = \frac{N}{k^2 + \alpha^2}$$

calculate the form of $f(x)$. Again, plot the two functions and show that

$$\Delta k \, \Delta x > 1$$

independent of your choice of α.

3. Consider the problem of the spreading of a Gaussian wave packet for a free particle, where the relation

$$\omega = \frac{\hbar k^2}{2m}$$

holds. Use (2-17) to calculate the fractional change in the size of the wave packet in one second, if

(a) the packet represents an electron, with the wave packet having a size of 10^{-4} cm; 10^{-8} cm.

(b) the packet represents an object of mass 1 gm and has size 1 cm.

It will be convenient to express the width in units of $\hbar/mc$, where m is the mass of the particle represented by the packet.

4. A beam of electrons is to be fired over a distance of 10^4 km. If the size of the initial packet is 1 mm, what will be its size upon arrival, if its kinetic energy is (a) 13.6 eV, (b) 100 MeV?

(*Caution.* The relation between kinetic energy and momentum is not always K.E. $= p^2/2m$!)

5. The relation between the wavelength and the frequency in a wave guide is given by

$$\lambda = \frac{c}{\sqrt{\nu^2 - \nu_0^2}}$$

What is the group velocity of such waves?

6. For surface tension waves in shallow water, the relation between frequency and wavelength is given by

$$\nu = \left(\frac{2\pi T}{\rho \lambda^3}\right)^{1/2}$$

where T is the surface tension and ρ the density. What is the group velocity of the waves, and its relation to the phase velocity, defined to be $v_p = \lambda \nu$? For gravity waves (deep water), the relation is given by

$$\nu = \left(\frac{g}{2\pi\lambda}\right)^{1/2}$$

what are the group and phase velocities?

7. Use the uncertainty relation to estimate the ground state energy of a harmonic oscillator. The energy is given by

$$E = \frac{p^2}{2m} + \frac{1}{2} m\omega^2 x^2$$

8. Use the value of the "lifetime" of an electron in an $n = 2$ Bohr obit, calculated in Problem 21 of Chapter 1, to estimate the uncertainty in the energy of the $n = 2$ energy level. How does it compare with the energy of that level?

9. Nuclei, typically of size 10^{-12} cm, frequently emit electrons, with typical energies of 1–10 MeV. Use the uncertainty principle to show that electrons of energy 1 MeV could not be contained in the nucleus before the decay.

10. The apparatus sketched below appears to allow a violation of the uncertainty relation. The lateral location can be determined with accuracy $\Delta y \sim a$, and the transverse momentum of the incident beam can be made as small as possible by making L arbitrarily large. Analyze the apparatus in detail, point out the hidden assumptions made in the above, and show that the uncertainty relation is not violated.

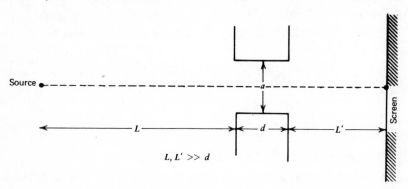

References

Wave packets are discussed in a number of textbooks. Most useful at this level are:

S. Borowitz, *Fundamentals of Wave Mechanics*, W. A. Benjamin, Inc., New York, 1967.

J. L. Powell and B. Crasemann, *Quantum Mechanics*, Addison-Wesley, Reading, Mass., 1961.

D. Bohm, *Quantum Theory*, Prentice-Hall, Englewood Cliffs, N.J., 1951.

All textbooks on quantum mechanics necessarily deal with the uncertainty relations. Very thorough discussions can be found in the book by Bohm cited above, and in W. Heisenberg, *The Physical Principles of the Quantum Theory*, Dover Publications, Inc., 1930.

Discussions of the uncertainty relations may also be found in any of the more advanced books listed at the end of this book.

chapter 3

The Schrödinger Wave Equation

In Chapter 2 we obtained a partial differential equation satisfied by a wave packet that, within certain approximations, described a freely moving "particle." From this point we shall take this equation

$$i\hbar \frac{\partial \psi(x,t)}{\partial t} = - \frac{\hbar^2}{2m} \frac{\partial^2 \psi(x,t)}{\partial x^2} \tag{3-1}$$

as the correct equation for the description of a free particle. Inverting the sequence that led to (2-23), we see that the most general solution of this equation is

$$\psi(x,t) = \frac{1}{\sqrt{2\pi\hbar}} \int dp \; \phi(p) \; e^{i[px-(p^2/2m)t]/\hbar} \tag{3-2}$$

[The reason for the normalization factor in front of the integral appears in (3.26).] Before turning to the crucial point of interpreting the meaning of the solution $\psi(x,t)$ of this equation, we draw attention to the fact that the equation is of first order in the time-derivative. This implies that once the initial value of ψ, namely, $\psi(x,0)$, is given, its values at all other times can be found. This is evident from the form of the equation as seen by a digital computer[1]

$$\psi(x,t + \Delta t) = \psi(x,t) + \frac{i\hbar}{2m} \frac{\partial^2 \psi(x,t)}{\partial x^2} \Delta t \tag{3-3}$$

or from the form of the most general solution. Given $\psi(x,0)$, the function $\phi(p)$ may be found from (3-2), with $t = 0$:

$$\psi(x,0) = \frac{1}{\sqrt{2\pi\hbar}} \int dp \; \phi(p) \; e^{ipx/\hbar} \tag{3-4}$$

may be inverted, and once $\phi(p)$ is known, the solution is known for all values

---

[1] For a discrete mesh, $\partial\psi(x, t)/\partial t$ must be replaced by $[\psi(x, t + \Delta t) - \psi(x,t)]/\Delta t$, with Δt small but not vanishing.

of t. Note that there is no "uncertainty" in the differential equation: once the initial state of the wave packet is specified—and there are, so far, no restrictions on $\psi(x,0)$—then that wave packet is completely specified at all later times.

In searching for an interpretation for $\psi(x, t)$ we must bear in mind (1) that $\psi(x, t)$ is in general a complex function [e.g. (2-16)], and (2) that the function $|\psi(x, t)|$ is large where the particle is supposed to be, and small elsewhere. It also has associated with it the feature of spreading, discussed in Chapter 2. The suggestion of Max Born that

$$P(x, t)\ dx = |\psi(x, t)|^2\ dx \tag{3-5}$$

define the *probability that the particle described by the wave function $\psi(x, t)$ may be found between x and $x + dx$ at time t* turns out to provide the correct interpretation of the wave function. The probability density $P(x, t)$ is real, is large where the particle is supposed to be, and its spreading does not imply that a particular particle is spreading; all it means is that as time goes by, one is less likely to find the particle where one put it at $t = 0$.

For this interpretation to hold, we must require that

$$\int_{-\infty}^{\infty} P(x, t)\ dx = 1 \tag{3-6}$$

since the particle must be somewhere. In a linear equation like (3-1), the solution $\psi(x, t)$ may be multiplied by a constant, and it still remains a solution. Thus (3-6) restricts the solutions $\psi(x, t)$ to a class of functions that are *square integrable*. We shall see below that it is enough to require that

$$\int_{-\infty}^{\infty} dx |\psi(x, 0)|^2 < \infty \tag{3-7}$$

that is, the *initial state wave functions must be square integrable*. With an infinite integration interval this means that $\psi(x, 0)$ must go to zero at infinity at least as fast as $x^{-1/2-\epsilon}$ where ϵ can be arbitrarily small, but must be positive. We shall also require that the wave functions $\psi(x, t)$ be *continuous* in x.

Since $|\psi(x, t)|^2$ is the physically significant quantity, it would appear that the *phase* of the solution of the equation is somehow unimportant. That is wrong! Since the equation (3-1) is linear, if $\psi_1(x, t)$ and $\psi_2(x, t)$ are solutions, so is

$$\psi(x, t) = \alpha_1\psi_1(x, t) + \alpha_2\psi_2(x, t) \tag{3-8}$$

where α_1 and α_2 are arbitrary complex numbers. Clearly the absolute square of $\psi(x, t)$ in (3-8) will depend crucially on the relative phases of the two parts. A more physical way of seeing this is to note that as in classical optics, the interference pattern is determined by the phase relation between the two parts of the wave function associated with the two slits in a two-slit experiment. It is, of course, true that an *overall* phase factor can be ignored.

We now show that the condition (3-6), imposed at $t = 0$, holds true for all times. We need Eq. 3-1 and its complex conjugate

$$-i\hbar \, \frac{\partial \psi^*(x,t)}{\partial t} = - \, \frac{\hbar^2}{2m} \, \frac{\partial^2 \psi^*(x,t)}{\partial x^2} \tag{3-9}$$

Now

$$\frac{\partial}{\partial t} \, P(x, t) = \frac{\partial \psi^*}{\partial t} \, \psi + \psi^* \, \frac{\partial \psi}{\partial t}$$

$$= \frac{1}{i\hbar} \left(\frac{\hbar^2}{2m} \, \frac{\partial^2 \psi^*}{\partial x^2} \, \psi - \frac{\hbar^2}{2m} \, \psi^* \, \frac{\partial^2 \psi}{\partial x^2} \right)$$

$$= - \, \frac{\partial}{\partial x} \left[\frac{\hbar}{2im} \left(\psi^* \, \frac{\partial \psi}{\partial x} - \frac{\partial \psi^*}{\partial x} \, \psi \right) \right]$$

If we define the *flux* by

$$j(x, t) = \frac{\hbar}{2im} \left(\psi^* \, \frac{\partial \psi}{\partial x} - \frac{\partial \psi^*}{\partial x} \, \psi \right) \tag{3-10}$$

we see that

$$\frac{\partial}{\partial t} \, P(x, t) + \frac{\partial}{\partial x} \, j(x, t) = 0 \tag{3-11}$$

Integrating, we find that

$$\frac{\partial}{\partial t} \int_{-\infty}^{\infty} dx \, P(x, t) = - \int_{-\infty}^{\infty} dx \, \frac{\partial}{\partial x} \, j(x, t) = 0 \tag{3-12}$$

since for square integrable functions, $j(x, t)$ vanishes at infinity. Incidentally, had we allowed discontinuities in $\psi(x)$, we would have been led to delta-function[2] singularities in the flux, and hence in the probability density, which is unacceptable in a physically observable quantity.

The relation (3-11) is a conservation law. It expresses the fact that a change in the density in a region in x is compensated by a net change in flux into that region

$$\frac{\partial}{\partial t} \int_{a}^{b} dx \, P(x, t) = - \int_{a}^{b} dx \, \frac{\partial}{\partial x} \, j(x, t)$$

$$= j(a, t) - j(b, t) \tag{3-13}$$

[2] See Appendix A for a discussion of delta functions.

The definition of $P(x, t)$, $j(x, t)$ and the conservation law are maintained if the equation (3-1) is changed to

$$i\hbar \frac{\partial \psi(x,t)}{\partial t} = -\frac{\hbar^2}{2m} \frac{\partial^2 \psi(x,t)}{\partial x^2} + V(x)\, \psi(x, t) \tag{3-14}$$

provided that $V(x)$ is real. This is important, since we will argue later that (3-14) is the Schrödinger equation for a particle in a potential $V(x)$. The generalization to three dimensions is straightforward. Eq. 3-14 becomes

$$i\hbar \frac{\partial \psi(x, y, z, t)}{\partial t} = -\frac{\hbar^2}{2m} \left(\frac{\partial^2}{\partial x^2} + \frac{\partial^2}{\partial y^2} + \frac{\partial^2}{\partial z^2} \right) \psi(x, y, z, t)$$
$$+ V(x, y, z)\, \psi(x, y, z, t)$$

that is,

$$i\hbar \frac{\partial \psi(\mathbf{r}, t)}{\partial t} = -\frac{\hbar^2}{2m} \nabla^2 \psi(\mathbf{r}, t) + V(\mathbf{r})\, \psi(\mathbf{r}, t) \tag{3-15}$$

and the generalization of (3-11) reads

$$\frac{\partial}{\partial t} P(\mathbf{r}, t) + \nabla \cdot \mathbf{j}(\mathbf{r}, t) = 0 \tag{3-16}$$

where

$$P(\mathbf{r}, t) = |\psi(\mathbf{r}, t)|^2 \tag{3-17}$$

and

$$\mathbf{j}(\mathbf{r}, t) = \frac{\hbar}{2im} [\psi^*(\mathbf{r}, t)\, \nabla \psi(\mathbf{r}, t) - \nabla \psi^*(\mathbf{r}, t)\, \psi(\mathbf{r}, t)] \tag{3-18}$$

Given the probability density $P(x, t)$, expectation values of functions of x may be calculated. In general, we have[3]

$$\langle f(x) \rangle = \int dx\, f(x)\, P(x, t) = \int dx\, \psi^*(x, t)\, f(x)\, \psi(x, t) \tag{3-19}$$

This only has meaning if the integral converges. The expression does not help us if we want to calculate the expectation value of the momentum, since we do not know how to write momentum in terms of x. We try the following: since classically,

$$p = mv = m \frac{dx}{dt} \tag{3-20}$$

---

[3] For a finite, discrete "sample space" with probabilities p_i so that $\Sigma p_i = 1$, the mean value of any variable over the space is $\langle f \rangle = \Sigma f_i p_i$.

we shall write

$$\langle p \rangle = m \frac{d}{dt} \langle x \rangle = m \frac{d}{dt} \int dx \, \psi^*(x, t) \, x\psi(x, t) \qquad (3\text{-}21)$$

This yields

$$\langle p \rangle = m \int_{-\infty}^{\infty} dx \left(\frac{\partial \psi^*}{\partial t} x\psi + \psi^* x \frac{\partial \psi}{\partial t} \right)$$

Note that there is no dx/dt under the integral sign. The only quantity that varies with time is $\psi(x, t)$, and it is this variation that gives rise to a change in x with time. Making use of (3-1) and its complex conjugate, we have

$$\langle p \rangle = \frac{\hbar}{2i} \int_{-\infty}^{\infty} dx \left(\frac{\partial^2 \psi^*}{\partial x^2} x\psi - \psi^* x \frac{\partial^2 \psi}{\partial x^2} \right)$$

Now

$$\frac{\partial^2 \psi^*}{\partial x^2} x\psi = \frac{\partial}{\partial x} \left(\frac{\partial \psi^*}{\partial x} x\psi \right) - \frac{\partial \psi^*}{\partial x} \psi - \frac{\partial \psi^*}{\partial x} x \frac{\partial \psi}{\partial x}$$

$$= \frac{\partial}{\partial x} \left(\frac{\partial \psi^*}{\partial x} x\psi \right) - \frac{\partial}{\partial x} (\psi^*\psi) + \psi^* \frac{\partial \psi}{\partial x}$$

$$- \frac{\partial}{\partial x} \left(\psi^* x \frac{\partial \psi}{\partial x} \right) + \psi^* \frac{\partial \psi}{\partial x} + \psi^* x \frac{\partial^2 \psi}{\partial x^2}$$

Hence the integrand has the form

$$\frac{\partial}{\partial x} \left(\frac{\partial \psi^*}{\partial x} x\psi - \psi^* x \frac{\partial \psi}{\partial x} - \psi^*\psi \right) + 2\psi^* \frac{\partial \psi}{\partial x}$$

so that

$$\langle p \rangle = \int dx \, \psi^*(x, t) \frac{\hbar}{i} \frac{\partial}{\partial x} \psi(x, t) \qquad (3\text{-}22)$$

since the integral of the derivatives vanishes for square integrable functions.

This suggests that the momentum is represented by the *operator*

$$\boxed{p = \frac{\hbar}{i} \frac{\partial}{\partial x}} \qquad (3\text{-}23)$$

and that, more generally

$$\langle f(p) \rangle = \int dx \, \psi^*(x, t) \, f\left(\frac{\hbar}{i} \frac{\partial}{\partial x} \right) \psi(x, t) \qquad (3\text{-}24)$$

Armed with this representation we can now discuss the physical significance of $\phi(p)$, which appears in (3-2). First, it is sufficient to consider that equation at $t = 0$, since $\phi(p)$ does not have any time dependence. With

$$\psi(x) = \frac{1}{\sqrt{2\pi\hbar}} \int dp \, \phi(p) \, e^{ipx/\hbar} = \sqrt{\frac{\hbar}{2\pi}} \int dk \, \phi(\hbar k) \, e^{ikx}$$

we find, using the inversion formula for a Fourier integral, that

$$\phi(\hbar k) = \frac{1}{\sqrt{2\pi\hbar}} \int dx \, \psi(x) \, e^{-ikx}$$

that is,

$$\phi(p) = \frac{1}{\sqrt{2\pi\hbar}} \int dx \, \psi(x) \, e^{-ipx/\hbar} \tag{3-25}$$

Now

$$\int dp \, \phi^*(p) \, \phi(p) = \int dp \, \phi^*(p) \, \frac{1}{\sqrt{2\pi\hbar}} \int dx \, \psi(x) \, e^{-ipx/\hbar}$$

$$= \int dx \, \psi(x) \, \frac{1}{\sqrt{2\pi\hbar}} \int dp \, \phi^*(p) \, e^{-ipx/\hbar}$$

$$= \int dx \, \psi(x) \, \psi^*(x) = 1 \tag{3-26}$$

This result is known as *Parseval's theorem* in the mathematical literature. It states that if a function is normalized to 1, so is its Fourier transform.

Next consider

$$\langle p \rangle = \int dx \, \psi^*(x) \, \frac{\hbar}{i} \, \frac{d\psi(x)}{dx}$$

$$= \int dx \, \psi^*(x) \, \frac{\hbar}{i} \, \frac{d}{dx} \, \frac{1}{\sqrt{2\pi\hbar}} \int dp \, \phi(p) \, e^{ipx/\hbar}$$

$$= \int dp \, \phi(p) \, p \, \frac{1}{\sqrt{2\pi\hbar}} \int dx \, \psi^*(x) \, e^{ipx/\hbar}$$

$$= \int dp \, \phi(p) \, p\phi^*(p) \tag{3-27}$$

This result, together with (3-26), strongly suggests that $\phi(p)$ should be interpreted as the wave function in momentum space, with $|\phi(p)|^2$ yielding the probability density for finding the particle with momentum p. When $\psi(x, t)$ is a solution of (3-14), we may define $\phi(p, t)$ by

$$\psi(x, t) = \frac{1}{\sqrt{2\pi\hbar}} \int dp \, \phi(p, t) \, e^{ipx/\hbar} \qquad (3\text{-}28)$$

The fact that in general $\phi(p, t)$ has a time dependence does not change (3-26), (3-27), or its interpretation. Lest the reader think that in spite of this symmetry between x- and p-space, $p = (\hbar/i)(\partial/\partial x)$ is an operator, and x is not, we note that x is in fact an operator too. It happens to have a particularly simple form in x-space, but if we want to calculate $\langle f(x) \rangle$ in momentum space, then, we can show by methods very similar to the ones used above that

$$\langle f(x) \rangle = \int dp \, \phi^*(p, t) \, f\left(i\hbar \frac{\partial}{\partial p}\right) \phi(p, t) \qquad (3\text{-}29)$$

In other words, the operator x has the representation

$$\boxed{x = i\hbar \frac{\partial}{\partial p}} \qquad (3\text{-}30)$$

in momentum space.

We will find that operators play a central role in quantum mechanics, and we will slowly learn a great deal about them. At this point we will indicate only that:

1. In contrast to ordinary numbers, operators do not always commute. If we define

$$[A, B] = AB - BA \qquad (3\text{-}31)$$

then

$$[p, x] \, \psi(x, t) = \frac{\hbar}{i} \frac{\partial}{\partial x} x\psi(x, t) - x \frac{\hbar}{i} \frac{\partial \psi(x, t)}{\partial x}$$

$$= \frac{\hbar}{i} \, \psi(x, t) \qquad (3\text{-}32)$$

that is, we have the *commutation relation*

$$[p, x] = \frac{\hbar}{i} \qquad (3\text{-}33)$$

This leads to an ambiguity in transcribing a classical function $f(x, p)$ into operator form, and we shall adopt the rule that $f(x, p)$ be symmetrized in x and p. Thus

$$xp \rightarrow \tfrac{1}{2}(xp + px)$$
$$x^2p \rightarrow \tfrac{1}{4}(x^2p + 2xpx + px^2) \qquad (3\text{-}34)$$

and so on.

Later we will see that it is the lack of commutativity of x and p that stands behind the uncertainty relations connecting these two variables.

2. The appearance of the operator p, with its i, might lead us to worry about the reality of the expectation value of p. We can, however, check the fact that p is real. We have

$$
\begin{aligned}
\langle p \rangle - \langle p \rangle^* &= \int dx\, \psi^*(x)\, \frac{\hbar}{i}\, \frac{\partial \psi}{\partial x} - \int dx\, \psi(x) \left(-\frac{\hbar}{i}\, \frac{\partial \psi^*}{\partial x} \right) \\
&= \frac{\hbar}{i} \int dx \left(\psi^* \frac{\partial \psi}{\partial x} + \frac{\partial \psi^*}{\partial x} \psi \right) \\
&= \frac{\hbar}{i} \int dx\, \frac{\partial}{\partial x}\, (\psi^* \psi) \qquad\qquad (3\text{-}35) \\
&= 0
\end{aligned}
$$

provided the wave function vanishes at infinity, which it does for a square integrable function. Sometimes one has occasion to use functions that are not square integrable but that have certain periodicity conditions, for example,

$$\psi(x) = \psi(x + L) \qquad\qquad (3\text{-}36)$$

If one restricts onself to working in the region $0 \leq x \leq L$, then $\hbar/i \; d/dx$ is still a hermitian operator, since in (3-35),

$$
\begin{aligned}
\langle p \rangle - \langle p \rangle^* &= \frac{\hbar}{i} \int_0^L dx\, \frac{\partial}{\partial x}\, (\psi^*(x)\, \psi(x)) \\
&= \frac{\hbar}{i}\, |\psi(L)|^2 - \frac{\hbar}{i}\, |\psi(0)|^2 = 0 \qquad\qquad (3\text{-}37)
\end{aligned}
$$

An operator whose expectation value for all admissible wave functions is real is called a *hermitian operator*, and hence p, like x, is a hermitian operator.[4]

We conclude this chapter by noting that the equation

$$i\hbar\, \frac{\partial \psi(x, t)}{\partial t} = -\frac{\hbar^2}{2m}\, \frac{\partial^2 \psi(x, t)}{\partial x^2}$$

may, with the identification $(\hbar/i)(\partial/\partial x) = p_{op}$ be written in the form

$$i\hbar\, \frac{\partial \psi(x, t)}{\partial t} = \frac{p_{op}^2}{2m}\, \psi(x, t) \qquad\qquad (3\text{-}38)$$

The operator on the right is just the energy for a free particle. If we generalize this to a particle in a potential, we write

$$i\hbar\, \frac{\partial \psi(x, t)}{\partial t} = \left[\frac{p_{op}^2}{2m} + V(x) \right] \psi(x, t) \qquad\qquad (3\text{-}39)$$

---

[4] Some mathematical background on operators is discussed in Appendix B.

or, more explicitly

$$i\hbar \frac{\partial \psi(x, t)}{\partial t} = - \frac{\hbar^2}{2m} \frac{\partial^2 \psi(x, t)}{\partial x^2} + V(x)\, \psi(x, t) \qquad (3\text{-}40)$$

This equation, generalizing (3-1), is the *basic equation of nonrelativistic quantum mechanics*, and it was first proposed by Schrödinger. The Schrödinger equation, obtained above, can also be written in the form

$$i\hbar \frac{\partial \psi(x, t)}{\partial t} = H\, \psi(x, t) \qquad (3\text{-}41)$$

where H is the *energy operator*. H is commonly called the *Hamiltonian*, because it is an operator version of the classical mechanical Hamiltonian function. Since[5] p is a hermitian operator, so is p^2, and therefore so is

$$H = \frac{p^2}{2m} + V(x) \qquad (3\text{-}42)$$

if $V(x)$ is a real potential.

In summary:

1. The time dependence of wave functions is given by the first order partial differential equation

$$i\hbar \frac{\partial \psi(x, t)}{\partial t} = H\, \psi(x, t)$$

where H is the operator $p^2/2m + V(x)$.

2. Wave functions are restricted to square integrable functions.

3. The probability density for finding the particle at x is

$$P(x, t) = |\psi(x, t)|^2$$

4. The function $\phi(p, t)$ defined by

$$\psi(x, t) = \frac{1}{\sqrt{2\pi\hbar}} \int dp\, \phi(p, t)\, e^{ipx/\hbar}$$

is the wave function in momentum space, and the probability density for finding the particle with momentum p is $|\phi(p, t)|^2$.

5. The momentum p and the position x are *operators*, that is, they are quantities that differ from numbers because of their lack of commutativity. In x-space, the momentum operator takes the form

$$p = \frac{\hbar}{i} \frac{\partial}{\partial x}$$

---

[5] From now on we will drop the subscript op on p_{op}. We will use it only when there is danger of confusion with a number described by the letter p.

and in p-space, the x operator takes the form

$$x = i\hbar \frac{\partial}{\partial p}$$

both consistent with the fundamental commutation relation for x with p,

$$[p, x] = \frac{\hbar}{i}$$

We are now ready for a quantitative discussion of quantum mechanics. We have abandoned the notion of a wave packet as representing a particle. This notion was helpful to us in making the Schrödinger equation plausible, but now it is $\psi(x, t)$ and its probabilistic interpretation that tell us where the particle is, without the particle being thought of as "made up out of waves."

Problems

1. Use (3-2) and (3-4) to write the solution of the free particle Schrödinger equation in the form

$$\psi(x, t) = \int dx' \, K(x, x'; t) \, \psi(x', 0)$$

Obtain a representation for $K(x, x'; t)$ in the form of an integral, and evaluate the integral. Show that

$$K(x, x'; 0) = \delta(x - x')$$

2. Show that the conservation law (3-11) holds when $\psi(x, t)$ is a solution of the Schrödinger equation with a potential $V(x)$, (3-14), provided that $V(x)$ is real.

3. Suppose that $V(x)$ is complex. Obtain an expression for $\partial P(x, t)/\partial t$ and $d/dt \int dx \, P(x, t)$. For absorption, the last must be negative. What does this tell us about $V(x)$?

4. Consider the Klein-Gordon equation

$$\frac{1}{c^2} \frac{\partial^2 \psi(x, t)}{\partial t^2} - \frac{\partial^2 \psi(x, t)}{\partial x^2} + \left(\frac{\mu c}{\hbar}\right)^2 \psi(x, t) = 0$$

Show that there is a conservation law of the form (3-11) given that $j(x, t)$ has the form

$$j(x, t) = -\frac{\hbar}{2i\mu}\left(\psi \frac{\partial \psi^*}{\partial x} - \psi^* \frac{\partial \psi}{\partial x}\right)$$

What is the form of $P(x, t)$? Can you give an argument for why the Klein-Gordon equation is not a good candidate for a one-particle equation (i.e., an alternative for the Schrödinger equation)?

5. Given that

$$\psi(x) = \left(\frac{\pi}{\alpha}\right)^{-1/4} e^{-\alpha x^2/2}$$

calculate

(a) $\langle x^n \rangle$

(b) $\sqrt{\langle x^2 \rangle - \langle x \rangle^2} \equiv \Delta x$

6. Calculate the momentum space wave function for the system described by the wave function in problem 5. Use it to calculate

(a) $\langle p^n \rangle$

(b) $\sqrt{\langle p^2 \rangle - \langle p \rangle^2} \equiv \Delta p$

Calculate the value of $\Delta x \, \Delta p$ using the above, and the result of problem 5(b).

7. Given the wave function

$$\psi(x) = \frac{N}{x^2 + a^2}$$

(a) Calculate N needed to normalize $\psi(x)$.

(b) Use the above wave function to calculate $\langle x^n \rangle$. What values of n lead to convergent integrals?

(c) Calculate $\langle p^2 \rangle$ directly, and using the momentum space wave function.

(d) Use the definitions

$$\Delta x = \sqrt{\langle x^2 \rangle - \langle x \rangle^2}$$

$$\Delta p = \sqrt{\langle p^2 \rangle - \langle p \rangle^2}$$

to calculate $\Delta x \, \Delta p$ for this problem.

8. Show that the operator relation

$$e^{ipa/\hbar} x e^{-ipa/\hbar} = x + a$$

holds. The operator e^A is defined to be

$$e^A = \sum_{n=0}^{\infty} A^n/n!$$

[*Hint.* Calculate $e^{ipa/\hbar} x e^{-ipa/\hbar} f(p)$ where $f(p)$ is any function of p, and use the representation $x = i\hbar \, d/dp$.]

9. Consider the functions $\psi(\theta)$ of the angular variable θ, restricted to the interval $-\pi \le \theta \le \pi$.

If the wave functions satisfy the condition $\psi(\pi) = \psi(-\pi)$, show that the operator

$$L = \frac{\hbar}{i} \frac{d}{d\theta}$$

is hermitian.

10. Consider $\phi(p)$, the momentum space wave function of a particle. If this function is only defined for positive values of p, what condition must $\phi(p)$ satisfy in order that x be a hermitian operator? [Use (3-30).]

chapter 4

Eigenfunctions and Eigenvalues

Let us consider the time-dependent Schrödinger equation obtained in Chapter 3,

$$i\hbar \frac{\partial \psi(x,t)}{\partial t} = -\frac{\hbar^2}{2m} \frac{\partial^2 \psi(x,t)}{\partial x^2} + V(x)\, \psi(x,t) \tag{4-1}$$

and attempt to solve it by reducing it to a pair of ordinary differential equations in one variable. Write

$$\psi(x,t) = T(t)u(x) \tag{4-2}$$

which implies that

$$i\hbar u(x) \frac{dT(t)}{dt} = \left[-\frac{\hbar^2}{2m} \frac{d^2 u(x)}{dx^2} + V(x)u(x) \right] T(t)$$

Dividing by $u(x)\,T(t)$ we get

$$i\hbar \frac{dT(t)/dt}{T(t)} = \frac{-(\hbar^2/2m)\,(d^2 u(x)/dx^2) + V(x)\,u(x)}{u(x)} \tag{4-3}$$

This can only be satisfied if both sides are equal to a constant, which we call E. The solution of

$$i\hbar \frac{dT(t)}{dt} = ET(t) \tag{4-4}$$

is

$$T(t) = C\,e^{-iEt/\hbar} \tag{4-5}$$

where C is a constant. The other equation is

$$-\frac{\hbar^2}{2m} \frac{d^2 u(x)}{dx^2} + V(x)u(x) = Eu(x) \tag{4-6}$$

This equation is frequently called the *time-independent Schrödinger equation*. Its character is really different from that of (4-1). Equation 4-1 describes the time

57

development of $\psi(x,t)$; Eq. 4-6 is an *eigenvalue equation*. To explain what this means, we must return to the notion of an operator, which was briefly mentioned but not defined in the last chapter.

Most generally, an operator acting on a function maps it into another function. Let us consider some examples

$$Of(x) = f(x) + x^2$$
$$Of(x) = [f(x)]^2$$
$$Of(x) = f(3x^2 + 1)$$
$$Of(x) = [df(x)/dx]^3$$
$$Of(x) = df(x)/dx - 2f(x)$$
$$Of(x) = \lambda f(x) \tag{4-7}$$

All of these examples share the property that given a function $f(x)$, there is a rule that determines $Of(x)$ for us. There is a special class of operators, called *linear operators* (we denote these operators by L to distinguish them from the general operators O). These have the property that

$$L[f_1(x) + f_2(x)] = Lf_1(x) + Lf_2(x) \tag{4-8}$$

and,[1] with c an arbitrary complex number,

$$Lcf(x) = cLf(x) \tag{4-9}$$

Thus, in our list only the last two are linear operators.

A linear operator will map one function into another, as in the example

$$Lf(x) = \frac{df(x)}{dx} - 2f(x)$$

It is instructive to think of the functions as analogous to vectors in a three-dimensional space. The action of an operator is to transform a vector into another vector. In the special case that the vectors are all of unit length, an operator will transform one point on a unit sphere into another. An operator, in this special (but very relevant) example, may be a rotation about an axis (Fig. 4.1). Let the operator be a rotation of, say, 30° about the z-axis. It is easy to visualize what happens to various vectors under this operation. There will be two vectors that have a special property: the unit vectors to the north and south poles will be mapped into themselves under the rotation. This is a special example of an operator equation like (4-6), which may be written as

$$Hu_E(x) = Eu_E(x) \tag{1-10}$$

---

[1] There are also *antilinear* operators, for which (4-9) is replaced by $Lcf(x) = c*Lf(x)$.

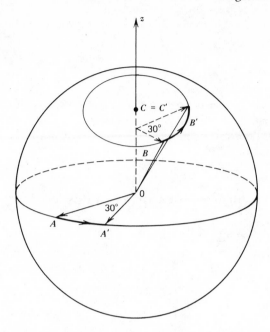

Fig. 4-1. An illustration of the operator rotating all vectors by 30° with the vectors lying on the unit sphere: for vectors on the equator $(A \rightarrow A')$, at an intermediate latitude $(B \rightarrow B')$, and at a pole $(C \rightarrow C' = C)$.

This equation states that H, the Hamiltonian operator acting on a special class of functions, will give back the function that it is acting on, multiplied by a constant. The constant is called the *eigenvalue*. The solution of the equation depends on E, and we have therefore labeled it with an E. The solution $u_E(x)$ is called the *eigenfunction*, corresponding to the eigenvalue E, of the operator H. We shall see that eigenvalues can form a continuum or be discrete.

The solution (4-2) is of the form $u_E(x) \, e^{-iEt/\hbar}$. Since (4-1) is a linear equation, a sum of solutions of the above form, with permissible values of E, is also a solution. Thus the most general solution of (4-1) is

$$\psi(x,t) = \left(\sum + \int dE \right) C(E) \, u_E(x) \, e^{-iEt/\hbar} \qquad (4\text{-}11)$$

where $C(E)$ is an arbitrary function of the eigenvalues, and the sum extends over the discrete values of E, the integral over the continuous range of eigenvalues. The eigenvalues of the operator H are called the energy eigenvalues, as is suggested by the form of

$$H = \frac{p_{op}^2}{2m} + V(x) \tag{4-12}$$

Before discussing a very simple but instructive example, we note that the separation of the equation would fail if the potential V depended explicitly on time. We will see later that when this is the case, energy is not a constant of the motion.

A. The Eigenvalue Problem for a Particle in a Box

We consider Eq. 4-6 with

$$V(x) = 0 \qquad |x| < a$$
$$= \infty \qquad \text{elsewhere} \tag{4-13}$$

This implies that the wave function must vanish for $|x| > a$, that is,

$$u(a) = u(-a) = 0 \tag{4-14}$$

Inside the box

$$\frac{d^2 u(x)}{dx^2} + \frac{2mE}{\hbar^2} u(x) = 0 \tag{4-15}$$

First we notice that if $E < 0$, then (4-15) takes the form

$$\frac{d^2 u(x)}{dx^2} - \kappa^2 u(x) = 0 \tag{4-16}$$

with $\kappa^2 = 2m|E|/\hbar^2$. The most general solution is a linear combination of $e^{\kappa x}$ and $e^{-\kappa x}$, and there is no way of satisfying the boundary conditions (4-14). Thus the energy eigenvalues must be positive. We write

$$k^2 = \frac{2mE}{\hbar^2} \tag{4-17}$$

so that the equation (4-15) takes the form

$$\frac{d^2 u(x)}{dx^2} + k^2 u(x) = 0 \tag{4-18}$$

whose solutions are $\sin kx$ and $\cos kx$. The boundary conditions imply that for the sine solution, which we denote by $u_n^{(-)}(x)$,

$$ka = n\pi \qquad n = 1, 2, 3, \ldots \tag{4-19}$$

so that

$$E_n^{(-)} = \frac{n^2 \pi^2 \hbar^2}{2ma^2} \qquad (4\text{-}20)$$

It is easy to check that the normalized solution is

$$u_n^{(-)}(x) = \frac{1}{\sqrt{a}} \sin \frac{n\pi x}{a} \qquad (4\text{-}21)$$

The cosine solution, denoted by $u_n^{(+)}(x)$ must be such that

$$ka = \left(n - \tfrac{1}{2}\right) \pi \qquad n = 1, 2, 3, \ldots \qquad (4\text{-}22)$$

that is,

$$E_n^{(+)} = \frac{[n - (1/2)]^2 \pi^2 \hbar^2}{2ma^2} \qquad (4\text{-}23)$$

The normalized solution is therefore

$$u_n^{(+)}(x) = \frac{1}{\sqrt{a}} \cos \frac{[n - (1/2)] \pi x}{a} \qquad (4\text{-}24)$$

We see that the ($\pm$) signs refer to the even/odd property under the reflection $x \rightarrow -x$.

The solutions have the property that

$$\int_{-a}^{a} dx u_m^{(+)}(x)^* \, u_n^{(+)}(x) = \int_{-a}^{a} dx u_m^{(-)}(x)^* \, u_n^{(-)}(x) = \delta_{mn}$$

$$\int_{-a}^{a} dx u_m^{(+)}(x)^* \, u_n^{(-)}(x) = 0 \qquad (4\text{-}25)$$

that is, they satisfy what are called *orthonormality conditions*. Since the solutions are real, the complex conjugation is not really necessary, but is inserted for consistency with future usage.

The state of lowest energy, the *ground state* is represented by $u_1^{(+)}(x)$, and its energy is

$$E_1^{(+)} = \frac{\pi^2 \hbar^2}{8ma^2} \qquad (4\text{-}26)$$

The solutions are real. It therefore follows that

$$\langle p \rangle = 0 \qquad (4\text{-}27)$$

This can be done by direct calculation, or by a symmetry argument: for any one of the solutions, which are real, $\langle p \rangle$ is of the form $(\hbar/i) \times$ (integral). Since $\langle p \rangle$ must be real, the integral, involving only real functions, must vanish; equivalently, the integral involves a product of two even or two odd functions, with

d/dx inserted between them. The total integrand is thus an odd function of x, and upon integration over a symmetric interval must yield a vanishing integral.

We can calculate $\langle p^2 \rangle$ for the various solutions. In fact, since inside the box $p^2 = 2mE$, we have

$$\langle p^2 \rangle = 2mE_n^{(\pm)} \tag{4-28}$$

Notice that

$$2a \sqrt{\langle p^2 \rangle} \sim 2n\pi\hbar > \hbar \tag{4-29}$$

is consistent with the uncertainty relation.[2] We also note that the larger the number of nodes in a solution, the higher is its energy (Fig. 4.2). This is understandable, since the kinetic energy is larger for a solution with a larger curvature, a measure of which is d^2u/dx^2. Specifically

$$-\frac{\hbar^2}{2m} \int dx u^*(x) \frac{d^2u}{dx^2} = \frac{\hbar^2}{2m} \int dx \frac{du^*}{dx}\frac{du}{dx} = \frac{\hbar^2}{2m} \int dx \left|\frac{du}{dx}\right|^2$$

is large when the function has a lot of variation in it.

B. The Expansion Postulate

An arbitrary function $\psi(x)$, satisfying the boundary conditions $\psi(a) = \psi(-a) = 0$, can be constructed from our solutions. It will be a superposition of all of them

$$\psi(x) = \sum_{n=1}^{\infty} [A_n^{(+)}u_n^{(+)}(x) + A_n^{(-)}u_n^{(-)}(x)] \tag{4-30}$$

The orthonormality relations can be used to determine the coefficients $A_n^{(\pm)}$ With the help of (4-25) we can calculate, for example,

$$\int dx u_n^{(+)*}(x)\psi(x)$$

$$= \sum_{m=1}^{\infty} \left[A_m^{(+)} \int u_n^{(+)*}(x)u_m^{(+)}(x)\, dx + A_m^{(-)} \int u_n^{(+)*}(x)u_m^{(-)}(x)\, dx \right]$$

$$= A_n^{(+)}$$

so that

$$A_n^{(\pm)} = \int dx u_n^{(\pm)*}(x)\, \psi(x) \tag{4-31}$$

[2] It is a general feature that for higher eigenfunctions $\Delta x \Delta p$ grows with the eigenvalue.

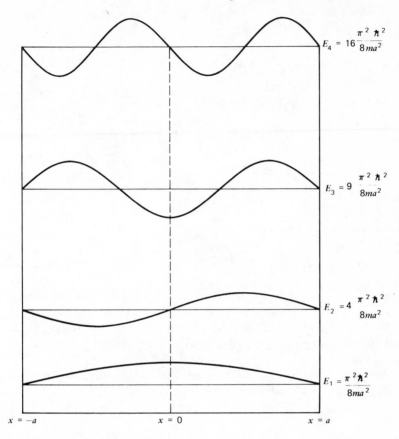

Fig. 4-2. Eigensolutions for particle in a box.

As in our discussion of the free wave packet, we can calculate the time development of this arbitrary initial packet. Since each of the solutions $u_n^{(\pm)}(x)$ acquires the time dependence $e^{-iEn(\pm)t/\hbar}$ [see (4-11)], we have quite generally

$$\psi(x,t) = \sum_{n=1}^{\infty} [A_n^{(+)}u_n^{(+)}(x)\,e^{-iEn(+)t/\hbar} + A_n^{(-)}u_n^{(-)}(x)\,e^{-iEn(-)t/\hbar}] \quad (4\text{-}32)$$

To get an idea of the physical meaning of the coefficients $A_n^{(\pm)}$, we calculate the expectation value of the energy in an arbitrary state. Since inside the box $H = p^2/2m$, and outside the box nothing contributes, and since

$$Hu_n^{(\pm)}(x) = E_n^{(\pm)}u_n^{(\pm)}(x) \quad (4\text{-}33)$$

we have, using the orthonormality relations (4-25),

$$\langle H \rangle = \int dx \psi^*(x) \, H\psi(x)$$

$$= \int dx \left\{ \sum_{n=1}^{\infty} \left[A_n^{(+)*} u_n^{(+)}(x)^* + A_n^{(-)*} u_n^{(-)}(x)^* \right] \right\}$$

$$\times \left\{ \sum_{m=1}^{\infty} \left[E_m^{(+)} A_m^{(+)} u_m^{(+)}(x) + E_m^{(-)} A_m^{(-)} u_m^{(-)}(x) \right] \right\}$$

$$= \sum_{m=1}^{\infty} (E_m^{(+)} |A_m^{(+)}|^2 + E_m^{(-)} |A_m^{(-)}|^2) \tag{4-34}$$

In exactly the same way we show that

$$\int dx \psi^*(x) \, \psi(x) = 1$$

implies that

$$\sum_{n=1}^{\infty} (|A_n^{(+)}|^2 + |A_n^{(-)}|^2) = 1 \tag{4-35}$$

Equation 4-34, together with the normalization condition (4-35), strongly suggests that $|A_n^{(\pm)}|^2$ *be interpreted as the probability that a measurement of the energy for the arbitrary state yields the value $E_n^{(\pm)}$*. Note that only the values $E_n^{(\pm)}$ are possible for the energy, so that a given measurement can only yield one of the values $E_n^{(\pm)}$.

For what packet will the measurement always yield an energy $E_k^{(+)}$ (an eigenvalue)? Clearly this will be so only when

$$|A_n^{(+)}|^2 = \delta_{nk} \tag{4-36}$$

that is, when $\psi(x) = u_k^{(+)}(x)$, the eigenfunction corresponding to the eigenvalue $E_k^{(+)}$. This leads us to a very important conclusion:

Suppose that we have a general packet described by $\psi(x)$. If an energy measurement is carried out, only an eigenvalue of the Hamiltonian operator H can result, with probability

$$P(E_n) = |\int dx u_n^*(x) \, \psi(x)|^2 \tag{4-37}$$

(where we have left off the ($\pm$) label for generality). *Furthermore, after the measurement that has yielded the eigenvalue E_n, the state of the system is described by the eigenfunction $u_n(x)$*, since otherwise a repetition of the measurement would not necessarily give the same result, and reproducibility of a measurement for a given system is essential for the measurement to have any meaning. These statements are not peculiar to the problem of a particle in a box. They hold for more general systems [with a $V(x)$], and also for hermitian operators other than the Hamiltonians, as will be seen again and again, and these statements lie at the heart of quantum mechanics.

C. Parity

The eigenfunctions for the particle in a box were divided into two classes: those even in x, denoted with a $(+)$ and those odd in x, denoted with $(-)$. If we start with a wave packet $\psi(x)$ that is even in x, say, then in (4-30) all the $A_n^{(-)}$ must vanish. Equation (4-32) then shows that the packet remains even in x for all time. The same holds for a packet that is initially odd. Thus for our box, which was symmetrically centered about $x = 0$, we find that "evenness" and "oddness" are time independent. Since any constant of the motion is of interest to us, we will formalize the discussion somewhat.

We do this by introducing the *parity operator* P, whose rule of operation is to reflect $x \rightarrow -x$. Thus for any packet $\psi(x)$, we have

$$P\psi(x) = \psi(-x) \tag{4-38}$$

For an even packet we have

$$P\psi^{(+)}(x) = \psi^{(+)}(x) \tag{4-39}$$

and for an odd packet

$$P\psi^{(-)}(x) = -\psi^{(-)}(x) \tag{4-40}$$

These two equations are eigenvalue equations, and what we have shown is that even functions are eigenfunctions of P with eigenvalue $+1$, while odd functions are eigenfunctions of P with eigenvalue -1. In the problem of the particle in a box, the functions $u_n^{(\pm)}(x)$ are not only eigenfunctions of H; *they are simultaneously eigenfunctions of P.*

The eigenvalues ± 1 are the only possible ones. Suppose we have

$$Pu(x) = \lambda u(x) \tag{4-41}$$

Applying P again, we would get

$$P^2 u(x) = P\lambda u(x) = \lambda^2 u(x) \tag{4-42}$$

However $P^2 u(x) = u(x)$, since two reflections should not change anything. Hence $\lambda^2 = 1$, that is, $\lambda = \pm 1$. An arbitrary function $\psi(x)$ can always be written as a sum of an even and an odd function

$$\psi(x) = \tfrac{1}{2}[\psi(x) + \psi(-x)] + \tfrac{1}{2}[\psi(x) - \psi(-x)] \tag{4-43}$$

that is, just as with the eigenfunctions of H discussed in our example, any function can be expanded in terms of the eigenfunctions of this new operator. This too is a general feature of hermitian operators: *the eigenfunctions of any hermitian operator are said to form a complete set, in terms of which any function can be expanded.* We leave it to the reader to show that $\langle P \rangle$ is real for any state $\psi(x)$, which implies that the operator is hermitian.

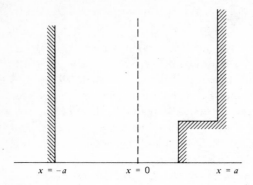

$$x = -a \qquad x = 0 \qquad x = a$$

Fig. 4-3. Box for which there is no symmetry under reflections.

The explicit appearance of evenness and oddness came about because we centered the box at $x = 0$. Had we taken it to lie between 0 and $2a$, nothing would have changed, and there would still be symmetry under reflections about $x = a$. Such symmetry would, however, be much less apparent. The lesson to be learned here is that in setting up a quantum mechanical problem one should always pay attention to the symmetries in the Hamiltonian, and choose the coordinates in a way that exhibits the symmetries most explicitly. If the box were uneven (Fig. 4-3), no amount of changing coordinates would bring about a symmetry. The important fact is that *the symmetry be in the Hamiltonian*.[3] This may be seen more clearly by asking under what circumstances an even function will remain even for all time. Let

$$\psi(x,0) = \psi(-x,0) \equiv \psi^{(+)}(x) \tag{4-44}$$

The time development is given by

$$i\hbar \frac{\partial \psi(x,t)}{\partial t} = H\psi(x,t) \tag{4-45}$$

If we operate with P on this equation, we get

$$i\hbar \frac{\partial}{\partial t} P\psi(x,t) = PH\psi(x,t) \tag{4-46}$$

Under the special circumstances that

$$PH\psi(x,t) = HP\psi(x,t) \tag{4-47}$$

---

[3] When dealing with the box, we consider the walls as part of the potential, that is, the Hamiltonian. That is why we do not speak of boundary conditions instead of the Hamiltonian.

which holds when H is even under $x \to -x$, that is, when $V(x)$ is an even function (since d^2/dx^2 is even), we have

$$i\hbar \frac{\partial}{\partial t} [P\psi(x,t)] = H[P\psi(x,t)] \tag{4-48}$$

Hence

$$\psi^{(+)}(x,t) = \tfrac{1}{2}(1 + P)\,\psi(x,t) \tag{4-49}$$

and

$$\psi^{(-)}(x,t) = \tfrac{1}{2}(1 - P)\,\psi(x,t) \tag{4-50}$$

separately obey the Schrödinger equation, and do not mix, if the initial state is even (or odd). The condition for the time-independence of parity only holds if

$$(PH - HP)\,\psi(x,t) = 0 \tag{4-51}$$

for all possible states, that is, if the operatores P and H commute

$$[P,H] = 0 \tag{4-52}$$

This important condition will be seen to be quite general: *any operator that does not have an explicit time dependence and that commutes with the Hamiltonian H is a constant of the motion.* In particular, if the potential changes with time, that is, we have $V(x,t)$, then the energy itself is not a constant of the motion, just as in classical mechanics. Note that when V depends on t, the separation of the equation into an equation for the time dependence and an energy eigenvalue equation is not possible.

D. Momentum Eigenfunction and the Free Particle

Our discussion of parity showed that it is not only the energy operator H that has eigenfunctions and eigenvalues. Let us now solve the eigenvalue equation for the *momentum operator*

$$p_{op}u_p(x) = pu_p(x) \tag{4-53}$$

Since $p_{op} = (\hbar/i)(d/dx)$, this reads

$$\frac{du_p(x)}{dx} = \frac{ip}{\hbar}\,u_p(x) \tag{4-54}$$

The solution to this equation is

$$u_p(x) = C\,e^{ipx/\hbar} \tag{4-55}$$

with C a constant to be determined by normalization, and the eigenvalue p real, so that the eigenfunction does not blow up at either $+\infty$ or $-\infty$. This is the only constraint on p: we say that p_{op} has a *continuous spectrum*. We might, by analogy with (4-25), expect that the eigenfunctions obey orthonormality conditions. We see that

$$\int dx\, u_{p'}^*(x)\, u_p(x) = |C|^2 \int dx\, e^{i(p-p')x/\hbar}$$

$$= 2\pi |C|^2 \hbar \delta(p - p') \qquad (4\text{-}56)$$

With the choice

$$u_p(x) = \frac{1}{\sqrt{2\pi\hbar}}\, e^{ipx/\hbar} \qquad (4\text{-}57)$$

(4-56) reads

$$\int_{-\infty}^{\infty} dx\, u_{p'}^*(x)\, u_p(x) = \delta(p - p') \qquad (4\text{-}58)$$

This differs from (4-25) only in that the Kroenecker δ_{mn}, appropriate for discrete indices is replaced by a Dirac delta function $\delta(p - p')$ for the continuous indices.

The statement that any wave packet $\psi(x)$ may be expanded in terms of a complete set of eigenfunctions can also be established here. The analog of (4-30) must take into account that we are summing over a continuous index p, so that we write

$$\psi(x) = \int_{-\infty}^{\infty} dp\, \phi(p)\, \frac{e^{ipx/\hbar}}{\sqrt{2\pi\hbar}} \qquad (4\text{-}59)$$

According to the interpretation implicit in (4-37), $|\phi(p)|^2$, where

$$\phi(p) = \int dx \left(\frac{e^{ipx/\hbar}}{\sqrt{2\pi\hbar}}\right)^* \psi(x) \qquad (4\text{-}60)$$

gives the probability that a measurement of the momentum for an arbitrary packet $\psi(x)$ yields the eigenvalue p. In this way we justify the conjecture made about $\phi(p)$ in Chapter 3 (cf. Eq. 3-30).

Let us now turn to the free particle Hamiltonian. When $V(x)$ is zero everywhere, the energy eigenvalue equation reads

$$\frac{d^2u(x)}{dx^2} + k^2u(x) = 0 \qquad (4\text{-}61)$$

where $k^2 = 2mE/\hbar^2$. The solutions are e^{ikx} and e^{-ikx}, or linear combinations of these, for example, $\cos kx$ and $\sin kx$. The trouble with all of them is that they are not square integrable, since $\int_{-\infty}^{\infty} dx\, |Ae^{ikx} + Be^{-ikx}|^2$ diverges for all values of A and B.

There are three ways of getting around this difficulty.

(a) We may consider the problem defined by (4-61) as the limiting case of a particle in a box, with the walls receding to infinity, that is, $a \to \infty$. In this limit the solutions (4-21) and (4-24), even aside from the normalization factors $1/\sqrt{a}$ will become trivial, unless n becomes very large, so that

$$\frac{n\pi}{a} = k \tag{4-62}$$

becomes finite.[4] We can then neglect the $\frac{1}{2}$ in the $(n - \frac{1}{2})$ in the even solutions (4-24), and obtain the solutions

$$\frac{1}{\sqrt{a}} \sin kx \qquad \frac{1}{\sqrt{a}} \cos kx \tag{4-63}$$

We may keep the $1/\sqrt{a}$ factors: they will drop out of the answer to any physical question that we may ask about the system.[5] It is sometimes useful to keep them, since their presence in a final result indicates that an error has been made.

(b) We may work with wave packets. A solution of the form

$$\psi(x) = e^{ikx} \tag{4-64}$$

is a special case of (4-59) with

$$\phi(p) = \sqrt{2\pi\hbar}\, \delta(p - \hbar k) \tag{4-65}$$

that is, an infinitely peaked momentum-space distribution. Suppose we replace this limiting $\phi(p)$ by a very sharply peaked function $\sqrt{2\pi\hbar}\, g(p - \hbar k)$. Then e^{ikx} will be replaced by

$$\psi(x) = \int dp\, e^{ipx/\hbar} g(p - \hbar k)$$

$$= e^{ikx} \int dq\, e^{iqx/\hbar} g(q) \tag{4-66}$$

which is a plane wave, e^{ikx}, multiplied by a very broad function of x. We may make this function so broad that it is essentially constant over the region of physical interest. The uncertainty in the momentum will now be of the order of magnitude $\hbar/$(size of x-packet), and if the denominator is of macroscopic size, this uncertainty is negligible. We thus satisfy the mathematical requirements without changing any of the physics. The wave packet description is actually the one that is closest to what really happens physically, since any way of pre-

[4] We also keep x finite, and are not particularly interested in values of x that are a finite fraction of a.

[5] A question that is not meaningful physically is one that depends on the existence of the walls. For example, "How long will it take for a wave packet to go to the walls and return to $x = 0$?" is a question that we classify as not physically relevant.

paring the initial state, for example, firing an electron gun, can never, in practice, create an exact momentum eigenstate.

(c) The difficulty stems from the fact that for a wave function like e^{ikx}, the particle is not confined to any region of space, so that the probability of finding it anywhere is zero. If we do not ask questions that involve the probability of finding the particle in any finite region, no problems arise. One way of avoiding the normalization difficulty is to deal with the probability current, or *flux*

$$j(x) = \frac{\hbar}{2im}\left[\psi^*(x)\frac{d\psi(x)}{dx} - \frac{d\psi^*(x)}{dx}\psi(x)\right] \qquad (4\text{-}67)$$

discussed at the beginning of Chapter 3. For a wave function $Ce^{ipx/\hbar}$, the flux is $|C|^2 p/m$; for the wave function $Ce^{-ipx/\hbar}$, the flux is $-|C|^2 p/m$. If we note that for a one-dimensional problem, the flux of particles with a density of 1 particle/cm, moving with velocity $v = p/m$ is just v—that is the number crossing a point $x = x_0$ per second—we see that $|C|^2$ represents the density of particles per cm. Thus (4-57) represents particles with a density $1/2\pi\hbar$ per cm.

In three dimensions, with

$$u_{\mathbf{p}}(\mathbf{r}) = C\,e^{i\mathbf{p}\cdot\mathbf{r}/\hbar} \qquad (4\text{-}68)$$

the flux will be $|C|^2\,\mathbf{p}/m$, and this corresponds to a flow of particles, with density $|C|^2$ per cm^3 crossing a unit area perpendicular to $\mathbf{p}$, when the particles are moving with velocity $\mathbf{v} = \mathbf{p}/m$ (Fig. 4.4).

The energy eigenvalue equation (4-61) has two independent solutions, e^{ikx} and e^{-ikx}; equivalently, the pair of real solutions $\cos kx$ and $\sin kx$ is also independent. Whichever pair we choose, we notice that in contrast to the problem of a particle in a box, there are *two* solutions that have the same energy associated with them. This is an example of something that happens quite frequently: *there may be more than one independent eigenfunction that corresponds to the same eigenvalue of a hermitian operator.* When this occurs, we have a *degeneracy*.

In the two cases that we have above, the two solutions are orthogonal:

$$\int_{-\infty}^{\infty} dx(e^{-ikx})^* e^{ikx} = \int_{-\infty}^{\infty} dx\, e^{2ikx} = 0$$

$$\int_{-\infty}^{\infty} dx\, \sin kx \cos kx = 0 \qquad (4\text{-}69)$$

for $k \neq 0$. It is always possible to make linear combinations such that this is true. Such linear combinations are, of course, orthogonal to eigenfunctions that correspond to different values of the eigenvalue, for example, the energy.[6]

What distinguishes the two degenerate eigenfunctions? For the set

[6] See Appendix B.

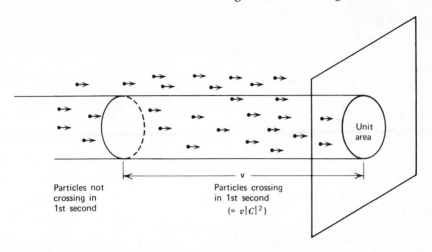

Fig. 4-4. The relation between velocity of particles and flux, that is, number of particles crossing a unit area perpendicular to velocity, per unit time.

(e^{-ikx}, e^{ikx}), the difference is that they are eigenfunctions of the momentum operator

$$p_{op} e^{\pm ikx} = \frac{\hbar}{i} \frac{d}{dx} e^{\pm ikx} = \pm \hbar k \, e^{\pm ikx} \qquad (4\text{-}70)$$

corresponding to *different* eigenvalues of the momentum. Similarly the pair $(\cos kx, \sin kx)$ are eigenfunctions of the parity operator, corresponding to different eigenvalues

$$P \cos kx = \cos kx$$
$$P \sin kx = - \sin kx \qquad (4\text{-}71)$$

In both cases, what differentiates the degenerate eigenfunctions is that they are simultaneous eigenfunctions of another hermitian operator. Both the operators p_{op} and P have the property that they commute with the Hamiltonian $p_{op}^2/2m$ in this problem. We shall show later that this is a necessary condition for the existence of simultaneous eigenfunctions. For example p_{op} and P do not commute, [since $(\hbar/i)(d/dx)$ changes sign under $x \rightarrow -x$], and therefore the eigenfunctions of one of the operators cannot all be simultaneous eigenfunctions of the other.

We have learned an enormous amount about quantum mechanics from the two simple problems that we have considered. We shall return to these matters in later chapters and generalize them. In Chapter 5 we will again consider some very simple problems, but this time we will concentrate not on the mathematical features, but rather on the physical systems that they are simple models of.

Problems

✓ 1. You are given the following operators

(a) $O_1\psi(x) = x^3\psi(x)$ (b) $O_2\psi(x) = x\dfrac{d}{dx}\psi(x)$

(c) $O_3\psi(x) = \lambda\psi^*(x)$ (d) $O_4\psi(x) = e^{\psi(x)}$

(e) $O_5\psi(x) = \dfrac{d\psi(x)}{dx} + a$ (f) $O_6\psi(x) = \displaystyle\int_{-\infty}^{x} dx'\,(\psi(x')x')$

Which of these are linear operators?

✓ 2. Solve the eigenvalue problem

$$O_6\psi(x) = \lambda\psi(x)$$

What values of the eigenvalue λ lead to square integrable eigenfunctions?
(*Hint.* Differentiate both sides of the equation with respect to x.)

✓ 3. Calculate the following commutators

(a) $[O_2, O_6]$
(b) $[O_1, O_2]$

The procedure is to calculate $[A,B]$ by expressing $A(B\psi) - B(A\psi)$ in the form $C\psi$.

4. Calculate

$$\Delta x = \sqrt{\langle x^2 \rangle}$$

for the $u_n^{(\pm)}(x)$ given by (4-21) and (4-24). Using $\langle p^2 \rangle$ given by (4-28) calculate

$$\Delta p\,\Delta x$$

It is characteristic that for the higher states the uncertainty increases with n.

5. Solve the Schrödinger equation for a particle in a box with sides at $x = 0$ and $x = L$ with the boundary condition that

$$\psi(0) = \psi(L)$$

What are the eigenvalues and the normalized eigenfunctions?

6. A particle is in the ground state of a box with sides at $x = \pm a$. Very suddenly the sides of the box are moved to $x = \pm b$ ($b > a$). What is the probability that the particle will be found in the ground state for the new potential? What is the probability that it will be found in the first excited state? In the latter case, the simple answer has a simple explanation. What is it?

7. A particle is known to be localized in the left half of a box with sides at $x = \pm a$. If all values of x in the left half side are equally probable, what wave

function describes the particle at $t = 0$? Will the particle remain localized at later times?

Calculate the probability that an energy measurement yields the ground state energy; the energy of the first excited state.

8. A particle is in the ground state of a box with sides at $x = 0$ and $x = L$. Suddenly the walls of the box are moved to $\pm \infty$, respectively, so that the particle is free. What is the probability that the particle has momentum in the range $(p, p + dp)$? After the removal of the walls, the energy of the particle is $p^2/2m$, which need not be equal to the ground state energy. Can you give an explanation for the apparent lack of energy conservation?

9. Repeat the above calculation for a particle initially in the nth eigenstate. Show that the corresponding probability is given by

$$\frac{2n^2\pi}{\hbar L^3} \frac{1 - (-1)^n \cos pL/\hbar}{[(p/\hbar)^2 - (n\pi/L)^2]^2}$$

Sketch the distribution. Show that it conforms with the uncertainty relation, and that the result is in agreement with the correspondence principle when n is large.

10. A particle in free space is initially in a wave packet described by

$$\psi(x) = \left(\frac{\alpha}{\pi}\right)^{1/4} e^{-\alpha x^2/2}$$

(a) What is the probability that its momentum is in the range $(p, p + dp)$?

(b) What is the expectation value of the energy? Can you give a rough argument, based on the "size" of the wave function and the uncertainty principle, for why the answer should be roughly what it is?

11. The wave function for a particle is given by

$$\psi(x) = A e^{ikx} + B e^{-ikx}$$

What flux does this represent?

12. What is the flux associated with a particle described by the wave function

$$\psi(x) = u(x) e^{ikx}$$

where $u(x)$ is a real function?

13. Consider the eigenfunctions for a box with sides at $x = \pm a$. Without working out the integral, prove that the expectation value of the quantity

$$x^2p^3 + 3xp^3x + p^3x^2$$

vanishes for all the eigenfunctions.

14. Prove that the parity operator, defined by

$$P\psi(x) = \psi(-x)$$

is a hermitian operator. Also prove that the eigenfunctions of P, corresponding to the eigenvalues $+1$ and -1 are orthogonal.

References

A detailed discussion of the properties of second order differential equations as related to quantum mechanics may be found in J. L. Powell and B. Crasemann, *Quantum Mechanics*, Addison-Wesley, Inc., Reading, Mass., 1961, and D. S. Saxon, *Elementary Quantum Mechanics*, Holden-Day, San Francisco (1968).

See also any of the more advanced textbooks listed at the end of the book.

chapter 5

One-Dimensional Potentials

Here we solve some simple problems of one-dimensional motion. They are of interest because they illustrate some nonclassical effects, and because many physical situations are effectively one-dimensional even though we live in a three-dimensional world.

A. The Potential Step

For this problem we take (Fig. 5-1) the form of $V(x)$ to be

$$V(x) = 0 \qquad x < 0$$
$$ = V_0 \qquad x > 0 \tag{5-1}$$

The Schrödinger equation

$$-\frac{\hbar^2}{2m}\frac{d^2u(x)}{dx^2} + V(x)\,u(x) = Eu(x) \tag{5-2}$$

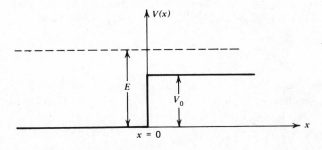

Fig. 5-1. The potential step.

takes the form

$$\frac{d^2u(x)}{dx^2} + \frac{2m}{\hbar^2} [E - V(x)]\, u(x) = 0 \qquad (5\text{-}3)$$

We write, as usual

$$\frac{2mE}{\hbar^2} = k^2 \qquad (5\text{-}4)$$

and we also introduce

$$\frac{2m(E - V_0)}{\hbar^2} = q^2 \qquad (5\text{-}5)$$

The most general solution of (5-3) for $x < 0$, where $V(x) = 0$ is

$$u(x) = e^{ikx} + R\, e^{-ikx} \qquad (5\text{-}6)$$

This corresponds to a flux moving in the positive x direction, of magnitude

$$j = \frac{\hbar}{2im} [(e^{-ikx} + R^* e^{ikx})\, (ik\, e^{ikx} - ikR\, e^{-ikx}) - \text{complex conjugate}]$$

$$= \frac{\hbar k}{m} (1 - |R|^2) \qquad (5\text{-}7)$$

We may view e^{ikx} with flux $\hbar k/m$ as an *incoming wave*. If there were no potential, we could choose e^{ikx} as the solution for all x, so that we attribute R to the presence of the potential. This potential gives rise to a reflected wave, $R\, e^{-ikx}$, with a reflected flux $\hbar k |R|^2/m$.

For $x > 0$, we write the solution

$$u(x) = T\, e^{iqx} \qquad (5\text{-}8)$$

The most general solution for $x > 0$ is a linear combination of e^{iqx} and e^{-iqx}, but a term involving the latter would describe a wave coming from $+\infty$ in the negative direction, and with the "experiment" that we have set up, the only wave on the right can be a transmitted wave. The flux corresponding to (5-8) is

$$j = \frac{\hbar q}{m} |T|^2 \qquad (5\text{-}9)$$

Since there is no time dependence in the problem, the conservation law (3-11) implies that $j(x)$ is independent of x. Hence the flux on the left must be equal to the flux on the right, that is, we expect that

$$\frac{\hbar k}{m} (1 - |R|^2) = \frac{\hbar q}{m} |T|^2 \qquad (5\text{-}10)$$

The continuity of the wave function implies that

$$1 + R = T \tag{5-11}$$

obtained by matching the two solutions at $x = 0$. In spite of the fact that the potential is discontinuous, the slope of the wave function is also continuous, as can be seen by integrating (5-3) from $-\epsilon$ to $+\epsilon$ (with ϵ arbitrarily small and positive) and using the continuity of the wave function:

$$\left(\frac{du}{dx}\right)_{\epsilon} - \left(\frac{du}{dx}\right)_{-\epsilon} = \int_{-\epsilon}^{\epsilon} dx \, \frac{d}{dx} \frac{du}{dx}$$

$$= \int_{-\epsilon}^{\epsilon} dx \, \frac{2m}{\hbar^2} \left[V(x) - E \right] u(x) = 0 \tag{5-12}$$

We note, for future reference, that if the potential contains a term like $V_0 \delta(x - a)$ then integration of the equation from $a - \epsilon$ to $a + \epsilon$ gives

$$\left(\frac{du}{dx}\right)_{a+\epsilon} - \left(\frac{du}{dx}\right)_{a-\epsilon} = \frac{2m}{\hbar^2} \int_{a-\epsilon}^{a+\epsilon} dx \, V_0 \delta(x - a) \, u(x)$$

$$= \frac{2m}{\hbar^2} V_0 u(a) \tag{5-13}$$

The continuity of the derivative for our potential implies that

$$ik(1 - R) = iqT \tag{5-14}$$

We can therefore solve for R and T to obtain

$$R = \frac{k - q}{k + q}$$

$$T = \frac{2k}{k + q} \tag{5-15}$$

From this we can calculate the reflected and transmitted fluxes:

$$\frac{\hbar k}{m} |R|^2 = \frac{\hbar k}{m} \left(\frac{k - q}{k + q}\right)^2$$

$$\frac{\hbar q}{m} |T|^2 = \frac{\hbar k}{m} \frac{4kq}{(k + q)^2} \tag{5-16}$$

We note the following:

1. In contrast to classical mechanics, according to which a particle going over a potential step would slow down (to conserve energy) but would never be reflected, here we do have a certain fraction of the incident particles reflected.

This is, of course, a consequence of the wave properties of the particle; partial reflection of light from an interface between two media is a familiar phenomenon.

2. With the help of (5-16) we can easily check that the conservation law (5-10) is indeed satisfied.

3. For $E \gg V_0$, that is, for $q \rightarrow k$ from below, the ratio of the reflected flux to the incident flux, that is, $|R|^2$ approaches zero. This agrees with intuition, which tells us that at very high energies, the presence of the step is but a small perturbation on the propagation of the wave.

4. If the energy E is less than V_0, then q becomes imaginary. If we note that now the solution for $x > 0$ must be of the form

$$u(x) = T e^{-|q|x} \tag{5-17}$$

so as not to blow up at $+\infty$, we see that now

$$|R|^2 = \left(\frac{k - i|q|}{k + i|q|}\right)\left(\frac{k - i|q|}{k + i|q|}\right)^* = 1 \tag{5-18}$$

Thus, as in classical mechanics, there is now total reflection. Note, however, that

$$T = \frac{2k}{k + i|q|} \tag{5-19}$$

does not vanish, and a part of the wave penetrates into the forbidden region. This penetration phenomenon again is characteristic of waves, and we shall see a little later that it permits a "tunneling" through barriers that would totally block particles in a classical description. There is no flux to the right, since $j(x)$ vanishes for a real solution even if the coefficient in front of it is taken to be complex.

B. The Potential Well

We next consider the potential (Fig. 5-2)

$$\begin{aligned} V(x) &= 0 & x < -a \\ &= -V_0 & -a < x < a \\ &= 0 & a < x \end{aligned} \tag{5-20}$$

We again write

$$k^2 = \frac{2mE}{\hbar^2} \tag{5-21}$$

and

$$q^2 = \frac{2m(E + V_0)}{\hbar^2} \tag{5-22}$$

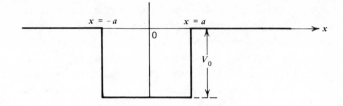

Fig. 5-2.   The potential well.

We can immediately write down the solutions

$$u(x) = e^{ikx} + R \, e^{-ikx} \qquad\qquad x < -a$$
$$u(x) = A \, e^{iqx} + B \, e^{-iqx} \qquad -a < x < a$$
$$u(x) = T \, e^{ikx} \qquad\qquad\qquad a < x \qquad\qquad (5\text{-}23)$$

These correspond to an incoming flux $\hbar k/m$ from the left, a reflected flux $\hbar k |R|^2/m$ and a transmitted flux $\hbar k |T|^2/m$ to the right. Inside the well, there are waves going in both directions because of the reflections at both discontinuities at $\pm a$. According to flux conservation we should get

$$\frac{\hbar k}{m} (1 - |R|^2) = \frac{\hbar q}{m} (|A|^2 - |B|^2) = \frac{\hbar k}{m} |T|^2 \qquad (5\text{-}24)$$

Matching wave functions and derivatives gives the four equations

$$e^{-ika} + R \, e^{ika} = A \, e^{-iqa} + B \, e^{iqa}$$
$$ik(e^{-ika} - R \, e^{ika}) = iq(A \, e^{-iqa} - B \, e^{iqa})$$
$$A \, e^{iqa} + B \, e^{-iqa} = T \, e^{ika}$$
$$iq(A \, e^{iqa} - B \, e^{-iqa}) = ikT \, e^{ika} \qquad\qquad (5\text{-}25)$$

A little algebra yields the results

$$R = i \, e^{-2ika} \, \frac{(q^2 - k^2) \sin 2qa}{2kq \cos 2qa - i(q^2 + k^2) \sin 2qa}$$

$$T = e^{-2ika} \, \frac{2kq}{2kq \cos 2qa - i(q^2 + k^2) \sin 2qa} \qquad (5\text{-}26)$$

Again, if $E \gg V_0$, there is practically no reflection, since $q^2 - k^2 \ll 2kq$, and as $E \to 0$, the transmission goes to zero. There is an item of special interest: in the special case that $\sin 2qa = 0$, that is, for the energies given by

$$E = -V_0 + \frac{n^2\pi^2\hbar^2}{8ma^2} \qquad n = 1, 2, 3, \ldots \qquad (5\text{-}27)$$

there is no reflection. This is actually a model of what happens in the scattering of low energy electrons (0.1 eV) by noble gas atoms, for example, neon and argon, in which there is anomalously large transmission. The effect, first observed by Ramsauer and Townsend, is described as a transmission resonance. A more accurate discussion must, of course, involve three-dimensional considerations. In wave language, the effect is due to a destructive interference between the wave reflected at $x = -a$ and the wave reflected once, twice, thrice, . . . , at the edge $x = a$. The resonance condition $2qa = n\pi$, which may be written in the form

$$\lambda = \frac{2\pi}{q} = \frac{4a}{n} \tag{5-28}$$

is just the one that describes the Fabry-Perot interferometer.

In addition to the above solutions for $E > 0$, there are, remarkably, also solutions for $E < 0$ provided the potential is negative, that is, $V_0 > 0$ in (5-20). They will turn out to be discrete. Let us write

$$\frac{2mE}{\hbar^2} = -\kappa^2 \tag{5-29}$$

The solutions outside the well that are bounded at infinity are

$$u(x) = C_1 e^{\kappa x} \qquad x < -a$$
$$u(x) = C_2 e^{-\kappa x} \qquad a < x \tag{5-30}$$

Since we are dealing with real functions, it is more convenient to write the solution inside the well in the form

$$u(x) = A \cos qx + B \sin qx \qquad -a < x < a \tag{5-31}$$

Note that

$$q^2 = \frac{2m}{\hbar^2} (V_0 - |E|) > 0 \tag{5-32}$$

Matching solutions and derivatives at the edges $x = \pm a$ yields

$$C_1 e^{-\kappa a} = A \cos qa - B \sin qa$$
$$\kappa C_1 e^{-\kappa a} = q(A \sin qa + B \cos qa)$$
$$C_2 e^{-\kappa a} = A \cos qa + B \sin qa$$
$$-\kappa C_2 e^{-\kappa a} = -q(A \sin qa - B \cos qa) \tag{5-33}$$

These may be combined to yield

$$\kappa = q \frac{A \sin qa - B \cos qa}{A \cos qa + B \sin qa}$$
$$= q \frac{A \sin qa + B \cos qa}{A \cos qa - B \sin qa} \tag{5-34}$$

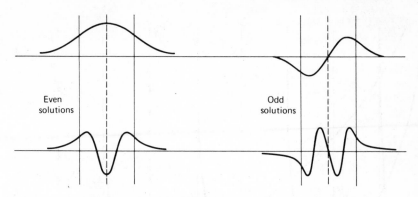

Fig. 5-3. Solutions for discrete spectrum in attractive potential well.

Together these imply that $AB = 0$, that is, the solutions are either even in $x(B = 0)$ or odd in $x(A = 0)$, a situation encountered in the case of the infinite box. The wave functions are roughly of the shape shown in Fig. 5-3. The ground state, with no nodes, is even. This is a general property of simple systems. The conditions that determine the energy are from (5-34)

$$\kappa = q \tan qa \qquad \text{even solutions}$$

$$\kappa = -q \cot qa \qquad \text{odd solutions} \qquad (5\text{-}35)$$

Let us examine these separately.

(a) The even solutions:

With the notation

$$\lambda = \frac{2mV_0 a^2}{\hbar^2}$$

$$y = qa \qquad (5\text{-}36)$$

the first of the relations (5-35) reads

$$\frac{\sqrt{\lambda - y^2}}{y} = \tan y \qquad (5\text{-}37)$$

If we plot $\tan y$ and $\sqrt{\lambda - y^2}/y$ as functions of y (Fig. 5-4), the points of intersection determine the eigenvalues. These form a discrete set. The larger λ is, the further the curves for $\sqrt{\lambda - y^2}/y$ go, that is, *when the potential is deeper and/or broader, there are more bound states*. The figure also shows that no matter how small λ is, there will always be at least one bound state. This is characteristic of one-dimensional attractive potentials, and is *not* true for three-dimensional potentials, which behave much more like the odd-solution problem that we will

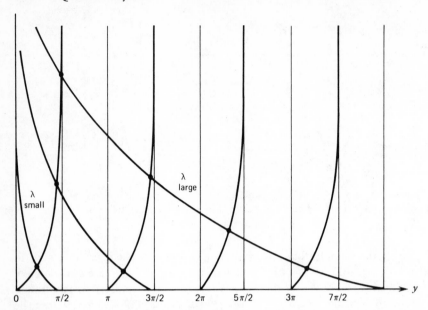

Fig. 5-4. Location of discrete eigenvalues for even solutions in square well. The rising curves represent $\tan y$; the falling curves are $\sqrt{\lambda - y^2}/y$ for different values of λ.

discuss below. As λ becomes large, the eigenvalues tend to become equally spaced in y, with the intersection points given approximately by

$$y \simeq (n + \tfrac{1}{2}) \pi \qquad n = 0, 1, 2, \ldots \qquad (5\text{-}38)$$

This is just the eigenvalue condition for the even solutions of the infinite box, and this is as might be expected, since for the deep-lying states in the potential, the fact that it is not really infinitely deep does not matter very much.

(b) The odd solutions:

Here the eigenvalue condition reads

$$\frac{\sqrt{\lambda - y^2}}{y} = - \cot y \qquad (5\text{-}39)$$

Since $-\cot y = \tan (\pi/2 + y)$, the plot in Fig. 5-5 is the same as in Fig. 5.4 with the tangent curves shifted by $\pi/2$. The large λ behavior is more or less the same, with (5-38) replaced by

$$y \simeq n\pi \qquad n = 1, 2, 3, \ldots \qquad (5\text{-}40)$$

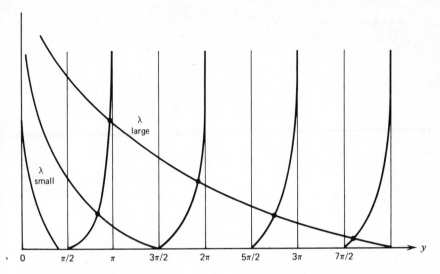

Fig. 5-5. Location of discrete eignevlues for odd solutions in square well. The rising curves represent $-\cot y$; the falling curves are $\sqrt{\lambda - y^2}/y$ for different values of λ. Note that there is no eigenvalue for $\lambda < (\pi/2)^2$.

In contrast to the even solutions, there will only be an intersection if $\sqrt{\lambda - \pi^2/4} > 0$, that is, if

$$\frac{2mV_0a^2}{\hbar^2} > \frac{\pi^2}{4} \tag{5-41}$$

The odd solutions all vanish at $x = 0$, and hence the bound-state problem for the odd solutions will be the same as for the potential well shown in Fig. 5-6, since in the latter, the condition $u(0) = 0$ would be imposed. We shall see that such conditions are imposed on wave functions in the three-dimensional world.

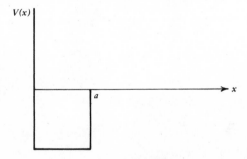

Fig. 5-6. Equivalent potential for odd solutions of square well bound state problem.

C. The Potential Barrier

We now consider

$$V(x) = 0 \qquad x < -a$$
$$= V_0 \qquad -a < x < a$$
$$= 0 \qquad a < x \qquad (5\text{-}42)$$

We will limit our discussion to energies such that $E < V_0$, that is, energies such that no penetration of the barrier would occur in classical physics (Fig. 5-7). Inside the barrier we have the equation

$$\frac{d^2u(x)}{dx^2} + \frac{2m}{\hbar^2}\,(E - V_0)\,u(x) = 0$$

that is

$$\frac{d^2u(x)}{dx^2} - \kappa^2 u(x) = 0 \qquad (5\text{-}43)$$

The general solution

$$u(x) = A\,e^{\kappa x} + B\,e^{-\kappa x} \qquad |x| < a \qquad (5\text{-}44)$$

is to be matched onto

$$u(x) = e^{ikx} + R\,e^{-ikx} \qquad x < -a$$
$$= T\,e^{ikx} \qquad x > a \qquad (5\text{-}45)$$

Actually we need not go through the trouble of solving this since the results can be read off from (5-26) with the substitution

$$q \rightarrow i\kappa = i\sqrt{(2m/\hbar^2)(V_0 - E)} \qquad (5\text{-}46)$$

Thus, for example,

$$T = e^{-2ika}\,\frac{2k\kappa}{2k\kappa\cosh 2\kappa a - i(k^2 - \kappa^2)\sinh 2\kappa a} \qquad (5\text{-}47)$$

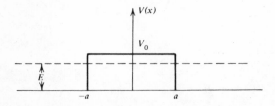

Fig. 5-7. Potential barrier. Energy is such that a classical particle would be totally reflected by the barrier.

and this implies that

$$|T|^2 = \frac{(2k\kappa)^2}{(k^2 + \kappa^2)^2 \sinh^2 2\kappa a + (2k\kappa)^2} \tag{5-48}$$

There is transmission, even though the energy lies below the top of the barrier. This is a wave phenomenon, and in quantum mechanics it is also one exhibited by particles. This *tunneling* of a particle through a barrier is frequently encountered, and we shall discuss some applications. We also note that when κa is large, the ratio of transmitted flux to incident flux is

$$|T|^2 \simeq \left(\frac{2k\kappa}{k^2 + \kappa^2}\right)^2 e^{-4\kappa a} \tag{5-49}$$

This becomes an extremely sensitive function of the width of the barrier, and of the amount by which the barrier exceeds the incident energy, since

$$\kappa a = \left[\frac{2ma^2}{\hbar^2}(V_0 - E)\right]^{1/2} \tag{5-50}$$

In general, the barriers that occur in physical phenomena are not square, and to discuss some applications, we must first obtain an approximate expression for the transmission coefficient $|T|^2$ through an irregularly shaped barrier. The proper way to do this, given the fact that there is no exact solution available for most potentials, is through the Wentzel-Kramers-Brillouin (WKB) approximation technique.[1] Our discussion will be less mathematical.

We observe that (5-49) consists of a product of two terms, the second of which is by far the more important one. If we write

$$\log|T|^2 \simeq -2\kappa(2a) + 2\log\frac{2(ka)(\kappa a)}{(ka)^2 + (\kappa a)^2}$$

we see that under most circumstances the first term dominates the second for any reasonable size of κa. The procedure we adopt is to treat a smooth, curved barrier as a juxtaposition of square barriers (Fig. 5-8). Since transmission coefficients are multiplicative[2] when they are small (in effect, with most of the flux reflected, the transmission through each slice is an independent, improbable event), we may write, approximately

$$\log|T|^2 \approx \sum_{\substack{\text{partial} \\ \text{barriers}}} \log|T_{\text{partial}\atop\text{barrier}}|^2$$

$$\approx -2\sum \Delta x \langle\kappa\rangle$$

[1] See the WKB approximation in Special Topics section 3.

[2] This statement is only correct for the most important exponential part, as can be seen from the fact that doubling the width will only approximately square the transmission coefficient $|T|^2$.

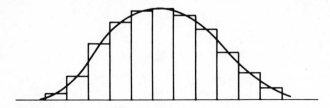

Fig. 5-8. Approximation of smooth barrier by a juxtaposition of square potential barriers.

$$\approx -2 \int_{\text{barrier}} dx \sqrt{(2m/\hbar^2)[V(x) - E]} \qquad (5\text{-}51)$$

In the partial barriers, Δx is the width and $\langle \kappa \rangle$ the average value of κ for that barrier. In the last step a limit of infinitely narrow barriers was taken. It is clear from the expression that the approximation is least accurate near the "turning points" where the energy and potential are nearly equal, since there (5-49) is not a good approximation to (5-48). It is also important that $V(x)$ be a slowly varying function of x, since otherwise the approximation of a curved barrier by a stack of square ones is only possible if the latter are narrow, and there again (5-49) is a poor approximation. A proper treatment, using the WKB approximation includes a discussion of the behavior near the turning points. For most purposes, it is still a fair approximation to write

$$|T|^2 \approx e^{-2 \int dx \sqrt{(2m/\hbar^2)[V(x) - E]}} \qquad (5\text{-}52)$$

with the integration over the region in which the square root is real.

D. Tunneling Phenomena

The phenomenon of particle tunneling is quite common in atomic and nuclear physics, and we discuss two examples at this point.

(a) Consider electrons in a metal. As noted in our discussion of the photo-electric effect in Chapter 1, these electrons are held in a metal by a potential, which, to first approximation, may be described by a box of finite depth, as shown in Fig. 5-9a. The electrons are actually stacked up in energy levels that are very dense, since the box is very wide. It is a property of electrons[3] that no more than two of them can occupy any given energy level; thus for the lowest energy state of the metal, all the levels up to a certain energy, called the *Fermi energy*

---

[3] This property of electrons is described by the Pauli exclusion principle, which will be discussed in Chapter 8.

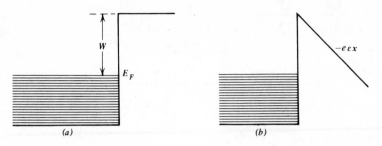

Fig. 5-9. (*a*) Electronic energy levels in metal. E_F is the Fermi energy and W is the work function. (*b*) Potential altered by an external electric field.

(which depends on the density of free electrons) are filled. When the temperature is above 0°K, a few electrons are thermally excited to higher levels, but even at room temperature, the number is small. The difference between the Fermi energy and the top of the well is what is required to bring an electron out; it is the *work function* discussed in connection with the photoelectric effect. Electrons can be removed by transferring energy to them, either by photons, or by heating them. They can also be removed by the application of an external electric field $\mathcal{E}$. *Cold emission* occurs because the external field changes the potential seen by an electron from W to $(W - e\mathcal{E}x)$ (Fig. 5-9), if the electron is at the top of the "sea" of levels. The transmission coefficients is

$$|T|^2 = e^{-2\int_0^a dx\, [2m(W - e\mathcal{E}x)/\hbar^2]^{1/2}} \tag{5-53}$$

Since

$$\int dx(A + Bx)^{1/2} = \frac{(A + Bx)^{3/2}}{3B/2}$$

this leads to

$$|T|^2 = e^{-(4\sqrt{2}/3)\,\sqrt{mW}/\hbar^2\,(W/e\mathcal{E})} \tag{5-54}$$

The Fowler-Nordheim formula, as (5-54) is called, describes the emission only qualitatively. One effect, which is easily included, is the additional attraction of the electron back to the plate, caused by the image charge. The other effect, much harder to handle, is that there are surface imperfections in the metal surface, which change the electric field locally, and since $\mathcal{E}$ appears in the exponent, this can make a large difference. Incidentally, we see that the exponent may be written in terms of the barrier thickness at the top of the Fermi sea, since that thickness is given by

$$a = \frac{W}{e\mathcal{E}} \tag{5-55}$$

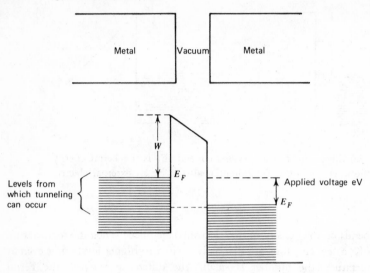

Fig. 5-10. Energy diagram for tunneling between two metals separated by vacuum. Tunneling between metals is possible only when there are empty states on the right. Such empty states are created when eV is applied to lower the Fermi level on the right.

The same effect appears if we bring two metal plates close together. Figure 5-10 shows the situation both without a potential difference, and with a potential difference. Without the potential difference, tunneling is not possible because the levels on both sides of the barrier are filled. The effect of even a weak electric field is to change the shape of the barrier a little (Fig. 5-10)—an effect that we can neglect—and to lower the Fermi sea on one side of the barrier. This, in effect, brings some empty levels in correspondence with the filled ones on the other side of the barrier, and now tunneling can proceed, with transmission coefficient

$$|T|^2 \simeq e^{-2\sqrt{(2mW/\hbar^2)}\,a} \qquad (5\text{-}56)$$

Such a factor acts as a resistance. Unfortunately this expression is very sensitive to the gap separation a, and since for a work function of the order of electron volts, the separation has to be of the order of angstroms, it has not proved possible to make metal plates sufficiently flat and parallel. The formula has been applied to the interpretation of currents flowing between two plates with an oxide between them (Ni-NiO-Pb), where the gap can be made as small as 50 Å, and it is qualitatively correct.

An interesting effect occurs when the metal on the right is in a superconducting state. A characteristic of such a state is that above the Fermi level

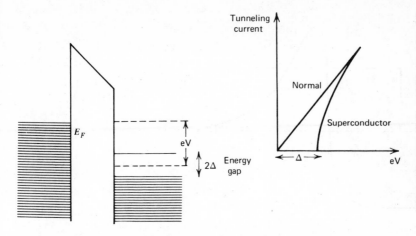

Fig. 5-11. Energy diagram for tunneling from metal to superconductor. In contrast to the metal-metal tunneling shown in Fig. 5-10, no tunneling is allowed into the energy gap. This affects the current-voltage characteristic as shown.

there is a gap in the level density, that is, there are no allowed states between an energy $E_F - \Delta$ and $E_F + \Delta$ with Δ of the order of 10^{-3} eV compared with the Fermi energy E_F of order 10 eV. These levels do not disappear, but are squeezed up and down, so that the level density just below and just above the gap is very large. If the electric field is small enough, that is, $a\mathcal{E} \leq \Delta/e$, there will be no tunneling, since there is no place for the electrons to go. The qualitative features of the current-voltage relation and the energetics are shown in Fig. 5-11. These features are in good agreement with experiment.

(b) Tunneling is also important in nuclear physics. Nuclei are very complicated objects, but under certain circumstances it is appropriate to view them as independent particles occupying levels in a potential well. With this picture in mind, the decay of a nucleus into an α-particle (a He nucleus with $Z = 2$) and a daughter nucleus may be described as the tunneling of an α-particle through a barrier caused by the Coulomb potential between the daughter and the α-particle. The α-particle is not viewed as being in a bound state: if it were, the nucleus could not decay. Rather, the α-particle is taken to have positive energy, and its decay is only inhibited by the existence of the barrier.[4]

---

[4] If you find it difficult to imagine why a repulsion would keep two objects from separating, think of the inverse process, α capture. It is clear that the barrier will tend to keep the α-particle out.

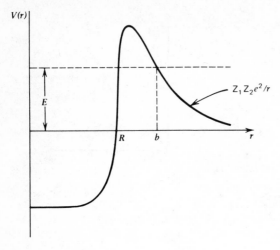

Fig. 5-12. Potential barrier for α decay.

If we write

$$|T|^2 = e^{-G} \qquad (5\text{-}57)$$

then

$$G = 2 \left(\frac{2m}{\hbar^2}\right)^{1/2} \int_R^b dr \left(\frac{Z_1 Z_2 e^2}{r} - E\right)^{1/2} \qquad (5\text{-}58)$$

where R is the nuclear radius[5] and b is the turning point, determined by the vanishing of the integrand (Fig. 5-12). Z_1 is the charge of the daughter nucleus, and Z_2 ($= 2$ here) is the charge of the particle being emitted. The integral can be done exactly

$$\int_R^b dr \left(\frac{1}{r} - \frac{1}{b}\right)^{1/2} = \sqrt{b} \left[\cos^{-1}\left(\frac{R}{b}\right)^{1/2} - \left(\frac{R}{b} - \frac{R^2}{b^2}\right)^{1/2}\right] \qquad (5\text{-}59)$$

At low energies (relative to the height of the Coulomb barrier at $r = R$, we have $b \gg R$, and then

$$G \simeq 2 \left(\frac{2m Z_1 Z_2 e^2 b}{\hbar^2}\right)^{1/2} \left[\frac{\pi}{2} - \left(\frac{R}{b}\right)^{1/2}\right] \qquad (5\text{-}60)$$

[5] If fact, early estimations of the nuclear radius came from the study of α-decay. Nowadays one uses the size of the charge distribution, as measured by scattering electrons off nuclei to get nuclear radii. It is not clear that the two should be expected to give exactly the same answer.

with $b = Z_1 Z_2 e^2 / E$. If we write for the α-particle energy $E = mv^2/2$, where v is its final velocity, then

$$G \simeq \frac{2\pi Z_1 Z_2 e^2}{\hbar v} = 2\pi \alpha Z_1 Z_2 \left(\frac{c}{v}\right) \tag{5-61}$$

The time taken for an α-particle to get out of the nucleus may be estimated as follows: the probability of getting through the barrier on a single encounter is e^{-G}. Thus the number of encounters needed to get through is $n \simeq e^{G}$. The time between encounters is of the order of $2R/v$, where R is again the nuclear radius, and v is the α velocity inside the nucleus. Thus the lifetime is

$$\tau \simeq \frac{2R}{v} e^{G} \tag{5-62}$$

The velocity of the α inside the nucleus is a rather fuzzy concept, and the whole picture is very classical, so that the factor in front of the e^{G} cannot really be predicted without a much more adequate theory. Our considerations do give us an order of magnitude for it. For a 1 MeV α-particle,

$$v = \sqrt{\frac{2E}{m}} = c\sqrt{\frac{2E}{mc^2}} = 3 \times 10^{10} \sqrt{\frac{2}{4 \times 940}} \simeq \frac{3}{43} \times 10^{10} \text{ cm/sec}$$

Also, for R we take

$$R \simeq 1.5 \times 10^{-13} A^{1/3} \text{ cm} \tag{5-63}$$

and for $A = 216$ we get, for the factor in front, 2.6×10^{-21}. We can also rewrite G in the form

$$G \simeq 4 \frac{Z_1}{\sqrt{E(\text{MeV})}} \tag{5-63}$$

so that one predicts, for low energy α's, the straight-line plot

$$\log_{10} \frac{1}{\tau} \simeq \text{const} - 1.73 \frac{Z_1}{\sqrt{E(\text{MeV})}} \tag{5-64}$$

with the constant in front of the order of magnitude 27–28 when τ is measured in years instead of seconds, Figure 5-13 shows that a good fit to the lifetime data of a large number of α emitters is obtained with the formula

$$\log_{10} \frac{1}{\tau} = C_2 - C_1 \frac{Z_1}{\sqrt{E}}$$

where $C_1 = 1.61$ and $C_2 = 28.9 + 1.6 Z_1^{2/3}$. Thus the very simple considerations give a rather remarkable fit to the data.

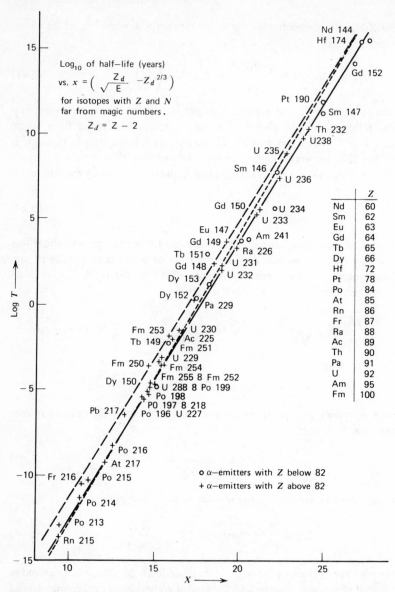

Fig. 5-13. Plot of $\log_{10} 1/\tau$ versus $C_2 - C_1 Z_1/\sqrt{E}$ with $C_1 = 1.61$ and a slowly varying $C_2 = 28.9 + 1.6\, Z_1^{2/3}$. (From E. K. Hyde, I. Perlman and G. T. Seaborg, *The Nuclear Properties of the Heavy Elements*, Vol. 1, Prentice-Hall, Inc. (1964), reprinted by permission.)

With more energetic α emission, the G factor depends on R, and with $R = r_0 A^{1/3}$, one finds that r_0 is a constant, that is, that the notion of a Coulomb barrier taking over the role of the potential beyond the nuclear radius has some validity. Again, simple qualitative considerations explain the data.

The fact that the probability of a reaction (e.g., capture) between nuclei is attenuated by the factor

$$e^{-2(Z_1 Z_2 / \sqrt{E})} \tag{5-65}$$

implies that at low energies and/or for high Z's, such reactions are rare. That is why all attempts to make thermonuclear reactors concentrate on the burning of hydrogen (actually heavy hydrogen—deuterium).

$$_1H^2 + {}_1H^2 \rightarrow {}_2He^3 + n \qquad (3.27 \text{ MeV})$$

$$_1H^2 + {}_1H^2 \rightarrow {}_1H^3 + p \qquad (4.03 \text{ MeV})$$

$$_1H^2 + {}_1H^3 \rightarrow {}_2He^4 + n \qquad (17.6 \text{ MeV})$$

since reactions involving higher Z elements would require much higher energies, that is, much higher temperatures, with correspondingly greater confinement problems. For the same reason, neutrons are used in nuclear reactors to fission the heavy elements. Protons, at the low energies available, would not be able to get near enough to the nuclei to react with them.

E. One-Dimensional Model of Molecule

Some aspects of what gives rise to molecules are exhibited by the example of a particle in a double potential well (Fig. 5-14). The algebraic work is greatly simplified if we consider a square well in the limit of great depth with the width going to zero such that $V_0 a$ remains a constant. In that case we get a delta-function well, which is very easy to handle. Just to show this, consider first a single attractive potential well

$$(2m/\hbar^2) \, V(x) = -\frac{\lambda}{a} \, \delta(x) \tag{5-66}$$

The equation to be solved is, when $E < 0$,

$$\frac{d^2 u(x)}{dx^2} - \kappa^2 u(x) = -\frac{\lambda}{a} \, \delta(x) \, u(x) \tag{5-67}$$

where $\kappa^2 = 2m|E|/\hbar^2$.

The solution everywhere, except at $x = 0$, must satisfy the equation $d^2 u/dx^2 - \kappa^2 u = 0$, and if it is to vanish at $x \rightarrow \pm \infty$, we must have

$$u(x) = e^{-\kappa x} \qquad x > 0$$

$$= e^{\kappa x} \qquad x < 0 \tag{5-68}$$

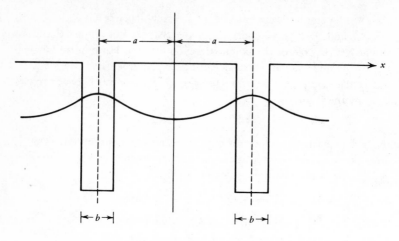

Fig. 5-14. Double one-dimensional potential well. The shape of the wave function for a bound state is sketched in.

The coefficients in front are the same (and here chosen to be unity—we can normalize afterwards) because of the continuity of the wave function. The derivative of the wave function is no longer continuous. As argued before (Eq. 5-13) we have

$$\left(\frac{du}{dx}\right)_{x=0+} - \left(\frac{du}{dx}\right)_{x=0-} = -\frac{\lambda}{a} u(0) \qquad (5\text{-}69)$$

The last relation gives the eigenvalue condition

$$-\kappa - \kappa = -\frac{\lambda}{a}$$

that is

$$\kappa = \frac{\lambda}{2a} \qquad (5\text{-}70)$$

The double square well will be replaced by

$$(2m/\hbar^2)\, V(x) = -\frac{\lambda}{a} [\delta(x - a) + \delta(x + a)] \qquad (5\text{-}71)$$

Because the potential is symmetric under the interchange $x \rightarrow -x$, we expect that there will be solutions of definite parity, and we will first consider the even solutions.

1. For the even solution we write

$$
\begin{aligned}
u(x) &= e^{-\kappa x} & x > a \\
&= A \cosh \kappa x & a > x > -a \\
&= e^{\kappa x} & x < -a
\end{aligned}
\tag{5-72}
$$

and continuity of the wave function gives

$$
e^{-\kappa a} = A \cosh \kappa a \tag{5-73}
$$

Because of the symmetry, it is sufficient to apply the discontinuity condition for the derivative at $x = a$. Nothing new will come of the application at $x = -a$. We get

$$
-\kappa e^{-\kappa a} - \kappa A \sinh \kappa a = -\frac{\lambda}{a} e^{-\kappa a} \tag{5-74}
$$

and the eigenvalue condition is

$$
\tanh \kappa a = \frac{\lambda}{\kappa a} - 1 \tag{5-75}
$$

Figure 5-15 shows this graphically. There is only one intersection point of the curve $\tanh y$ with $(\lambda/y) - 1$. It is obvious that when $y = \lambda$, the right side is zero, whereas $\tanh y > 0$. Thus the intersection point occurs for $y < \lambda$. On the other hand, since $\tanh y < 1$, we must have $(\lambda/y) < 2$ at the intersection point, that is,

$$
\kappa > \frac{\lambda}{2a} \tag{5-76}
$$

If we compare this with (5-70), we see that the energy for the double well is a *larger negative number*, that is, the energy for the double potential is lower. Note that this is not because somehow the strength of a pair of potentials is larger than that of a single potential, as might be the case if one compared an electron

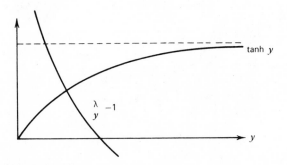

Fig. 5-15. Solution of the eigenvalue condition $\tanh y = \lambda/y - 1$.

bound to two protons with an electron bound to one proton. The larger binding is there because, as Fig. 5-16 indicates, it is easier to accommodate a sharply dropping exponential to a symmetric function (here cosh x) with a discontinuity in slope as given, than it is to accommodate it to an equally sharply dropping exponential on the other side of the potential. In the real world, a single electron bound to two protons separated by a small distance will have a lower energy than a single proton plus a hydrogen atom far away, even though in the first case there is a more effective repulsion between the protons. Again it is the way in which the wave function can accommodate itself to the geometrical situation that is the dominant effect.

2. The odd solution will have the form

$$
\begin{aligned}
u(x) &= e^{-\kappa x} & x > a \\
&= A \sinh \kappa x & a > x > -a \\
&= -e^{\kappa x} & x < -a
\end{aligned}
\tag{5-77}
$$

Again, because of the antisymmetry, it is sufficient to apply the conditions at $x = a$, say. Continuity of the wave function gives

$$
A \sinh \kappa a = e^{-\kappa a}
\tag{5-78}
$$

and the discontinuity equation reads

$$
-\kappa \, e^{-\kappa a} - \kappa A \cosh \kappa a = -\frac{\lambda}{a} e^{-\kappa a}
\tag{5-79}
$$

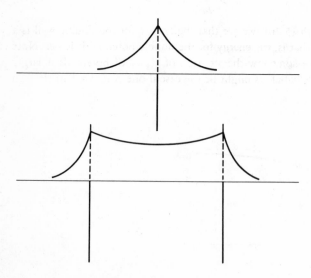

Fig. 5-16. Bound state wave functions for single and double delta function attractive potentials.

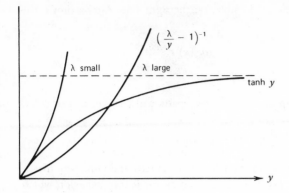

Fig. 5-17. Solution of the eigenvalue condition $\tanh y = (\lambda/y) - 1)^{-1}$.

Combining the two yields the eigenvalue condition

$$\coth \kappa a = \frac{\lambda}{\kappa a} - 1 \qquad (5\text{-}80)$$

Figure 5-17 shows a plot of the reciprocal of this equation, that is, $\tanh y$ against $(\lambda/y - 1)^{-1}$. There will only be an intersection if the slope of the former at the origin is larger than that of the second, that is, if

$$\lambda > 1 \qquad (5\text{-}81)$$

At $y = \lambda/2$ the term $(\lambda/y - 1)^{-1}$ is already at 1, so that the intersection had to occur for $y < \lambda/2$, that is,

$$\kappa < \frac{\lambda}{2a} \qquad (5\text{-}82)$$

Thus the odd solution, if there is a bound state, is less strongly bound than the even solution. The wave function, which has to go through zero, is forced to be steep between the wells, and thus can only accommodate to a less rapidly falling exponential. Depending on the size of λ, there may or may not exist an *excited state*.

Let us now consider a superposition of the ground state $u_e(x)$, with energy E_e and the excited state $u_o(x)$, with energy E_o (e and o stand for even and odd)

$$\psi(x) = u_e(x) + \alpha u_o(x) \qquad (5\text{-}83)$$

with α chosen so as to make $\int_{-\infty}^{0} dx |\psi(x)|^2$ as small as possible, that is, with the "electron" localized, as far as possible, on the right side. After a time t, the wave function will be

$$
\begin{aligned}
\psi(x,t) &= u_e(x)\, e^{-iE_e t/\hbar} + \alpha u_o(x)\, e^{-iE_o t/\hbar} \\
&= e^{-iE_e t/\hbar}\, [u_e(x) + \alpha\, e^{-i(E_o - E_e)t/\hbar}\, u_o(x)]
\end{aligned}
\tag{5-84}
$$

that is, the phase relationship between the two parts will change. In particular, after a time such that

$$
e^{-i(E_o - E_e)t/\hbar} = -1
\tag{5-85}
$$

the "electron" will be localized on the left side in exactly the same way that it was localized on the right at $t = 0$. Thus there is an oscillatory behavior, which may be described by the electron going back and forth between the two potentials, with frequency

$$
\omega = 2\omega_{oe} = 2\, \frac{E_o - E_e}{\hbar}
\tag{5-86}
$$

We shall leave it to the reader to convince himself that the period associated with the frequency ω_{oe} is, for large λ, approximately equal to the "tunneling time" across the barrier separating the two wells, as might be determined from the material presented in Sections C and D. This is a model for the ammonia molecule. There are ways of measuring such a frequency with high precision, and thus we have at our disposal a very accurate "clock."[6]

F. The Kronig-Penney Model

Metals generally have a crystalline structure, that is, the ions are arranged in a way that exhibits a spatial periodicity. This periodicity has an effect on the motion of the free electrons in the metal, and this effect is exhibited in the simple model that we will now discuss.

The periodicity will be built into the potential, for which we require that

$$
V(x + a) = V(x).
\tag{5-87}
$$

Since the kinetic energy term $-(\hbar^2/2m)(d^2/dx^2)$ is unaltered by the change $x \to x + a$, the whole *Hamiltonian is invariant under displacements by a*. For the case of zero potential, when the solution corresponding to a given energy $E = \hbar^2 k^2/2m$ is

$$
\psi(x) = e^{ikx}
\tag{5-88}
$$

<hr>

[6] For a discussion of the ammonia molecule, see R. P. Feynman, R. B. Leighton, and M. Sands, *The Feynman Lectures on Physics*, Vol. III, Addison-Weesley, Reading, Mass., 1965.

the displacement yields

$$\psi(x + a) = e^{ik(x+a)} = e^{ika} \psi(x) \tag{5-89}$$

that is, the original solution multiplied by a phase factor, so that

$$|\psi(x + a)|^2 = |\psi(x)|^2 \tag{5-90}$$

The observables will therefore be the same at x as at $x + a$, that is, we cannot tell whether we are at x or at $x + a$. In our example we shall also insist that $\psi(x)$ and $\psi(x + a)$ differ only by a phase factor, which need not, however, be of the form e^{ika}.

To simplify the algebra, we will take a series of repulsive delta-function potentials,

$$V(x) = \frac{\hbar^2}{2m} \frac{\lambda}{a} \sum_{n=-\infty}^{\infty} \delta(x - na) \tag{5-91}$$

Away from the points $x = na$, the solution will be that of the free particle equation, that is, some linear combination of $\sin kx$ and $\cos kx$ (we deal with real functions for simplicity). Let us assume that in the region R_n defined by $(n - 1)\, a \leq x \leq na$, we have

$$\psi(x) = A_n \sin k(x - na) + B_n \cos k(x - na) \tag{5-92}$$

and in the region R_{n+1}, defined by $na \leq x \leq (n + 1)\, a$ we have

$$\psi(x) = A_{n+1} \sin k[x - (n + 1)\, a] + B_{n+1} \cos k[x - (n + 1)\, a] \tag{5-93}$$

Continuity of the wave function implies that $(x = na)$

$$-A_{n+1} \sin ka + B_{n+1} \cos ka = B_n \tag{5-94}$$

and the discontinuity condition (5-13) here reads

$$kA_{n+1} \cos ka + kB_{n+1} \sin ka - kA_n = \frac{\lambda}{a} B_n \tag{5-95}$$

A little manipulation yields

$$A_{n+1} = A_n \cos ka + (g \cos ka - \sin ka) B_n$$
$$B_{n+1} = (g \sin ka + \cos ka) B_n + A_n \sin ka \tag{5-96}$$

where $g = \lambda/ka$.

The requirement that the wave functions (5-92) and (5-93) be related by

$$\psi(R_{n+1}) = e^{i\phi} \psi(R_n) \tag{5-97}$$

is satisfied if

$$A_{n+1} = e^{i\phi} A_n$$
$$B_{n+1} = e^{i\phi} B_n \tag{5-98}$$

When this is inserted into (5-96), we find a consistency condition that reads

$$(e^{i\phi} - \cos ka)(e^{i\phi} - g \sin ka - \cos ka) = \sin ka(g \cos ka - \sin ka)$$

that is,

$$e^{2i\phi} - e^{i\phi}(2 \cos ka + g \sin ka) + 1 = 0$$

Multiplication by $e^{-i\phi}$ gives

$$\cos \phi = \cos ka + \tfrac{1}{2} g \sin ka \tag{5-99}$$

If we take periodic boundary conditions for our "crystal" so that

$$\psi(R_{n+N}) = \psi(R_n) \tag{5-100}$$

then it follows from (5-98) that $e^{iN\phi} = 1$, that is,

$$\phi = \frac{2\pi}{N} m \qquad m = 0, \pm 1, \pm 2, \ldots \tag{5-101}$$

We denote ϕ by qa, where q is the wave number of an electron in a box of length Na, with periodic boundary conditions and without any potential, that is, without any ions present. Thus (5-99) should be rewritten in the form

$$\cos qa = \cos ka + \tfrac{1}{2} \lambda \frac{\sin ka}{ka} \tag{5-102}$$

This is a very interesting result, because the left side is always bounded by 1, that is, there are restrictions on the possible ranges of the energy $E = \hbar^2 k^2/2m$ that depend on the parameters of our "crystal." Figure 5-18 shows a plot of the function $\cos x + \lambda \sin x/2x$ as a function of $x = ka$. The horizontal line represents the bounds on $\cos qa$, and the regions of x, for which the curve lies outside the strip, are forbidden regions. Thus there are *allowed energy bands* separated by regions that are forbidden. Note that the onset of a forbidden band corresponds to the condition

$$ka = n\pi \qquad n = \pm 1, \pm 2, \pm 3, \ldots \tag{5-103}$$

This, however, is just the condition for Bragg reflection with normal incidence.

The Kronig-Penney model has some relevance to the theory of metals, insulators, and semiconductors if we take into account the fact (to be studied later) that energy levels occupied by electrons cannot accept more electrons. Thus a metal may have an energy band partially filled. If an external field is applied, the electrons are accelerated, and if there are momentum states available to them, the electrons will occupy the momentum states under the influence of the electric field. Insulators have completely filled bands, and an electric field cannot accelerate electrons, since there are no neighboring empty states. If the electric field is strong enough, the electrons can "jump" across a forbidden energy gap and go into an empty allowed energy band. This corresponds

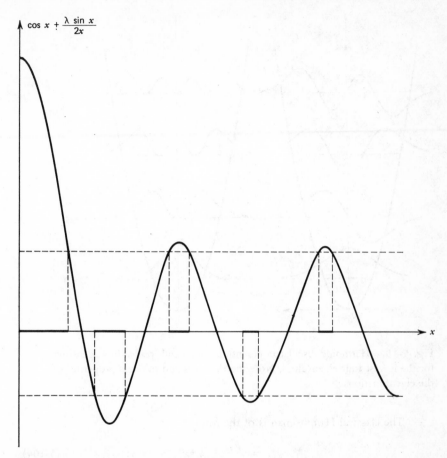

Fig. 5-18.   Plot of $\cos x + (\lambda/2)(\sin x/x)$ as a function of x. The horizontal lines represent the bounds ± 1. The regions of x for which the curve lines outside the strip are forbidden.

to the breakdown of an insulator. The semiconductor is an insulator with a very narrow forbidden gap. There, small changes of conditions, for example, a rise in temperature, can produce the "jump" and the insulator becomes a conductor.

G. The Harmonic Oscillator

As our last example we consider the harmonic oscillator (Fig. 5.19). In contrast to the examples dealt with until now, the differential equation that needs to be solved is not so trivial, and one reason for discussing this problem is to learn something about the technique for solving such equations.

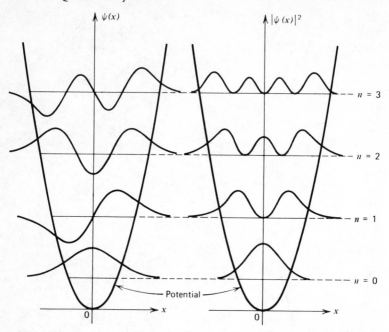

Fig. 5-19. Harmonic oscillator eigenfunctions, and probability densities for the lowest four eigenvalues. Note the evenness and oddness properties of the eigenfunctions.

The classical Hamiltonian is of the form

$$H = \frac{p^2}{2m} + \tfrac{1}{2}\,kx^2 \tag{5-104}$$

so that the eigenvalue equation is

$$-\frac{\hbar^2}{2m}\frac{d^2u(x)}{dx^2} + \tfrac{1}{2}kx^2u(x) = Eu(x) \tag{5-105}$$

We introduce the frequency of the oscillator

$$\omega = \sqrt{k/m} \tag{5-106}$$

write

$$\epsilon = \frac{2E}{\hbar\omega} \tag{5-107}$$

and change variables to

$$y = \sqrt{\frac{m\omega}{\hbar}}\,x \tag{5-108}$$

to finally get the simpler form of the equation

$$\frac{dy^2}{dy^2} + (\epsilon - y^2)\, u = 0 \tag{5-109}$$

All quantities that appear are dimensionless.

For any eigenvalue ϵ, as $y^2 \to \infty$, the term involving ϵ is negligible, and we must therefore require that $u(y)$ asymptotically satisfy the equation

$$\frac{d^2 u_0(y)}{dy^2} - y^2 u_0(y) = 0 \tag{5-110}$$

We multiply by $2 du_0/dy$, which allows us to rewrite this in the form

$$\frac{d}{dy}\left(\frac{du_0}{dy}\right)^2 - y^2\,\frac{d}{dy}\,(u_0{}^2) = 0 \tag{5-111}$$

or, equivalently,

$$\frac{d}{dy}\left[\left(\frac{du_0}{dy}\right)^2 - y^2 u_0{}^2\right] = -2y u_0{}^2 \tag{5-112}$$

This simplifies a great deal if we neglect the term on the right side of the equation. We assume that this can be done, and then check that the assumption was correct. If we drop the right side, we find that

$$\frac{du_0}{dy} = (C + y^2 u_0{}^2)^{1/2}$$

where C is a constant of integration. Since both $u_0(y)$ and du_0/dy must vanish at infinity, we must have $C = 0$. Thus

$$\frac{du_0}{dy} = \pm y u_0 \tag{5-113}$$

whose solution, acceptable at infinity, is

$$u_0(y) = e^{-y^2/2} \tag{5-114}$$

We can now check that $2y u_0{}^2 = 2y\, e^{-y^2}$ is indeed negligible compared with

$$\frac{d}{dy}\,(y^2 u_0{}^2) = \frac{d}{dy}\,(y^2\, e^{-y^2}) \simeq -4y^3\, e^{-y^2}$$

for large y. If we now introduce a new function $h(y)$, such that

$$u(y) = h(y)\, e^{-y^2/2} \tag{5-115}$$

then the differential equation is easily seen to take the form

$$\frac{d^2 h(y)}{dy^2} - 2y\,\frac{dh(y)}{dy} + (\epsilon - 1)\, h(y) = 0 \tag{5-116}$$

This may not seem like much of a simplification, but we have accounted for the behavior at infinity, and we can now look at the behavior near $y = 0$. Let us attempt a power series expansion

$$h(y) = \sum_{m=0}^{\infty} a_m y^m \tag{5-117}$$

When this is inserted into the equation, we find that the coefficients of y^m satisfy the recursion relation

$$(m + 1)(m + 2) a_{m+2} = (2m - \epsilon + 1) a_m \tag{5-118}$$

Thus, given a_0 and a_1, the even and odd series can be generated separately. That they do not mix is a consequence of the invariance of the Hamiltonian under reflections. For arbitrary ϵ, we find that for large m (say $m > N$)

$$a_{m+2} \simeq \frac{2}{m} a_m \tag{5-119}$$

This means that the solution is approximately
$h(y) = $ (a polynomial in y)

$$+ a_N \left[y^N + \frac{2}{N} y^{N+2} + \frac{2^2}{N(N + 2)} y^{N+4} + \frac{2^3}{N(N + 2)(N + 4)} y^{N+6} + \cdots \right]$$

where, for simplicity, we have only taken the even solution. The series may be written in the form

$$a_N y^2 \left(\frac{N}{2} - 1 \right)! \left[\frac{(y^2)^{N/2-1}}{(N/2 - 1)!} + \frac{(y^2)^{N/2}}{(N/2)!} + \frac{(y^2)^{N/2+1}}{(N/2 + 1)!} + \cdots \right]$$

which is of the form of a polynomial $+$ a constant $\times y^2 e^{y^2}$. When this is inserted into (5-115), we get a solution that does not vanish at infinity. An acceptable solution can be found if the recursion relation terminates, that is, if

$$\epsilon = 2N + 1 \tag{5-120}$$

For that particular value of ϵ the recursion relations yield

$$a_{2k} = (-2)^k \frac{N(N - 2) \ldots (N - 2k + 4)(N - 2k + 2)}{(2k)!} a_0 \tag{5-121}$$

and

$$a_{2k+1} = (-2)^k \frac{(N + 1)(N - 1) \ldots (N - 2k + 3)(N - 2k + 1)}{(2k + 1)!} a_1 \tag{5-122}$$

Thus the results are:

1. There are discrete, equally spaced eigenvalues. (5-120) translates into

$$E = \hbar\omega(n + \tfrac{1}{2}) \tag{5-123}$$

a form that looks familiar, since the relation between energy and frequency is the same as that discovered by Planck for the radiation field modes. This is no accident, since a decomposition of the electromagnetic field into normal modes is essentially a decomposition into harmonic oscillators that are decoupled.

2. The polynomials $h(y)$ are, except for normalization constants, the Hermite polynomials $H_n(y)$, whose properties may be found in many textbooks We are not really interested in these details, and we will solve the harmonic oscillator problem again, so that we do not pursue these matters. It is, however, worth pointing out that the reason for the importance of the harmonic oscillator in quantum mechanics, as in classical mechanics, is that any small perturbation of a system from its equilibrium state will give rise to small oscillations, which are ultimately decomposable into normal modes, that is, independent oscillators.

3. As (5-123) shows, even the lowest state has some energy, the *zero-point energy*. Its presence is a purely quantum mechanical effect, and can be interpreted in terms of the uncertainty principle. It is the zero-point energy that is responsible for the fact that helium does not "freeze" at extremely low temperatures, but remains liquid down to temperatures of the order of 10^{-3} degrees Kelvin, at normal pressures. The frequency ω is larger for lighter atoms, which is why the effect is not seen for nitrogen, say. It also depends on detailed features of the interatomic forces, which is why liquid hydrogen does freeze.

Problems

1. Consider an arbitrary potential localized on a finite part of the x-axis. The solutions of the Schrödinger equation to the left and to the right of the potential region are given by

respectively. Show that if we write

$$C = S_{11}A + S_{12}D$$

$$B = S_{21}A + S_{22}D$$

that is, relate the "outgoing" waves to the "ingoing" waves by

$$\begin{pmatrix} C \\ B \end{pmatrix} = \begin{pmatrix} S_{11} & S_{12} \\ S_{21} & S_{22} \end{pmatrix} \begin{pmatrix} A \\ D \end{pmatrix}$$

that the following relations hold

$$|S_{11}|^2 + |S_{12}|^2 = 1$$
$$|S_{21}|^2 + |S_{22}|^2 = 1$$
$$S_{11}S_{12}^* + S_{21}S_{22}^* = 0$$

This is equivalent to the statement that the *matrix*

$$S = \begin{pmatrix} S_{11} & S_{12} \\ S_{21} & S_{22} \end{pmatrix}$$

is unitary.

(*Hint.* Use flux conservation and the possibility that A and D are arbitrary complex numbers.)

2. Calculate the elements of the scattering matrix, S_{11}, S_{12}, S_{21}, and S_{22} for the potential

$$V(x) = 0 \qquad x < -a$$
$$= V_0 \qquad -a < x < a$$
$$= 0 \qquad x < a$$

and show that the general conditions proved in Problem 1 are indeed satisfied.

3. The elements $S_{11} \ldots S_{22}$ are functions of k. Show that

$$S_{11}(-k) = S_{11}^*(k)$$
$$S_{22}(-k) = S_{22}^*(k)$$
$$S_{12}(-k) = S_{21}^*(k)$$

that is, that the matrix has the property

$$S(-k) = S^+(k)$$

4. Consider the odd solution to the potential well (e.g., Eq. 5-39), which can be used as a model for a three-dimensional potential well with zero angular momentum. If the range of the potential is given to be 1.4×10^{-13} cm and the binding energy of a system is -2.2 MeV, and if the mass to be used is 0.8×10^{-24} gm, find the depth of the potential in MeV.

[*Hints.* (1) First, convert distances and masses into units of some mass, so that the range is $d(\hbar/\mu c)$ and the binding energy is of the form $\epsilon(\mu c^2)$. A convenient mass might be the one given. (2) The binding energy is very small, so that it is almost zero. If it were zero, condition (5-41) would yield V_0. Expand about this value.]

5. Without actually solving the Schrödinger equation, set up the solutions

so that only the matching of eigenfunctions and their derivatives remain to be done for the following situations:

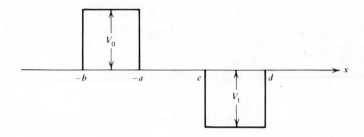

with the following conditions (a) flux $\hbar k/m$ would be incident from the left if the potentials were absent; take $E < V_0$.

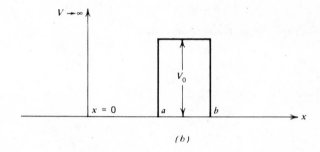

(b)

with flux of magnitude $\hbar k/m$ incident from the right if the potential were absent, $E < V_0$.

6. Show that the conditions for a bound state (5-35) may be obtained by requiring the vanishing of the denominators in (5-25) at $k = i\kappa$. Can you give an argument for why this is not an accident?

7. Consider the scattering matrix for the potential

$$V(x) = \frac{\lambda}{a} \delta(x - b)$$

Show that it has the form

$$\begin{pmatrix} \dfrac{2ika}{2ika - \lambda} & \dfrac{\lambda}{2ika - \lambda} e^{-2ikb} \\ \dfrac{\lambda}{2ika - \lambda} e^{2ikb} & \dfrac{2ika}{2ika - \lambda} \end{pmatrix}$$

Prove that it is unitary, and that it will yield the condition for bound states when the elements of that matrix become infinite. (This will only occur for $\lambda < 0$.)

8. Calculate α in Eq. 5-83, which will localize the particle as far as possible on the right side of the origin.

9. Work out in detail the wave functions for the three lowest eigenfunctions of the harmonic oscillator.

10. Consider the harmonic oscillator potential perturbed by a small cubic term, so that

$$V(x) = \tfrac{1}{2}m\omega^2 \left(x^2 - \frac{1}{a} x^3 \right)$$

If a is large (compared to the characteristic dimension $(\hbar/m\omega)^{1/2}$, estimate how long it takes a particle in the ground state to "leak out" to the region on the far right. Note that with this perturbation alone, there is no lowest energy state, since for large enough x the potential becomes arbitrarily deep.

11. Consider the potential shown below

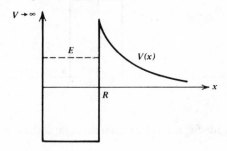

with

$$V(x) = \frac{\hbar^2 l(l + 1)}{2mx^2} \qquad x > R_0$$

Estimate the lifetime of a particle of energy E in this potential. (The outside potential represents a centrifugal barrier in a three-dimensional world.) Express your result in terms of the dimensionless ratio l/kR, where $E = \hbar^2 k^2/2m$. Take $l \gg 1$.

12. Consider the Kronig-Penney potential with

$$\lambda = 3\pi$$

(a) Make a detailed plot of

$$\cos x + \frac{\lambda}{2} \frac{\sin x}{x}$$

as a function of $x = ka$.

(b) Show that forbidden energy bands start just above $ka = n\pi$.

(c) Show that the allowed energy bands get narrower as λ increases.

(d) Plot the energy $\hbar^2 k^2/2m$ as a function of q.

13. Consider the model of a molecule defined by (5.71). Show that when λ is large,

$$\omega_{oe} \simeq \frac{\hbar}{2ma^2} \lambda^2 e^{-\lambda}$$

References

The Kronig-Penney model is also discussed in detail in:

E. Merzbacher, *Quantum Mechanics* (2nd edition), John Wiley and Son, New York, 1970.

For a more detailed discussion of "band theory" see:

C. Kittel, *Introduction to Solid State Physics* (4th edition), John Wiley and Sons, Inc., New York, 1971, Chapter 9.

For a more complete discussion of barrier penetration, using the WKB approximation, see any of the more advanced textbooks listed at the end of the book.

chapter 6

The General Structure of Wave Mechanics

In Chapter 5 the energy eigenfunctions, that is, the solutions of the equation

$$Hu_E(x) = Eu_E(x) \qquad (6\text{-}1)$$

with

$$H = \frac{p_{op}^2}{2m} + V(x) = \frac{1}{2m}\left(\frac{\hbar}{i}\frac{d}{dx}\right)^2 + V(x) \qquad (6\text{-}2)$$

were obtained for a number of physically interesting Hamiltonians. The Hamiltonian operator H was emphasized, because it is this operator that determines the time development of a system. The initial state of a system can be described by any wave function $\psi(x)$, which is only constrained by the requirement that

$$\int dx\, \psi^*(x)\, \psi(x) < \infty \qquad (6\text{-}3)$$

that is, that it be square integrable. The $\psi(x)$ can, without loss of generality, be multiplied by a constant, so that it is normalized

$$\int dx\, \psi^*(x)\, \psi(x) = 1 \qquad (6\text{-}4)$$

The time-dependent Schrödinger equation

$$\frac{\hbar}{i}\frac{\partial}{\partial t}\psi(x,t) = H\psi(x,t) \qquad (6\text{-}5)$$

describes the time development of the wave function, and given the energy eigenfunctions $u_E(x)$, this problem can be solved immediately. What goes into

111

the solution is a general theorem that states: an arbitrary function $\psi(x)$ can be expanded in a complete set of eigenfunctions of H, that is,

$$\psi(x) = \sum_E C_E u_E(x) \tag{6-6}$$

If we choose the eigenfunctions of H to be normalized, and if we take into account that the eigenfunctions corresponding to different values of E are orthogonal, so that

$$\int u_{E'}^*(x)\, u_{E''}(x)\, dx = \delta_{E'E''} \tag{6-7}$$

a result proved in Appendix B, then

$$\int dx\, u_{E'}^*(x)\, \psi(x) = \sum_E C_E \int u_{E'}^*(x)\, u_E(x)\, dx$$

$$= \sum_E C_E\, \delta_{EE'}$$

$$= C_{E'} \tag{6-8}$$

that is, the expansion coefficients are determined. Now the time dependence for each energy eigenfunction is

$$u_E(x,t) = u_E(x)\, e^{-iEt/\hbar} \tag{6-9}$$

as can easily be seen when the above is substituted into (6-5), and hence

$$\psi(x,t) = \sum_E C_E\, e^{-iEt/\hbar}\, u_E(x) \tag{6-10}$$

It should be noted, as we have learned from considering a large number of examples, that the energy eigenvalues may take on discrete values and/or continuous values. We speak of the *spectrum* of eigenvalues being discrete and/or continuous. Thus (6-6) really reads

$$\psi(x) = \sum_n C_n u_{E_n}(x) + \int dE C(E)\, u_E(x) \tag{6-11}$$

corresponding to the two possibilities, and (6-7) reads

$$\int u_{E_m}^*(x)\, u_{E_n}(x)\, dx = \delta_{mn} \tag{6-12}$$

for the discrete values, and

$$\int u_E^*(x)\, u_{E'}(x)\, dx = \delta(E - E') \tag{6-13}$$

for the continuous ones. This is not the only possible choice. As we saw in our solutions of problems with potential wells or barriers, the solutions of the energy eigenvalue equation can be made up of functions that far from the potential are momentum eigenfunctions. There is a relation between the energy and the momentum ($E = p^2/2m$ away from the potential), and it turns out to be possible to normalize the solutions so that the right side of (6-13) is replaced by $\delta(p - p')$ or, in three dimensions by $\delta(\mathbf{p} - \mathbf{p'})$.

We also postulated an interpretation for the expansion coefficients: $|C_E|^2$ is the probability that an energy measurement of the state described by $\psi(x)$ yields the particular eigenvalue E. Any particular measurement can only yield an eigenvalue, but in contrast to classical physics, we cannot predict which one it will be: we only have the probability that it will be a particular value E. In quantum mechanics, as in classical theory, a measurement must be reproducible to have any meaning. Thus if an observer, upon making a single measurement on a system finds that the energy is, say, E_1, then a subsequent energy measurement for that system must again yield E_1. Hence, after the first measurement, the state of the system is described by a new wave function, namely the eigenfunction $u_{E_1}(x)$; only then will a repeated measurement yield E_1 with probability 1. The expression "a measurement projects a state into an eigenstate of the observable" is sometimes used.

The expansion theorem may be viewed as a generalization of the expansion of a vector $\mathbf{A}$ in terms of orthonormal unit vectors in an N-dimensional vector space

$$\mathbf{A} = a_1 \hat{\imath}_1 + a_2 \hat{\imath}_2 + \ldots + a_N \hat{\imath}_N \tag{6-14}$$

The unit vectors $\hat{\imath}_k$ satisfy

$$\hat{\imath}_k \cdot \hat{\imath}_l = \delta_{kl} \tag{6-15}$$

and are the analogs of $u_E(x)$. The coefficients a_k are given by

$$a_k = \hat{\imath}_k \cdot \mathbf{A} \tag{6-16}$$

and they are the analogs of the C_E. We shall often use the language of vector spaces in talking about quantum mechanics. Thus we shall often refer to the coefficients C_E as the "projections" of $\psi(x)$ "along" $u_E(x)$, and the quantity

$$C_E = \int u_E^*(x)\, \psi(x)\, dx \tag{6-17}$$

will often be called a *scalar product*. We will, following Dirac, introduce a convenient notation for the scalar product:

$$\int \phi^*(x)\, \psi(x)\, dx = \langle \phi | \psi \rangle \tag{6-18}$$

The similarity between the acceptable wave functions, and the collection of all N-dimensional vectors is actually quite deep. Just as the sum of any two vectors yields a vector

$$\mathbf{A} + \mathbf{B} = \mathbf{C} \tag{6-19}$$

and the product of a vector with a number is again a vector, so will the sum of any two square integrable functions again be a square integrable function, as will the product of a square integrable function with an arbitrary (complex) number. In both cases, provided we define the notion of a scalar product

$$\langle A|B\rangle = \mathbf{A\cdot B} \tag{6-20}$$

in one case,

$$\langle\phi|\psi\rangle = \int dx\ \phi^*(x)\ \psi(x) \tag{6-21}$$

in the other, we have a linear vector space. The only difference is that in quantum mechanics, the vector space is infinite dimensional. In fact, since in (6-21) it is the continuous label x that plays the role that the index i plays in

$$\mathbf{A\cdot B} = \sum_{i=1}^{N} a_i b_i \tag{6-22}$$

we see that the space is continuously infinite. This does mean that a proper mathematical treatment of such vector spaces is much more complicated, since questions of convergence of integrals like (6-21) have to be faced, and in contrast to a finite dimensional space, proving completeness is much more difficult. In mathematical parlance, the square integrable functions form a Hilbert space, and the energy eigenfunctions form a *complete set of basis vectors*.

In vector spaces, be they finite dimensional or more general, an operator is defined to be something that transforms a vector into another vector, or in this case, a square integrable function into another square integrable function. We are actually interested in *linear operators* that have the property that

$$H(\alpha\psi_1 + \beta\psi_2) = \alpha H\psi_1 + \beta H\psi_2 \tag{6-23}$$

The simple example discussed in Chapter 4 showed that the expectation value of H, defined by

$$\langle H\rangle_\psi = \int \psi^*(x)\ H\psi(x)\ dx \tag{6-24}$$

was real. This is to be expected for a physically measurable quantity, and it generalizes to the statement that the expectation value, for all $\psi(x)$, of an operator representing an observable quantity, has to be real. We called operators that had this property, *hermitian*.

For an arbitrary linear operator A, we have

$$\langle A\rangle_\psi = \int \psi^*(x)\ A\psi(x)\ dx \tag{6-25}$$

and

$$\langle A\rangle_\psi{}^* = \int [A\psi(x)]^*\ \psi(x)\ dx \tag{6-26}$$

The operator $A^\dagger$ (pronounced A-dagger) is defined by the relation

$$\int [A\psi(x)]^* \, \psi(x) \, dx = \int \psi^*(x) \, A^\dagger\psi(x) \, dx \qquad (6\text{-}27)$$

and is called the *hermitian conjugate operator*. For example, the relation

$$\int dx \left(\frac{d\psi}{dx}\right)^* \psi = \int dx \, \frac{d}{dx} (\psi^*\psi) - \int dx \, \psi^* \, \frac{d\psi}{dx} = -\int dx \, \psi^* \, \frac{d\psi}{dx}$$

shows that

$$\left(\frac{d}{dx}\right)^\dagger = -\frac{d}{dx} \qquad \longleftarrow$$

Similarly, the hermitian conjugate of the operator

$$\left(\frac{d^2}{dx^2} - ae^{ix}\right)$$

is easily shown to be

$$\left(\frac{d^2}{dx^2} - a^* e^{-ix}\right)$$

For a hermitian operator

$$\langle H \rangle_\psi^* = \int [H\psi(x)]^* \, \psi(x) \, dx$$

$$= \int \psi^*(x) \, H^\dagger\psi(x) \, dx$$

$$= \langle H \rangle_\psi$$

$$= \int \psi^*(x) \, H\psi(x) \, dx \qquad (6\text{-}28)$$

and since this is true for all $\psi(x)$, we say that

$$H^\dagger = H \qquad (6\text{-}29)$$

The Dirac notation for scalar products that involve operators is

$$\int \phi^*(x) \, A\psi(x) \, dx = \langle \phi | A | \psi \rangle \qquad (6\text{-}30)$$

Thus[1]

$$\langle \phi | A | \psi \rangle^* = \int [A\psi(x)]^* \, \phi(x) \, dx$$

---

[1] The definition of $A^\dagger$ in (6-27) only involved the expectation value of A. It is easy to check, by writing $\psi(x) = u(x) + \lambda \, v(x)$ with λ an arbitrary complex number, that (6-27) implies the step between line one and line two in (6-31).

$$= \int \psi^*(x) \, A^\dagger \phi(x) \, dx$$

$$= \langle \psi | A^\dagger | \phi \rangle \qquad (6\text{-}31)$$

The reason for our drive toward generality is, as already seen, the fact that H is not the only operator of interest. Other physical observables, such as the momentum operator p_{op}, parity, position, and so on are represented by hermitian operators. We shall use the letters A, B, C, ... for operators, and since we are only dealing with operators that represent observables, they are all hermitian, that is,

$$A = A^\dagger$$

$$B = B^\dagger$$

and so on.

All hermitian operators have eigenfunctions, that is, there exists a set of vectors that have the property that the operator acting on them reproduces them, except for a proportionality constant, the eigenvalue

$$A u_a(x) = a u_a(x) \qquad (6\text{-}33)$$

The spectrum of eigenvalues, as for the Hamiltonian, may be discrete and/or continuous. The spectrum of momentum eigenvalues was found to be continuous; that of the parity eigenvalues, ± 1 was discrete. As for the energy eigenfunctions, those corresponding to different values of a are orthogonal, and the others may be chosen to be normalized, so that we have

$$\int u_a^*(x) \, u_{a'}(x) \, dx = \delta(a,a') \qquad (6\text{-}34)$$

or, in our new notation

$$\langle u_a | u_{a'} \rangle = \delta(a,a') \qquad (6\text{-}35)$$

Here $\delta(a,a')$ may be a Kroenecker delta $\delta_{aa'}$ if the eigenvalues are discrete, or a Dirac delta function $\delta(a - a')$ if they are continuous. It follows from (6-33) and (6-34) that

$$a = \int u_a^*(x) \, A u_a(x) \, dx \qquad (6\text{-}36)$$

that is,

$$a = \langle u_a | A | u_a \rangle \qquad (6\text{-}37)$$

Thus the eigenvalues of a hermitian operator must be real. Since the results of an individual measurement of the observable described by A must be one of the eigenvalues, this must be so.

Just as for the Hamiltonian, we found in Chapter 4 that the eigenfunctions of other hermitian operators also form a *complete set*, so that the expansion theorem

$$\psi(x) = \sum_a C_a u_a(x) \qquad (6\text{-}38)$$

where

$$C_a = \int u_a^*(x)\, \psi(x)\, dx = \langle u_a | \psi \rangle \qquad (6\text{-}39)$$

holds. The interpretation of C_a is again that of a *probability amplitude*, that is, $|C_a|^2$ is the probability of finding the eigenvalue a in making a measurement of A on a system described by $\psi(x)$. Again, after a measurement, reproducibility requires that the system be found in the eigenstate $u_a(x)$.

In both the problems discussed in Chapter 4, the particle in the box, and the free particle, we found that the eigenfunctions were simultaneous eigenfunctions of H and another operator, parity in the first case, momentum in the second, and we saw that in both cases the additional operators commuted with H. Let us now examine the general conditions under which this happens.

The eigenfunctions u_a, corresponding to the eigenvalue a of the operator A,

$$A u_a(x) = a u_a(x) \qquad (6\text{-}40)$$

will be simultaneous eigenfunctions of another operator B, when

$$B u_a(x) = b u_a(x) \qquad (6\text{-}41)$$

This, however, implies that

$$AB u_a(x) = Ab u_a(x) = bA u_a(x) = ab u_a(x)$$

and

$$BA u_a(x) = Ba u_a(x) = aB u_a(x) = ab u_a(x)$$

that is, that

$$(AB - BA)\, u_a(x) = 0 \qquad (6\text{-}42)$$

If this were to hold for just one u_a, it would not be very interesting, but if it holds for the complete set u_a, then it means that for all square integrable functions $\psi(x) = \sum_a C_a u_a(x)$,

$$\sum_a C_a (AB - BA)\, u_a(x) = (AB - BA) \sum_a C_a u_a(x)$$

$$= (AB - BA)\, \psi(x) = 0 \qquad (6\text{-}43)$$

that is, the operators commute

$$[A, B] = 0 \qquad (6\text{-}44)$$

Conversely, if we have two hermitian operators A and B that commute, so that (6-44) holds, then

$$ABu_a(x) = BAu_a(x)$$
$$= aBu_a(x) \qquad (6\text{-}45)$$

that is,

$$A[Bu_a(x)] = a[Bu_a(x)] \qquad (6\text{-}46)$$

Thus the function $Bu_a(x)$ is also an eigenfunction of A with eigenvalue a. If there is only one eigenfunction of A corresponding to the eigenvalue a, then this implies that $Bu_a(x)$ must be proportional to $u_a(x)$, that is,

$$Bu_a(x) = bu_a(x) \qquad (6\text{-}47)$$

Then $u_a(x)$ is a simultaneous eigenfunction of A and B. This situation, in which the eigenfunctions of A are *not degenerate*, is the one that we saw for the particle in the box. If, on the other hand, there are two eigenfunctions of A corresponding to the eigenvalue a, that is, we have a twofold *degeneracy*

$$Au_a^{(1)}(x) = au_a^{(1)}(x)$$
$$Au_a^{(2)}(x) = au_a^{(2)}(x) \qquad (6\text{-}48)$$

a situation illustrated in the free particle example, then we can only assert that $Bu_a^{(1)}(x)$ and $Bu_a^{(2)}(x)$ must be linear combinations of $u_a^{(1)}(x)$ and $u_a^{(2)}(x)$:

$$Bu_a^{(1)}(x) = b_{11}u_a^{(1)}(x) + b_{12}u_a^{(2)}(x)$$
$$Bu_a^{(2)}(x) = b_{21}u_a^{(1)}(x) + b_{22}u_a^{(2)}(x) \qquad (6\text{-}49)$$

It is evident, however, that we can take linear combinations of these equations to obtain equations of the type

$$Bv_a^{(1)}(x) = b_+v_a^{(1)}(x)$$
$$Bv_a^{(2)}(x) = b_-v_a^{(2)}(x) \qquad (6\text{-}50)$$

For example,

$$B(u_a^{(1)} + \lambda u_a^{(2)}) = (b_{11} + \lambda b_{21})\, u_a^{(1)} + (b_{12} + \lambda b_{22})\, u_a^{(2)}$$
$$= b_\pm(u_a^{(1)} + \lambda u_a^{(2)})$$

provided we choose λ such that

$$\frac{b_{12} + \lambda b_{22}}{b_{11} + \lambda b_{21}} = \lambda$$

This is a quadratic equation and there will be two values of λ, corresponding to the two eigenvalues $b_\pm$. It is more appropriate to denote the simultaneous eigenfunctions of A and B in (6-50) by $u_{ab}^{(1)}(x)$ and $u_{ab}^{(2)}(x)$. Since these correspond to different eigenvalues of the operator B, they will be orthogonal to each other. In practice, for twofold degeneracy, the degenerate eigenfunctions of A, if they are taken to be orthogonal to each other (e.g., e^{ikx} and e^{-ikx} for the free particle case), will automatically be eigenfunctions of B.

Even after finding eigenfunctions of A and then making linear combinations that are eigenfunctions of a commuting operator B, there may still be some degeneracy, that is, there are several eigenfunctions of A and B simultaneously, with the same a and b. This means that there must be a third operator C that commutes with both A and B, and the functions can be recombined to be simultaneous eigenfunctions of A, B, and C whose eigenvalues distinguish the degenerate eigenfunctions of A and B. This will go on until there is no more degeneracy. The set of mutually commuting operators $A, B, C, \ldots, M$ of which our set of functions is a set of common eigenfunctions is called a *complete set of commuting observables*. We have

$$[A, \ B] = [A, \ C] = \ldots = [A, \ M] = 0$$
$$[B, \ C] = [B, \ D] = \ldots = (B, \ M] = 0 \qquad (6\text{-}51)$$

and so on.

$$Au_{ab\ldots m}(x) = au_{ab\ldots m}(x)$$
$$Bu_{ab\ldots m}(x) = bu_{ab\ldots m}(x)$$
$$Mu_{ab\ldots m}(x) = mu_{ab\ldots m}(x) \qquad (6\text{-}52)$$

The state described by $u_{abc\ldots m}(x)$ has definite values of the observables A, B, $C, \ldots, M$. This is the largest possible amount of information that we can have about a system all at once. The reason is that if we consider another operator that is not some function of the operators $A, B, \ldots, M$ (since these commute, such a function is unambiguously defined), then a measurement of it will not give a sharp value for the state $u_{ab\ldots m}(x)$. In general, if two operators do not commute, then a type of uncertainty relation connects the precision with which the two observables can be determined.

To demonstrate this, we must first agree on a definition of uncertainty. A natural definition is

$$(\Delta A)^2 = \langle A^2 \rangle - \langle A \rangle^2 \qquad (6\text{-}53)$$

also called the *dispersion*. It has the advantage that it does not vanish even if $\langle A \rangle = 0$, and it vanishes if the expectation value is taken in an eigenstate of A. Note that we may also write this as

$$(\Delta A)^2 = \langle (A - \langle A \rangle)^2 \rangle \qquad (6\text{-}54)$$

since

$$\langle A^2 - 2A\langle A \rangle + \langle A \rangle^2 \rangle = \langle A^2 \rangle - \langle A \rangle \langle A \rangle$$

This shows that $(\Delta A)^2$ deals with the magnitude of fluctuations about the mean. It is straightforward to show (see Appendix B) that

$$(\Delta A)^2 (\Delta B)^2 - \tfrac{1}{4} \langle i[A,B] \rangle^2 \geq 0 \tag{6-55}$$

Thus for x and p, for which $[x,p] = i\hbar$, it follows that

$$(\Delta x)^2 (\Delta p)^2 \geq \frac{\hbar^2}{4} \tag{6-56}$$

Notice that in the derivation no use was made of wave properties, x-space or p-space functions, or particle-wave duality. Our result depends entirely on the operator properties of the observables A and B.

Let us now turn to the important question of the classical limit of quantum theory. To do this we must first study the time development of expectation values of operators. In general, the expectation value of an operator changes with time. It may change with time because the operator has an explicit time dependence, for example, the operator $x + pt/m$, and it also changes with time because the expectation value is taken with respect to a wave function that itself changes with time. If we write

$$\langle A \rangle_t = \int \psi^*(x,t)\, A\psi(x,t)\, dx \tag{6-57}$$

then

$$\frac{d}{dt} \langle A \rangle_t = \int \psi^*(x,t)\, \frac{\partial A}{\partial t}\, \psi(x,t)\, dx$$

$$+ \int \frac{\partial \psi^*(x,t)}{\partial t}\, A\psi(x,t)\, dx$$

$$+ \int \psi^*(x,t)\, A\, \frac{\partial \psi(x,t)}{\partial t}\, dx$$

$$= \left\langle \frac{\partial A}{\partial t} \right\rangle_t + \int \left(\frac{1}{i\hbar}\, H\psi(x,t) \right)^* A\psi(x,t)$$

$$+ \int \psi^*(x,t)\, A \left(\frac{1}{i\hbar}\, H\psi(x,t) \right)$$

$$= \left\langle \frac{\partial A}{\partial t} \right\rangle_t + \frac{i}{\hbar} \int \psi^*(x,t)\, HA\psi(x,t)\, dx$$

$$- \frac{i}{\hbar} \int \psi^*(x,t)\, AH\psi(x,t)\, dx$$

that is,

$$\frac{d}{dt} \langle A \rangle_t = \left\langle \frac{\partial A}{\partial t} \right\rangle_t + \frac{i}{\hbar} \langle [H,A] \rangle_t \qquad (6\text{-}58)$$

In the derivation we made use of the fact that H is a hermitian operator. We observe that if A has no explicit time dependence, then the change of the expectation value *for any state* is

$$\frac{d}{dt} \langle A \rangle_t = \frac{i}{\hbar} \langle [H,A] \rangle_t \qquad (6\text{-}59)$$

If the operator commutes with H, then its expectation value is always constant, that is, we may say that *the observable is a constant of the motion.* If the Hamiltonian is one of the complete set of commuting observables, then all the others are constants of the motion.

Let us consider successively $A = x$ and $A = p$. We first have

$$\frac{d}{dt} \langle x \rangle = \frac{i}{\hbar} \langle [H,x] \rangle$$

$$= \frac{i}{\hbar} \left\langle \left[\frac{p^2}{2m} + V(x),x \right] \right\rangle$$

Now x commutes with any function of x,

$$[V(x),x] = 0 \qquad (6\text{-}60)$$

so that we only have to calculate

$$[p^2,x] = p[p,x] + [p,x] p$$

$$= \frac{2\hbar}{i} p \qquad (6\text{-}61)$$

Thus we obtain

$$\frac{d}{dt} \langle x \rangle = \left\langle \frac{p}{m} \right\rangle \qquad (6\text{-}62)$$

Next we have

$$\frac{d}{dt} \langle p \rangle = \frac{i}{\hbar} \left\langle \left[\frac{p^2}{2m} + V(x),p \right] \right\rangle$$

$$= -\frac{i}{\hbar} \langle [p,V(x)] \rangle \qquad (6\text{-}63)$$

since p^2 and p evidently commute. To evaluate the last commutator, we note that

$$pV(x)\,\psi(x) - V(x)\,p\psi(x) = \frac{\hbar}{i}\frac{d}{dx}[V(x)\,\psi(x)] - \frac{\hbar}{i}\,V(x)\,\frac{d}{dx}\,\psi(x)$$

$$= \frac{\hbar}{i}\frac{dV(x)}{dx}\,\psi(x) \tag{6-64}$$

so that

$$[p,V(x)] = \frac{\hbar}{i}\frac{dV(x)}{dx} \tag{6-65}$$

and thus

$$\frac{d}{dt}\langle p\rangle_t = -\left\langle\frac{dV(x)}{dx}\right\rangle_t \tag{6-66}$$

We may combine (6-62) and (6-66) to obtain

$$m\frac{d^2}{dt^2}\langle x\rangle_t = -\left\langle\frac{dV(x)}{dx}\right\rangle_t \tag{6-67}$$

This looks very much like the equation of motion of a classical point particle in a potential $V(x)$

$$m\frac{d^2x_{cl}}{dt^2} = -\frac{dV(x_{cl})}{dx_{cl}} \tag{6-68}$$

The only thing that keeps us from making the identification

$$x_{cl} = \langle x\rangle \tag{6-69}$$

is that

$$\left\langle\frac{dV}{dx}\right\rangle \neq \frac{d}{d\langle x\rangle}V(\langle x\rangle) \tag{6-70}$$

Under circumstances where the above inequality becomes an approximate equality, the motion is essentially classical, as was first noted by Ehrenfest. This requires that the potential be a slowly varying function of its argument. If we write

$$F(x) = -\frac{dV(x)}{dx} \tag{6-71}$$

then

$$F(x) = F(\langle x\rangle) + (x - \langle x\rangle)\,F'(\langle x\rangle) + \frac{(x - \langle x\rangle)^2}{2!}\,F''(\langle x\rangle) + \cdots$$

If the uncertainty $(\Delta x)^2 = \langle (x - \langle x \rangle)^2 \rangle$ is small, and the higher terms in the expansion can be neglected, then we have

$$\langle F(x) \rangle \cong F(\langle x \rangle) + \langle x - \langle x \rangle \rangle F'(\langle x \rangle)$$
$$\cong F(\langle x \rangle) \tag{6-72}$$

It is indeed true that even for electrons and other subatomic particles, (6-72) can be valid. For macroscopic fields (6-72) is a good approximation, and this allows us to describe electron or proton orbits in an accelerator by means of classical equations of motion.

Problems

1. If A and B are hermitian operators, prove that (1) the operator AB is only hermitian if A and B commute, that is, if $AB = BA$, and (2) the operator $(A + B)^n$ is hermitian.

2. Prove that $A + A^\dagger$ and $i(A - A^\dagger)$ are hermitian for any operator, as is $AA^\dagger$.

3. Prove that if H is a hermitian operator, then the hermitian conjugate operator of e^{iH} (defined to be $\sum_{n=0}^{\infty} i^n H^n / n!$) is the operator e^{-iH}.

4. Prove the Schwartz inequality

$$\langle \psi | \psi \rangle \langle \phi | \phi \rangle \geq | \langle \psi | \phi \rangle |^2$$

Note that this is equivalent to $\cos^2 \theta \leq 1$ for three dimensional vectors.
(*Hint.* Consider $\langle \psi + \lambda \phi | \psi + \lambda \phi \rangle \geq 0$ and calculate the value of λ that minimizes the l.h.s.)

5. Consider Eq. 6-38 and 6-39. Calculate $\langle \phi | \psi \rangle$ for an arbitrary ϕ in terms of $\langle \phi | u_a \rangle$, and show that it is possible to write

$$\langle \phi | \psi \rangle = \sum_a \langle \phi | u_a \rangle \langle u_a | \psi \rangle$$

In a sense the sum over a complete set of

$$\sum_a | u_a \rangle \langle u_a |$$

is equivalent to the unit operator.

7. If A is hermitian, show that $\langle A^2 \rangle \geq 0$.

8. Consider the hermitian operator H that has the property that

$$H^4 = 1$$

What are the eigenvalues of the operator H? What are the eigenvalues if H is not restricted to being hermitian?

9. An operator is said to be unitary if it has the property that

$$UU^\dagger = U^\dagger U = 1$$

Show that if $\langle \psi | \psi \rangle = 1$ then $\langle U\psi | U\psi \rangle = 1$.

10. Show that if A is hermitian, then e^{iA} is unitary.

11. Show that if the $\{u_a\}$ form an orthonormal complete set, with

$$\langle u_a | u_b \rangle = \delta_{ab}$$

then the set

$$|v_a\rangle = U|u_a\rangle$$

with U unitary is also orthonormal. (The meaning of the above is a unitary operator acting on a set of "basis" states yields another set of "basis" states.)

12. Use the definition of Δx and Δp given in (6-54) to show that

$$\Delta p \, \Delta x \sim \hbar n$$

for a particle in an infinite box in the state characterized by the quantum number n.

13. Show that if the hermitian conjugate operator $A^\dagger$ is defined by (6-27), then

$$\int dx [A\phi(x)]^* \, \psi(x) = \int dx \phi^*(x) \, A^\dagger \psi(x)$$

(*Hint.* See footnote 1, p. 115.)

14. Use the commutation relations between the momentum p and the position x to obtain the equations describing the time dependence of $\langle x \rangle$ and $\langle p \rangle$ given the Hamiltonians

(a) $$H = \frac{p^2}{2m} + \tfrac{1}{2}m(\omega_1^2 x^2 + \omega_2 x + \epsilon)$$

(b) $$H = \frac{p^2}{2m} + \tfrac{1}{2}m\omega^2 x^2 - \frac{A}{x^2}$$

Solve the first set of equations (Hamiltonian (a)).

References

The general structure of wave mechanics is discussed in all books on quantum mechanics. See, for example, among the more introductory books,

D. Bohm, *Quantum Theory*, Prentice Hall, Inc., 1951.

R. H. Dicke and J. P. Wittke, *Introduction to Quantum Mechanics*, Addison-Wesley Publishing Co., Inc., 1960.

J. L. Powell and B. Crasemann, *Quantum Mechanics*, Addison-Wesley Publishing Co., Inc., 1961.

E. Merzbacher, *Quantum Mechanics*, John Wiley and Sons, Inc. (1970), is among the more advanced textbooks.

chapter 7

Operator Methods in Quantum Mechanics

The discussion of the general structure of wave mechanics placed equal weight on the operators that represent the observables, and on their eigenfunctions. Although the latter were at one point described as analogous to an orthonormal basis of unit vectors in an N-dimensional vector space—which would certainly downgrade them in importance—they, rather than the operators, seemed to play the leading role in our discussion of the physical problems in Chapter 5. In this chapter we will show, using a simple example, (a) that one can go very far toward finding the eigenvalue spectrum using the operators alone, and (b) that the description of eigenfunctions as a basis can be made a little more abstract. The latter is important because so far we have only considered functions that depend on x or on p. We shall see later that there exist observables that cannot be associated with x-space in any direct way, and for these a more abstract notion of *eigenstate* must be developed. These remarks will become somewhat clearer in the course of the solution of our example, the *harmonic oscillator problem*.[1]

The Hamiltonian has the form

$$H = \frac{p^2}{2m} + \tfrac{1}{2}m\omega^2 x^2 \qquad (7\text{-}1)$$

where x and p are operators. We do not insist that p be represented by $(\hbar/i)(d/dx)$. The only vestige of the explicit representation that we obtained in Chapter 3 is the statement of the fundamental commutation relation

$$[p,x] = \frac{\hbar}{i} \qquad (7\text{-}2)$$

---

[1] There are few problems that are exactly soluble, whether as differential equations or in operator form. This example is the simplest and thus most suitable for our purposes.

Classically, the Hamiltonian could be factored into

$$H = \omega \left(\sqrt{\frac{m\omega}{2}} \, x - i \, \frac{p}{\sqrt{2m\omega}} \right) \left(\sqrt{\frac{m\omega}{2}} \, x + i \, \frac{p}{\sqrt{2m\omega}} \right)$$

but because p and x do not commute, we have

$$\omega \left(\sqrt{\frac{m\omega}{2}} \, x - i \, \frac{p}{\sqrt{2m\omega}} \right) \left(\sqrt{\frac{m\omega}{2}} \, x + i \, \frac{p}{\sqrt{2m\omega}} \right)$$

$$= \frac{p^2}{2m} + \frac{m\omega^2}{2} \, x^2 - \frac{i\omega}{2} \, (px - xp)$$

$$= H - \tfrac{1}{2}\hbar\omega \tag{7-3}$$

Let us now introduce the notation

$$A = \sqrt{\frac{m\omega}{2}} \, x + i \, \frac{p}{\sqrt{2m\omega}}$$

$$A^\dagger = \sqrt{\frac{m\omega}{2}} \, x - i \, \frac{p}{\sqrt{2m\omega}} \tag{7-4}$$

Since x and p are hermitian operators, the labeling of the second operator with a dagger is appropriate. The two operators do not commute. We may compute

$$[A, A^\dagger] = \left[\sqrt{\frac{m\omega}{2}} \, x, \, -i \, \frac{p}{\sqrt{2m\omega}} \right] + \left[i \, \frac{p}{\sqrt{2m\omega}}, \, \sqrt{\frac{m\omega}{2}} \, x \right]$$

$$= \hbar \tag{7-5}$$

and rewrite the Hamiltonian in terms of the new operators,

$$H = \tfrac{1}{2}\hbar\omega + \omega A^\dagger A \tag{7-6}$$

The simplicity of the Hamiltonian is reflected in the simplicity of the commutation relations of A and $A^\dagger$ with H. We have[2]

$$[H, A] = [\omega A^\dagger A, A] = \omega[A^\dagger, A] \, A$$

$$= -\hbar\omega A \tag{7-7}$$

and similarly

$$[H, A^\dagger] = [\omega A^\dagger A, A^\dagger] = \omega A^\dagger[A, A^\dagger]$$

$$= \hbar\omega A^\dagger \tag{7-8}$$

---

[2] We shall make repeated use of the rules for commutators exhibited in Appendix B:

$$[A + B, C] = [A, C] + [B, C] \text{ and } [AB, C] = A[B, C] + [A, C] B$$

It is, of course, essential that the order of operators not be disturbed.

Incidentally, it is a useful technical trick in deriving commutation relations involving hermitian adjoint operators to recall that

$$[A,B]^\dagger = (AB - BA)^\dagger = B^\dagger A^\dagger - A^\dagger B^\dagger = [B^\dagger, A^\dagger] \qquad (7\text{-}9)$$

In particular

$$[H,A]^\dagger = [A^\dagger, H] = -[H, A^\dagger]$$
$$= (-\hbar\omega A)^\dagger \qquad (7\text{-}10)$$

from which (7-8) follows.

Let us now write down the eigenvalue equation, which reads

$$Hu_E = Eu_E \qquad (7\text{-}11)$$

In the past, whenever we wrote down such an equation, the implication was that H contained some differential operators like d/dx and that u_E was a function of x. That was appropriate when our operators were specifically tied to the space defined by all square integrable functions of x, but in what we are doing now, we are not being very specific about what our operators operate on. We shall assume that they are defined in some abstract vector space, and relate that abstract vector space to the space of functions of x later. To translate this abstraction into the language that we use to describe the equations, we shall not speak of eigenfunctions but of *eigenstates*, and what we called wave functions or wave packets, we shall now call *state vectors*. Thus the eigenfunction $u_{ab...m}(x)$ of the maximal commuting set of observables can be replaced by the eigenvector or eigenstate of this maximal commuting set, $u_{ab...m}$; the labels $a, b, \ldots, m$ give the values of the eigenvalues of the observables $A, B, \ldots, M$, and this description, without the x, does explicitly show the maximum information content.

Let us now take (7-7) and have it act on u_E.

$$HAu_E - AHu_E = -\hbar\omega Au_E$$

With the help of (7-11) this becomes

$$HAu_E = (E - \hbar\omega)\, Au_E \qquad (7\text{-}12)$$

This equation states that if u_E is an eigenstate of H with eigenvalue E, then Au_E is also an eigenstate of H but with eigenvalue $E - \hbar\omega$, that is, with energy lowered by one unit of

$$\epsilon = \hbar\omega \qquad (7\text{-}13)$$

We may therefore write

$$Au_E = c(E)\, u_{E-\epsilon} \qquad (7\text{-}14)$$

The constant $c(E)$ is necessary, since even if u_E is normalized to 1, Au_E need not be. In our emphasis on separation from x-dependence, the normalization condition that was always written as

$$\int u_E^*(x)\, u_E(x)\, dx = 1$$

is now, using the notation defined in (6-18), written as

$$\langle u_E | u_E \rangle = 1 \qquad (7\text{-}15)$$

We shall always normalize all eigenstates to 1, unless they belong to the continuum, in which case

$$\langle u_E | u_{E'} \rangle = \begin{cases} \delta(E - E') \\ \text{or } \delta(p - p') \end{cases} \qquad (7\text{-}16)$$

If we now apply (7-7) to the state $u_{E-\epsilon}$, we find, in exactly the same way, that $A u_{E-\epsilon}$, or, equivalently, $A^2 u_E$ gives a state of energy $E - 2\epsilon$. Thus by repeated application of the operator A to any u_E we can generate states of lower and lower energy. A is appropriately called a *lowering operator*. There is a limit to how many times it can be applied, since it is a consequence of (7-1) that H must always have positive expectation value. For an arbitrary wave function

$$\langle \psi | p^2 | \psi \rangle = \int \psi^*(x)\, p^2 \psi(x)\, dx = \int [p^\dagger \psi(x)]^* \, (p\psi)\, dx$$

$$= \int [p\psi(x)]^* \, [p\psi(x)]\, dx$$

$$= \hbar^2 \int |d\psi(x)/dx|^2\, dx \geq 0 \qquad (7\text{-}17)$$

which we rewrite in our coordinate-deemphasizing way as

$$\langle \psi | p^2 | \psi \rangle = \langle p^\dagger \psi | p \psi \rangle$$

$$= \langle p\psi | p\psi \rangle \geq 0 \qquad (7\text{-}18)$$

Similarly, since x is also a hermitian operator

$$\langle \psi | x^2 | \psi \rangle = \langle x^\dagger \psi | x\psi \rangle$$

$$= \langle x\psi | x\psi \rangle \geq 0 \qquad (7\text{-}19)$$

and all scalar products of vectors with themselves yield the square of their length, that is, a positive number. Thus our lowering procedure must end somewhere, and there is a *ground state*, which we will now denote by u_0, beyond which lowering ends. This must mean that

$$A u_0 = 0 \qquad (7\text{-}20)$$

The energy of the ground state is

$$Hu_0 = (\omega A^\dagger A + \tfrac{1}{2}\hbar\omega)\, u_0 = \tfrac{1}{2}\hbar\omega\, u_0 \qquad (7\text{-}21)$$

Let us apply (7-8) to the ground state:

$$HA^\dagger u_0 - A^\dagger H u_0 = \hbar\omega A^\dagger u_0$$

that is,

$$HA^\dagger u_0 = (\hbar\omega + \tfrac{1}{2}\hbar\omega)\, A^\dagger u_0 \qquad (7\text{-}22)$$

The energy has been raised by one unit of $\hbar\omega$, and $A^\dagger$ is aptly described as a *raising operator*. We will change our notation a little, namely, label the state by the number of energy units $\epsilon = \hbar\omega$ it has over the ground state energy $\tfrac{1}{2}\hbar\omega$. Thus we write

$$A^\dagger u_0 = c u_1 \qquad (7\text{-}23)$$

Note that (7-12) implies that

$$A u_1 = c' u_0 \qquad (7\text{-}24)$$

so that $A^\dagger$ and A move up and down the same "ladder." All the states may be generated by repeated application of $A^\dagger$ to u_0. One consequence is that the energy spectrum is given by

$$E = (n + \tfrac{1}{2})\, \hbar\omega \qquad n = 0, 1, 2, \dots \qquad (7\text{-}25)$$

We have succeeded in obtaining the energy spectrum without solving any differential equation. We have also a general representation of the eigenvectors

$$u_n = \frac{1}{\sqrt{n!}} \left(\frac{A^\dagger}{\sqrt{\hbar}} \right)^n u_0 \qquad (7\text{-}26)$$

where we have put in the correct normalization constant.[3] With the help of this representation, we can prove the orthogonality of eigenstates corresponding to different energies. What is involved is an evaluation of an expression of the form

$$\langle u_0 | A^m (A^\dagger)^n | u_0 \rangle$$

and this is done by commuting A's through the $A^\dagger$'s to the right, where, at the

---

[3] For the algebraically oriented reader, we describe briefly the way in which this is derived. With $(A^\dagger)^n u_0 = c_n u_n$ we have $|c_n|^2 \langle u_n | u_n \rangle = |c_n|^2 = \langle (A^\dagger)^n u_0 | (A^\dagger)^n u_0 \rangle = \langle u_0 | A^n (A^\dagger)^n u_0 \rangle$.

Now we may use (7-5) to derive the relation $A^n (A^\dagger)^n = A^{n-1}[n\hbar(A^\dagger)^{n-1} + (A^\dagger)^n A]$. When this is put between $\langle u_0 | \dots u_0 \rangle$, the second term on the r.h.s. vanishes and we get the recursion relation $|c_n|^2 = n\hbar |c_{n-1}|^2$.

$\therefore\ |c_n|^2 = n!(\hbar)^n |c_0|^2 = n!(\hbar)^n$. We can, without loss of generality, choose c_n real.

end, they act on u_0 giving a vanishing result. Using (7-5) we see, for example, that

$$A^2(A^\dagger)^3 = A(A^\dagger A + \hbar)(A^\dagger)^2 = \hbar A(A^\dagger)^2$$
$$+ AA^\dagger(A^\dagger A + \hbar) A^\dagger = 3\hbar A(A^\dagger)^2 + A(A^\dagger)^3 A$$

The last term, sandwiched between the u_0 gives zero because $Au_0 = 0$, and the first term can be manipulated in the same way to finally give $6\hbar^2 A^\dagger$. Now

$$\langle u_0 | A^\dagger u_0 \rangle = \langle Au_0 | u_0 \rangle = 0 \tag{7-27}$$

so that we have proved $\langle u_2 | u_3 \rangle = 0$. This procedure, when followed in the general case, allows us to prove that

$$\langle u_m | u_n \rangle = 0 \qquad m \neq n \tag{7-28}$$

The statement that an arbitrary state vector can be expanded in eigenstates of H now reads in the coordinate independent way

$$\psi = \sum_{n=0}^{\infty} C_n u_n \tag{7-29}$$

and since $\langle u_m | u_n \rangle = \delta_{mn}$, we have

$$C_m = \langle u_m | \psi \rangle \tag{7-30}$$

We digress briefly from the main thrust of this chapter to point out that the raising and lowering operators may also be used to advantage in solving the harmonic oscillator equation. In x-space, (7-20) reads

$$\left(\sqrt{\frac{m\omega}{2}} x + i \frac{p}{\sqrt{2m\omega}} \right) u_0(x) = 0 \tag{7-31}$$

Using the x-representation of the operator p, $p = (\hbar/i(d/dx))$ this is

$$\left(m\omega x + \hbar \frac{d}{dx} \right) u_0(x) = 0 \tag{7-32}$$

This is a simple differential equation, whose solution is

$$u_0(x) = C e^{-m\omega x^2/2\hbar} \tag{7-33}$$

The constant C is determined by the requirement that $u_0(x)$ be normalized to unity:

$$1 = C^2 \int_{-\infty}^{\infty} dx \, e^{-m\omega x^2/\hbar}$$
$$= C^2 \left(\frac{\hbar\pi}{m\omega} \right)^{1/2}$$

that is,

$$C = \left(\frac{m\omega}{\hbar\pi}\right)^{1/4} \tag{7-34}$$

We may also obtain the excited states by working out in detail

$$u_n(x) = \frac{\hbar^{-n/2}}{\sqrt{n!}} (A^\dagger)^n u_0(x)$$

$$= \frac{1}{\sqrt{n!}}\left(\frac{m\omega}{\hbar\pi}\right)^{1/4} \left(\sqrt{\frac{m\omega}{2\hbar}}\, x - \sqrt{\frac{\hbar}{2m\omega}}\,\frac{d}{dx}\right)^n e^{-m\omega x^2/2\hbar} \tag{7-35}$$

This is, in fact, a very compact way of writing out the general solution of the differential equation.

We have succeeded in making the point that one can solve for the eigenvalues of the harmonic oscillator using operator methods alone. For this problem all that is needed to specify the eigenstates is the energy, that is, the integer $n = 0, 1, 2, \ldots$ appearing in

$$E = (n + \tfrac{1}{2})\, \hbar\omega$$

and thus the complete set of commuting observables consists of H alone.[4] Thus the label n on the eigenstate u_n describes its whole content. We would therefore be quite willing to give up the privileged role of the eigenfunction in x-space, $u_n(x)$, except for one point: $u_n(x)$ does provide us with more information in that it gives us the probability density [via $|u_n(x)|^2$] of finding the particle at x. Does this additional content single out the x-space wave function after all? Let us recall the role of the wave function in momentum space $\phi(p)$ that appears in Chapter 3 for example. As the Fourier transform of the x-space function it might have had some claim to a privileged role, but later, in (4-59), for example, we explained that $\phi(p)$ was "merely" an expansion coefficient of an arbitrary $\psi(x)$ in eigenstates of the momentum operator, and that is why its absolute square yielded the probability of finding a momentum p for that state. Similarly, the fact that $|\psi(x)|^2$ yields the probability density of finding x for the position of the system could be interpreted by the statement that $\psi(x)$ is the expansion coefficient of an arbitrary abstract state in eigenstates of the position operator x_{op}. We write the eigenvalue equation abstractly as

$$x_{op}\phi_x = x\phi_x \tag{7-36}$$

keeping x as a subscript to stress that it is a label of the eigenstate, just as n is the label for u_n. The spectrum of x_{op}, a hermitian operator, is continuous, so that

---

[4] The parity is contained in the label n. States with n even are positive parity states, and those with n odd have negative parity. This follows from the fact that under reflection A and $A^\dagger$ are odd.

the expansion theorem, instead of taking a form like (7-29), really reads

$$\psi = \int dx \, C(x) \, \phi_x \qquad (7\text{-}37)$$

Since the eigenstates defined in (7-36) form an orthonormal set,

$$\langle \phi_x | \phi_{x'} \rangle = \delta(x - x') \qquad (7\text{-}38)$$

we can derive

$$C(x) = \langle \phi_x | \psi \rangle \qquad (7\text{-}39)$$

and this quantity is the probability amplitude for finding a particle at x—more specifically, the measurement of the observable x will yield the eigenvalue x with probability $|C(x)|^2$. All we have to do is change the notation, rewriting (7-37) as

$$\psi = \int dx \psi(x) \, \phi_x \qquad (7\text{-}40)$$

to show that the wave function in x-space has no privileged role, and we use it only as a matter of convenience. The basic principles deal with operators and their eigenvectors and eigenvalues in an abstract space, and the rest is a matter of *representation*. The latter is, of course, crucial in obtaining numbers, which is what physics is all about. That is why we will not lay too much stress on the formal structure of the theory, and continue using wave functions. Later we will have to deal with operators that have no classical analog, such as the intrinsic spin of electrons and other particles, and there we will exercise our freedom to use other representations.

We conclude this chapter by discussing the time development of a system in our representation-independent way. The time-dependent Schrödinger equation

$$i\hbar \, \frac{d\psi(t)}{dt} = H\psi(t) \qquad (7\text{-}41)$$

is now an operator equation in an abstract space. $\psi(t)$ is a vector, and it points in a direction that depends on time. The equation can easily be solved. The solution is

$$\psi(t) = e^{-iHt/\hbar} \, \psi(0) \qquad (7\text{-}42)$$

where $\psi(0)$ is the vector at time $t = 0$ and the operator $e^{-iHt/\hbar}$ is defined by

$$e^{-iHt/\hbar} = \sum_{n=0}^{\infty} \frac{(-iHt/\hbar)^n}{n!} \qquad (7\text{-}43)$$

The solution (7-42) allows us to describe the change with time of the expectation value of some operator A that does not have any explicit time dependence:

$$\langle A \rangle_t = \langle \psi(t) | A\psi(t) \rangle$$

$$= \langle e^{-iHt/\hbar}\,\psi(0) | A\ e^{-iHt/\hbar}\,\psi(0) \rangle$$

$$= \langle \psi(0) |\ e^{iHt/\hbar}\ A\ e^{-iHt/\hbar}\,\psi(0) \rangle$$

$$= \langle \psi(0) | A(t)\,\psi(0) \rangle$$

$$= \langle A(t) \rangle_0 \tag{7-44}$$

We used

$$(e^{-iHt/\hbar})^{\dagger} = e^{iH^{\dagger}t/\hbar} = e^{iHt/\hbar} \tag{7-45}$$

along the way, and defined

$$A(t) = e^{iHt/\hbar}\ A\ e^{-iHt/\hbar} \tag{7-46}$$

What (7-44) says is that the expectation value of a time independent operator A on a state that varies with time as (7-42) may be written as the expectation value of a time-varying operator $A(t)$ [given by (7-46)] in the time-independent state $\psi(0)$. This is very useful in the formal discussion of quantum mechanics, since it is convenient to set up a basis of orthonormal eigenvectors in the abstract vector space once for all, and not worry about how the basis vectors change with time. When we do this, we are working in the *Heisenberg picture*, whereas keeping A without time dependence means that we are working in the *Schrödinger picture*. The result is the same, whatever picture we use: this is analogous to the option of describing a rotating body relative to a fixed set of axes, or of describing the body at rest in a rotating coordinate system. The choice is one of convenience. If we do work in the Heisenberg picture, then state vectors are fixed, and we need not refer to them. How an observable varies with time is determined by (7-46), which yields

$$\frac{d}{dt}\,A(t) = \frac{iH}{\hbar}\ e^{iHt/\hbar}\ A\ e^{-iHt/\hbar} - \frac{i}{\hbar}\ e^{iHt/\hbar}\ AH\ e^{-iHt/\hbar}$$

$$= \frac{i}{\hbar}\ HA(t) - \frac{i}{\hbar}\ A(t)\ H$$

$$= \frac{i}{\hbar}\ [H,A(t)] \tag{7-47}$$

a form remarkably like (6-59). That equation was an equation for expectation values, but since its form was independent of the state in which the expectation value was taken, it had to reflect operator properties, and (7-47) shows that explicitly.

For the harmonic oscillator

$$H = \omega A^{\dagger}A + \tfrac{1}{2}\hbar\omega$$

and since H is a constant of the motion, we have

$$H = \omega A^\dagger(t)\, A(t) + \tfrac{1}{2}\hbar\omega \qquad (7\text{-}48)$$

We can also show, using (7-46), that

$$[A(t),\, A^\dagger(t)] = \hbar \qquad (7\text{-}49)$$

Hence (7-7) and (7-8) still have the same form, and we get

$$\frac{d}{dt}\, A(t) = -i\omega A(t)$$

$$\frac{d}{dt}\, A^\dagger(t) = i\omega A^\dagger(t) \qquad (7\text{-}50)$$

Thus the time dependence of $A(t)$ and $A^\dagger(t)$ is obtained by solving (7-50), with the result that

$$A(t) = e^{-i\omega t}\, A(0)$$

$$A^\dagger(t) = e^{i\omega t}\, A^\dagger(0) \qquad (7\text{-}51)$$

Using the relation (7-4) it is easy to show that

$$p(t) = p(0) \cos \omega t - m\omega x(0) \sin \omega t$$

$$x(t) = x(0) \cos \omega t + \frac{p(0)}{m\omega} \sin \omega t \qquad (7\text{-}52)$$

expressing the operators $x(t)$ and $p(t)$ in terms of the operators $x(0)$ and $p(0)$.

Problems

1. Use the commutation relation (7-5) and the definition of the state u_n given in (7-26) to prove that

$$A u_n = \sqrt{n\hbar}\, u_{n-1}$$

(*Hint.* Use induction, that is, show that if this relation is true for n it is true for $n + 1$, and establish it directly for $n = 1$.

2. Use the above relation to show that if $f(A^\dagger)$ is any polynomial in $A^\dagger$, then

$$A f(A^\dagger)\, u_0 = \hbar \frac{df(A^\dagger)}{dA^\dagger}\, u_0$$

Note that representing A in the form

$$A = \hbar \frac{d}{dA^\dagger}$$

is consistent with the commutation relation (7-5) and is quite analogous to the representation

$$p = \frac{\hbar}{i} \frac{d}{dx}$$

3. Calculate the form of $\langle u_n | x | u_m \rangle$, and show that it vanishes unless $n = m \pm 1$.

(*Hint.* It is sufficient to calculate $\langle u_n | A | u_m \rangle$ since $\langle u_n | A^\dagger | u_m \rangle = \langle A u_n | u_m \rangle = \langle u_m | A | u_n \rangle^*$. Use the results of Problem 1.)

4. Use the results of Problem 2 to show that

$$e^{\lambda A} f(A^\dagger) \, u_0 = f(A^\dagger + \lambda \hbar) \, u_0$$

(*Hint.* Expand the exponential in a series, and use the fact that

$$f(x + a) = \sum \frac{a^n}{n!} f^{(n)}(x)$$

$$f^{(n)}(x) = \frac{d^n}{dx^n} f(x)$$

to work out this problem.)

5. Use the results of Problem 4 to establish the operator relation

$$e^{\lambda A} f(A^\dagger) \, e^{-\lambda A} = f(A^\dagger + \lambda \hbar)$$

Note that an operator relation must hold when it acts on an arbitrary state. Let an arbitrary state be of the form $g(A^\dagger) \, u_0$. Thus what must be proved is that

$$e^{\lambda A} f(A^\dagger) \, e^{-\lambda A} g(A^\dagger) \, u_0 = f(A^\dagger + \lambda \hbar) \, g(A^\dagger) \, u_0$$

This can also be proved from the general relation

$$e^{\lambda A} A^\dagger e^{-\lambda A} = A^\dagger + \lambda [A, A^\dagger] + \frac{\lambda^2}{2!} [A, [A, A^\dagger]] + \cdots$$

6. Use the above relation to prove that

$$e^{aA + bA^\dagger} = e^{aA} \, e^{bA^\dagger} \, e^{-(1/2)ab\hbar}$$

The procedure is the following. Let

$$e^{\lambda(aA + bA^\dagger)} \equiv e^{\lambda aA} \, F(\lambda)$$

Differentiation with respect to λ yields

$$(aA + bA^\dagger) \, e^{\lambda(aA + bA^\dagger)} = aA \, e^{\lambda aA} \, F(\lambda) + e^{\lambda aA} \frac{dF}{d\lambda}$$

that is,

$$(aA + bA^\dagger) e^{\lambda aA} F(\lambda) = aA e^{\lambda aA} (F\lambda) + e^{\lambda aA} \frac{dF}{d\lambda}$$

Use Problem 5 to show that

$$\frac{dF}{d\lambda} = (bA^\dagger - \lambda ab\hbar) F(\lambda)$$

so that

$$F(\lambda) = e^{\lambda bA^\dagger} e^{-(1/2)\lambda^2 ab\hbar}$$

7. Use the procedure of problem 5 to show that

$$e^{\lambda A^\dagger} f(A) e^{-\lambda A^\dagger} = f(A - \lambda\hbar)$$

Show from this that,

$$e^{aA+bA^\dagger} = e^{bA^\dagger} e^{aA} e^{(1/2)ab\hbar}$$

using the method outlined in Problem 6.

8. Use the above result to show that

$$e^{ikx} = e^{(ik/\sqrt{2m\omega})A^\dagger} e^{(ik/\sqrt{2m\omega})A} e^{-(\hbar k^2/4m\omega)}$$

Note that $x = (1/\sqrt{2m\omega})(A + A^\dagger)$

Use this expression to calculate

$$\langle u_0 | e^{ikx} | u_0 \rangle$$

9. Show that the result obtained above is the same as the one obtained from

$$\int dx\, u_0^*(x)\, e^{ikx}\, u_0(x)$$

10. Use the general operator equation of motion (7-47) to solve for the time dependence of the operator $x(t)$ given that

$$H = \frac{p^2(t)}{2m} + mgx(t)$$

11. Consider the Hamiltonian describing a one-dimensional oscillator in an external electric field.

$$H = \frac{p^2(t)}{2m} + \tfrac{1}{2}m\omega^2 x^2(t) - e\mathcal{E}x(t)$$

Calculate the equation of motion for the operators $p(t)$ and $x(t)$ using Eq. 7-47 and the commutation relation

$$[p(t),x(t)] = \frac{\hbar}{i}$$

Show that the equation of motion is just the classical equation of motion. Solve for $p(t)$ and $x(t)$ in terms of $p(0)$ and $x(0)$. Show that

$$[x(t_1),x(t_2)] \neq 0 \qquad \text{for } t_1 \neq t_2$$

This shows that operators that commute at the same time need not commute at different times.

12. Use Eq. 7-35 to calculate the eigenfunctions for $n = 1, 2, 3$. *Note.* Be sure to keep track of the ordering of x and d/dx in the expansion of the binomial series.

References

The material discussed in this chapter is also treated in almost all of the books in the reference list at the end of the book. The student is encouraged to look up some of them, since it is always useful to see the same basic material presented from different points of view.

chapter 8

N-Particle Systems

Our discussion of a single particle is easily generalized to an N-particle system. The N particles are described by a wave function $\psi(x_1, x_2, \ldots, x_N)$ that is normalized such that

$$\int \ldots \int dx_1 dx_2 \ldots dx_N |\psi(x_1, x_2, \ldots, x_N)|^2 = 1 \tag{8-1}$$

The interpretation of $|\psi(x_1, x_2, \ldots, x_N)|^2$ is a generalization of the interpretation of $|\psi(x)|^2$, that is, it yields the probability density for finding particle 1 at x_1, particle 2 at x_2, $\ldots$, particle N at x_N. The time development of such a wave function is given by the solution of the differential equation

$$i\hbar \frac{\partial}{\partial t} \psi(x_1, \ldots, x_N; t) = H\psi(x_1, \ldots, x_N; t) \tag{8-2}$$

where the Hamiltonian is again constructed in correspondence with the classical form

$$H = \sum_{i=1}^{N} \frac{p_i^2}{2m_i} + V(x_1, x_2, \ldots, x_N) \tag{8-3}$$

as

$$H = -\hbar^2 \left(\frac{1}{2m_1} \frac{\partial^2}{\partial x_1^2} + \ldots + \frac{1}{2m_N} \frac{\partial^2}{\partial x_N^2} \right) + V(x_1, \ldots, x_N) \tag{8-4}$$

The whole formalism of quantum mechanics developed before is easily generalized, with the proviso that operators describing single particle observables commute when they refer to different particles, for example,

$$[p_i, x_j] = \frac{\hbar}{i} \delta_{ij} \tag{8-5}$$

141

If there are no external fields, such as the common gravitational field of the earth, or externally imposed electric or magnetic fields, then the potential energy can only depend on the relative separation of the particles, that is,

$$V = V(x_1 - x_2, x_1 - x_3, \ldots, x_{N-1} - x_N) \tag{8-6}$$

This must be the case, because in the absence of any external agency that somehow determines an "origin," the displacement of the whole system should not change any physical properties of the system. In other words, the form of the potential (8-6) is a consequence of the invariance of all physically significant quantities under the transformation

$$x_i \rightarrow x_i + a \tag{8-7}$$

A very important special case of (8-6) is the case of two-body forces, in which case

$$V = \sum_{i>j} V(x_i - x_j) \tag{8-8}$$

The summation is over all indices i and j, subject to the condition $i > j$ to avoid double counting, and the counting of $i = j$. Actually, in the description of electrons in an atom, we will be dealing with the common Coulomb potential, as well as the electron-electron repulsion, and there the nucleus provides an origin. The potential in that case is a three-dimensional generalization of

$$\sum_{i=1}^{N} W(x_i) + \sum_{i>j} V(x_i - x_j) \tag{8-9}$$

When there are no external forces, then in classical mechanics the total momentum is conserved. This follows from the equations of motion

$$m_i \frac{d^2x_i}{dt^2} = -\frac{\partial}{\partial x_i} V(x_1 - x_2, x_1 - x_3, \ldots, x_{N-1} - x_N) \tag{8-10}$$

a consequence of which is that

$$\frac{d}{dt} \sum_i m_i \frac{dx_i}{dt} = -\sum_i \frac{\partial}{\partial x_i} V(x_1 - x_2, \ldots, x_{N-1} - x_N) \tag{8-11}$$

$$= 0$$

The reason for the vanishing of the right side of the above equation is that for every argument in V, there are equal and opposite contributions that come from $\sum_i \partial/\partial x_i$ acting on it. Thus

$$P = \sum_i m_i \frac{dx_i}{dt} \tag{8-12}$$

is a constant of the motion.

In quantum mechanics the same conclusion holds. We shall demonstrate it by using the invariance of the Hamiltonian under the transformation (8-7). The invariance implies that both

$$Hu_E(x_1, x_2, \ldots, x_N) = Eu_E(x_1, x_2, \ldots, x_N) \tag{8-13}$$

and

$$Hu_E(x_1 + a, x_2 + a, \ldots, x_N + a) = Eu_E(x_1 + a, x_2 + a, \ldots, x_N + a) \tag{8-14}$$

hold. Let us take a infinitesimal, so that terms of $0(a^2)$ can be neglected. Then

$$u(x_1 + a, \ldots, x_N + a) \simeq u(x_1, \ldots, x_N) + a \frac{\partial}{\partial x_1} u(x_1, \ldots, x_N)$$

$$+ a \frac{\partial}{\partial x_2} u(x_1, \ldots, x_N) + \ldots$$

$$\simeq u(x_1, \ldots, x_N) + a \sum_i \frac{\partial}{\partial x_i} u(x_1, \ldots, x_N)$$

and hence, subtracting (8-13) from (8-14)

$$aH \left(\sum_{i=1}^N \frac{\partial}{\partial x_i} \right) u_E(x_1, \ldots, x_N) = aE \left(\sum_{i=1}^N \frac{\partial}{\partial x_i} \right) u_E(x_1, \ldots, x_N)$$

$$= a \left(\sum_{i=1}^N \frac{\partial}{\partial x_i} \right) Eu_E(x_1, \ldots, x_N)$$

$$= a \left(\sum_{i=1}^N \frac{\partial}{\partial x_i} \right) Hu_E(x_1, \ldots, x_N) \tag{8-15}$$

If we now define

$$P = \frac{\hbar}{i} \sum_{i=1}^N \frac{\partial}{\partial x_i} \equiv \sum_{i=1}^N p_i \tag{8-16}$$

we see that we have demonstrated that

$$(HP - PH) u_E(x_1, \ldots, x_N) = 0 \tag{8-17}$$

Since the energy eigenstates for N-particles presumably form a complete set of states, in the sense that any function of $x_1, x_2, \ldots, x_N$ can be expanded in terms of all the $u_E(x_1, \ldots, x_N)$ the above equation can be translated into

$$[H,P] \psi(x_1, \ldots, x_N) = 0 \tag{8-18}$$

for all $\psi(x_1, \ldots, x_N)$, that is, into the operator relation

$$[H,P] = 0 \tag{8-19}$$

This, however, implies that P, the total momentum of the system, is a *constant of the motion*. This is a very deep consequence of what is really a statement about the nature of space. The statement that there is no origin, that is, that the laws of physics are invariant under displacement by a fixed distance, leads to a conservation law. In particle physics there are no potentials of the form that we consider here; nevertheless the invariance principle, as stated above, still leads to a conserved total momentum.

Our main interest will be in the *two-particle system*, which we discuss next. For two noninteracting particles we have the simple Hamiltonian

$$H = \frac{p_1^2}{2m_1} + \frac{p_2^2}{2m_2} \tag{8-20}$$

We might expect that since the two particles are totally uncorrelated, the probability of finding one at x_1 and the other at x_2 is the product of two independent probabilities

$$P(x_1, x_2) = P(x_1)\, P(x_2) \tag{8-21}$$

Thus we expect that the solution of

$$\left(-\frac{\hbar^2}{2m_1}\frac{\partial^2}{\partial x_1^2} - \frac{\hbar^2}{2m_1}\frac{\partial^2}{\partial x_2^2} \right) u(x_1, x_2) = Eu(x_1, x_2) \tag{8-22}$$

should be separable into

$$u(x_1, x_2) = \phi_1(x_1)\, \phi_2(x_2) \tag{8-23}$$

Substituting this into (8-22) and dividing by $u(x_1, x_2)$ we get

$$\frac{-(\hbar^2/2m_1)(d^2\phi_1(x_1)/dx_1^2)}{\phi_1(x_1)} + \frac{-(\hbar^2/2m_2)(d^2\phi_2(x_2)/dx_2^2)}{\phi_2(x_2)} = E \tag{8-24}$$

The two terms in the equation depend on different variables, and that is why we set both of them equal to the constants E_1 and E_2 ,respectively:

$$E = E_1 + E_2$$

$$-\frac{\hbar^2}{2m_1}\frac{d^2\phi_1(x_1)}{dx_1^2} = E_1\phi_1(x_1)$$

$$-\frac{\hbar^2}{2m_1}\frac{d^2\phi_2(x_2)}{dx_2^2} = E_2\phi_2(x_2) \tag{8-25}$$

The two equations are easily solved, and we get

$$u(x_1, x_2) = C\, e^{ik_1x_1 + ik_2x_2} \tag{8-26}$$

with

$$k_1^2 = \frac{2m_1E_1}{\hbar^2} \qquad k_2^2 = \frac{2m_2E_2}{\hbar^2} \tag{8-27}$$

Let us now rewrite the solution using the coordinates

$$x = x_1 - x_2$$

$$X = \frac{m_1 x_1 + m_2 x_2}{m_1 + m_2} \tag{8-28}$$

that is, the separation between the particles, and the center of mass coordinate. If we write

$$k_1 x_1 + k_2 x_2 = \alpha(x_1 - x_2) + \beta \frac{m_1 x_1 + m_2 x_2}{m_1 + m_2}$$

we find that

$$\beta = k_1 + k_2 \equiv K$$

$$\alpha = \frac{m_2 k_1 - m_1 k_2}{m_1 + m_2} \equiv k$$

so that the solution has the form

$$u(x_1, x_2) = C\, e^{iKX}\, e^{ikx} \tag{8-29}$$

where $K = k_1 + k_2$ is the wave number corresponding to the total momentum, and k is the wave number corresponding to the relative momentum. The first factor represents the motion of the center of mass, and the second factor is the "internal" wave function. The energy may be written as

$$E = \frac{\hbar^2 K^2}{2(m_1 + m_2)} + \frac{\hbar^2 k^2}{2}\left(\frac{1}{m_1} + \frac{1}{m_2}\right) \tag{8-30}$$

The first factor is the energy of the two-particle system, with mass $m_1 + m_2$ moving freely with the total momentum; the second term is the internal energy. If we introduce the *reduced mass* μ, defined by

$$\frac{1}{\mu} = \frac{1}{m_1} + \frac{1}{m_2} \tag{8-31}$$

then the term is $\hbar^2 k^2 / 2\mu$, which is effectively a one-particle energy, namely, that of a free particle with mass μ and momentum $\hbar k$.

When the Hamiltonian in (8-20) is altered by the addition of a potential that depends on $x_1 - x_2$ only, then we have

$$\left(-\frac{\hbar^2}{2m_1}\frac{\partial^2}{\partial x_1{}^2} - \frac{\hbar^2}{2m_2}\frac{\partial^2}{\partial x_2{}^2}\right) u(x_1, x_2) + V(x_1 - x_2)\, u(x_1, x_2) = E u(x_1, x_2) \tag{8-32}$$

Using the coordinates

$$x = x_1 - x_2$$

$$X = \frac{m_1 x_1 + m_2 x_2}{m_1 + m_2} = \frac{\mu}{m_2} x_1 + \frac{\mu}{m_1} x_2 \tag{8-33}$$

so that

$$x_1 = X + \frac{\mu}{m_1} x$$

$$x_2 = X - \frac{\mu}{m_2} x \tag{8-34}$$

a little algebra shows that the equation takes the form

$$\left(- \frac{\hbar^2}{2(m_1 + m_2)} \frac{\partial^2}{\partial X^2} - \frac{\hbar^2}{2\mu} \frac{\partial^2}{\partial x^2} + V(x) \right) u(x,X) = Eu(x,X) \tag{8-35}$$

If we write

$$u(x,X) = e^{iKX} \phi(x) \tag{8-36}$$

we find that the equation for $\phi(x)$ is

$$- \frac{\hbar^2}{2\mu} \frac{d^2\phi(x)}{dx^2} + V(x) \phi(x) = \epsilon\phi(x) \tag{8-37}$$

that is, a one-particle Schrödinger equation with reduced mass, and energy

$$\epsilon = E - \frac{\hbar^2 K^2}{2(m_1 + m_2)} \tag{8-38}$$

In Chapter 9 we will obtain the separation in a somewhat more sophisticated way. We now turn to the problem of *identical particles*.

There is compelling evidence that electrons are indistinguishable. If this were not so, then the spectrum of an atom, say, helium, would vary from experiment to experiment, depending on "what kind" of electrons were contained in it. No such variation has ever been observed. Similarly, nuclear spectra are always the same, indicating that protons are indistinguishable, as are neutrons. Similar evidence from high energy physics experiments indicates very strongly that other particles, for example, pi-mesons, are also indistinguishable. This is a purely quantum-mechanical property: in classical mechanics it is possible to follow the orbits of all particles (in principle) so that they are never really indistinguishable.

We shall learn that electrons are characterized by an internal quantum number, called the *spin*, and thus their states must include in their description the spin label. This has a further effect on the consequences of indistinguishability, which we discuss next.

A Hamiltonian for indistinguishable particles must be completely symmetric in the coordinates of the particles. For a two-particle system, if there is no dependence on the spin labels, the Hamiltonian is

$$H = \frac{p_1^2}{2m} + \frac{p_2^2}{2m} + V(x_1,x_2) \tag{8-39}$$

with

$$V(x_1, x_2) = V(x_2, x_1) \tag{8-40}$$

We write this symmetry symbolically as

$$H(1,2) = H(2,1) \tag{8-41}$$

and it is understood that if the Hamiltonian does depend on the spins of the particles, then the spins are to be included in the labeling "1," "2." A wave function for an N-particle system, with all the particles identical, will be denoted by $\psi(1, 2, \ldots, N)$, and this stands for the more explicit $\psi(x_1, \sigma_1; x_2, \sigma_2; \ldots; x_N, \sigma_N)$ where the σ_is describe the spin states.

For a two-particle system the energy eigenvalue equation reads

$$H(1,2)\, u_E(1,2) = E u_E(1,2) \tag{8-42}$$

Since the labeling does not matter, we may write this as

$$H(2,1)\, u_E(2,1) = E u_E(2,1) \tag{8-43}$$

On the other hand, using (8-41) we also have

$$H(1,2)\, u_E(2,1) = E u_E(2,1) \tag{8-44}$$

If we now follow the formal approach that we used in our discussion of parity, we will introduce an *exchange operator* P_{12}, which, acting on a state, interchanges all coordinates (space and spin) of particles 1 and 2. The definition of P_{12} implies that

$$P_{12}\psi(1,2) = \psi(2,1) \tag{8-45}$$

Eq. 8-44 may be written as follows

$$\begin{aligned} HP_{12}\, u_E(1,2) &= E u_E(2,1) \\ &= E P_{12}\, u_E(1,2) \\ &= P_{12}\, E u_E(1,2) \\ &= P_{12}\, H u_E(1,2) \end{aligned} \tag{8-46}$$

and this, as usual, implies the operator relation

$$[H, P_{12}] = 0 \tag{8-47}$$

Thus P_{12}, like parity, is a constant of the motion. Also, like parity

$$(P_{12})^2\, \psi(1,2) = \psi(1,2) \tag{8-48}$$

so that the eigenvalues of P_{12} are ± 1. The eigenstates are the symmetric and antisymmetric combinations

$$\psi^{(S)}(1,2) = \frac{1}{\sqrt{2}} [\psi(1,2) + \psi(2,1)]$$

$$\psi^{(A)}(1,2) = \frac{1}{\sqrt{2}} [\psi(1,2) - \psi(2,1)] \tag{8-49}$$

The fact that P_{12} is a constant of the motion implies that a state that is symmetric at an initial time will always be symmetric, and an antisymmetric state will always be antisymmetric.

It is an important *law of nature* that the symmetry or antisymmetry under the interchange of two particles is a characteristic of the particles, and not something that can be arranged in the preparation of the initial state. The law, which was discovered by Pauli, states that

1. Systems consisting of identical particles of half-odd-integral spin (i.e., spin 1/2, 3/2, . . .) are described by antisymmetric wave functions. Such particles are called fermions, and are said to obey Fermi-Dirac statistics.

2. Systems consisting of identical particles of integral spin (spin 0, 1, 2, . . .) are described by symmetric wave functions. Such particles are called bosons, and are said to obey Bose-Einstein statistics.

The law extends to N-particle states. For a system of N identical fermions, the wave function is antisymmetric under the interchange of any pair of particles. For example, a three-particle wave function, properly antisymmetrized, has the form

$$\psi^{(A)}(1,2,3) = \frac{1}{\sqrt{6}} [\psi(1,2,3) - \psi(2,1,3) + \psi(2,3,1)$$
$$- \psi(3,2,1) + \psi(3,1,2) - \psi(1,3,2)] \tag{8-50}$$

whereas the three identical boson wave function has the form

$$\psi^{(S)}(1,2,3) = \frac{1}{\sqrt{6}} [\psi(1,2,3) + \psi(2,1,3) + \psi(2,3,1)$$
$$+ \psi(3,2,1) + \psi(3,1,2) + \psi(1,3,2)] \tag{8-51}$$

Let us now consider a very interesting special case, in which N fermions do not interact with each other, but do interact with a common potential. In that case

$$H = \sum_{i=1}^{N} H_i \tag{8-52}$$

where

$$H_i = \frac{p_i^2}{2m} + V(x_i) \tag{8-53}$$

The eigenstates of the one-particle potential are denoted by $u_{E_k}(x)$ where

$$H_k u_{E_k}(x_k) = E_k u_{E_k}(x_k) \tag{8-54}$$

A solution of

$$Hu_E(1, 2, \ldots, N) = Eu_E(1, 2, \ldots, N) \tag{8-55}$$

is

$$u_E(1, 2, \ldots, N) = u_{E_1}(x_1)\, u_{E_2}(x_2) \ldots u_{E_N}(x_N) \tag{8-56}$$

where

$$E_1 + E_2 + \ldots + E_N = E \tag{8-57}$$

In (8-56) we suppressed the σ_i labels that go with the x_i. Our task now is to antisymmetrize (8-56). If there are only two particles, we evidently have

$$u^{(A)}(1,2) = \frac{1}{\sqrt{2}} [u_{E_1}(x_1)\, u_{E_2}(x_2) - u_{E_1}(x_2)\, u_{E_2}(x_1)] \tag{8-58}$$

With three particles, the form is

$$u^{(A)}(1,2,3) = \frac{1}{\sqrt{6}} [u_{E_1}(x_1)\, u_{E_2}(x_2)\, u_{E_3}(x_3) - u_{E_1}(x_2)\, u_{E_2}(x_1)\, u_{E_3}(x_3)$$

$$+ u_{E_1}(x_2)\, u_{E_2}(x_3)\, u_{E_3}(x_1) - u_{E_1}(x_3)\, u_{E_2}(x_2)\, u_{E_3}(x_1)$$

$$+ u_{E_1}(x_3)\, u_{E_2}(x_1)\, u_{E_3}(x_2) - u_{E_1}(x_1)\, u_{E_2}(x_3)\, u_{E_3}(x_2)] \tag{8-59}$$

For N particles, the answer is a determinant, the so-called *Slater determinant*:[1]

$$u^{(A)}(1,2, \ldots, N) = \frac{1}{\sqrt{N!}} \begin{vmatrix} u_{E_1}(x_1) & u_{E_1}(x_2) \ldots u_{E_1}(x_N) \\ u_{E_2}(x_1) & u_{E_2}(x_2) \ldots u_{E_2}(x_N) \\ \cdot \\ \cdot \\ \cdot \\ u_{E_N}(x_1) & u_{E_N}(x_2) \ldots u_{E_N}(x_N) \end{vmatrix} \tag{8-60}$$

Clearly the interchange of two particles involves the interchange of two columns in the determinant, and this changes the sign. If two electrons are in the same energy eigenstate, for example, $E_1 = E_2$, and if they are in the same spin state, that is, the spin labels are the same $\sigma_1 = \sigma_2$, then the determinant vanishes when $x_1 = x_2$, that is, the electrons cannot be at the same place. Thus the requirement of antisymmetry introduces an effective interaction between two fermions: qualitatively we see that two particles in the same state tend to stay away from each other, since the joint wave function vanishes when their separation goes to zero. Thus even noninteracting particles behave as if there were a repulsive

---

[1] The wave function for N identical bosons is totally symmetric, and the general form is obtained by expanding the determinant in (8-60) and making all the signs positive.

interaction between them. We will see that a complete set of commuting observables for electrons includes an additional two-valued observable associated with the spin. Thus a state of given energy, angular momentum, parity, and so on, can be occupied by two electrons (of opposite spin variable), but by no more than two electrons. This is a restricted version of the *Pauli exclusion principle*.

The statement "no two electrons can be in the same quantum state" strikes one by its global nature. Suppose we have a hydrogen atom in the ground state on earth and another hydrogen atom in its ground state on the moon. Does this mean that the two electrons must be in opposite spin states? To answer this, we note that a specification of the state of the two electrons requires not just a statement that the electrons have spin "up" or spin "down" and that they are in the ground states of their respective atoms, but it also requires a specification of the energy of the atoms. How well do we know these? Suppose we consider a box of width L, and suppose the atoms are localized in $0 \leq x \leq L/4$ and $3L/4 \leq x \leq L$ respectively. Then the momentum of the atoms can be determined with an accuracy that is restricted by the uncertainty principle. The possible values of the energy are given by

$$E_n = \frac{\hbar^2 \pi^2 n^2}{2ML^2} \tag{8-61}$$

from which we deduce that possible values of the momentum are

$$p = \frac{\hbar \pi n}{L}$$

Measurements of the momenta of atoms are restricted by the uncertainty relation

$$\Delta p \sim \frac{n\hbar}{\Delta x} \sim \frac{n\hbar}{L} \tag{8-62}$$

and hence their energies can only be determined with an accuracy

$$\Delta E \simeq \frac{p\Delta p}{M} \sim \frac{\hbar^2 \pi n^2}{ML^2} \tag{8-63}$$

This, however, is larger than

$$E_n - E_{n-1} \simeq \frac{\hbar^2 \pi^2 n}{ML^2} \tag{8-64}$$

In fact, for atoms separated by 1 meter, say, moving with velocity 10^6 cm/sec $n \sim 10^{11}$, so that there is no possibility that in a macroscopic situation there will be conflict with classical intuition. In effect, if the two atoms are labeled A and B, the question is whether there is a difference between using the wave function $\psi_A(x_1)\,\psi_B(x_2)$ and

$$\frac{[\psi_A(x_1)\,\psi_B(x_2) - \psi_A(x_2)\,\psi_B(x_1)]}{\sqrt{2}}$$

to describe the two electron state. We shall see in our discussion of molecules that the overlap between wave functions falls off exponentially with the distance between the two atoms. When the atoms are far apart, it does not make any difference which wave function we use. When the atoms are close, as in a H_2 molecule, for example, the value of n is of order 1, the wave functions do overlap and it does make a difference whether one uses the uncorrelated wave functions or the antisymmetrized wave functions. Experiment tells us that it is the latter that should be used.

An interesting consequence of the Pauli exclusion principle is that the ground state for N electrons in a potential is very different from the ground state for N bosons or N distinguishable particles. Consider, for example, the infinite potential box,

$$
\begin{aligned}
V(x) &= \infty & x < 0 \\
&= 0 & 0 < x < b \\
&= \infty & b < x
\end{aligned}
\tag{8-65}
$$

The solution of the Schrödinger equation that vanishes at $x = 0$ and $x = b$ is given by

$$u_n(x) = \sin n\pi x/b \tag{8-66}$$

with $n = 1, 2, 3, \ldots$, and the energy eigenvalues are

$$E_n = \frac{\hbar^2 \pi^2 n^2}{2mb^2} \tag{8-67}$$

For N noninteracting bosons, the ground state has all the particles in the $n = 1$ state, and thus the energy is given by

$$E = N \frac{\hbar^2 \pi^2}{2mb^2} \tag{8-68}$$

so that the energy per particle is

$$\frac{E}{N} = \frac{\hbar^2 \pi^2}{2mb^2} \tag{8-69}$$

For N noninteracting fermions the situation is quite different. Only two electrons can go into each of the states $n = 1, 2, 3, \ldots$, so that $N/2$ states are filled. Thus the total energy is given by

$$E = 2 \sum_{n=1}^{N/2} \frac{\hbar^2 \pi^2 n^2}{2mb^2} = \frac{\hbar^2 \pi^2}{2mb^2} \frac{N^3}{24} \tag{8-70}$$

In obtaining the last result we have assumed that N is large, so that it does not matter whether the last level is filled with one or two electrons, and we have used

$$\sum_{n=1}^{N/2} n^2 \approx \int_1^{N/2} n^2 \, dn \simeq \frac{1}{3}\left(\frac{N}{2}\right)^3$$

Thus the energy per particle

$$\frac{E}{N} = \frac{\hbar^2 \pi^2}{24mb^2} N^2 \tag{8-71}$$

grows with N^2. Equivalently, for a given energy, the number of bosons filling the well is proportional to E, while the number of fermions filling the well is proportional to $E^{1/3}$. The highest level to be filled in the fermion case is the one for which $n = N/2$, and its energy is

$$E_F = \frac{\hbar^2 \pi^2 N^2}{8mb^2} \tag{8-72}$$

The subscript F has been put in because this energy is called the *Fermi energy*. We may write it in terms of the density of fermions, which in the one-dimensional problem is $N/b \equiv \rho$, as

$$E_F = \frac{\hbar^2 \pi^2}{8m} \rho^2 \tag{8-73}$$

We shall return to the significance of these remarks in Chapter 9.

The exclusion principle plays an extremely important role in the structure of atoms. The enormous richness in the variety of chemical properties of the various elements is directly traceable to the fact that only a limited number of electrons can occupy a given energy eigenstate.

Problems

1. What is the reduced mass of an electron-proton system? How does it differ from the reduced mass of an electron-deuteron system? What is the reduced mass of a system of two identical particles?

2. Prove that the exchange operator P_{12} is hermitian.

3. Consider two noninteracting electrons in an infinite potential well. What is the ground state wave function if the two electrons are in the *same* spin state?

4. Consider two electrons in the same spin state, interacting with a potential

$$V(|x_1 - x_2|) = -V_0 \qquad |x_1 - x_2| \leq a$$
$$= 0 \qquad \text{elsewhere}$$

What is the lowest energy of the two-electron state, assuming that the total momentum of the two electrons is zero?

[*Hint.* Separate the equation in a manner leading to (8-37) and then apply the Pauli principle.]

5. Consider two identical particles, each of spin 0 interacting with potential energy

$$V(x_1,x_2) = K[(x_1 - x_0) - (x_2 + x_0)]^2$$

where x_0 and $-x_0$ are the equilibrium positions of the particles.

What is the spectrum of the two-particle system? What is the spectrum of the system when the identical particles have spin 1/2?

6. Consider two identical particles described by the energy operator

$$H = H(p_1,x_1) + H(p_2,x_2)$$

where

$$H(p,x) = \frac{p^2}{2m} + \tfrac{1}{2}m\omega^2x^2$$

Separate out the center of mass motion, and obtain the energy spectrum for this system. Show that it agrees with that obtained by solving

$$H\psi(x_1,x_2) = E\psi(x_1,x_2)$$

with

$$\psi(x_1,x_2) = u_1(x_1)\,u_2(x_2)$$

Discuss the degeneracy of the energy spectrum.

References

See any of the references listed at the end of Chapter 6 and also

D. S. Saxon, *Elementary Quantum Mechanics*, Holden-Day, Inc., 1968.
D. Park, *Introduction to the Quantum Theory*, McGraw-Hill Co., 1964.

chapter 9

The Schrödinger Equation in Three Dimensions

The Hamiltonian for a single particle moving in three-dimensional space reads

$$H = \frac{p_x^2 + p_y^2 + p_z^2}{2m} + V(x,y,z) \tag{9-1}$$

which we write in the form

$$H = \frac{\mathbf{p}^2}{2m} + V(\mathbf{r}) \tag{9-2}$$

The three-dimensional momentum $\mathbf{p}$ has the representation

$$\mathbf{p} = \frac{\hbar}{i} \boldsymbol{\nabla} \tag{9-3}$$

For two particles in three dimensions, the general form of the Hamiltonian is

$$H = \frac{\mathbf{p}_1^2}{2m_1} + \frac{\mathbf{p}_2^2}{2m_2} + V(\mathbf{r}_1,\mathbf{r}_2) \tag{9-4}$$

If the potential depends on the separation between the particles alone, that is, if

$$V(\mathbf{r}_1,\mathbf{r}_2) = V(|\mathbf{r}_1 - \mathbf{r}_2|) \tag{9-5}$$

then the Hamiltonian is invariant under the displacement of the whole system, $\mathbf{r}_1 \to \mathbf{r}_1 + \mathbf{a}$, $\mathbf{r}_2 \to \mathbf{r}_2 + \mathbf{a}$, and, as we saw in Chapter 8, this implies the conservation of the total momentum and a separation of variables. In what follows we will achieve the separation by finding functions that are simultaneous eigenfunctions of the commuting operators H and $\mathbf{P} = \mathbf{p}_1 + \mathbf{p}_2$. The momentum eigenvalue equation reads

$$\mathbf{P}_{op}f(\mathbf{r}_1,\mathbf{r}_2) = \mathbf{P}\, f(\mathbf{r}_1,\mathbf{r}_2) \tag{9-6}$$

that is,

$$\frac{\hbar}{i} \, (\boldsymbol{\nabla}_1 + \boldsymbol{\nabla}_2) \, f(\mathbf{r}_1,\mathbf{r}_2) = \mathbf{P} \, f(\mathbf{r}_1,\mathbf{r}_2) \qquad (9\text{-}7)$$

If we write

$$f(\mathbf{r}_1,\mathbf{r}_2) = \psi(\mathbf{r}_1 - \mathbf{r}_2, \, \alpha\mathbf{r}_1 + \beta\mathbf{r}_2) \qquad (9\text{-}8)$$

then with $\mathbf{R} = \alpha\mathbf{r}_1 + \beta\mathbf{r}_2$, (9-7) reads

$$\frac{\hbar}{i} \, (\alpha + \beta) \, \boldsymbol{\nabla}_R \psi(\mathbf{r},\mathbf{R}) = \mathbf{P}\psi(\mathbf{r},\mathbf{R}) \qquad (9\text{-}9)$$

that is, the variable $\mathbf{r} = \mathbf{r}_1 - \mathbf{r}_2$ is a constant parameter as far as this equation is concerned. Thus the solution of this equation is

$$\psi(\mathbf{r},\mathbf{R}) = u(\mathbf{r}) \, e^{i\mathbf{P}\cdot\mathbf{R}/\hbar(\alpha+\beta)} \qquad (9\text{-}10)$$

We will now choose α and β to simplify the energy eigenvalue equation, which reads

$$\left[-\frac{\hbar^2}{2m_1} \boldsymbol{\nabla}_1{}^2 - \frac{\hbar^2}{2m_2} \boldsymbol{\nabla}_2{}^2 + V(|\mathbf{r}|) - E_{\text{tot}} \right] u(\mathbf{r}) \, e^{i\mathbf{P}\cdot\mathbf{R}/\hbar(\alpha+\beta)} = 0 \qquad (9\text{-}11)$$

Since

$$\boldsymbol{\nabla}_1 = \boldsymbol{\nabla}_r + \alpha\boldsymbol{\nabla}_R$$

$$\boldsymbol{\nabla}_2 = -\boldsymbol{\nabla}_r + \beta\boldsymbol{\nabla}_R \qquad (9\text{-}12)$$

this equation takes the form

$$-\frac{\hbar^2}{2m_1} \left[\boldsymbol{\nabla}_r{}^2 \, u(\mathbf{r}) + \frac{2i\alpha}{(\alpha + \beta) \, \hbar} \, \mathbf{P}\cdot\boldsymbol{\nabla}_r \, u(\mathbf{r}) - \frac{\alpha^2 \mathbf{P}^2}{(\alpha + \beta)^2 \, \hbar^2} \, u(\mathbf{r}) \right]$$

$$-\frac{\hbar^2}{2m_2} \left[\boldsymbol{\nabla}_r{}^2 \, u(\mathbf{r}) - \frac{2i\beta}{(\alpha + \beta) \, \hbar} \, \mathbf{P}\cdot\boldsymbol{\nabla}_r \, u(\mathbf{r}) - \frac{\beta^2 \mathbf{P}^2}{(\alpha + \beta)^2 \, \hbar^2} \, u(\mathbf{r}) \right]$$

$$+ V(|\mathbf{r}|) \, u(\mathbf{r}) = E_{\text{tot}} \, u(\mathbf{r}) \qquad (9\text{-}13)$$

after the exponential factor has been divided out, following the differentiation with respect to $\mathbf{R}$. This equation simplifies if the cross terms are eliminated with the choice

$$\alpha = \gamma m_1$$

$$\beta = \gamma m_2 \qquad (9\text{-}14)$$

It then reads

$$-\frac{\hbar^2}{2\mu} \boldsymbol{\nabla}_r{}^2 \, u(\mathbf{r}) + V(|\mathbf{r}|) \, u(\mathbf{r}) = \left(E_{\text{tot}} - \frac{\mathbf{P}^2}{2(m_1 + m_2)} \right) u(\mathbf{r}) \qquad (9\text{-}15)$$

where we have introduced the reduced mass μ by

$$\frac{1}{\mu} = \frac{1}{m_1} + \frac{1}{m_2} \tag{9-16}$$

This is really a one-particle Schrödinger equation with energy

$$E = E_{\text{tot}} - \frac{\mathbf{P}^2}{2(m_1 + m_2)} \tag{9-17}$$

Thus the energy that enters into the effective one-particle equation is the total energy, less the kinetic energy of the two-particle system, whose center of mass moves with momentum $\mathbf{P}$ and whose total mass is $m_1 + m_2$.

The quantity γ is not specified by the above equation. If, however, we require that the variable $\mathbf{R}$ be canonically conjugate to the total momentum $\mathbf{P}$, that is, if we require that

$$[P_x, R_x] = \frac{\hbar}{i} \tag{9-18}$$

and so on, then we see that

$$[p_{1x} + p_{2x}, \alpha x_1 + \beta x_2] = \frac{\hbar}{i}(\alpha + \beta) = \frac{\hbar}{i} \tag{9-19}$$

implies that

$$\alpha + \beta = 1 \tag{9-20}$$

that is,

$$\gamma = \frac{1}{m_1 + m_2} \tag{9-21}$$

The reason for carrying out what is after all a very simple separation of variables in this seemingly complicated way is that this procedure will serve as an example of how to proceed in the further separations of the one-particle Schrödinger equation. Such a separation is possible when the potential depends on the separation between the particles, $|\mathbf{r}|$ alone. With $|\mathbf{r}| = r$, the Hamiltonian

$$H = \frac{\mathbf{p}^2}{2\mu} + V(r) \tag{9-22}$$

is invariant under rotations; $V(r)$ is certainly a function of the distance from the origin alone and does not depend on the angular variables that locate the direction of the vector $\mathbf{r}$; $\mathbf{p}^2$ is also a scalar quantity, the length of the vector $\mathbf{p}$, and thus independent of the orientation of $\mathbf{p}$. Equivalently $p^2 = -\hbar^2\boldsymbol{\nabla}^2$ is invariant under rotations. The sceptical reader can check this explicitly by considering the special case of a rotation through an angle θ about the z-axis: with

$$x' = x \cos \theta - y \sin \theta$$

$$y' = x \sin \theta + y \cos \theta \qquad (9\text{-}23)$$

it is easy to see that

$$r' = (x'^2 + y'^2 + z'^2)^{1/2} = (x^2 + y^2 + z^2)^{1/2} = r$$

and

$$\left(\frac{\partial}{\partial x'}\right)^2 + \left(\frac{\partial}{\partial y'}\right)^2 = \left(\cos\theta\,\frac{\partial}{\partial x} - \sin\theta\,\frac{\partial}{\partial y}\right)^2 + \left(\sin\theta\,\frac{\partial}{\partial x} + \cos\theta\,\frac{\partial}{\partial y}\right)^2$$

$$= \left(\frac{\partial}{\partial x}\right)^2 + \left(\frac{\partial}{\partial y}\right)^2$$

Since the Hamiltonian has an invariance property, we expect a conservation law, as we saw in the case of parity and invariance under displacements. To identify the operators that commute with H, let us consider an infinitesimal rotation about the z-axis. Keeping terms of order θ only so that

$$x' = x - \theta y$$

$$y' = y + \theta x \qquad (9\text{-}24)$$

we require that

$$H u_E(x - \theta y, y + \theta x, z) = E u_E(x - \theta y, y + \theta x, z) \qquad (9\text{-}25)$$

If we expand this to first order in θ and subtract from it

$$H u_E(x,y,z) = E u_E(x,y,z) \qquad (9\text{-}26)$$

we obtain

$$H\left(x\,\frac{\partial}{\partial y} - y\,\frac{\partial}{\partial x}\right) u_E(x,y,z) = E\left(x\,\frac{\partial}{\partial y} - y\,\frac{\partial}{\partial x}\right) u_E(x,y,z) \qquad (9\text{-}27)$$

Since the right side of this may be written as

$$\left(x\,\frac{\partial}{\partial y} - y\,\frac{\partial}{\partial x}\right) H u_E(x,y,z)$$

and since the $u_E(\mathbf{r})$ form a complete set, we find that with

$$L_z = \frac{\hbar}{i}\left(x\,\frac{\partial}{\partial y} - y\,\frac{\partial}{\partial x}\right) = x p_y - y p_x \qquad (9\text{-}28)$$

the commutation relation

$$[H, L_z] = 0 \qquad (9\text{-}29)$$

holds. L_z is the z-component of the operator

$$\mathbf{L} = \mathbf{r} \times \mathbf{p} \qquad (9\text{-}30)$$

which is the angular momentum. Had we taken rotations about the x- and y-axes, we would have found, in addition, that

$$[H, L_x] = 0$$
$$[H, L_y] = 0 \qquad (9\text{-}31)$$

Thus the three components of the angular momentum operators commute with the Hamiltonian, that is, the angular momentum is a constant of the motion. This parallels the classical result that central forces imply conservation of the angular momentum.

We might be tempted to look for simultaneous eigenfunctions of H, L_x, L_y, and L_z, but these do not form a complete set of commuting variables. For example

$$[L_x, L_y] = [yp_z - zp_y, zp_x - xp_z]$$
$$= [yp_z, zp_x] - [zp_y, zp_x] - [yp_z, xp_z] + [zp_y, xp_z]$$
$$= y\,[p_z, z]\,p_x + x[z, p_z]\,p_y$$
$$= \frac{\hbar}{i}\,(yp_x - xp_y)$$
$$= i\hbar L_z \qquad (9\text{-}32)$$

Similarly

$$[L_y, L_z] = i\hbar L_x$$
$$[L_z, L_x] = i\hbar L_y \qquad (9\text{-}33)$$

Thus only one component of $\mathbf{L}$ may be chosen with H to form the commuting set of observables. We can do a little better, however, since (9-32) and (9-33) imply that $\mathbf{L}^2$ commutes with all three components of $\mathbf{L}$:

$$[L_z, \mathbf{L}^2] = [L_z, L_x{}^2 + L_y{}^2 + L_z{}^2] = [L_z, L_x{}^2] + [L_z, L_y{}^2]$$
$$= L_x[L_z, L_x] + [L_z, L_x]\,L_x + L_y[L_z, L_y] + [L_z, L_y]\,L_y$$
$$= i\hbar L_x L_y + i\hbar L_y L_x - i\hbar L_y L_x - i\hbar L_x L_y$$
$$= 0 \qquad (9\text{-}34)$$

and so on. We thus choose as our complete set of commuting observables the operators H, L_z (a purely conventional choice) and $\mathbf{L}^2$. We could also have included parity, since the Hamiltonian is manifestly invariant under $x \to -x$, $y \to -y$ and $z \to -z$, but, as we shall see later, specification of $\mathbf{L}^2$ determines the parity.

In Chapter 10 we will determine the eigenvalues and eigenfunctions of L_z and L^2; here we merely note that their use greatly simplifies the solution of the Schrödinger equation. This follows from a relation derived below.

$$L^2 = (r \times p)^2 = [(r \times p)_x]^2 + [(r \times p)_y]^2 + [(r \times p)_z]^2$$

$$= -\hbar^2 \left(y\,\frac{\partial}{\partial z} - z\,\frac{\partial}{\partial y} \right) \left(y\,\frac{\partial}{\partial z} - z\,\frac{\partial}{\partial y} \right)$$

$$-\hbar^2 \left(z\,\frac{\partial}{\partial x} - x\,\frac{\partial}{\partial z} \right) \left(z\,\frac{\partial}{\partial x} - x\,\frac{\partial}{\partial z} \right)$$

$$-\hbar^2 \left(x\,\frac{\partial}{\partial y} - y\,\frac{\partial}{\partial x} \right) \left(x\,\frac{\partial}{\partial y} - y\,\frac{\partial}{\partial x} \right)$$

$$= -\hbar^2 \left[x^2 \left(\frac{\partial^2}{\partial y^2} + \frac{\partial^2}{\partial z^2} \right) + y^2 \left(\frac{\partial^2}{\partial z^2} + \frac{\partial^2}{\partial x^2} \right) \right.$$

$$+ z^2 \left(\frac{\partial^2}{\partial x^2} + \frac{\partial^2}{\partial y^2} \right) - 2xy\,\frac{\partial^2}{\partial x \partial y} - 2yz\,\frac{\partial^2}{\partial y \partial z}$$

$$\left. - 2zx\,\frac{\partial^2}{\partial z \partial x} - 2x\,\frac{\partial}{\partial x} - 2y\,\frac{\partial}{\partial y} - 2z\,\frac{\partial}{\partial z} \right] \qquad (9\text{-}35)$$

as a little algebra shows. Similarly

$$(r \cdot p)^2 = -\hbar^2 \left(x\,\frac{\partial}{\partial x} + y\,\frac{\partial}{\partial y} + z\,\frac{\partial}{\partial z} \right) \left(x\,\frac{\partial}{\partial x} + y\,\frac{\partial}{\partial y} + z\,\frac{\partial}{\partial z} \right)$$

$$= -\hbar^2 \left(x^2\,\frac{\partial^2}{\partial x^2} + y^2\,\frac{\partial^2}{\partial y^2} + z^2\,\frac{\partial^2}{\partial z^2} + 2xy\,\frac{\partial^2}{\partial x \partial y} + 2yz\,\frac{\partial^2}{\partial y \partial z} \right.$$

$$\left. + 2zx\,\frac{\partial^2}{\partial z \partial x} + x\,\frac{\partial}{\partial x} + y\,\frac{\partial}{\partial y} + z\,\frac{\partial}{\partial z} \right) \qquad (9\text{-}36)$$

The sum of the two yields

$$-\hbar^2(x^2 + y^2 + z^2) \left(\frac{\partial^2}{\partial x^2} + \frac{\partial^2}{\partial y^2} + \frac{\partial^2}{\partial z^2} \right) + \hbar^2 \left(x\,\frac{\partial}{\partial x} + y\,\frac{\partial}{\partial y} + z\,\frac{\partial}{\partial z} \right)$$

$$(9\text{-}37)$$

We therefore get the identity

$$L^2 + (r \cdot p)^2 = r^2 p^2 + i\hbar r \cdot p \qquad (9\text{-}38)$$

Since we are dealing with operators, keeping track of the order of the terms is crucial. It follows from the identity that

$$p^2 = \frac{1}{r^2} \left[L^2 + (r \cdot p)^2 - i\hbar r \cdot p \right]$$

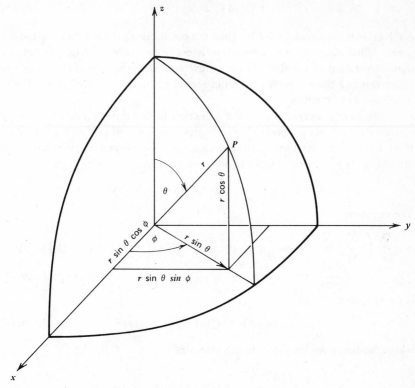

Fig. 9-1. The definition of the spherical coordinates used in the text and the relation between the cartesian coordinates (x,y,z) and the spherical coordinates (r,θ,ϕ).

$$= \frac{1}{r^2} \mathbf{L}^2 - \hbar^2 \frac{1}{r^2} \left(r \frac{\partial}{\partial r} \right)^2 - \hbar^2 \frac{1}{r} \frac{\partial}{\partial r} \qquad (9\text{-}39)$$

Thus the Schrödinger equation takes the form

$$-\frac{\hbar^2}{2\mu} \left[\frac{1}{r^2} \left(r \frac{\partial}{\partial r} \right) \left(r \frac{\partial}{\partial r} \right) + \frac{1}{r} \frac{\partial}{\partial r} - \frac{1}{\hbar^2 r^2} \mathbf{L}^2 \right] u_E(\mathbf{r})$$
$$+ V(r) \, u_E(\mathbf{r}) = E u_E(\mathbf{r}) \qquad (9\text{-}40)$$

If we work in spherical coordinates (Fig. 9-1), which is the natural thing to do, then the only operator that involves the spherical angles θ and ϕ is $\mathbf{L}^2$. If we therefore pick eigenfunctions of the form

$$u_E(\mathbf{r}) = Y_\lambda(\theta,\phi) \, R_{E\lambda}(r) \qquad (9\text{-}41)$$

where

$$\mathbf{L}^2 Y_\lambda(\theta,\phi) = \lambda Y_\lambda(\theta,\phi) \qquad (9\text{-}42)$$

is the eigenvalue equation for $\mathbf{L}^2$, then the equation separates into (9-42) and a purely radial equation. Our procedure is really no different than the conventional separation of variables. It does, however, stress the role of the symmetry in determining the complete commuting set of operators, and with this help the separation can be effected.

We have concentrated on the reduction of the three-dimensional energy eigenvalue equation in spherical coordinates, since central potentials, for which $V = V(r)$, are by far the most interesting ones. One other situation that is of interest to us is the case when the potential is of the form

$$V(x,y,z) = V_1(x) + V_2(y) + V_3(z)$$

The equation

$$-\frac{\hbar^2}{2m}\left(\frac{\partial^2}{\partial x^2} + \frac{\partial^2}{\partial y^2} + \frac{\partial^2}{\partial z^2}\right) u_E(x,y,z) + [V_1(x) + V_2(y) + V_3(z)]\, u_E(x,y,z)$$
$$= E u_E(x,y,z) \qquad (9\text{-}43)$$

is easily seen to be solved by

$$u_E(x,y,z) = u_{\epsilon_1}(x)\, v_{\epsilon_2}(y)\, w_{\epsilon_3}(z) \qquad (9\text{-}44)$$

where the functions on the right are solutions of

$$\left[-\frac{\hbar^2}{2m}\frac{d^2}{dx^2} + V_1(x)\right] u_{\epsilon_1}(x) = \epsilon_1 u_{\epsilon_1}(x)$$

$$\left[-\frac{\hbar^2}{2m}\frac{d^2}{dy^2} + V_2(y)\right] v_{\epsilon_2}(y) = \epsilon_2 v_{\epsilon_2}(y)$$

$$\left[-\frac{\hbar^2}{2m}\frac{d^2}{dz^2} + V_3(z)\right] w_{\epsilon_3}(z) = \epsilon_3 w_{\epsilon_3}(z) \qquad (9\text{-}45)$$

and

$$E = \epsilon_1 + \epsilon_2 + \epsilon_3$$

A particularly interesting example is the three-dimensional generalization of the potential hole with infinite walls. If the three-dimensional box is cubical in shape, with side L, then

$$\begin{aligned} V_1(x) &= \infty & x &< 0 \\ &= 0 & 0 &< x < L \\ &= \infty & L &< x \end{aligned} \qquad (9\text{-}46)$$

and so on. Thus, aside from a normalizing factor, the general solution is

$$u_E(x,y,z) = \sin \frac{n_1\pi x}{L} \sin \frac{n_2\pi y}{L} \sin \frac{n_3\pi z}{L} \qquad (9\text{-}47)$$

and

$$E = \frac{\hbar^2\pi^2}{2mL^2}\,(n_1^2 + n_2^2 + n_3^2) \qquad (9\text{-}48)$$

Note that there is quite a lot of degeneracy in the problem: there are as many solutions for a given E as there are sets of integers $\{n_1,n_2,n_3\}$ that satisfy (9-48). The degeneracy is usually associated with the existence of mutually commuting operators, and this example is no exception. Here these operators are H_x, H_y, and H_z, defined by

$$H_x = \frac{p_x^2}{2m} + V_1(x)$$

$$H_y = \frac{p_y^2}{2m} + V_2(y) \qquad (9\text{-}49)$$

$$H_z = \frac{p_z^2}{2m} + V_3(z)$$

so that

$$H_x + H_y + H_z = H \qquad (9\text{-}50)$$

It is interesting to ask for the ground state energy of N noninteracting identical fermions, for example, electrons, in the box of volume L^3. For each triplet of integers, $(1,1,1)$, $(2,1,1)$, $(1,2,1)$, . . . , two electrons can be accommodated. It is easier to ask the question in a different way: How many triplets of integers $\{n_1,n_2,n_3\}$ are there such that E given by (9-48) is less than the Fermi energy E_F? Each triplet forms a lattice point in a three-dimensional space, and if there are very many of them, then it is a very good approximation to say that they must lie inside a sphere of radius R given according to (9-48) by

$$n_1^2 + n_2^2 + n_3^2 = R^2 = \frac{2mE_F}{\hbar^2\pi^2}\,L^2 \qquad (9\text{-}51)$$

and their number is given by the volume of the octant of the sphere for which all the n_i are positive. Thus the number of lattice points is

$$\frac{1}{8} \cdot \frac{4\pi}{3}\,R^3 = \frac{1}{8}\frac{4\pi}{3}\left(\frac{2mE_F}{\hbar^2\pi^2}\,L^2\right)^{3/2} \qquad (9\text{-}52)$$

and hence the number of electrons with energy less than the Fermi energy E_F is twice that, that is,

$$N = \frac{\pi}{3}\,L^3\left(\frac{2mE_F}{\hbar^2\pi^2}\right)^{3/2} \qquad (9\text{-}53)$$

The number of electrons is proportional to the volume of the box L^3, which is to be expected. In terms of the density of electrons,

$$n = \frac{N}{L^3} \tag{9-54}$$

we have

$$E_F = \frac{\hbar^2 \pi^2}{2m} \left(\frac{3n}{\pi} \right)^{2/3} \tag{9-55}$$

To calculate the total energy, the number of lattice points may be written as

$$\frac{1}{8} \int_{|\mathbf{n}| \leqslant R} d^3\mathbf{n} \tag{9-56}$$

The factor $1/8$ comes from our restriction to positive integers in (9-48); in the above integration this restriction is removed and must be compensated for by the factor in front. At each lattice point the energy is given by

$$E = \frac{\hbar^2 \pi^2}{2mL^2} \mathbf{n}^2 \tag{9-57}$$

so that the total energy is

$$
\begin{aligned}
E_{\text{tot}} &= \frac{\hbar^2 \pi^2}{2mL^2} \cdot \frac{1}{8} \int_{|\mathbf{n}| \leqslant R} \mathbf{n}^2 \, d^3\mathbf{n} \\
&= \frac{\hbar^2 \pi^2}{2mL^2} \cdot \frac{1}{8} \cdot 4\pi \int_0^R n^4 \, dn \\
&= \frac{\pi^3 \hbar^2}{20mL^2} R^5
\end{aligned} \tag{9-58}
$$

Since R is related to the number of electrons by

$$N = 2 \cdot \frac{1}{8} \cdot \frac{4\pi}{3} R^3 \tag{9-59}$$

we finally get

$$E_{\text{tot}} = \frac{\pi^3 \hbar^2}{20mL^2} \left(\frac{3N}{\pi} \right)^{5/3} \tag{9-60}$$

If we write this in terms of $n = N/L^3$ we get

$$E_{\text{tot}} = \frac{\pi^3 \hbar^2}{20m} \left(\frac{3n}{\pi} \right)^{5/3} L^3 \tag{9-61}$$

The fact that the ground state of a many-electron system in a potential consists of a large number of filled levels has many ramifications. Typical values of E_F are of the order of 5–10 eV. Thus, at ordinary temperatures very few electrons can be thermally excited; most of them could only be excited to states that are already occupied. The implication of this is that for a metal, which is quite well described as a crystal lattice of ions with one or two free electrons per atom, only the ions contribute to the specific heat. If an electric field is applied to the metal, only the electrons near the top of the "Fermi sea" can be accelerated, since those that lie deeper cannot find available energy states. Those that are accelerated have long mean free paths. Collisons with ions that would reduce their energies below E_F are inhibited because there are no available empty states. These matters are discussed more fully in books on solid state physics.

Problems

1. Consider a particle moving in a cylindrically symmetric potential $V(\rho)$, where $\rho^2 = x^2 + y^2$. What is the complete set of commuting observables that you would use to specify the state of the system?

2. Use your conclusions from Problem 1 to separate the Schrödinger equation in cylindrical coordinates.

3. Given that the number density of free electrons in copper is 8.5×10^{22} cm^{-3}, calculate (1) the Fermi energy in electron volts; (2) the velocity of an electron moving with kinetic energy equal to the Fermi energy.

4. A nucleus consists of N neutrons and Z protons, with $N + Z = A$. If the radius of the nucleus is given by $R = r_0 A^{1/3}$, with $r_0 = 1.1$ fm (1 fm = 10^{-13} cm), and if the neutron and proton masses are both very nearly 1.6×10^{-24} gm, write expressions for the Fermi energy of the proton "gas" and the neutron "gas," assuming that the protons and the neutrons move freely. What are the Fermi energies if $N = 126$ and $Z = 82$?

5. Consider a neutron gas in its ground state, with mass density ρ varying from 10^{11} to 10^{16} gm cm^{-3}. Calculate the Fermi energy as a function of ρ. Note that at some point the neutron gas becomes relativistic, that is, the relation between energy and momentum is relativistic. In what range of densities should one begin to use the relativistic formula?

6. The mean electron energy in a degenerate electron gas is given by

$$\langle E \rangle = \frac{\dfrac{1}{8} \displaystyle\int d^3\mathbf{n}\, \dfrac{p^2}{2m_e}}{\dfrac{1}{8} \displaystyle\int d^3\mathbf{n}}$$

for nonrelativistic electrons, and

$$\langle E \rangle = \frac{\dfrac{1}{8}\displaystyle\int d^3\mathbf{n}[(p^2c^2 + m_e{}^2c^4)^{1/2} - m_ec^2]}{\dfrac{1}{8}\displaystyle\int d^3\mathbf{n}}$$

more generally. Calculate the general expression for the mean energy as a function of $\xi = p_F/m_ec$. Use this to calculate the pressure, defined by the thermodynamic formula

$$P = -\frac{\partial \langle E \rangle}{\partial (1/n)}$$

in the nonrelativistic formula and in the ultrarelativistic domain, where $\xi \gg 1$.

References

For a good discussion of angular momentum in the context used here, see J. L. Powell and B. Crasemann, *Quantum Mechanics*, Addison-Wesley, Inc., 1961.

This book, as well as every introductory book, works out the separation of the three-dimensional Schrödinger equation.

chapter 10

Angular Momentum

Our task in this chapter is to find the eigenvalues and the eigenfunctions of the operators L_z and $\mathbf{L}^2$. Since the angular momentum has the dimensions of $\hbar$, we may write the eigenvalue equations in the form

$$L_z Y_{lm} = m\hbar Y_{lm}$$

$$\mathbf{L}^2 Y_{lm} = l(l+1)\,\hbar^2 Y_{lm} \tag{10-1}$$

where m and $l(l+1)$ are real numbers. The peculiar way of writing the eigenvalue of $\mathbf{L}^2$ will prove its convenience later. There are several ways of proceeding. The conventional way is to write out the operators $\mathbf{L}$ in spherical coordinates. We have

$$x = r \sin\theta \cos\phi$$

$$y = r \sin\theta \sin\phi$$

$$z = r \cos\theta \tag{10-2}$$

so that

$$dx = \sin\theta \cos\phi \, dr + r \cos\theta \cos\phi \, d\theta - r \sin\theta \sin\phi \, d\phi$$

$$dy = \sin\theta \sin\phi \, dr + r \cos\theta \sin\phi \, d\theta + r \sin\theta \cos\phi \, d\phi$$

$$dz = \cos\theta \, dr - r \sin\theta \, d\theta \tag{10-3}$$

These can be solved to give

$$dr = \sin\theta \cos\phi \, dx + \sin\theta \sin\phi \, dy + \cos\theta \, dz$$

$$d\theta = \frac{1}{r}\,(\cos\theta \cos\phi \, dx + \cos\theta \sin\phi \, dy - \sin\theta \, dz)$$

$$d\phi = \frac{1}{r \sin\theta}\,(-\sin\phi \, dx + \cos\phi \, dy) \tag{10-4}$$

168 Quantum Physics

With the help of this equation we can obtain

$$\frac{\partial}{\partial x} = \frac{\partial r}{\partial x}\frac{\partial}{\partial r} + \frac{\partial \theta}{\partial x}\frac{\partial}{\partial \theta} + \frac{\partial \phi}{\partial x}\frac{\partial}{\partial \phi}$$

$$= \sin\theta\cos\phi\,\frac{\partial}{\partial r} + \frac{1}{r}\cos\theta\cos\phi\,\frac{\partial}{\partial \theta} - \frac{\sin\phi}{r\sin\theta}\frac{\partial}{\partial \phi}$$

$$\frac{\partial}{\partial y} = \sin\theta\sin\phi\,\frac{\partial}{\partial r} + \frac{1}{r}\cos\theta\sin\phi\,\frac{\partial}{\partial \theta} + \frac{\cos\phi}{r\sin\theta}\frac{\partial}{\partial \phi}$$

$$\frac{\partial}{\partial z} = \cos\theta\,\frac{\partial}{\partial r} - \frac{\sin\theta}{r}\frac{\partial}{\partial \theta} \tag{10-5}$$

and thus we finally obtain

$$L_z = \frac{\hbar}{i}\left(x\frac{\partial}{\partial y} - y\frac{\partial}{\partial x}\right) = \frac{\hbar}{i}\frac{\partial}{\partial \phi} \tag{10-6}$$

The other two components of the angular momentum are more compactly expressed, if we introduce

$$L_\pm = L_x \pm iL_y \tag{10-7}$$

Then

$$L_\pm = \frac{\hbar}{i}\left[y\frac{\partial}{\partial z} - z\frac{\partial}{\partial y} \pm i\left(z\frac{\partial}{\partial x} - x\frac{\partial}{\partial z}\right)\right]$$

$$= \frac{\hbar}{i}\left[\pm iz\left(\frac{\partial}{\partial x} \pm i\frac{\partial}{\partial y}\right) \mp i(x \pm iy)\frac{\partial}{\partial z}\right]$$

$$= \pm\hbar r\cos\theta\left(\sin\theta\, e^{\pm i\phi}\frac{\partial}{\partial r} + \frac{1}{r}\cos\theta\, e^{\pm i\phi}\frac{\partial}{\partial \theta} \pm \frac{i\, e^{\pm i\phi}}{r\sin\theta}\frac{\partial}{\partial \phi}\right)$$

$$\mp \hbar r\sin\theta\, e^{\pm i\phi}\left(\cos\theta\frac{\partial}{\partial r} - \frac{\sin\theta}{r}\frac{\partial}{\partial \theta}\right) \tag{10-8}$$

that is,

$$L_\pm = \hbar\, e^{\pm i\phi}\left(\pm\frac{\partial}{\partial \theta} + i\cot\theta\frac{\partial}{\partial \phi}\right) \tag{10-9}$$

One can then construct the $\mathbf{L}^2$ operator by observing that

$$L_+L_- = (L_x + iL_y)(L_x - iL_y)$$

$$= L_x^2 + L_y^2 - i[L_x, L_y] \tag{10-10}$$

so that

$$\mathbf{L}^2 = L_z^2 + L_+L_- + i[L_x, L_y]$$

$$= L_+L_- + L_z^2 - \hbar L_z \tag{10-11}$$

$(9\text{-}32) = [L_x, L_y] = i\hbar L_z$

In the last line we used (9-32). We thus get a second order differential operator involving θ and ϕ, and there remains the task of solving the differential equations that (10-1) represent. This is discussed in many textbooks on quantum mechanics or classical electrodynamics. We will proceed algebraically but digress for a moment to discuss the eigenvalue equation

$$L_z Y_{lm} = m\hbar Y_{lm} \tag{10-12}$$

and some applications. The equation, using (10-6), reads

$$\frac{\partial}{\partial\phi} Y_{lm}(\theta,\phi) = im Y_{lm}(\theta,\phi) \tag{10-13}$$

so that the solution is of the form $Y_{lm}(\theta,\phi) = \Theta_{lm}(\theta)\, \Phi_m(\phi)$ where

$$\frac{d\Phi_m(\phi)}{d\phi} = im\Phi_m(\phi) \tag{10-14}$$

The solution to this, normalized such that

$$\int_0^{2\pi} d\phi\, |\Phi_m|^2 = 1 \tag{10-15}$$

is

$$\Phi_m(\phi) = \frac{1}{\sqrt{2\pi}}\, e^{im\phi} \tag{10-16}$$

It is sometimes argued that since a rotation through 360°, that is, a transformation $\phi \to \phi + 2\pi$, leaves the system invariant, it is necessary that

$$e^{2\pi im} = 1 \tag{10-17}$$

so that m is an integer. This is not quite correct, since the quantities that enter into physical observables are of the type $\int_0^{2\pi} d\phi\psi_1{}^*(\phi)\, A\psi_2(\phi)$, with wave functions $\psi(\phi)$ of the form

$$\psi(\phi) = \sum_{m=-\infty}^{\infty} C_m \frac{e^{im\phi}}{\sqrt{2\pi}} \tag{10-18}$$

If we require that these arbitrary wave packets do not change (except for an overall phase factor) under the transformation $\phi \to \phi + 2\pi$, then we are led to the conclusion that the most general allowed values of m are $m = c +$ integer where c is a constant. It is only if we view the operator L_z as part of the total set (L_x, L_y, L_z) that we can say something about the constant c. We shall argue below that the eigenvalues are distributed symmetrically about zero, so that $c = 0$ or $c = 1/2$, and for the operators considered in this chapter, we shall restrict ourselves to $c = 0$, that is, the condition that m is an integer.

The eigenvalue equation for L_z appears in another context. Consider a classical rotator, rotating in the x-y plane. If the moment of inertia is I, then the energy is

$$E = \frac{L_z^2}{2I} \tag{10-19}$$

and thus the Hamiltonian is

$$H = \frac{L_z^2}{2I} \tag{10-20}$$

The eigenvalues of the Hamiltonian are now immediately seen to be

$$E_m = \frac{\hbar^2 m^2}{2I} \tag{10-21}$$

and the eigenfunctions are $e^{\pm im\phi}$. There is a degeneracy, since H commutes with L_z, and the two eigenfunctions for a given E_m correspond to the two senses of rotation. If we have N particles rigidly fixed on a circle, with equal angles $2\pi/N$ between neighboring particles, and *if the particles are identical*, then the solution of the energy eigenvalue equation

$$H\Phi_E(\phi) = E\Phi_E(\phi) \tag{10-22}$$

will again be $e^{\pm i\lambda\phi}$. The physical system is unaltered under a rotation of $2\pi/N$ radians (or an integral multiple of the angle), and the solutions should reflect this. The same kind of arguments that forced m to be an integer now imply that $\lambda = N \times$ (an integer).[1] The energy is therefore

$$E = \frac{\hbar^2(Nm)^2}{2I} \tag{10-23}$$

Let us now return to our equations (10-1), and try to obtain the eigenvalues in a manner reminiscent of our treatment of the harmonic oscillator in Chapter 7. The eigenfunctions of the hermitian operators L_z and $\mathbf{L}^2$ will be orthogonal, if the eigenvalues are different, and with proper normalization, we will write

$$\langle Y_{l'm'} | Y_{lm} \rangle = \delta_{ll'}\delta_{mm'} \tag{10-24}$$

Since

$$\langle Y_{lm} | (L_x{}^2 + L_y{}^2 + L_z{}^2) Y_{lm} \rangle$$
$$= \langle L_x Y_{lm} | L_x Y_{lm} \rangle + \langle L_y Y_{lm} | L_y Y_{lm} \rangle + m^2\hbar^2$$
$$\geq 0 \tag{10-25}$$

[1] The reader might look back to the Dicke-Wittke Gedankenexperiment discussed in Chapter 1.

it follows that

$$l(l + 1) \geq 0 \tag{10-26}$$

The operators $L_\pm$ introduced in (10-7) are very useful in what follows, and we shall see that they play the role of raising and lowering operators. First, we already saw that

$$\mathbf{L}^2 = L_+L_- + L_z{}^2 - \hbar L_z \tag{10-27}$$

In the same way we see that

$$\mathbf{L}^2 = L_-L_+ + L_z{}^2 + \hbar L_z \tag{10-28}$$

It follows from the above, as well as directly from (9-32) that

$$[L_+, L_-] = 2\hbar L_z \tag{10-29}$$

The remaining commutation relations are

$$[L_+, L_z] = [L_x + iL_y, L_z] = -i\hbar L_y - \hbar L_x$$
$$= -\hbar L_+ \tag{10-30}$$

and

$$[L_-, L_z] = \hbar L_- \tag{10-31}$$

From the fact that $[\mathbf{L}^2,\mathbf{L}] = 0$, it also follows that

$$[\mathbf{L}^2, L_\pm] = 0$$
$$[\mathbf{L}^2, L_z] = 0 \tag{10-32}$$

This implies that

$$\mathbf{L}^2 L_\pm Y_{lm} = L_\pm \mathbf{L}^2 Y_{lm} = l(l + 1)\,\hbar^2 L_\pm Y_{lm} \tag{10-33}$$

that is, $L_\pm Y_{lm}$ are also eigenfunctions of $\mathbf{L}^2$ with the eigenvalue characterized by l. On the other hand,

$$L_z L_+ Y_{lm} = (L_+L_z + \hbar L_+)\,Y_{lm}$$
$$= m\hbar L_+ Y_{lm} + \hbar L_+ Y_{lm}$$
$$= \hbar(m + 1)\,L_+ Y_{lm} \tag{10-34}$$

so that $L_+ Y_{lm}$ is also an eigenfunction of L_z, but with m-value increased by unity. Similarly we can show that

$$L_z L_- Y_{lm} = \hbar(m - 1)\,L_- Y_{lm} \tag{10-35}$$

so that $L_- Y_{lm}$ is an eigenfunction of L_z with m-value lowered by unity. Thus we call $L_\pm$ raising and lowering operators, respectively. We may write

$$L_\pm Y_{lm} = C_\pm(l,m)\,Y_{l,m\pm1} \tag{10-36}$$

It follows from the hermiticity of L_x and L_y that

$$L_\pm^\dagger = (L_x \pm iL_y)^\dagger = L_x \mp iL_y = L_\mp \qquad (10\text{-}37)$$

Hence, a consequence of

$$\langle L_\pm Y_{lm} | L_\pm Y_{lm} \rangle \geq 0 \qquad (10\text{-}38)$$

is that

$$\langle Y_{lm} | L_\mp L_\pm Y_{lm} \rangle \geq 0 \qquad (10\text{-}39)$$

and therefore (10-27) and (10-28) imply that

$$\langle Y_{lm} | (\mathbf{L}^2 - L_z^2 \pm \hbar L_z)\, Y_{lm} \rangle \geq 0 \qquad (10\text{-}40)$$

that is,

$$l(l + 1) \geq m^2 + m$$
$$l(l + 1) \geq m^2 - m \qquad (10\text{-}41)$$

Since $l(l + 1) \geq 0$, we can take $l \geq 0$ without loss of generality.[2] Then (10-41) shows that

$$-l \leq m \leq l \qquad (10\text{-}42)$$

If there is a minimum value of $m(= m_-)$ then for the corresponding eigenstate

$$L_- Y_{lm_-} = 0 \qquad (10\text{-}43)$$

We may then calculate m_- by using (10-27) and applying it to Y_{lm_-}: we get

$$l(l + 1)\, \hbar^2 = m_-^2 \hbar^2 - m_- \hbar^2 \qquad (10\text{-}44)$$

Similarly, if there is a maximum value of $m(= m_+)$ then

$$L_+ Y_{lm_+} = 0 \qquad (10\text{-}45)$$

and an application of (10-28) to the maximum eigenstate gives

$$l(l + 1)\, \hbar^2 = m_+^2 \hbar^2 + m_+ \hbar^2 \qquad (10\text{-}46)$$

Hence

$$m_- = -l$$
$$m_+ = +l \qquad (10\text{-}47)$$

Since the maximum value is to be reached from the minimum value by unit steps (repeated application of L_+), we find (Fig. 10.1) (a) that there are $(2l + 1)$ states, that is, $2l + 1$ is an integer, and (b) that m can take on the values

$$m = -l, -l + 1, -l + 2, \dots, l - 1, l$$

The possibility that l is half-odd integral, that is, $l = 1/2, 3/2, \dots$ will be dis-

---

[2] If we were to find that $l \leq -1$, we would merely define $L = -l - 1$, and replace the old l, with the new, positive L. Nothing would change, since $L(L + 1) = l(l + 1)$.

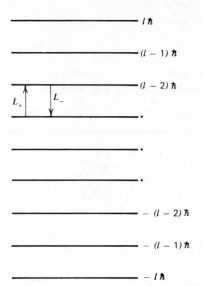

$l\,\hbar$

$(l-1)\,\hbar$

$(l-2)\,\hbar$

L_+ L_-

·

·

·

$-(l-2)\,\hbar$

$-(l-1)\,\hbar$

$-l\,\hbar$

Fig. 10-1. Spectrum of the operator L_z for a given value of l.

cussed in Chapter 14 where we discuss *spin*. In this chapter we restrict ourselves to *integral values of l*.

We may also calculate the coefficients $C_\pm(l,m)$ defined in (10-36). We have

$$
\begin{aligned}
|C_\pm(l,m)|^2 \langle Y_{l,m\pm1} | Y_{l,m\pm1}\rangle &= \langle L_\pm Y_{lm} | L_\pm Y_{lm}\rangle \\
&= \langle Y_{lm} | L_\mp L_\pm Y_{lm}\rangle \\
&= \langle Y_{lm} | (\mathbf{L}^2 - L_z^2 \mp \hbar L_z)\, Y_{lm}\rangle \\
&= \hbar^2[l(l+1) - m(m\pm1)]
\end{aligned}
$$

so that, with a convenient choice of phase, we get

$$
C_\pm(l,m) = \hbar[l(l+1) - m(m\pm1)]^{1/2} \tag{10-48}
$$

This is as far as operator methods can take us. We shall now use the explicit form of the operators L_z and $L_\pm$ to obtain convenient expressions for the eigenfunctions in terms of the spherical angles θ and ϕ. This development will parallel that of Eq. (7-31) to (7-35). We write, as already suggested

$$
Y_{lm}(\theta,\phi) = \Theta_{lm}(\theta)\, e^{im\phi} \tag{10-49}
$$

The condition (10-45) reads

$$
\hbar\, e^{i\phi}\left(\frac{\partial}{\partial\theta} + i\cot\theta\,\frac{\partial}{\partial\phi}\right)\Theta_{ll}(\theta)\, e^{il\phi}
$$

$$
= \hbar\, e^{i(l+1)\phi}\left(\frac{\partial}{\partial\theta} - l\cot\theta\right)\Theta_{ll}(\theta) = 0 \tag{10-50}
$$

The solution to this equation is easily found to be

$$\Theta_{ll}(\theta) = (\sin\theta)^l \tag{10-51}$$

The appropriate multiplicative constant will be obtained later from the normalization condition. An arbitrary state is obtained by the lowering procedure

$$Y_{lm}(\theta,\phi) = C(L_-)^{l-m} (\sin\theta)^l \, e^{il\phi} \tag{10-52}$$

Consider first

$$L_- Y_{ll}(\theta,\phi) = \hbar\, e^{-i\phi} \left(-\frac{\partial}{\partial\theta} + i\cot\theta \,\frac{\partial}{\partial\phi} \right) (\sin\theta)^l \, e^{il\phi}$$

$$= \hbar\, e^{i(l-1)\phi} \left(-\frac{\partial}{\partial\theta} - l\cot\theta \right) (\sin\theta)^l$$

Since one can show that for an arbitrary function $f(\theta)$

$$\left(\frac{d}{d\theta} + l\cot\theta \right) f(\theta) = \frac{1}{(\sin\theta)^l} \frac{d}{d\theta} \left[(\sin\theta)^l f(\theta) \right] \tag{10-53}$$

we have obtained

$$Y_{l,l-1} = C' \frac{e^{i(l-1)\phi}}{(\sin\theta)^l} \left(-\frac{d}{d\theta} \right) [(\sin\theta)^l(\sin\theta)^l] \tag{10-54}$$

The next step is the same, except that l is replaced by $l-1$ and the operation in (10-53) acts on the form obtained in (10-54). Thus

$$Y_{l,l-2} = C'' \frac{e^{i(l-2)\phi}}{(\sin\theta)^{l-1}} \left(-\frac{d}{d\theta} \right) \left[(\sin\theta)^{l-1} \frac{1}{(\sin\theta)^l} \left(-\frac{d}{d\theta} \right) (\sin\theta)^{2l} \right]$$

$$= C''(-1)^2 \frac{e^{i(l-2)\phi}}{(\sin\theta)^{l-1}} \frac{d}{d\theta} \left[\frac{1}{\sin\theta} \frac{d}{d\theta} (\sin\theta)^{2l} \right] \tag{10-55}$$

In terms of the variable $u = \cos\theta$, $-1/(\sin\theta)(d/d\theta) = d/du$, and (10-54), (10-55), respectively, read

$$Y_{l,l-1} = C' \frac{e^{i(l-1)\phi}}{(\sin\theta)^{l-1}} \frac{d}{du} [(1-u^2)^l]$$

$$Y_{l,l-2} = C'' \frac{e^{i(l-2)\phi}}{(\sin\theta)^{l-2}} \frac{d^2}{du^2} [(1-u^2)^l] \tag{10-56}$$

The general form is

$$Y_{lm} = C \frac{e^{im\phi}}{(\sin\theta)^m} \left(\frac{d}{du} \right)^{l-m} [(1-u^2)^l] \tag{10-57}$$

The eigenfunctions are to be normalized. Since we are dealing with spherical angles whose range of integration is $0 \le \phi \le 2\pi$, $0 \le \theta \le \pi$ (see Fig. 9.1) and where the integral over the surface of the sphere ($r =$ constant) is

$$\int d\Omega = \int_0^{2\pi} d\phi \int_0^{\pi} \sin\theta \, d\theta$$

we must impose

$$\langle Y_{lm} | Y_{lm} \rangle = 1 = \int_0^{2\pi} d\phi \int_{-1}^{1} du \, |c|^2 \left[\frac{1}{(1-u^2)^{m/2}} \left(\frac{d}{du} \right)^{l-m} (1-u^2)^l \right]^2$$

(10-58)

The integration is tedious. We content ourselves with writing down the appropriately normalized $Y_{lm}(\theta,\phi)$ with the phases that are conventionally established:

$$Y_{lm}(\theta,\phi) = (-1)^m \left[\frac{2l+1}{4\pi} \frac{(l-m)!}{(l+m)!} \right]^{1/2} P_l^m(\cos\theta) \, e^{im\phi}$$

(10-59)

with

$$Y_{l,-m} = (-1)^m Y_{lm}^*$$

(10-60)

The associated Legendre polynomials are given by

$$P_l^m(u) = (-1)^{l+m} \frac{(l+m)!}{(l-m)!} \frac{(1-u^2)^{-m/2}}{2^l l!} \left(\frac{d}{du} \right)^{l-m} (1-u^2)^l$$

(10-61)

with the value for negative m obtained from

$$P_l^{-m}(u) = (-1)^m \frac{(l-m)!}{(l+m)!} P_l^m(u)$$

(10-62)

It will be enough, for our purposes, to list a few of the eigenfunctions:

$$Y_{0,0} = \frac{1}{\sqrt{4\pi}}$$

$$Y_{1,1} = -\sqrt{\frac{3}{8\pi}} \, e^{i\phi} \sin\theta$$

$$Y_{1,0} = \sqrt{\frac{3}{4\pi}} \cos\theta$$

$$Y_{2,2} = \sqrt{\frac{15}{32\pi}} \, e^{2i\phi} \sin^2\theta$$

$$Y_{2,1} = -\sqrt{\frac{15}{8\pi}} \, e^{i\phi} \sin\theta \cos\theta$$

$$Y_{2,0} = \sqrt{\frac{5}{16\pi}} \, (3 \cos^2 \theta - 1) \tag{10-63}$$

With the knowledge that $\mathbf{L}^2$, acting on an eigenfunction, as in (9-40), is to be replaced by $l(l + 1) \, \hbar^2$, we can now write the radial differential equation that determines the energy eigenvalues and eigenfunctions. The equation, which we will discuss in great detail for a variety of potentials, is

$$-\frac{\hbar^2}{2\mu} \left[\frac{1}{r} \frac{d}{dr} \left(r \frac{d}{dr} \right) + \frac{1}{r} \frac{d}{dr} - \frac{l(l + 1)}{r^2} \right] R_{Elm}(r)$$
$$+ V(r) \, R_{Elm}(r) = E R_{Elm}(r) \tag{10-64}$$

We note that there is no dependence on m in the equation. Thus, for a given l, there will always be a $(2l + 1)$-fold degeneracy, since all the possible m-values will have the same energy.

Problems

1. A molecule consists of two identical atoms, each of which, in its ground state, has spin 0. The molecule has, among its possible excitations, rotational excitations. If only rotations about the z-axis are considered, so that $H = L_z^2/2I$ and the separation between the atoms is considered fixed, what is the rotational spectrum? If the atoms have spin $1/2$ and they are both in the same spin state, what is the spectrum?

2. Express the spherical harmonics listed in (10-63) in terms of $x = r \sin \theta \cos \phi$, $y = r \sin \theta \sin \phi$, and $z = r \cos \theta$.

3. The Legendre polynomials $P_l(u) = P_l^0(u)$ can be defined in terms of the expression (10-61). Use this definition to show that $P_l(u)$ satisfies the equation

$$(1 - u^2) \, P_l''(u) - 2u P_l'(u) + l(l + 1) \, P_l(u) = 0$$

4. Show that the Legendre polynomials $P_l(u)$ satisfy the recurrence relations

$$l P_{l-1} = u l P_l + (1 - u^2) \, P_l'$$
$$(l + 1)P_{l+1} = (l + 1) \, u P_l - (1 - u^2) \, P_l'$$
$$(l + 1) \, P_{l+1} - (2l + 1) \, u P_l + l P_{l-1} = 0$$

5. Use (10-61) to show that

$$\sum_{l=0}^{\infty} z^l P_l(u) = (1 - 2uz + z^2)^{-1/2} \qquad z < 1$$

6. Use the procedure outlined in this chapter to discuss rotations in four dimensions. The generalization of L is now the set of operators that may be written as

$$L_{ij} = -i(x_i \partial_j - x_j \partial_i)$$

$(i,j = 1,2,3,4)$. Introduce

$$(J_1, J_2, J_3) = (L_{23}, L_{31}, L_{12})$$

and

$$(K_1, K_2, K_3) = (L_{14}, L_{24}, L_{34})$$

(a) Find the commutation relations of all six operators among themselves.

(b) Show that the operators

$$\mathbf{J}^{(+)} = (\mathbf{J} + \mathbf{K}); \ \mathbf{J}^{(-)} = (\mathbf{J} - \mathbf{K})$$

each obey angular momentum commutation relations and that they commute with each other. Use the final result to determine the maximal set of mutually commuting observables, and thus the quantum numbers that would be used to label an eigenfunction.

7. Consider an electron in an arbitrary potential $V(r)$ and a state of angular momentum l. Show that the probability of finding it at the point $\mathbf{r}$ is only a function of $|\mathbf{r}|$.

[*Hint.* Note that the solutions for the $(2l + 1)$ m-values are degenerate, and that if no special alignment is prepared, all m-values are equally probable. Use the formula

$$\sum_{m=-l}^{l} |Y_{lm}(\theta,\phi)|^2 = \frac{2l + 1}{4\pi}$$

8. A particle in a spherically symmetric potential is in a state described by the wave packet

$$\psi(x,y,z) = C(xy + yz + zx)\, e^{-\alpha r^2}$$

What is the probability that a measurement of the square of the angular momentum yields 0? What is the probability that it yields $6\hbar^2$? If the value of l is found to be 2, what are the relative probabilities for $m = 2,1,0,-1,-2$?

9. Consider the following model of a perfectly smooth cylinder. It is a ring of equally spaced, identical particles, with mass M/N so that the mass the ring is M and its moment of inertia is MR^2, with R the radius of the ring. Calculate the possible values of the angular momentum. Calculate the energy eigenvalues. What is the energy difference between the ground state of zero angular momentum, and the first rotational state? Show that this approaches infinity as $N \to \infty$. Contrast this with the comparable energy for a "nicked" cylinder, which lacks the symmetry under the rotation through $2\pi/N$ radians. This example implies that it is impossible to set a perfectly smooth cylinder in

rotation, which is consistent with the fact that for a perfectly smooth cylinder such a rotation would be unobservable.

10. Express $\mathbf{L}^2$ in terms of $\partial/\partial\theta$ and $\partial/\partial\phi$. Write down the differential equation obeyed by Θ_{lm} defined in Eq. 10-49.

[*Hint.* Use the variable $z = \cos\theta$). Show that $(\sin\theta)^l$ is the solution of the equation for $l = m$.

References

This is standard material found in any of the books listed on page 501. For a deeper look into the consequences of invariance under rotation see especially

K. Gottfried, *Quantum Mechanics, Vol. 1*, W. A. Benjamin, Inc., 1966.

M. E. Rose, *Elementary Theory of Angular Momentum*, John Wiley and Sons, Inc., 1957.

chapter 11

The Radial Equation

The radial Schrödinger equation (10-64) may be written as

$$\left(\frac{d^2}{dr^2} + \frac{2}{r} \frac{d}{dr} \right) R_{nlm}(r) - \frac{2\mu}{\hbar^2} \left[V(r) + \frac{l(l+1)\hbar^2}{2\mu r^2} \right] R_{nlm}(r)$$

$$+ \frac{2\mu E}{\hbar^2} R_{nlm}(r) = 0 \qquad (11\text{-}1)$$

where we have replaced the label E by n in the subscripts of the eigenfunction $R_{nlm}(r)$. We will examine the solutions to this equation for a variety of potentials restricted by the condition that they go to zero at infinity faster than $1/r$, except for the important special case of the Coulomb potential. We will also assume that the potentials are not as singular as $1/r^2$ at the origin, so that

$$\underset{r \to 0}{\text{Lim}} \ r^2 V(r) = 0 \qquad (11\text{-}2)$$

It is sometimes convenient to introduce the function

$$u_{nlm}(r) = r R_{nlm}(r) \qquad (11\text{-}3)$$

Since

$$\left(\frac{d^2}{dr^2} + \frac{2}{r} \frac{d}{dr} \right) \frac{u_{nlm}(r)}{r} = \frac{1}{r} \frac{d^2}{dr^2} u_{nlm}(r) \qquad (11\text{-}4)$$

it follows that

$$\frac{d^2 u_{nlm}(r)}{dr^2} + \frac{2\mu}{\hbar^2} \left[E - V(r) - \frac{l(l+1)\hbar^2}{2\mu r^2} \right] u_{nlm}(r) = 0 \qquad (11\text{-}5)$$

This looks very much like a one-dimensional equation, except that

(a) the potential $V(r)$ is altered by the addition of a repulsive centrifugal barrier,

$$V(r) \to V(r) + \frac{l(l+1)\hbar^2}{2\mu r^2} \qquad (11\text{-}6)$$

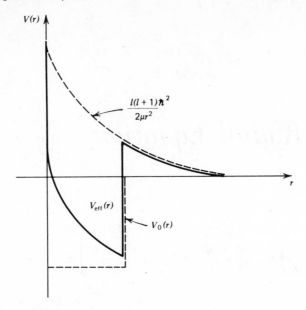

Fig. 11-1. Effective potential acting in radial equation for $u = rR(r)$ when the real potential is a square well.

(b) the definition of $u_{nlm}(r)$ and the finiteness of the wave function at the origin require that

$$u_{nlm}(0) = 0 \tag{11-7}$$

which makes it more like the one-dimensional problem for which $V = +\infty$ in the left-hand region (Fig. 11.1).

First we consider the radial equation near the origin, dropping all subscripts for convenience. As $r \to 0$, the leading terms in our equation are

$$\frac{d^2u}{dr^2} - \frac{l(l+1)}{r^2} u \simeq 0 \tag{11-8}$$

because the potential does not contribute for small enough r when (11-2) is satisfied. If we make the *Ansatz*

$$u(r) \sim r^s \tag{11-9}$$

we find that the equation will be satisfied, provided that

$$s(s-1) - l(l+1) = 0 \tag{11-10}$$

that is, $s = l + 1$ or $s = -l$. The solution that satisfies the condition $u(0) = 0$, that is, the solution that behaves like r^{l+1} is called the *regular solution*; the solution that behaves like r^{-l} is the *irregular solution*.

For large r we can drop the potential terms, and the equation becomes

$$\frac{d^2u}{dr^2} + \frac{2\mu E}{\hbar^2} u \simeq 0 \tag{11-11}$$

The square integrability condition implies that

$$1 = \int d^3r |\psi(\mathbf{r})|^2 = \int_0^\infty r^2 dr \int d\Omega |R_{nlm}(r) Y_{lm}(\theta,\phi)|^2$$

$$= \int_0^\infty r^2 dr |R_{nlm}(r)|^2 \tag{11-12}$$

that is,

$$\int_0^\infty dr |u_{nlm}(r)|^2 = 1 \tag{11-13}$$

so that the wave function should vanish at infinity. If $E < 0$, so that

$$\frac{2\mu E}{\hbar^2} = -\alpha^2 \tag{11-14}$$

the asymptotic solution is

$$u(r) \sim e^{-\alpha r} \tag{11-15}$$

If $E > 0$, we have solutions that are only normalizable in a box (see discussion in Chapter 4). With

$$\frac{2\mu E}{\hbar^2} = k^2 \tag{11-16}$$

the solution will be a linear combination of e^{ikr} and e^{-ikr}, the proper combination being determined by the requirement that the asymptotic solution tie on continuously to the solution that is regular at the origin. We now consider some examples.

A. The Free Particle

In this example $V(r) = 0$, but there is still a centrifugal barrier present. The radial equation (11-1) takes the form

$$\left[\frac{d^2}{dr^2} + \frac{2}{r}\frac{d}{dr} - \frac{l(l+1)}{r^2} \right] R(r) + k^2 R(r) = 0 \tag{11-17}$$

If we introduce the variable $\rho = kr$, we get

$$\frac{d^2R}{d\rho^2} + \frac{2}{\rho}\frac{dR}{d\rho} - \frac{l(l+1)}{\rho^2} R + R = 0 \tag{11-18}$$

This equation can actually be solved in terms of simple functions. The solutions are known as *spherical Bessel* functions. The regular solution is

$$j_l(\rho) = (-\rho)^l \left(\frac{1}{\rho} \frac{d}{d\rho} \right)^l \left(\frac{\sin \rho}{\rho} \right) \tag{11-19}$$

and the irregular one is

$$n_l(\rho) = -(-\rho)^l \left(\frac{1}{\rho} \frac{d}{d\rho} \right)^l \left(\frac{\cos \rho}{\rho} \right) \tag{11-20}$$

The first few functions are listed below.

$$j_0(\rho) = \frac{\sin \rho}{\rho} \qquad n_0(\rho) = - \frac{\cos \rho}{\rho}$$

$$j_1(\rho) = \frac{\sin \rho}{\rho^2} - \frac{\cos \rho}{\rho} \qquad n_1(\rho) = - \frac{\cos \rho}{\rho^2} - \frac{\sin \rho}{\rho}$$

$$j_2(\rho) = \left(\frac{3}{\rho^3} - \frac{1}{\rho} \right) \sin \rho - \frac{3}{\rho^2} \cos \rho$$

$$n_2(\rho) = - \left(\frac{3}{\rho^3} - \frac{1}{\rho} \right) \cos \rho - \frac{3}{\rho^2} \sin \rho \tag{11-21}$$

The combinations that will be of interest for large ρ are the *spherical Hankel functions*

$$h_1^{(1)}(\rho) = j_l(\rho) + in_l(\rho) \tag{11-22}$$

and

$$h_1^{(2)}(\rho) = [h_l^{(1)}(\rho)]^* \tag{11-23}$$

Again the first few spherical Hankel functions are

$$h_0^{(1)}(\rho) = \frac{e^{i\rho}}{i\rho}$$

$$h_1^{(1)}(\rho) = - \frac{e^{i\rho}}{\rho} \left(1 + \frac{i}{\rho} \right)$$

$$h_2^{(1)}(\rho) = \frac{i\,e^{i\rho}}{\rho} \left(1 + \frac{3i}{\rho} - \frac{3}{\rho^2} \right) \tag{11-24}$$

Of special interest are
(a) the behavior near the origin: for $\rho \ll l$, it turns out that

$$j_l(\rho) \approx \frac{\rho^l}{1 \cdot 3 \cdot 5 \cdot \ldots (2l + 1)} \tag{11-25}$$

and

$$n_l(\rho) \simeq -\frac{1 \cdot 3 \cdot 5 \cdot \ldots (2l - 1)}{\rho^l} \qquad (11\text{-}26)$$

For $\rho \gg l$, we have the asymptotic expressions

$$j_l(\rho) \simeq \frac{1}{\rho} \sin\left(\rho - \frac{l\pi}{2}\right) \qquad (11\text{-}27)$$

and

$$n_l(\rho) \simeq -\frac{1}{\rho} \cos\left(\rho - \frac{l\pi}{2}\right) \qquad (11\text{-}28)$$

so that

$$h_l^{(1)}(\rho) \simeq -\frac{i}{\rho} e^{i(\rho - l\pi/2)} \qquad (11\text{-}29)$$

The solution that is regular at the origin is

$$R_l(r) = j_l(kr) \qquad (11\text{-}30)$$

Its asymptotic form is, using (11-27)

$$R_l(r) \simeq -\frac{1}{2ikr} [e^{-i(kr - l\pi/2)} - e^{i(kr - l\pi/2)}] \qquad (11\text{-}31)$$

We describe this as a sum of an "incoming" and an "outgoing" spherical wave. The nomenclature is arrived at in the following way. The generalization of the one-dimensional flux is

$$\mathbf{j} = \frac{\hbar}{2i\mu} [\psi^*(\mathbf{r}) \, \nabla\psi(\mathbf{r}) - \nabla\psi^*(\mathbf{r}) \, \psi(\mathbf{r})] \qquad (11\text{-}32)$$

We shall see that it is only the flux in the radial direction that is of interest for large r. Thus the radial flux, integrated over all angles, is

$$\int d\Omega \hat{r} \cdot \mathbf{j}(\mathbf{r}) = \frac{\hbar}{2i\mu} \int d\Omega \left(\psi^* \frac{\partial}{\partial r} \psi - \frac{\partial \psi^*}{\partial r} \psi\right) \qquad (11\text{-}33)$$

For a solution of the form

$$\psi(\mathbf{r}) = C \frac{e^{\pm ikr}}{r} Y_{lm}(\theta, \phi) \qquad (11\text{-}34)$$

with

$$\int d\Omega |Y_{lm}(\theta, \phi)|^2 = 1 \qquad (11\text{-}35)$$

we get

$$\int d\Omega j_r = \frac{\hbar}{2i\mu}|C|^2\left[\frac{e^{\mp ikr}}{r}\left(\pm ik\frac{e^{\pm ikr}}{r} - \frac{e^{\pm ikr}}{r^2}\right) - \text{complex conjugate}\right]$$

$$= \pm\frac{\hbar k|C|^2}{\mu}\frac{1}{r^2} \quad (11\text{-}36)$$

The $\pm$ signs describe outgoing/incoming flux. The factor $1/r^2$ that emerges from our calculation is actually necessary for flux conservation, since the flux going through the spherical surface at radius r is

$$\int r^2 d\Omega j_r = (\text{independent of } r) \quad (11\text{-}37)$$

For our solution (11-31), the incoming flux is, aside from $1/r^2$,

$$-\frac{\hbar k}{\mu}\left|\frac{i}{2k}e^{il\pi/2}\right|^2 = -\frac{\hbar k}{\mu}\frac{1}{4k^2} \quad (11\text{-}38)$$

and this is equal in magnitude to the outgoing flux. The net flux is therefore zero, as it should be, since there are no sources of flux.

In general, flux conservation demands that any solution—and this includes solutions for which $V(r) \neq 0$—whose form for r very large must [by the arguments following (11.16)] be

$$R_l(r) \simeq -\frac{1}{2ikr}[e^{-i(kr-l\pi/2)} - S_l(k)\,e^{i(kr-l\pi/2)}] \quad (11\text{-}39)$$

requires

$$|S_l(k)|^2 = 1 \quad (11\text{-}40)$$

as otherwise the outgoing flux would differ from the incoming one. A function whose absolute square is unity can always be written in the form

$$S_l(k) = e^{2i\delta_l(k)} \quad (11\text{-}41)$$

The real function $\delta_l(k)$ is called the *phase shift* because the radial function in the asymptotic region (11-39) may be rewritten as

$$R_l(r) \simeq e^{i\delta_l(k)}\frac{\sin[kr - l\pi/2 + \delta_l(k)]}{kr} \quad (11\text{-}42)$$

Aside from the phase factor in front, this differs from the free particle solution $j_l(kr)$, whose asymptotic form is $[\sin(kr - l\pi/2)]/kr$, only by the shift in phase, $\delta_l(k)$.

We note that the flux in the $\hat{\imath}_\theta$ direction involves

$$\hat{\imath}_\theta \cdot \mathbf{j} = \frac{\hbar}{2i\mu}\left(\psi^*\frac{1}{r}\frac{\partial}{\partial\theta}\psi - \text{complex conjugate}\right) \sim \frac{1}{r^3}(\ldots)$$

and at large distances, such a flux, when multiplied by the area factor $r^2 d\Omega$, still vanishes as $1/r$ relative to the dominant term in the radial flux. This is the justification for ignoring all but the radial flux at large distances.

B. The Square Well, Bound States

Consider the potential

$$V(r) = -V_0 \qquad r < a$$
$$= 0 \qquad r > a \qquad (11\text{-}43)$$

Then the radial equation has the form

$$\frac{d^2 R}{dr^2} + \frac{2}{r}\frac{dR}{dr} - \frac{l(l+1)}{r^2} R + \frac{2\mu}{\hbar^2}(V_0 + E) R = 0 \qquad r < a$$

$$\frac{d^2 R}{dr^2} + \frac{2}{r}\frac{dR}{dr} - \frac{l(l+1)}{r^2} R + \frac{2\mu E}{\hbar^2} R = 0 \qquad r > a \qquad (11\text{-}44)$$

We look for bound state solutions, for which $E < 0$. We write

$$\frac{2\mu}{\hbar^2}(V_0 + E) = \kappa^2$$

$$\frac{2\mu}{\hbar^2} E = -\alpha^2 \qquad (11\text{-}45)$$

The solution for $r < a$, which must be regular at the origin, is

$$R(r) = A j_l(\kappa r) \qquad (11\text{-}46)$$

The solution for $r > a$ must vanish as $r \to \infty$. The second of the equations (11-44) is just the equation for the spherical Bessel function, except that k is replaced by $i\alpha$. The solution that behaves like e^{ikr} now becomes the exponentially falling one, that is, we have

$$R(r) = B h_l^{(1)}(i\alpha r) \qquad (11\text{-}47)$$

for $r > a$. The two solutions must match at $r = a$ and so must the derivatives. This leads to the condition

$$\kappa \left[\frac{dj_l(\rho)/d\rho}{j_l(\rho)} \right]_{\rho = \kappa a} = i\alpha \left[\frac{dh_l^{(1)}(\rho)/d\rho}{h_l^{(1)}(\rho)} \right]_{\rho = i\alpha a} \qquad (11\text{-}48)$$

This is a very complicated transcendental equation involving l, V_0, and E. For $l = 0$ it simplifies greatly if one uses the function $u(r) = rR(r)$. The eigenvalue is obtained by matching $A \sin \kappa r$ and $B e^{-\alpha r}$ at $r = a$. The details are left as an

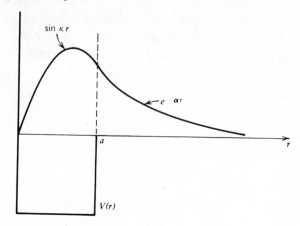

Fig. 11-2. The shape of the wave function $u(r) = rR(r)$ for an attractive square well when there exists one bound state ($l = 0$).

exercise for the reader; the shape of the radial wave function $u(r)$ for the first and second bound states is exhibited in Figs. 11.2 and 11.3.

Let us return to Eq. 11-48 for the case of a very deep potential for which $\kappa a \gg l$. In that case the left side of the equation simplifies, since we are justified in using the asymptotic form of $j_l(\rho)$. Computation shows that (11-48) takes the form

$$-\frac{1}{a} + \kappa \cot\left(\kappa a - \frac{l\pi}{2}\right) = \text{(right-hand side)} \qquad (11\text{-}49)$$

The right-hand side does not contain V_0, and if $|E| \ll V_0$ the largeness of κa

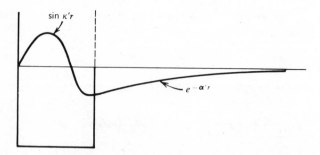

Fig. 11-3. The shape of the wave function $u(r) = rR(r)$ for an attractive square well when there exist two bound states ($l = 0$). Only the wave function for the second bound state is sketched in this figure.

implies that the cotangent must be close to zero. Thus we have approximately

$$\kappa a - \frac{l\pi}{2} \simeq (n + \tfrac{1}{2})\,\pi \tag{11-50}$$

Since for $|E| \ll V_0$ we have

$$\kappa \simeq \kappa_0 \left(1 + \frac{E}{2V_0}\right) \tag{11-51}$$

where

$$\kappa_0{}^2 = \frac{2\mu V_0}{\hbar^2} \tag{11-52}$$

(11-50) reads

$$\frac{E}{2V_0} = -1 + \frac{[n + (l+1)/2]\,\pi}{\kappa_0 a} \tag{11-53}$$

Thus the levels that are far from the bottom of the well are approximately equally spaced, for all $l \ll \kappa_0 a$, with the spacing

$$\frac{\Delta E}{2V_0} \simeq \frac{\pi}{\kappa_0 a} \tag{11-54}$$

A related problem is the infinite box in three dimensions. Here

$$V(r) = 0 \qquad r < a$$
$$= \infty \qquad r > a \tag{11-55}$$

In this case, writing

$$\frac{2\mu E}{\hbar^2} = k^2 \tag{11-56}$$

the solution that is regular at $r = 0$ is

$$R(r) = A j_l(kr) \tag{11-57}$$

with the eigenvalues determined by the condition that the solution vanish at $r = a$, that is, by

$$j_l(ka) = 0 \tag{11-58}$$

The roots for a few values of l are listed below.

$l = 0$	1	2	3	4	5
3.14	4.49	5.76	6.99	8.18	9.36
6.28	7.73	9.10	10.42		
9.42					

If the first root for a given l is labeled $n = 1$, the second root $n = 2$, and so on, and if we use the accepted spectroscopic notation for the l values,[1]

$$S : l = 0$$
$$P : l = 1$$
$$D : l = 2$$
$$F : l = 3$$
$$G : l = 4$$
$$\cdots$$

then the order in which the levels occur is

$$1S;\ 1P;\ 1D;\ 2S;\ 1F;\ 2P;\ 1G;\ 2D;\ 1H;\ 3S;\ \ldots$$

Suppose we consider a model of the nucleus that consists of protons and neutrons inside such an infinite box. Since neutrons and protons are spin $\frac{1}{2}$ particles, that is, fermions, no more than two neutrons and two protons can occupy a given state. If we concentrate on protons, we observe that in the $1S$ state only *two* protons can appear. In the next level we have $l = 1$, so that there are three states, and hence *six* protons will fill it. For the $1D$ level, with five possible m-values (since $l = 2$), *ten* protons are required to fill this "shell." Thus levels will be filled when the number of protons is 2, 8 ($= 2 + 6$), 18 ($= 2 + 6 + 10$), 20 ($= 18 + 2$), 34 ($= 20 + 14$), 40, 58, 68, 90, 92, 106, . . . , and similarly for the neutrons. A study of real nuclei shows that for the "magic" number of protons and neutrons, 2, 8, 20, 28, 50, 82, 126, . . . , these nuclei exhibit special characteristics that can be associated with filled levels, that is, closed shells. The difference between the real "magic" numbers, and those obtained in our primitive model comes about because there is an additional potential that depends on the spin and that shifts the levels about somewhat, thus reordering the numbers. The shell model of the nucleus, when properly constructed, explains many of the properties of nuclei. What is not obvious is why nuclei should behave like a collection of particles in a box.

C. The Square Well, Continuum Solutions

With $E > 0$ we write

$$\frac{2\mu E}{\hbar^2} = k^2 \tag{11-59}$$

---

[1] The historical origin of this notation was the description of spectral lines as Sharp, Principal, Diffuse, . . . , and their subsequent identification. It does not make sense, but it stuck. The notation here differs from that used in atomic physics, where the conventional notation adds the l-value to the index, so that the order would be written in the form,

$$1S,\ 2P,\ 3D,\ 2S,\ 4F,\ 3P,\ 5G,\ 4D,\ 6H,\ 3S,\ \ldots$$

The solution for $r > a$ will be a combination of the regular and irregular solutions of the free field equation

$$R_l(r) = Bj_l(kr) + Cn_l(kr) \qquad (11\text{-}60)$$

while the solution for $r < a$ must be the regular solution, that is,

$$R_l(r) = Aj_l(\kappa r) \qquad (11\text{-}61)$$

where

$$\kappa^2 = \frac{2\mu(E + V_0)}{\hbar^2} \qquad (11\text{-}62)$$

as before.

The matching of $\dfrac{1}{R_l}\dfrac{dR_l}{dr}$ at $r = a$ gives

$$\kappa \left[\frac{dj_l(\rho)/d\rho}{j_l(\rho)}\right]_{\rho=\kappa a} = k \left[\frac{Bdj_l/d\rho + Cdn_l/d\rho}{Bj_l(\rho) + Cn_l(\rho)}\right]_{\rho=ka} \qquad (11\text{-}63)$$

from which the ratio C/B can be calculated. This ratio can be related to the phase shift that appeared in (11-42). We do this by looking at the asymptotic (large r) form of (11-60)

$$R_l(r) \simeq \frac{B}{kr}\left[\sin\left(kr - \frac{l\pi}{2}\right) - \frac{C}{B}\cos\left(kr - \frac{l\pi}{2}\right)\right] \qquad (11\text{-}64)$$

which is to be compared with (11-42), rewritten as

$$R_l(r) \simeq \frac{1}{kr}\left[\sin\left(kr - \frac{l\pi}{2}\right)\cos\delta_l(k) + \cos\left(kr - \frac{l\pi}{\cdot 2}\right)\sin\delta_l(k)\right]$$

We see that the relation

$$\frac{C}{B} = -\tan\delta_l(k) \qquad (11\text{-}65)$$

holds.

The actual computation of C/B from (11-63) is tedious, except for $l = 0$. As for the bound state problem, the use $u(r) = rR(r)$ simplifies the calculation greatly. One only needs to match $A\sin\kappa r$ to $B\sin kr + C\cos kr$ at $r = a$ to obtain an expression for $\tan\delta_0$. The results for this case are schematically drawn in Figs. 11.4 and 11.5. They show that an attractive potential tends to "draw in" the wave function, while a repulsive potential tends to push it out. We will return to these matters in Chapter 24, when we discuss collision theory.

Before concluding this chapter, we focus on an important relation that can be obtained by solving the free particle equation in two ways. One solution is

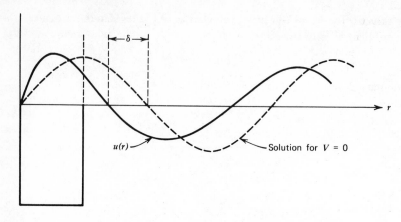

Fig. 11-4. Continuum solution $u(r) = rR(r)$ for attractive potential ($l = 0$).

obtained as a superposition of our separated solutions (11-30) multiplied by the appropriate spherical harmonic $Y_{lm}(\theta,\phi)$:

$$\psi(\mathbf{r}) = \sum_{l=0}^{\infty} \sum_{m=-l}^{l} A_{lm} j_l(kr)\ Y_{lm}(\theta,\phi) \tag{11-66}$$

Another solution of the free particle equation, which reads

$$(\nabla^2 + k^2)\ \psi(\mathbf{r}) = 0 \tag{11-67}$$

before the separation into angular and radial parts is made, is the plane wave

$$\psi(\mathbf{r}) = e^{i\mathbf{k}\cdot\mathbf{r}} \tag{11-68}$$

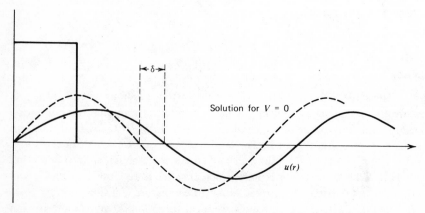

Fig. 11-5. Continuum solution $u(r) = rR(r)$ for repulsive potential ($l = 0$).

We may therefore find A_{lm} such that $\psi(\mathbf{r}) = e^{i\mathbf{k}\cdot\mathbf{r}}$ in (11-66). Note that the spherical angles (θ,ϕ) are the coordinates of the vector $\mathbf{r}$ relative to some arbitrarily chosen z-axis (see Fig. 9.1). If we define the z-axis by the direction of k (until now an arbitrary direction), then

$$e^{i\mathbf{k}\cdot\mathbf{r}} = e^{ikr\cos\theta} \tag{11-69}$$

Thus the left side of (11-66) has no azimuthal angle, ϕ, dependence, and thus on the right side only terms with $m = 0$ can appear; hence, making use of the fact that

$$Y_{l0}(\theta,\phi) = \left(\frac{2l+1}{4\pi}\right)^{1/2} P_l(\cos\theta) \tag{11-70}$$

where the $P_l(\cos\theta)$ are the Legendre polynomials, we get the relation

$$e^{ikr\cos\theta} = \sum_{l=0}^{\infty} \left(\frac{2l+1}{4\pi}\right)^{1/2} A_l j_l(kr)\, P_l(\cos\theta) \tag{11-71}$$

We may use the relation

$$\frac{1}{2}\int_{-1}^{1} d(\cos\theta)\, P_l(\cos\theta)\, P_{l'}(\cos\theta) = \frac{\delta_{ll'}}{2l+1} \tag{11-72}$$

which is a direct consequence of the orthonormality relation for the Y_{lm} and (11-70), to obtain

$$A_l j_l(kr) = \tfrac{1}{2}[4\pi(2l+1)]^{1/2} \int_{-1}^{1} dz\, P_l(z)\, e^{ikrz} \tag{11-73}$$

The integral can be looked up, or worked out by comparing both sides in the limit that $kr \to 0$. In any case, what results is the expansion

$$e^{ikr\cos\theta} = \sum_{l=0}^{\infty} (2l+1)\, i^l j_l(kr)\, P_l(\cos\theta) \tag{11-74}$$

which we will find exceedingly useful in discussions of collision theory.

Problems

1. Consider $l = 0$ bound states for an attractive square well. Find the eigenvalue condition for a bound state. What is the depth of the potential for a state that is barely bound?

2. Assume that the deuteron (consisting of a neutron and a proton, equal in mass) is a bound state with $l = 0$ and the potential square in shape and of range

$r_0 = 2.8 \times 10^{-13}$ cm. Given that the binding energy is -2.18 MeV, find the depth of the potential.

(*Hint.* Expand about the case of zero binding energy discussed in Problem 1).

3. Consider neutron-proton scattering, assumed to be interacting through a square well potential of range 2.8×10^{-13} cm and depth 20 MeV. Calculate the phase shift as a function of energy for very low energies, for $l = 0$.

4. Calculate the $l = 0$ phase shift for a square well potential. Use the procedure outlined following Eq. 11-65 to work out both the attractive and the repulsive potential case. Discuss various limits, such as E large and small, V_0 large and small.

5. Show that for $l = 0$ scattering by a square well of arbitrary range and depth V_0, it is always possible to write the phase shift as an expansion

$$ k \cot \delta_0 = -\frac{1}{a} + r_{\text{eff}} \, k^2/2 + 0(k^4) $$

Obtain an expression for a and r_{eff} in terms of the parameters of the well.

6. Consider a potential of arbitrary shape that vanishes for $r \geq a$. Let the logarithmic derivative of the radial function inside the potential,

$$ \frac{1}{R} \frac{dR(r)}{dr} \bigg|_{r=a} = f_l(E) $$

be a slowly varying function of the energy. Consider $l = 0$.

(a) If the potential has a bound state with energy, E_B, what is the value of $f_0(E_B)$?

(b) If $f_0(E)$ is independent of E, what is the phase shift as a function of energy?

(c) If $f_0(E) = f_0(E_B) + (E - E_B) f_0'$, how does f_0' enter into the phase shift?

It is simpler to work out (b) and (c) above in terms of $k \cot \delta_0(k)$, instead of the phase shift, and that is a preferable way to present your results.

7. Give a general argument for why $\delta_l(k)$ should be an odd function of k. Check that this is so for the square well [using (11-65), for example]. Show that

$$ S_l(-k) = S_l^*(k) $$

where $S_l(k)$ is defined in (11-41).

8. Calculate the function $S_l(k)$ for a potential

$$ V(r) = \infty \qquad r < a $$
$$ V(r) = 0 \qquad r > a $$

Consider the $l = 0$ phase shift. What is it for ka very small? What is it for ka very large? Note that this potential is a model for an impenetrable sphere.

9. Use the solution (11-63) together with the values of the spherical Bessel functions near the origin given in (11-25) and (11-26) to show that $\tan \delta_l(k) \to 0$ as $k \to 0$. How rapidly does it approach zero for a given l?

10. Consider the $l = 0$ radial equation for the potential

$$V(r) = V_0[e^{-2(r-r_0)/a} - 2\,e^{-(r-r_0)/a}]$$

(known as the Morse potential). Find the energy eigenvalues by simplifying the differential equation. Do this by defining a new variable $x = Ce^{-r/a}$ with C chosen to simplify the equation as much as possible, and then treating the equation in the manner that the simple harmonic oscillator problem was treated in Chapter 5.

Plot the potential. Show that for a deep, wide potential, the low-lying bound states approximate those of a harmonic oscillator, and explain why this is so.

References

The general properties of second-order differential equations in the context of quantum mechanics are discussed in

J. L. Powell and B. Crasemann, *Quantum Mechanics*, Addison-Wesley, Inc., 1961.

A comprehensive discussion of such equations may also be found in

P. M. Morse and H. Feshbach, *Methods of Theoretical Physics*, McGraw-Hill Book Co., Inc., 1953.

chapter 12

The Hydrogen Atom

The hydrogen atom is the simplest atom, since it contains only one electron. Thus the Schrödinger equation becomes a one-particle equation after the center of mass motion is separated out. We shall deal with hydrogenlike atoms, that is, atoms containing one electron only, but allowing for a nucleus more complicated than a single proton. The potential then is

$$V(r) = - \frac{Ze^2}{r} \tag{12-1}$$

and the radial Schrödinger equation is

$$\left(\frac{d^2}{dr^2} + \frac{2}{r} \frac{d}{dr} \right) R + \frac{2\mu}{\hbar^2} \left[E + \frac{Ze^2}{r} - \frac{l(l+1)\,\hbar^2}{2\mu r^2} \right] R = 0 \tag{12-2}$$

We will concentrate on the bound states, that is, $E < 0$ solutions. It is convenient to make a change of variables,

$$\rho = \left(\frac{8\mu|E|}{\hbar^2} \right)^{1/2} r \tag{12-3}$$

The equation then reads

$$\frac{d^2R}{d\rho^2} + \frac{2}{\rho} \frac{dR}{d\rho} - \frac{l(l+1)}{\rho^2} R + \left(\frac{\lambda}{\rho} - \frac{1}{4} \right) R = 0 \tag{12-4}$$

where we have introduced the dimensionless parameter

$$\lambda = \frac{Ze^2}{\hbar} \left(\frac{\mu}{2|E|} \right)^{1/2} = Z\alpha \left(\frac{\mu c^2}{2|E|} \right)^{1/2} \tag{12-5}$$

The second form makes it easier to compute with it, since $\alpha = 1/137$ and the energy is expressed in units of the rest mass; the first form does, however, make clear that the velocity of light c does not really appear in the equation, that is, that it is strictly a nonrelativistic equation.

We try to solve (12-4) in what is by now a familiar way. First, we extract the large ρ behavior. For large ρ, the only terms that remain in the equation are

$$\frac{d^2R}{d\rho^2} - \frac{1}{4} R \simeq 0 \qquad (12\text{-}6)$$

and the solution, which behaves properly at infinity, is $R \sim e^{-\rho/2}$. As in our treatment of the harmonic oscillator, we write

$$R(\rho) = e^{-\rho/2}\, G(\rho) \qquad (12\text{-}7)$$

substitute this into (12-4), and obtain the equation for $G(\rho)$. A little algebra, which we do not reproduce, leads to the equation

$$\frac{d^2G}{d\rho^2} - \left(1 - \frac{2}{\rho}\right)\frac{dG}{d\rho} + \left[\frac{\lambda - 1}{\rho} - \frac{l(l+1)}{\rho^2}\right] G = 0 \qquad (12\text{-}8)$$

We now write a power expansion for $G(\rho)$. This takes the form

$$G(\rho) = \rho^l \sum_{n=0}^{\infty} a_n \rho^n \qquad (12\text{-}9)$$

The fact that $R(\rho)$, and hence $G(\rho)$, behaves like ρ^l at the origin was established at the beginning of Chapter 11 for all potentials satisfying (11-2). When (12-9) is substituted into the differential equation, we find a relation between various coefficients a_n. The recursion relation is obtained from the differential equation obeyed by

$$H(\rho) = \sum_{n=0}^{\infty} a_n \rho^n \qquad (12\text{-}10)$$

which is

$$\frac{d^2H}{d\rho^2} + \left(\frac{2l+2}{\rho} - 1\right)\frac{dH}{d\rho} + \frac{\lambda - 1 - l}{\rho} H = 0 \qquad (12\text{-}11)$$

as can easily be obtained by substituting $G(\rho) = \rho^l H(\rho)$ into (12-8). We then have

$$\sum_{n=0}^{\infty} \left[n(n-1)\, a_n \rho^{n-2} + n a_n \rho^{n-1}\left(\frac{2l+2}{\rho} - 1\right) + (\lambda - 1 - l)\, a_n \rho^{n-1}\right] = 0$$

$$(12\text{-}12)$$

that is,

$$\sum_{n=0}^{\infty} \{(n+1)[n a_{n+1} + (2l+2)\, a_{n+1}] + (\lambda - 1 - l - n)\, a_n\}\, \rho^{n-1} = 0$$

Since this must vanish term by term, we get the recursion relation

$$\frac{a_{n+1}}{a_n} = \frac{n + l + 1 - \lambda}{(n + 1)(n + 2l + 2)} \qquad (12\text{-}13)$$

For large n this ratio is

$$\frac{a_{n+1}}{a_n} \simeq \frac{1}{n} \qquad (12\text{-}14)$$

and, as for the harmonic oscillator problem, we can show that we do not get a solution $R(\rho)$ that is well behaved at infinity, unless the series in (12-9) terminates. This means that for a given l, for some $n = n_r$ we must have

$$\lambda = n_r + l + 1 \qquad (12\text{-}15)$$

Let us introduce the *principal quantum number* n defined by

$$n = n_r + l + 1 \qquad (12\text{-}16)$$

Then, it follows from the fact that $n_r \geq 0$, that

1. $n \geq l + 1$
2. n is an integer
3. the relation

$$\lambda = n$$

implies that

$$E = -\frac{1}{2} \mu c^2 \frac{(Z\alpha)^2}{n^2} \qquad (12\text{-}17)$$

a result familiar from the old Bohr model. Notice that it is the *reduced mass* that appears in the expression; this, of course, is not peculiar to the differential equation approach. In the old Bohr theory, too, a proper treatment of the classical orbits, subsequently to be restricted by the quantization of angular momentum condition, would have introduced the reduced mass in the energy formula. The presence of the reduced mass

$$\mu = \frac{mM}{m + M} \qquad (12\text{-}18)$$

where m is the electron mass, and M the mass of the nucleus, means that the frequencies

$$\omega_{ij} = \frac{E_i - E_j}{\hbar} = \frac{mc^2/2\hbar}{1 + m/M} (Z\alpha)^2 \left(\frac{1}{n_i{}^2} - \frac{1}{n_j{}^2} \right) \qquad (12\text{-}19)$$

differ slightly for different hydrogenlike atoms. In particular, the difference between the hydrogen spectrum and the deuterium spectrum—where M, the

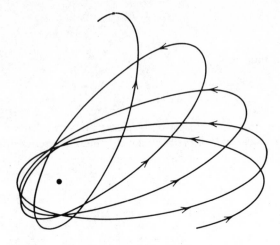

Fig. 12-1. Orbits for a potential that does not have the exact $1/r$ form do not close upon themselves and precess as shown here. The orbits remain planar as long as the potential is radial.

nuclear mass, is very close to being twice the proton mass—was responsible for the discovery of deuterium by Urey and collaborators in 1932.[1]

The energy does not depend on l, that is, for a given n the energies of all the states such that $l + 1 \leq n$ are *degenerate*. We did expect a $(2l + 1)$-fold degeneracy of the energy states for a given l, since the radial equation did not depend on m; here we find that although the radial equation does depend on l, there is an additional degeneracy. Such a degeneracy was formerly called "accidental," since there was no obvious reason for it. This, however, depends on what one means by "obvious." It is already known in classical mechanics that the potential $1/r$ has some special features: the orbits consist of ellipses that maintain their orientation in space, instead of forming precessing orbits (Fig. 12-1). Small modifications of the potential do cause a precession. Such modifications may come from a variety of sources, for example, the perturbations due to other planets, in the Kepler problem. In considering the planetary orbit of Mercury, it was found that after allowance was made for the effects of other planets, a precession of the perihelion in the amount of $42''$ per century remained unaccounted for, and this was finally explained by Einstein's general theory of relativity, which predicted just the right amount of $1/r^2$ potential to be added to the Newtonian $1/r$.

In quantum mechanics, too, there are perturbations, so that the l-degen-

---

[1] There are, of course, other shifts in spectral lines that arise from relativistic effects and from the existence of electron spin. These will be discussed later.

eracy is not really what is observed. In first approximation, however, we have, for a given n, the possible values of $l = 0, 1, 2, \ldots, (n-1)$, and for each there is the $(2l+1)$ degeneracy. Thus the total degeneracy is

$$\sum_{l=0}^{n-1} (2l+1) = n^2 \tag{12-20}$$

Strictly speaking, there are two possible states for the electron because of its spin, so that the true degeneracy is really $2n^2$.

Let us now return to the differential equation. If we set $\lambda = n$ in the recursion relation (12-13) so that

$$a_{k+1} = \frac{k+l+1-n}{(k+1)(k+2l+2)} a_k \tag{12-21}$$

we find that

$$a_{k+1} = (-1)^{k+1} \frac{n-(k+l+1)}{(k+1)(k+2l+2)} \cdot \frac{n-(k+l)}{k(k+2l+1)} \tag{12-22}$$
$$\cdots \frac{n-(l+1)}{1 \cdot (2l+2)} a_0$$

With the help of this we can obtain the power series expansion for $H(\rho)$. Equivalently, we observe that the equation for $H(\rho)$ is that for the *associated Laguerre polynomials*:

$$H(\rho) = L_{n-l-1}^{(2l+1)}(\rho) \tag{12-23}$$

The polynomials are tabulated and their various properties can be found in the mathematical literature.[2]

After conversion back to the radial coordinate r and after normalization, the first few radial functions can be computed. These are listed below. We use

$$a_0 = \frac{\hbar}{\mu c \alpha} \tag{12-24}$$

in the tabulation $R_{nl}(r)$:

$$R_{10}(r) = 2 \left(\frac{Z}{a_0}\right)^{3/2} e^{-Zr/a_0}$$

$$R_{20}(r) = 2 \left(\frac{Z}{2a_0}\right)^{3/2} \left(1 - \frac{Zr}{2a_0}\right) e^{-Zr/2a_0}$$

[2] An extremely useful book is M. Abramowitz and I. A. Stegun (eds.), *Handbook of Mathematical Functions*, National Bureau of Standards Publication, 1964.

$$R_{21}(r) = \frac{1}{\sqrt{3}} \left(\frac{Z}{2a_0}\right)^{3/2} \frac{Zr}{a_0} e^{-Zr/2a_0}$$

$$R_{30}(r) = 2\left(\frac{Z}{3a_0}\right)^{3/2}\left[1 - \frac{2Zr}{3a_0} + \frac{2(Zr)^2}{27a_0^2}\right]e^{-Zr/3a_0}$$

$$R_{31}(r) = \frac{4\sqrt{2}}{3}\left(\frac{Z}{3a_0}\right)^{3/2} \frac{Zr}{a_0}\left(1 - \frac{Zr}{6a_0}\right)e^{-Zr/3a_0}$$

$$R_{32}(r) = \frac{2\sqrt{2}}{27\sqrt{5}}\left(\frac{Z}{3a_0}\right)^{3/2}\left(\frac{Zr}{a_0}\right)^2 e^{-Zr/3a_0} \tag{12-25}$$

The following qualitative features emerge from the sampling of eigensolutions:

(a) The behavior of r^l for small r, which forces the wave function to stay small for a range of radii that increases with l, is a consequence of the centrifugal repulsive barrier that keeps the electrons from coming close to the nucleus.

(b) The recursion relation shows that $H(\rho)$ is a polynomial of degree $n_r = n - l - 1$, and thus it has n_r radial nodes (zeros). There will be $n - l$ "bumps" in the probability density distribution

$$P(r) = r^2[R_{nl})]^2 \tag{12-26}$$

When, for a given n, l has its largest value $l = n - 1$, then there is only one bump. As (12-25) suggests, and as can be seen from the solution to the differential equation,

$$R_{n,n-1}(r) \propto r^{n-1} e^{-Zr/a_0n} \tag{12-27}$$

Hence $P(r) \propto r^{2n} e^{-2Zr/a_0n}$ will peak at a value of r determined by

$$\frac{dP(r)}{dr} = \left(2nr^{2n-1} - \frac{2Z}{a_0n}r^{2n}\right)e^{-2Zr/a_0n} = 0 \tag{12-28}$$

that is, at

$$r = \frac{n^2a_0}{Z} \tag{12-29}$$

whick is the Bohr atom value for circular orbits. Smaller values of l give probability distributions with more bumps. One can show that they correspond to elliptical orbits in the large quantum number limit.

(c) Plots of the radial probability density $P(r)$ for finding the electron at a distance r from the origin can be constructed with the help of the wave functions. Figure 12-2 shows the general pattern. We must remember that the wave function also has an angular part, whose absolute square is $P_l^m (\cos \theta)^2$. Plots of the associated Legendre functions $P_l^m(\cos \theta)$ are given in Fig. 12.3. As m increases, the probability density is seen to shift from the z-axis toward the

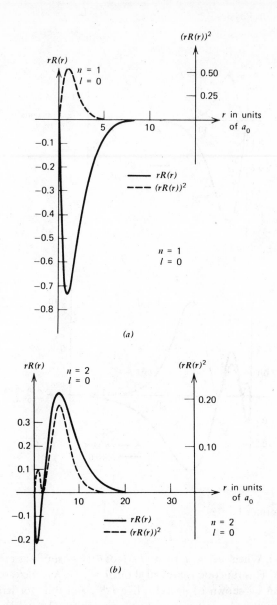

Fig. 12-2. The radial wave functions $u(r) = rR(r)$ and the radial probability density function $u^2(r)$ for values of $n = 1, 2, 3, 4$ and the values of l possible. The left abscissa measures $u(r)$ and the right abscissa measures $u^2(r)$. The wave functions are given by the solid lines, and the probability distributions by the dashed lines. The ordinate measures r in units of a_0.

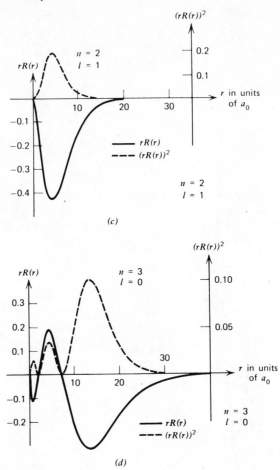

Fig. 12-2.   continued

equatorial plane. When $|m| = l$, then $|P_l^l (\cos \theta)|^2 \propto \sin^{2l} \theta$ as can be read off from Eq. 10-55. This function is peaked about $\theta = \pi/2$. As l increases, the width of the peak can be shown to decrease like $l^{-1/2}$, and thus for large quantum numbers we get the classical picture of planar orbits. The finite width of the peak can be understood from the following considerations. When $|m| = l$, we have $L_z^2 = l^2$ and consequently $L_x^2 + L_y^2 = l$. Thus the angular momentum vector can never be perfectly oriented along an axis. Incidentally, the degeneracy in m allows us to orient the "orbit" relative to some other axis, so that there really is no distinguished z-axis. Thus a state that is an eigenstate of L_x with eigenvalue l will be "oriented" in the x-direction. The wave function will now be

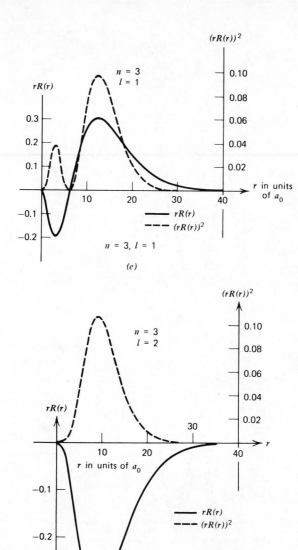

Fig. 12-2.   continued

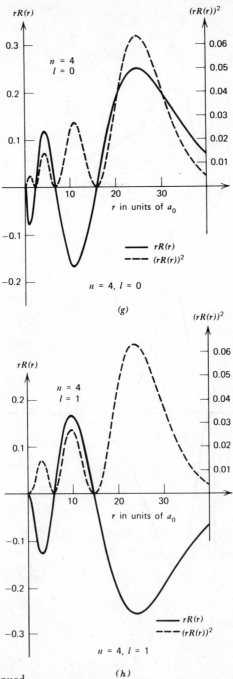

$n = 4$
$l = 0$

—— $rR(r)$
- - - $(rR(r))^2$

$n = 4, l = 0$

(g)

$n = 4$
$l = 1$

—— $rR(r)$
- - - $(rR(r))^2$

$n = 4, l = 1$

(h)

Fig. 12-2. continued

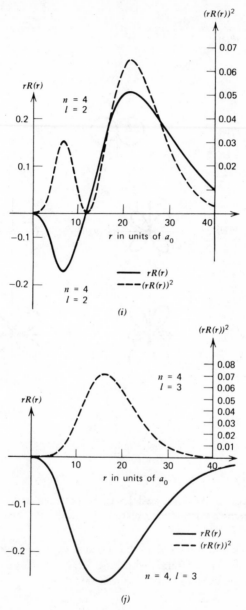

Fig. 12-2. continued

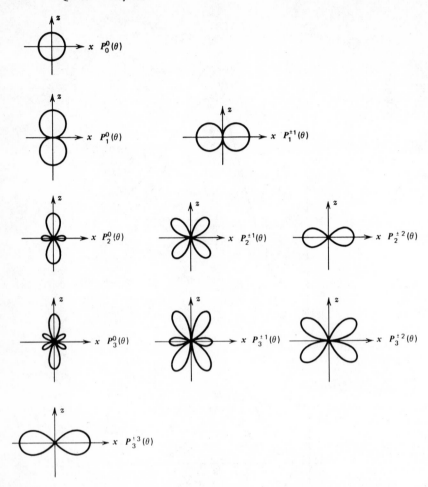

Fig. 12-3. Shapes of the associated Legendre polynomials as a function of θ, the angle between the z-axis and the equatorial plane, denoted here by the x-axis.

a linear combination of the $Y_{lm}(\theta,\phi)$, but because of the degeneracy, the energy will be the same as for the z-oriented orbits.

(d) Given the wave functions, we can calculate

$$\langle r^k \rangle = \int_0^\infty dr \ r^{2+k} \ [R_{nl}(r)]^2 \tag{12-30}$$

Some useful expectation values are given below:

$$\langle r \rangle = \frac{a_0}{2Z} \ [3n^2 - l(l+1)]$$

$$\langle r^2 \rangle = \frac{a_0^2 n^2}{2Z^2} [5n^2 + 1 - 3l(l+1)]$$

$$\left\langle \frac{1}{r} \right\rangle = \frac{Z}{a_0 n}$$

$$\left\langle \frac{1}{r^2} \right\rangle = \frac{Z^2}{a_0^2 n^3 (l + \frac{1}{2})} \tag{12-31}$$

Problems

1. Compare the wavelengths of the $2P \rightarrow 1S$ transitions in (1) hydrogen, (2) deuterium (nuclear mass = 2 × proton mass), (3) positronium (a bound state of an electron and a positron, whose mass is the same as that of an electron).

2. An electron is in the ground state of tritium, for which the nucleus consists of a proton and two neutrons. A nuclear reaction instantaneously changes the nucleus to He^3, that is, two protons and one neutron. Calculate the probability that the electron remains in the ground state of He^3. What is the probability that the electron is free, with momentum p?

[*Note.* The momentum eigenfunction for a free electron is $e^{i\mathbf{p}\cdot\mathbf{r}/\hbar}(2\pi\hbar)^{-3/2}$.]

3. The relativistic analog of the Schrödinger equation for a spin 0 electron (thus not applicable to the real electron) is the operator version of

$$(E - V)^2 = p^2 c^2 + m^2 c^4$$

that is,

$$\left(\frac{E}{\hbar c} - \frac{Ze^2}{\hbar c} \frac{1}{r} \right)^2 \psi = -\nabla^2 \psi + \left(\frac{mc}{\hbar} \right)^2 \psi$$

(a) Find the radial equation.

(b) Find the eigenvalue spectrum by noting the close relationship of the radial equation obtained in (a) with the radial equation for the hydrogen atom problem.

4. Using the expression for $\langle 1/r \rangle_{n,l}$ calculate the expression for

$$\langle T \rangle_{n,l} = \left\langle \frac{p^2}{2m} \right\rangle_{n,l}$$

for an arbitrary hydrogen atom eigenstate (with Z arbitrary). Show that generally for this potential

$$\langle T \rangle = -\tfrac{1}{2} \langle V \rangle$$

This is a special example of the *Virial theorem*.

5. An electron in the Coulomb field of a proton is in a state described by the wave function

$$\tfrac{1}{6}[4\psi_{100}(\mathbf{r}) + 3\psi_{211}(\mathbf{r}) - \psi_{210}(\mathbf{r}) + \sqrt{10}\,\psi_{21-1}(\mathbf{r})]$$

(a) What is the expectation value of the energy?
(b) What is the expectation value of $\mathbf{L}^2$?
(c) What is the expectation value of L_z?

6. An electron in the Coulomb field of a proton is in a state described by the wave function

$$\psi(\mathbf{r}) = \left(\frac{\alpha}{\sqrt{\pi}}\right)^{3/2} e^{-\alpha^2 r^2/2}$$

What is the probability that it will be found in the ground state of the hydrogen atom?

7. An electron is in the $n = 2$, $l = 1$, $m = 0$ state of the hydrogen atom. What is its wave function in momentum space?

8. The expectation value of $f(\mathbf{r},\mathbf{p})$ in any stationary state is a constant. Calculate

$$0 = \frac{d}{dt}\langle \mathbf{r}\cdot\mathbf{p}\rangle = \frac{i}{\hbar}\langle [H,\mathbf{r}\cdot\mathbf{p}]\rangle$$

for a Hamiltonian

$$H = \mathbf{p}^2/2m + V(r)$$

and show that

$$\left\langle \frac{\mathbf{p}^2}{m}\right\rangle = \langle \mathbf{r}\cdot\boldsymbol{\nabla}V(r)\rangle$$

Use this to establish the result of Problem 4.

9. Use the techniques developed in this chapter to discuss the three-dimensional harmonic oscillator problem, with

$$H = \frac{\mathbf{p}^2}{2m} + \tfrac{1}{2}m\omega^2 r^2$$

Note that the associated Laguerre polynomials also appear in this problem.

References

A very thorough discussion of the hydrogenlike atoms is to be found in

E. U. Condon and G. H. Shortley, *The Theory of Atomic Spectra*, Cambridge University Press, Cambridge, 1959.

The problem is discussed in every book on quantum mechanics.

chapter 13

Interaction of Electrons with Electromagnetic Field

In Chapter 12 we discussed the interaction of an electron with the static Coulomb field due to a point charge. To generalize this to the interaction with an external magnetic or electric field, we must first review the classical theory. Maxwell's equations in Gaussian units read, in the vacuum,

$$\nabla \cdot \mathbf{B}(\mathbf{r},t) = 0 \tag{13-1}$$

$$\nabla \times \mathbf{E}(\mathbf{r},t) + \frac{1}{c} \frac{\partial \mathbf{B}(\mathbf{r},t)}{\partial t} = 0 \tag{13-2}$$

$$\nabla \cdot \mathbf{E}(\mathbf{r},t) = 4\pi \rho(\mathbf{r},t) \tag{13-3}$$

$$\nabla \times \mathbf{B}(\mathbf{r},t) - \frac{1}{c} \frac{\partial \mathbf{E}(\mathbf{r},t)}{\partial t} = \frac{4\pi}{c} \mathbf{j}(\mathbf{r},t) \tag{13-4}$$

where $\rho(\mathbf{r},t)$ and $\mathbf{j}(\mathbf{r},t)$ are the charge and current densities that are the sources of the electromagnetic fields $\mathbf{E}(\mathbf{r},t)$ and $\mathbf{B}(\mathbf{r},t)$. The conservation of charge equation

$$\frac{\partial \rho(\mathbf{r},t)}{\partial t} + \nabla \cdot \mathbf{j}(\mathbf{r},t) = 0 \tag{13-5}$$

is automatically satisfied.

We may satisfy the first two equations by expressing the fields in terms of a scalar potential $\phi(\mathbf{r},t)$ and a vector potential $\mathbf{A}(\mathbf{r},t)$

$$\mathbf{B}(\mathbf{r},t) = \nabla \times \mathbf{A}(\mathbf{r},t)$$

$$\mathbf{E}(\mathbf{r},t) = -\frac{1}{c} \frac{\partial \mathbf{A}(\mathbf{r},t)}{\partial t} - \nabla \phi(\mathbf{r},t) \tag{13-6}$$

The fields $\mathbf{E}$ and $\mathbf{B}$ do not determine ϕ and $\mathbf{A}$ uniquely. New potentials, given by

$$\mathbf{A}'(\mathbf{r},t) = \mathbf{A}(\mathbf{r},t) - \boldsymbol{\nabla} f(\mathbf{r},t)$$

$$\phi'(\mathbf{r},t) = \phi(\mathbf{r},t) + \frac{1}{c}\frac{\partial f(\mathbf{r},t)}{\partial t} \qquad (13\text{-}7)$$

are easily seen to yield the same $\mathbf{E}$ and $\mathbf{B}$ fields. The transformation from the set $(\mathbf{A},\phi)$ to $(\mathbf{A}',\phi')$ is known as a *gauge transformation*, and the invariance of $\mathbf{E}$ and $\mathbf{B}$ allows us to choose the arbitrary function $f(\mathbf{r},t)$ in the most convenient way.

The source-dependent pair of equations (13-3) and (13-4) now read

$$-\nabla^2\phi(\mathbf{r},t) - \frac{1}{c}\frac{\partial}{\partial t}(\boldsymbol{\nabla}\cdot\mathbf{A}) = 4\pi\rho(\mathbf{r},t) \qquad (13\text{-}8)$$

and

$$\boldsymbol{\nabla}\times(\boldsymbol{\nabla}\times\mathbf{A}) + \frac{1}{c^2}\frac{\partial^2\mathbf{A}(\mathbf{r},t)}{\partial t^2} + \frac{1}{c}\frac{\partial}{\partial t}\boldsymbol{\nabla}\phi = \frac{4\pi}{c}\mathbf{j}(\mathbf{r},t)$$

which may be rewritten as

$$-\nabla^2\mathbf{A}(\mathbf{r},t) + \frac{1}{c^2}\frac{\partial^2\mathbf{A}(\mathbf{r},t)}{\partial t^2} + \boldsymbol{\nabla}\left(\boldsymbol{\nabla}\cdot\mathbf{A} + \frac{1}{c}\frac{\partial\phi}{\partial t}\right) = \frac{4\pi}{c}\mathbf{j}(\mathbf{r},t) \quad (13\text{-}9)$$

If the charge distribution is static, that is, $\rho(\mathbf{r})$ is independent of time, it is convenient to choose the gauge such that

$$\boldsymbol{\nabla}\cdot\mathbf{A}(\mathbf{r},t) = 0 \qquad (13\text{-}10)$$

This choice of $f(\mathbf{r},t)$ is given the name of *Coulomb gauge*. In that case we have

$$-\nabla^2\phi(\mathbf{r}) = 4\pi\rho(\mathbf{r}) \qquad (13\text{-}11)$$

that is, we have a time-independent scalar potential, and then the equation for $\mathbf{A}(\mathbf{r},t)$ reads

$$-\nabla^2\mathbf{A}(\mathbf{r},t) + \frac{1}{c^2}\frac{\partial^2\mathbf{A}(\mathbf{r},t)}{\partial t^2} = \frac{4\pi}{c}\mathbf{j}(\mathbf{r},t) \qquad (13\text{-}12)$$

When the charge distribution is not static, it is more convenient to choose the so-called *Lorentz gauge* for which

$$\boldsymbol{\nabla}\cdot\mathbf{A}(\mathbf{r},t) + \frac{1}{c}\frac{\partial\phi(\mathbf{r},t)}{\partial t} = 0 \qquad (13\text{-}13)$$

This leaves the equation for the vector potential unaltered, but now the scalar equation also obeys a wave equation. A technical point worth noting is that the relation

$$\boldsymbol{\nabla}\times(\boldsymbol{\nabla}\times\mathbf{A}) = -\nabla^2\mathbf{A} + \boldsymbol{\nabla}(\boldsymbol{\nabla}\cdot\mathbf{A})$$

used to obtain (13-9) is only valid in cartesian coordinates. Thus, $\nabla^2 \mathbf{A}(\mathbf{r},t)$, as it appears, must be calculated in terms of x, y, and z.

The equation describing the interaction of a point electron of mass μ with an electromagnetic field is the classical Lorentz force equation

$$\mu \frac{d^2\mathbf{r}}{dt^2} = -e\left[\mathbf{E}(\mathbf{r},t) + \frac{\mathbf{v}}{c} \times \mathbf{B}(\mathbf{r},t)\right] \tag{13-14}$$

We now assert that this equation will be obtained if the classical Hamiltonian for an electron in the absence of fields

$$H_0 = \frac{\mathbf{p}^2}{2\mu} \tag{13-15}$$

is changed by making the alteration

$$\mathbf{p} \rightarrow \mathbf{p} + \frac{e}{c}\mathbf{A}(\mathbf{r},t) \tag{13-16}$$

and adding the potential $e\phi(\mathbf{r})$ (we shall deal with static scalar potentials), so that

$$H = \frac{1}{2\mu}\left[\mathbf{p} + \frac{e}{c}\mathbf{A}(\mathbf{r},t)\right]^2 + e\phi(\mathbf{r}) \tag{13-17}$$

We shall leave the proof of this statement as an exercise for the reader.[1] The corresponding Schrödinger equation with the static potential taken over to the right side is

$$\frac{1}{2\mu}\left(\frac{\hbar}{i}\boldsymbol{\nabla} + \frac{e}{c}\mathbf{A}\right)^2 \psi(\mathbf{r},t) = [E + e\phi(\mathbf{r})]\,\psi(\mathbf{r},t) \tag{13-18}$$

The left side is

$$\frac{1}{2\mu}\left(\frac{\hbar}{i}\boldsymbol{\nabla} + \frac{e}{c}\mathbf{A}\right)\left(\frac{\hbar}{i}\boldsymbol{\nabla}\psi + \frac{e}{c}\mathbf{A}\psi\right)$$

$$= -\frac{\hbar^2}{2\mu}\nabla^2\psi - \frac{ie\hbar}{\mu c}\mathbf{A}\cdot\boldsymbol{\nabla}\psi - \frac{ie\hbar}{2\mu c}(\boldsymbol{\nabla}\cdot\mathbf{A})\,\psi + \frac{e^2}{2\mu c^2}\mathbf{A}^2\psi$$

$$= -\frac{\hbar^2}{2\mu}\nabla^2\psi - \frac{ie\hbar}{\mu c}\mathbf{A}\cdot\boldsymbol{\nabla}\psi + \frac{e^2}{2\mu c^2}\mathbf{A}^2\psi \tag{13-19}$$

For a constant uniform magnetic field, $\mathbf{B}$, we may take[2]

$$\mathbf{A} = -\tfrac{1}{2}\mathbf{r} \times \mathbf{B} \tag{13-20}$$

---

[1] See footnote 4, p. 216.

[2] Note that this choice is not unique, since we may add the gradient of any function to $\mathbf{A}$ without changing $\mathbf{B}$. This choice, however, is very convenient.

212   Quantum Physics

This means that the three components of **A** are

$$\mathbf{A} = -\tfrac{1}{2}(yB_z - zB_y, \; zB_x - xB_z, \; xB_y - yB_x)$$

and consequently

$$\boldsymbol{\nabla} \times \mathbf{A} = (\tfrac{1}{2}B_x + \tfrac{1}{2}B_x, \; B_y, \; B_z)$$
$$= \mathbf{B}$$

Hence the second term in (13-19) becomes

$$\frac{ie\hbar}{2\mu c} \mathbf{r} \times \mathbf{B} \cdot \boldsymbol{\nabla}\psi = -\frac{ie\hbar}{2\mu c} \mathbf{B} \cdot \mathbf{r} \times \boldsymbol{\nabla}\psi$$

$$= \frac{e}{2\mu c} \mathbf{B} \cdot \mathbf{r} \times \frac{\hbar}{i} \boldsymbol{\nabla}\psi = \frac{e}{2\mu c} \mathbf{B} \cdot \mathbf{L}\psi \qquad (13\text{-}21)$$

and the third term is

$$\frac{e^2}{8\mu c^2} (\mathbf{r} \times \mathbf{B})^2 \psi = \frac{e^2}{8\mu c^2} [r^2\mathbf{B}^2 - (\mathbf{r}\cdot\mathbf{B})^2] \, \psi = \frac{e^2 B^2}{8\mu c^2} (x^2 + y^2) \, \psi \qquad (13\text{-}22)$$

if **B** is the direction that defines the z-axis. This is of the form of a two-dimensional harmonic oscillator potential.

Let us compare the magnitudes of the two terms. The ratio is estimated with $\langle L_z \rangle$ taken of order $\hbar$ and $\langle x^2 + y^2 \rangle$ of order a_0^2, with a_0 the Bohr radius:

$$\frac{(e^2/8\mu c^2)\, a_0^2 B^2}{(e/2\mu c)\, \hbar B} \simeq \frac{1}{4} \frac{e^2}{\hbar c} \frac{B}{e/a_0^2} \simeq \frac{1}{548} \frac{B}{e/a_0^2}$$

$$\approx \frac{B}{548(4.8 \times 10^{-10})/(0.5 \times 10^{-8})^2}$$

$$\approx \frac{B}{9 \times 10^9 \text{ gauss}} \qquad (13\text{-}23)$$

Thus in atomic systems, with the kind of fields available in the laboratory, that is, $B \lesssim 10^4$ gauss, the quadratic term is certainly negligible. The term linear in B, compared with the Coulomb potential energy can be estimated in a similar way

$$\frac{(e/2\mu c)\, \hbar B}{e^2/a_0} \approx \frac{1}{2} \frac{\hbar/\mu c}{e/a_0} B \approx \frac{1}{274} \frac{B}{e/a_0^2} \approx \frac{B}{5 \times 10^9 \text{ gauss}} \qquad (13\text{-}24)$$

so that the linear term will only slightly perturb the atomic energy levels. The quadratic term can become very important under two conditions: if the magnetic field is very intense; it is believed that fields as large as 10^{12} gauss may exist on the surface of neutron stars, and this would radically alter the structure of atoms.[3] The quadratic term will also be important when we consider the macroscopic motion of an electron in an external field.

Let us first consider the linear term alone, and pick the z-direction to coincide with that of **B**. Then the Hamiltonian with **B** = 0 is altered by the addition of

$$H_1 = \frac{e}{2\mu c} BL_z \qquad (13\text{-}25)$$

If we define the frequency, called the Larmor frequency,

$$\frac{eB}{2\mu c} = \omega_L \qquad (13\text{-}26)$$

and deal with energy eigenstates that are simultaneously eigenstates of $\mathbf{L}^2$ and L_z, then the extra term (13-25), when acting on an eigenstate, yields a number, namely,

$$H_1 u_{nlm}(\mathbf{r}) = \hbar\omega_L m u_{nlm}(\mathbf{r}) \qquad (13\text{-}27)$$

where m is the z-component of the angular momentum eigenvalue, with $-l \leq m \leq l$. Thus the existing energy levels, with their $(2l + 1)$-fold degeneracy are split into $(2l + 1)$ components that are equally spaced, with energies given by

$$E = -\frac{1}{2} \mu c^2 \frac{(Z\alpha)^2}{n^2} + \hbar\omega_L m \qquad (13\text{-}28)$$

The size of the splitting is

$$\frac{eB\hbar}{2\mu c} = \frac{e\hbar}{2\mu c} \left(\frac{B}{e/a_0^2}\right) \frac{e}{a_0^2}$$

$$= \frac{e^2\hbar}{2\mu c} \left(\frac{\mu c\alpha}{\hbar}\right)^2 \left(\frac{B}{e/a_0^2}\right)$$

$$= (\tfrac{1}{2}\alpha^2\mu c^2)\, \alpha\, \frac{B}{e/a_0^2}$$

$$= \left(\frac{B}{2.4 \times 10^9}\right) \times 13.6 \text{ eV}$$

Since there are selection rules (to be discussed later) according to which only transitions in which the m-value changes by zero or unity are allowed, it turns out that the single line representing a transition with $B = 0$ splits into *three* lines, as can be seen in Fig. 13-1. This effect is the *normal Zeeman effect*. Actually, unless the electron spin state in the atom is one in which the spin is

[3] See R. Cohen, L. Lodenquai, and M. Ruderman, *Phys. Rev. Letters*, **25**, 467 (1970).

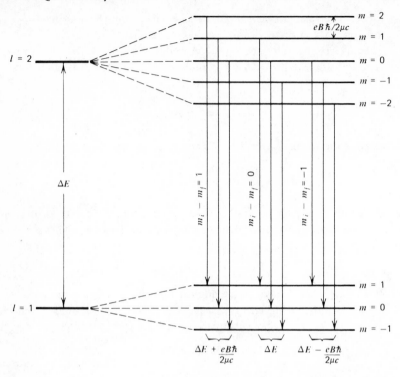

Fig. 13-1. Normal Zeeman effect: of the 15 possible transitions between the $l = 2$ and $l = 1$ states, split by the magnetic field, only 9, corresponding to $\Delta m = m_i - m_f = -1, 0, 1$ occur, in the form of three lines.

zero, the interactions of the electron spin with the magnetic field changes the pattern predicted above. The more common *anomalous Zeeman effect* will be discussed when we have learned about spin.

It is of some interest to discuss the solution of an electron in a constant magnetic field under conditions where the B^2 term is not negligible, and where the Coulomb potential can be neglected. Under those conditions, with **B** again chosen to define the z-direction, the Schrödinger equation reads

$$-\frac{\hbar^2}{2\mu} \nabla^2\psi + \frac{eB}{2\mu c} L_z\psi + \frac{e^2B^2}{8\mu c^2} (x^2 + y^2)\,\psi = E\psi \qquad (13\text{-}29)$$

where we have used (13-19), (13-21), and (13-22). The presence of the "potential" $(x^2 + y^2)$ suggests the use of cylindrical coordinates for the separation of the variables. Writing

$$x = \rho \cos \phi$$
$$y = \rho \sin \phi \qquad (13\text{-}30)$$

we follow the procedure outlined at the beginning of Chapter 10 to arrive at

$$\frac{\partial}{\partial x} = \cos\phi \, \frac{\partial}{\partial \rho} - \frac{\sin\phi}{\rho} \, \frac{\partial}{\partial \phi}$$

$$\frac{\partial}{\partial y} = \sin\phi \, \frac{\partial}{\partial \rho} + \frac{\cos\phi}{\rho} \, \frac{\partial}{\partial \phi} \tag{13-31}$$

and hence

$$\nabla^2 = \frac{\partial^2}{\partial z^2} + \frac{\partial^2}{\partial \rho^2} + \frac{1}{\rho} \frac{\partial}{\partial \rho} + \frac{1}{\rho^2} \frac{\partial^2}{\partial \phi^2} \tag{13-32}$$

If we now write

$$\psi(\mathbf{r}) = u_m(\rho) \, e^{im\phi} \, e^{ikz} \tag{13-33}$$

we find that the differential equation satisfied by $u_m(\rho)$ is

$$\frac{d^2u}{d\rho^2} + \frac{1}{\rho}\frac{du}{d\rho} - \frac{m^2}{\rho^2} u - \frac{e^2B^2}{4\hbar^2c^2}\rho^2 u + \left(\frac{2\mu E}{\hbar^2} - \frac{eB\hbar m}{\hbar^2 c} - k^2\right) u = 0 \tag{13-34}$$

If we introduce the variable

$$x = \sqrt{\frac{eB}{2\hbar c}} \, \rho \tag{13-35}$$

we can rewrite the equation in the form

$$\frac{d^2u}{dx^2} + \frac{1}{x}\frac{du}{dx} - \frac{m^2}{x^2} u - x^2 u + \lambda u = 0 \tag{13-36}$$

where

$$\lambda = \frac{4\mu c}{eB\hbar}\left(E - \frac{\hbar^2 k^2}{2\mu}\right) - 2m \tag{13-37}$$

It is fairly straightforward to determine that (a) the behavior of $u(x)$ at infinity, determined from

$$\frac{d^2u}{dx^2} - x^2 u \approx 0$$

is $u(x) \sim e^{-x^2/2}$, and (b) the behavior of $u(x)$ near $x = 0$, determined from

$$\frac{d^2u}{dx^2} + \frac{1}{x}\frac{du}{dx} - \frac{m^2}{x^2} u \approx 0$$

is $u(x) \sim x^{|m|}$. We thus write

$$u(x) = x^{|m|} \, e^{-x^2/2} \, G(x) \tag{13-38}$$

and determine the differential equation obeyed by $G(x)$.

A little algebra leads to

$$\frac{d^2G}{dx^2} + \left(\frac{2|m|+1}{x} - 2x\right)\frac{dG}{dx} + (\lambda - 2 - 2|m|)\,G = 0 \quad (13\text{-}39)$$

This can be brought into the same form as (12-11) if we change variables to

$$y = x^2 \quad (13\text{-}40)$$

The equation then takes the form

$$\frac{d^2G}{dy^2} + \left(\frac{|m|+1}{y} - 1\right)\frac{dG}{dy} + \frac{\lambda - 2 - 2|m|}{4y}\,G = 0 \quad (13\text{-}41)$$

We can now proceed as in Chapter 12. Comparison with (12-11) shows that we must have

$$\frac{1}{4}\lambda - \frac{1+|m|}{2} = n_r \quad (13\text{-}42)$$

as an eigenvalue condition, with $n_r = 0, 1, 2, 3, \ldots$. This implies that $E - \hbar^2 k^2/2\mu$, the energy with the kinetic energy of the free motion in the z-direction subtracted out, is given by

$$E - \frac{\hbar^2 k^2}{2\mu} = \frac{eB\hbar}{2\mu c}\,(2n_r + 1 + |m| + m) \quad (13\text{-}43)$$

and

$$G(y) = L_{n_r}^{|m|}(y) \quad (13\text{-}44)$$

Our discussion of this solution will be confined to the classical limit. To do this, we first review the classical theory. Given the Hamiltonian (13-17), without the scalar potential term, we have[4]

$$\mathbf{v} = \frac{\mathbf{p} + (e/c)\,\mathbf{A}}{\mu} \quad (13\text{-}45)$$

and with $\mathbf{A} = -\tfrac{1}{2}\mathbf{r} \times B$, we obtain

$$\mu\mathbf{r} \times \mathbf{v} = \mathbf{r} \times \mathbf{p} + \frac{e}{c}\,\mathbf{r} \times (-\tfrac{1}{2}\,\mathbf{r} \times \mathbf{B})$$

$$= \mathbf{L} - \frac{e}{2c}\,[\mathbf{r}(\mathbf{r}\cdot\mathbf{B}) - r^2\mathbf{B}] \quad (13\text{-}46)$$

[4] The reader who is not familiar with mechanics as formulated by Hamilton can convince himself that the equations $dx/dt = \partial H/\partial p_x$, $dp_x/dt = -\partial H/\partial x$, and so on, are equivalent to Newton's equations for $H = \mathbf{p}^2/2\mu + V(\mathbf{r})$. The equations also hold for the more complicated Hamiltonian $[\mathbf{p} + e\mathbf{A}(\mathbf{r})/c]^2/2\mu$.

with the help of the identity

$$\mathbf{a} \times (\mathbf{b} \times \mathbf{c}) = \mathbf{b}(\mathbf{a} \cdot \mathbf{c}) - \mathbf{c}(\mathbf{a} \cdot \mathbf{b}) \tag{13-47}$$

We take the z-component of this equation to obtain

$$\mu(\mathbf{r} \times \mathbf{v})_z = L_z + \frac{e}{2c} B \, (x^2 + y^2)$$

that is,

$$\mu\rho v = L_z + \frac{eB}{2c} \rho^2 \tag{13-48}$$

The expression for the force on the electron

$$\mathbf{F} = - \frac{e}{c} \mathbf{v} \times \mathbf{B} \tag{13-49}$$

yields the relation

$$\frac{\mu v^2}{\rho} = \frac{evB}{c} \tag{13-50}$$

for circular motion. This relation, together with (13-48), after a little algebra, yields,

$$\tfrac{1}{2}\mu v^2 = \frac{eB}{\mu c} L_z \tag{13-51}$$

and

$$\rho = \left[\frac{2c}{eB} L_z \right]^{1/2} \tag{13-52}$$

We now return to the expression for the energy, (13-43). Because of the smallness of $\hbar$, the energy can only be of macroscopic size for reasonable B, if $(2n_r + 1 + |m| + m)$ is very large. We have two cases: (a) If $m < 0$, this implies that n_r is very large. Now n_r determines the degree of the polynomial $L_{n_r}^{|m|} (y)$, that is, the number of the zeros in the function,[5] and if that is very large, the function cannot be large for some small range of y where the classical orbit would be located. (b) If $m > 0$, the coefficient is $(2n_r + 1 + 2m)$, and this can be large, with n_r small, provided that m is large. The energy now is

$$E - \frac{\hbar^2 k^2}{2\mu} \simeq \frac{eB}{\mu c} \hbar m \tag{13-53}$$

[5] See Eq. 12-23, the development leading up to it, and the discussion on p. 200.

in agreement with the classical result. Note that

$$L_z = \hbar m \tag{13-54}$$

is positive, as expected.

We can also show that the radius of the orbit, as determined by the peaking of the radial probability distribution, corresponds to the classical value. Let us take $n_r = 0$. In that case $L_{n_r}^{|m|}(y)$ is just a constant, and the square of the wave function is, according to (13-38),

$$P(x) = x^{2|m|} e^{-x2} \tag{13-55}$$

This has a maximum where

$$\frac{dP}{dx} = (2|m|x^{2|m|-1} - 2x^{2|m|+1}) e^{-x2} = 0$$

that is, at

$$x = \sqrt{|m|} \tag{13-56}$$

which yields

$$\rho = \left(\frac{2c}{eB} \hbar m\right)^{1/2} \tag{13-57}$$

This problem is a beautiful illustration of the correspondence principle.

There are several interesting quantum mechanical effects connected with the interaction with a magnetic field that we now turn to. The Schrödinger equation (13-18) appears to violate the principle of gauge invariance, since it is $\mathbf{A}(\mathbf{r},t)$ that appears in the equation, and under the transformation

$$\mathbf{A} \rightarrow \mathbf{A} + \mathbf{\nabla} f(\mathbf{r},t) \tag{13-58}$$

the Hamiltonian is changed according to

$$\frac{1}{2\mu}\left(\frac{\hbar}{i}\mathbf{\nabla} + \frac{e}{c}\mathbf{A}\right)^2 \rightarrow \frac{1}{2\mu}\left(\frac{\hbar}{i}\mathbf{\nabla} + \frac{e}{c}\mathbf{A} + \frac{e}{c}\mathbf{\nabla} f\right)^2 \tag{13-59}$$

It is possible to save gauge invariance by using the fact that a change of the wave function by a phase factor, which may depend on $\mathbf{r}$, has no physical consequences. Thus if we require that (13-58) *must* be accompanied by the transformation

$$\psi(\mathbf{r},t) \rightarrow e^{i\wedge(\mathbf{r},t)} \psi(\mathbf{r},t) \tag{13-60}$$

then the left hand side of eq. (13-18) becomes

$$\frac{1}{2\mu}\left(\frac{\hbar}{i}\mathbf{\nabla} + \frac{e}{c}\mathbf{A} + \frac{e}{c}\mathbf{\nabla} f\right) \cdot \left(\frac{\hbar}{i}\mathbf{\nabla} + \frac{e}{c}\mathbf{A} + \frac{e}{c}\mathbf{\nabla} f\right) e^{i\wedge}\psi$$

$$= \frac{1}{2\mu}\left(\frac{\hbar}{i}\boldsymbol{\nabla} + \frac{e}{c}\mathbf{A} + \frac{e}{c}\boldsymbol{\nabla}f\right) \cdot \left[e^{i\Lambda}\left(\frac{\hbar}{i}\boldsymbol{\nabla}\psi + \frac{e}{c}\mathbf{A}\psi + \frac{e}{c}\boldsymbol{\nabla}f\psi + \hbar\boldsymbol{\nabla}\Lambda\psi\right)\right]$$

$$= \frac{1}{2\mu}e^{i\Lambda}\left(\frac{\hbar}{i}\boldsymbol{\nabla} + \frac{e}{c}\mathbf{A} + \frac{e}{c}\boldsymbol{\nabla}f + \hbar\boldsymbol{\nabla}\Lambda\right)^2\psi \qquad (13\text{-}61)$$

Thus with the choice

$$\Lambda = -\frac{e}{\hbar c}f \qquad (13\text{-}62)$$

that is, with the transformation law

$$\psi(\mathbf{r},t) \longrightarrow e^{-(ie/\hbar c)f(r,t)}\,\psi(\mathbf{r},t) \qquad (13\text{-}63)$$

gauge invariance is restored.

In a field-free region, $\mathbf{B} = 0$, which implies that

$$\boldsymbol{\nabla}\times\mathbf{A} = 0 \qquad (13\text{-}64)$$

that is, $\mathbf{A}$ may be written as a gradient of a function

$$\mathbf{A} = \boldsymbol{\nabla}f \qquad (13\text{-}65)$$

In a field-free region, we may therefore describe the motion of an electron in two ways: either we do not consider the presence of a field at all, and write

$$\left[\frac{1}{2\mu}\left(\frac{\hbar}{i}\boldsymbol{\nabla}\right)^2 + V(\mathbf{r})\right]\psi = E\psi \qquad (13\text{-}66)$$

for the energy eigenfunction equation, or we write the equation with the vector potential given by (13-65)

$$\frac{1}{2\mu}\left(\frac{\hbar}{i}\boldsymbol{\nabla} + \frac{e}{c}\mathbf{A}\right)^2\psi' + V(\mathbf{r})\,\psi' = E\psi' \qquad (13\text{-}67)$$

and take

$$\psi' = e^{-(ie/\hbar c)f}\,\psi \qquad (13\text{-}68)$$

The function $f(\mathbf{r},t)$ may be written in terms of $\mathbf{A}(\mathbf{r},t)$ by solving (13-65):

$$f(\mathbf{r},t) = \int^{\mathbf{r}} d\mathbf{r}'\cdot\mathbf{A}(\mathbf{r}',t) \qquad (13\text{-}69)$$

where the path of integration is taken from an arbitrary fixed point, for example, the origin, or infinity, to the point $\mathbf{r}$. The integral only makes sense if $\mathbf{B} = 0$, that is, in a field-free region, since the difference in the integral along two different paths, labeled 1 and 2, is

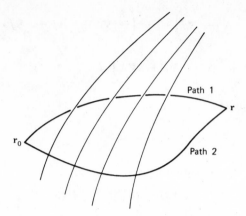

Fig. 13-2. The integrals $\int_{r_0}^{r} A(\mathbf{r}') \cdot d\mathbf{r}'$ along path 1 and path 2 are generally not the same, since the difference is equal to the magnetic flux Φ enclosed by the closed loop.

$$\int_1 d\mathbf{r}' \cdot \mathbf{A}(\mathbf{r}',t) - \int_2 d\mathbf{r}' \mathbf{A}(\mathbf{r}',t) = \oint d\mathbf{r}' \cdot \mathbf{A}(\mathbf{r}',t)$$
$$= \int_S \boldsymbol{\nabla}' \times \mathbf{A}(\mathbf{r}',t) \cdot d\mathbf{S} = \int_S \mathbf{B} \cdot d\mathbf{S} = \Phi \quad (13\text{-}70)$$

where we have used Stokes' theorem, and where Φ is the flux of magnetic field through the surface spanned by the two paths (Fig. 13.2). Thus only if $\Phi = 0$ will the phase factor in (13-68) be independent of the choice of path in the line integral. Such an independence is required if we insist that the wave function be single-valued.

If the two paths include flux, then the wave functions of electrons traveling along the two paths will acquire different phases. An interesting consequence is that if an electron moves in a field-free region that is not simply connected, but surrounds a "hole" containing flux Φ, then upon completing a circuit, the electron acquires an additional phase factor $e^{ie\Phi/\hbar c}$. The requirement that the electron wave function be single-valued, so that the phase factor is unity, implies that *the enclosed flux is quantized*

$$\Phi = \frac{2\pi\hbar c}{e} n \qquad n = 0, \pm1, \pm2, \ldots \quad (13\text{-}71)$$

Such a situation arises in the motion of electrons in a superconducting ring surrounding a region containing flux. The first experiments, done in 1961[6] were based on the following scheme: a ring, made of a superconductor, is

[6] B. S. Deaver and W. Fairbank, *Phys. Rev. Letters*, 7, 43 (1961); R. Döll and M. Nabauer, *ibid.*, 7, 51 (1961).

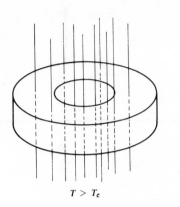
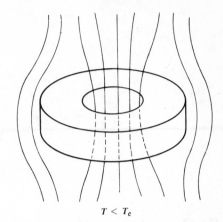

$T > T_c$

$T < T_c$

Fig. 13-3. A superconductor at temperature $T > T_c$ (the critical temperature) acts like any other metal, and magnetic flux lines can penetrate it. When the temperature is lowered until $T < T_c$, the ring becomes superconducting, and expels magnetic flux lines. Some of these become trapped inside the ring. It is the trapped flux that is found to be quantized.

placed in an external magnetic field at a temperature above the critical temperature, so that the metal is not superconducting. Since superconductors expel magnetic field lines, except for a thin surface layer, $\mathbf{B} = 0$ inside them. This is the *Meissner effect*.[7] When the ring is cooled below the critical temperature, it becomes superconducting, and magnetic flux is trapped inside the ring (Fig. 13.3). An ingenious measurement of the flux shows that (13-71) holds, with the modification that

$$\Phi = \frac{2\pi\hbar c}{(2e)}\, n \tag{13-72}$$

This is consistent with our present understanding of the phenomenon of superconductivity, according to which, "correlated states" of pairs of electrons (with charge 2e!) form the fundamental entities that one deals with in the superconductor.

Another manifestation of the dependence of the phase of the wave function on the flux, can, in principle, be seen in an interference experiment (Fig. 13.4) in which a solenoid confining magnetic flux is placed between the slits in a two-slit experiment. The interference pattern at the screen is due to the superposition of two parts of the wave function

$$\psi = \psi_1 + \psi_2 \tag{13-73}$$

[7] I strongly recommend Chapter 21 in the *Feynman Lectures on Physics*, Vol. III, for an excellent discussion of these macroscopic manifestations of quantum mechanics.

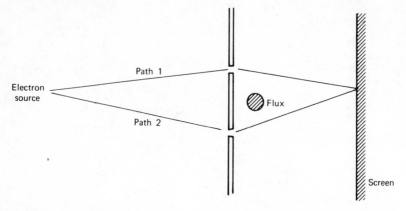

Fig. 13-4. Schematic sketch of experiment measuring shift of electron interference pattern by confined magnetic flux.

where ψ_1 denotes the part of the wave function that describes the electron following path 1, and ψ_2 the part appropriate to path 2. In the presence of the solenoid
we have

$$\psi = \psi_1 \, e^{ie/\hbar c \int_1 dr \cdot A} + \psi_2 \, e^{ie/\hbar c \int_2 dr \cdot A}$$

$$= (\psi_1 \, e^{ie\Phi/\hbar c} + \psi_2) \, e^{ie/\hbar c \int_2 dr \cdot A} \qquad (13\text{-}74)$$

The flux thus causes a *relative change in phase* between ψ_1 and ψ_2, and this will
change the interference pattern. This effect, first pointed out by Aharanov and
Bohm, has been observed experimentally.[8]

Problems

1. Show that with

$$H = \frac{\mathbf{p}^2}{2\mu} + V(r)$$

the equations

$$\frac{dx}{dt} = \frac{\partial H}{\partial p_x}, \cdots$$

$$\frac{dp_x}{dt} = -\frac{\partial H}{\partial x}, \cdots$$

[8] R. G. Chambers, *Phys. Rev. Letters*, 5, 3 (1960).

yield the equations of motion

$$\mu \frac{d^2 x}{dt^2} = - \frac{\partial V}{\partial x}, \ldots$$

2. Show that the Hamiltonian

$$H = \frac{1}{2\mu} \left[\mathbf{p} + \frac{e}{c} \mathbf{A}(\mathbf{r},t) \right]^2 - \frac{e^2}{r}$$

yields the Lorentz force equation

$$\mu \frac{d^2 \mathbf{r}}{dt^2} = - e \left[\mathbf{E}(\mathbf{r},t) + \frac{1}{c} \mathbf{v} \times \mathbf{B}(\mathbf{r},t) \right]$$

[*Note.* In your calculation use

$$\frac{d}{dt} \mathbf{A}(\mathbf{r},t) = \frac{\partial \mathbf{A}}{\partial t} + \frac{dx}{dt} \frac{\partial \mathbf{A}}{\partial x} + \frac{dy}{dt} \frac{\partial \mathbf{A}}{\partial y} + \frac{dz}{dt} \frac{\partial \mathbf{A}}{\partial z}$$

since the fields that enter into the equation of motion (and the Hamiltonian) must be evaluated at the position of the particle.]

3. Calculate the wavelengths of the three Zeeman lines in the $3D \to 2P$ transition in hydrogen, when the latter is in a field of 10^4 gauss.

4. Consider an electron confined to a region between two cylinders of radii a and b respectively ($b > a$). (a) Separate the Schrödinger equation in cylindrical coordinates (cf. Eq. 13-32), and show that the equation can be solved in terms of Bessel functions. What are the conditions for the determination of the energy eigenvalues? (b) Discuss the degeneracy of the energy eigenfunctions? What is it due to? For Bessel functions, see note after problem 8 below.

5. In this problem we work out an example showing how an enclosed magnetic flux changes the angular momentum of a particle in a region outside the flux tube. Consider a magnetic field confined in a cylindrical region $\rho < a$. Let the flux be Φ. In the region $\rho > a$ there is no magnetic field, and hence the vector potential is of the form

$$\mathbf{A}(\rho,\theta,z) = \nabla \Lambda(\rho,\theta,z)$$

(a) The choice of gauge $\nabla \cdot \mathbf{A} = 0$ implies that

$$\nabla^2 \Lambda = 0$$

Show that a solution of this equation, satisfying (13-70), is

$$\Lambda = \frac{1}{2\pi} \Phi \theta$$

(b) Calculate the angular momentum about the symmetry axis

$$(\mu \mathbf{r} \times \mathbf{v})_z = \tilde{L}_z = \left[\mathbf{r} \times \left(\frac{\hbar}{i} \boldsymbol{\nabla} + \frac{e}{c} \mathbf{A} \right) \right]_z$$

in cylindrical coordinates, and show that for the above Λ it is given by

$$\tilde{L}_z = \frac{\hbar}{i} \frac{\partial}{\partial \theta} + \frac{e}{c} \frac{\Phi}{2\pi}$$

(c) Solve the eigenvalue problem $\tilde{L}_z \psi = \lambda \psi$, and show that single-valuedness of the eigenfunctions leads to flux quantization.

6. Show that for a system described by the Hamiltonian

$$H = \frac{[\mathbf{p} + (e/c)\,\mathbf{A}(\mathbf{r},t)]^2}{2\mu}$$

the flux $\mathbf{j}$, which satisfies

$$\frac{\partial}{\partial t} \psi^* \psi + \boldsymbol{\nabla} \cdot \mathbf{j} = 0$$

is given by

$$\mathbf{j} = \frac{\hbar}{2i\mu} \left[\psi^* \boldsymbol{\nabla} \psi - \boldsymbol{\nabla} \psi^* \psi + \frac{2ie}{\hbar c} \mathbf{A}(\mathbf{r},t) \psi^* \psi \right]$$

Show also that the Hamiltonian equations of motion of Problem 1 imply that

$$\frac{d}{dt} \tilde{\mathbf{L}} = 0$$

where

$$\tilde{\mathbf{L}} = \mathbf{r} \times \left(\mathbf{p} + \frac{e}{c} \mathbf{A}(\mathbf{r},t) \right)$$

7. Consider the problem of a charged particle in an external magnetic field $\mathbf{B} = (0,0,B)$ with the gauge so chosen that $\mathbf{A} = (-yB, 0, 0)$. What are the constants of the motion? Go as far as you can in solving the equation of motion, and obtain the energy spectrum. Can you explain why the same problem in the gauges $\mathbf{A} = (-yB/2, xB/2, 0)$, $\mathbf{A} = (-yB, 0, 0)$, and $\mathbf{A} = (0, xB, 0)$ can still represent the same physical situation, even though the solutions look so different in all three cases?

8. Consider a charged particle in a magnetic field $\mathbf{B} = (0,0,B)$ and in a crossed electric field $\mathbf{E} = (E,0,0)$. Which of the three gauges mentioned in problem 7 would you use for this problem? Solve the eigenvalue problem.

Note. The solution of the equation

$$\frac{d^2u}{dz^2} + \frac{1}{z} \frac{du}{dz} + \left(1 - \frac{n^2}{z^2} \right) u = 0$$

with n integral, are known as Bessel functions, for the regular solutions

$$J_n(z) = \left(\frac{z}{2}\right)^n \sum_{l=0}^{\infty} \frac{(iz/2)^{2l}}{l!(n+l)!}$$

and Neumann functions for the irregular solutions

$$N_n(z) = \frac{2}{\pi} J_n(z) \log \frac{\gamma z}{2} - \frac{1}{\pi}\left(\frac{z}{2}\right)^n \sum_{l=0}^{\infty} \frac{(iz/2)^{2l}}{l!(n+l)!} a_{nl}$$

$$- \frac{1}{\pi}\left(\frac{z}{2}\right)^{-n} \sum_{l=0}^{n-1} \frac{(n-l-1)!}{l!}\left(\frac{z}{2}\right)^{2l}$$

$$(\log \gamma = 0.5772\ldots) \qquad a_{nl} = \left(\sum_{m=1}^{l} \frac{1}{m} + \sum_{m=1}^{l+n} \frac{1}{m}\right)$$

They have the asymptotic behavior

$$J_n(z) \sim \left(\frac{2}{\pi z}\right)^{1/2} \cos\left(z - \frac{n\pi}{2} - \frac{\pi}{4}\right)\left[1 + 0\left(\frac{1}{z^2}\right)\right]$$

$$N_n(z) \sim \left(\frac{2}{\pi z}\right)^{1/2} \sin\left(z - \frac{n\pi}{2} - \frac{\pi}{4}\right)\left[1 + 0\left(\frac{1}{z^2}\right)\right]$$

A detailed discussion of their properties may be found in any book on the special functions of mathematical physics.

References

The various aspects of electron motion in a magnetic field are very interestingly discussed in

R. P. Feynman, R. B. Leighton, and M. Sands, *The Feynman Lectures on Physics*, Vol. 3, Addison-Wesley, Inc., 1965.

chapter 14

Operators, Matrices, and Spin

A proper discussion of atoms is not possible without consideration of the spin of the electron. In spite of the suggestive name, this property of the electron has no classical analog, and, as will soon become evident, it must be treated by somewhat abstract methods. Fortunately we have some preparation for this further departure from a description closely tied to coordinate space, in that we discussed both the harmonic oscillator (Chapter 7) and the angular momentum eigenvalue problem

$$\mathbf{L}^2 Y_{lm} = \hbar^2 l(l + 1)\, Y_{lm}$$

$$L_z Y_{lm} = \hbar m Y_{lm} \tag{14-1}$$

by operator methods. For the harmonic oscillator we found states, defined by

$$u_n = \frac{1}{(n!\hbar^n)^{1/2}} (A^\dagger)^n\, u_0 \tag{14-2}$$

for which

$$Hu_n = \hbar\omega(n + \tfrac{1}{2})\, u_n \tag{14-3}$$

and we could also calculate the action of the raising and lowering operators on u_n,

$$A^\dagger u_n = \sqrt{(n + 1)\, \hbar}\; u_{n+1} \tag{14-4}$$

and

$$Au_n = \sqrt{n\hbar}\; u_{n-1} \tag{14-5}$$

We also showed that

$$\langle u_m | u_n \rangle = \delta_{mn} \tag{14-6}$$

a statement that can be made to hold for the eigenstates of any hermitian operator (H here). If we take the scalar product of (14-3) to (14-5) with u_m we find that

$$\langle u_m | Hu_n \rangle \equiv (u_m | H | u_n) = (n + \tfrac{1}{2})\, \hbar\omega\, \delta_{mn}$$

$$\langle u_m | A^\dagger u_n \rangle \equiv \langle u_m | A^\dagger | u_n \rangle = \sqrt{(n+1)\,\hbar}\; \delta_{m,n+1}$$

$$\langle u_m | A u_n \rangle \equiv \langle u_m | A | u_n \rangle = \sqrt{n\hbar}\; \delta_{m,n-1} \tag{14-7}$$

where we have introduced the more symmetric notation

$$\langle u_i | O u_j \rangle \equiv \langle u_i | O | u_j \rangle \tag{14-8}$$

These quantities may be arranged in arrays called *matrices*. The conventional notation for a matrix M_{ij} has the first index labeling the row, and the second labeling the column of the array. Thus if we convert the scalar product $\langle u_m | H | u_n \rangle$ into H_{mn} we find that

$$H = \hbar\omega \begin{pmatrix} 1/2 & 0 & 0 & 0 & \cdots \\ 0 & 3/2 & 0 & 0 & \cdots \\ 0 & 0 & 5/2 & 0 & \cdots \\ 0 & 0 & 0 & 7/2 & \cdots \\ \cdot & \cdot & \cdot & \cdot & \cdot \\ \cdot & \cdot & \cdot & \cdot & \quad\cdot \\ \cdot & \cdot & \cdot & \cdot & \qquad\cdot \end{pmatrix} \tag{14-9}$$

Similarly

$$A^\dagger = \sqrt{\hbar} \begin{pmatrix} 0 & 0 & 0 & 0 & \cdots \\ \sqrt{1} & 0 & 0 & 0 & \cdots \\ 0 & \sqrt{2} & 0 & 0 & \cdots \\ 0 & 0 & \sqrt{3} & 0 & \cdots \\ \cdot & \cdot & \cdot & \cdot \\ \cdot & \cdot & \cdot & \cdot \end{pmatrix} \tag{14-10}$$

and

$$A = \sqrt{\hbar} \begin{pmatrix} 0 & \sqrt{1} & 0 & 0 & \cdots \\ 0 & 0 & \sqrt{2} & 0 & \cdots \\ 0 & 0 & 0 & \sqrt{3} & \cdots \\ \cdot & \cdot & \cdot & \cdot \\ \cdot & \cdot & \cdot & \cdot \end{pmatrix} \tag{14-11}$$

We shall call the array $\langle u_m | F | u_n \rangle$, where F is any operator, and the u_i are any complete set, a matrix representation of F in the basis provided by the u_i. This appellation needs some justification. The product of two matrices, for example, satisfies

$$(FG)_{ij} = \sum_n (F)_{in}(G)_{nj} \tag{14-12}$$

and we need to verify this relation for the "matrix representations" of the operators F and G. To do this, let us consider the state Gu_j, and, using completeness, expand it in the form

$$Gu_j = \sum_n C_n u_n \tag{14-13}$$

The coefficients C_n are given by

$$C_n = \langle u_n | G | u_j \rangle \qquad (14\text{-}14)$$

Hence

$$\langle u_i | FG | u_j \rangle = \langle u_i | F (\sum_n C_n u_n) \rangle$$

$$= \sum_n C_n \langle u_i | F | u_n \rangle$$

$$= \sum_n \langle u_i | F | u_n \rangle \langle u_n | G | u_j \rangle \qquad (14\text{-}15)$$

which is the same as (14-12), provided we write

$$\langle u_i | F | u_n \rangle = F_{in} \qquad (14\text{-}16)$$

and so on. It is a useful mnemonic device to write the unit operator in the form

$$1 = \sum_n | u_n \rangle \langle u_n | \qquad (14\text{-}17)$$

and in that form it can be inserted between the two operators F and G in the matrix element $\langle u_i | FG | u_j \rangle$ to give (14-15).

Further justification for the matrix connection comes from the relation

$$\langle u_m | F | u_n \rangle^* = \langle F u_n | u_m \rangle = \langle u_n | F^\dagger | u_m \rangle \qquad (14\text{-}18)$$

which shows that if the operator F is represented by a matrix, then the hermitian conjugate operator $F^\dagger$ will be represented by the hermitian conjugate matrix, since the latter is defined by

$$(F^\dagger)_{nm} = F_{mn}^* \qquad (14\text{-}19)$$

Note that in our discussion we made no reference to the fact that we started out with eigenstates of the harmonic oscillator Hamiltonian. *The only thing that is special about them, is that they diagonalize the matrix representing H.* With another complete set, H would not be diagonal, and reading off its eigenvalues, that is, the matrix elements when it is diagonal, would not be easy.

The Y_{lm} were defined to be states that diagonalize $\mathbf{L}^2$ and L_z simultaneously. If we stay with a fixed l, that is, with states in which only the m-value is variable, then, with an abbreviated notation, the second of the relations (14-1) reads

$$\langle l,m' | L_z | lm \rangle = \hbar m \, \delta_{m'm} \qquad (14\text{-}20)$$

Furthermore (10-40) with (10-52) implies that

$$\langle l,m' | L_\pm | l,m \rangle = \hbar [l(l+1) - m(m \pm 1)]^{1/2} \, \delta_{m',m \pm 1} \qquad (14\text{-}21)$$

This leads to the matrix representations

$$L_z = \hbar \begin{pmatrix} 1 & 0 & 0 \\ 0 & 0 & 0 \\ 0 & 0 & -1 \end{pmatrix} \tag{14-22}$$

$$L_+ = \hbar \begin{pmatrix} 0 & \sqrt{2} & 0 \\ 0 & 0 & \sqrt{2} \\ 0 & 0 & 0 \end{pmatrix} \tag{14-23}$$

and

$$L_- = \hbar \begin{pmatrix} 0 & 0 & 0 \\ \sqrt{2} & 0 & 0 \\ 0 & \sqrt{2} & 0 \end{pmatrix} \tag{14-24}$$

for the $l = 1$ angular momentum operators. The rows and columns are labeled with $m = 1, 0, -1$ in order left to right and top to bottom. It is easy to check that the matrices satisfy the commutation relations. For example

$$[L_+, L_-] = \hbar^2 \begin{pmatrix} 0 & \sqrt{2} & 0 \\ 0 & 0 & \sqrt{2} \\ 0 & 0 & 0 \end{pmatrix} \begin{pmatrix} 0 & 0 & 0 \\ \sqrt{2} & 0 & 0 \\ 0 & \sqrt{2} & 0 \end{pmatrix} - \hbar^2 \begin{pmatrix} 0 & 0 & 0 \\ \sqrt{2} & 0 & 0 \\ 0 & \sqrt{2} & 0 \end{pmatrix} \begin{pmatrix} 0 & \sqrt{2} & 0 \\ 0 & 0 & \sqrt{2} \\ 0 & 0 & 0 \end{pmatrix}$$

$$= \hbar^2 \begin{pmatrix} 2 & 0 & 0 \\ 0 & 2 & 0 \\ 0 & 0 & 0 \end{pmatrix} - \hbar^2 \begin{pmatrix} 0 & 0 & 0 \\ 0 & 2 & 0 \\ 0 & 0 & 2 \end{pmatrix} = 2\hbar^2 \begin{pmatrix} 1 & 0 & 0 \\ 0 & 0 & 0 \\ 0 & 0 & -1 \end{pmatrix} = 2\hbar L_z \tag{14-25}$$

General relations between states can also be written in matrix representation. Consider, for example, a relation like

$$\psi = A\phi \tag{14-26}$$

If we take the scalar product of this with any member of a complete set u_i, we have

$$\langle u_i | \psi \rangle = \langle u_i | A\phi \rangle \tag{14-27}$$

Furthermore, the insertion of the unit operator, in the form (14-17) between A and ϕ yields

$$\langle u_i | \psi \rangle = \sum_n \langle u_i | A | u_n \rangle \langle u_n | \phi \rangle \tag{14-28}$$

If we write $\langle u_n | \phi \rangle$ as a column vector α_n

$$\langle u_n | \phi \rangle \rightarrow \begin{bmatrix} \langle u_1 | \phi \rangle \\ \langle u_2 | \phi \rangle \\ \langle u_3 | \phi \rangle \\ . \\ . \\ . \end{bmatrix} \equiv \begin{bmatrix} \alpha_1 \\ \alpha_2 \\ \alpha_3 \\ . \\ . \\ . \end{bmatrix} \tag{14-29}$$

and similarly

$$\langle u_n | \psi \rangle \rightarrow \begin{bmatrix} \langle u_1 | \psi \rangle \\ \langle u_2 | \psi \rangle \\ \langle u_3 | \psi \rangle \\ \cdot \\ \cdot \\ \cdot \end{bmatrix} \equiv \begin{bmatrix} \beta_1 \\ \beta_2 \\ \beta_3 \\ \cdot \\ \cdot \\ \cdot \end{bmatrix} \tag{14-30}$$

then the matrix representation of (14-26) is

$$\beta_i = \sum_n A_{in} \alpha_n \tag{14-31}$$

Thus matrices represent operators, and column vectors represent states. The scalar product $\langle \phi | u_n \rangle = \langle u_n | \phi \rangle^*$ is written conventionally in the form of a row

$$\langle \phi | u_n \rangle \rightarrow (\alpha_1^*, \alpha_2^*, \alpha_3^*, \ldots) \tag{14-32}$$

so that the scalar product $\langle \phi | \psi \rangle$, for example, can be written as

$$\langle \phi | \psi \rangle = \sum_n \langle \phi | u_n \rangle \langle u_n | \psi \rangle$$

$$= \sum_n \alpha_n^* \beta_n \tag{14-33}$$

An eigenvalue equation is a special case of (14-26). It reads

$$A\phi = a\phi \tag{14-34}$$

and it reads

$$\sum_n A_{in} \alpha_n = a\alpha_i \tag{14-35}$$

in matrix form. This is equivalent to

$$\begin{pmatrix} A_{11} - a & A_{12} & A_{13} & \cdots \\ A_{21} & A_{22} - a & A_{23} & \cdots \\ A_{31} & A_{32} & A_{33} - a & \cdots \\ \cdot & \cdot & \cdot & \\ \cdot & \cdot & \cdot & \\ \cdot & \cdot & \cdot & \end{pmatrix} \begin{pmatrix} \alpha_1 \\ \alpha_2 \\ \alpha_3 \\ \cdot \\ \cdot \\ \cdot \end{pmatrix} = 0 \tag{14-36}$$

and there will be a nontrivial solution of this equation only if the determinant of the matrix vanishes

$$\det | A_{in} - a\delta_{in} | = 0 \tag{14-37}$$

This is a good way of finding eigenvalues (and eigenvectors) for operators represented by finite matrices, but for infinite matrices this is unfortunately not so simple.

It is indeed fortunate that there is an alternative to representing operators by functions and differentials, since not all operators can be represented in that way. The simplest example is that corresponding to the angular momentum $l = \frac{1}{2}$. Eq. (10-51) and (10-60) tell us that

$$Y_{1/2,\pm1/2} = C_{\pm}\sqrt{\sin\theta}\, e^{\pm i\phi/2} \tag{14-38}$$

and (10-54) allows us to compute

$$L_- Y_{1/2,1/2} \propto \frac{\cos\theta}{\sqrt{\sin\theta}} e^{-i\phi/2} \tag{14-39}$$

This, however, is not proportional to $Y_{1/2,-1/2}$, and furthermore it is singular at $\theta = 0$ and π. Thus for $l = \frac{1}{2}$ there are troubles, and we must turn to matrix representations. Instead of talking about $l = \frac{1}{2}$, we shall talk about spin, $S = \frac{1}{2}$, reserving the letter l for the orbital angular momentum associated with $\mathbf{r} \times \mathbf{p}$. The spin operators are S_x, S_y, and S_z, and they are defined by their commutation relations

$$[S_x, S_y] = i\hbar S_z \tag{14-40}$$

and so on.

We wish to represent them by 2 x 2 matrices. (14-20) yields

$$S_z = \hbar \begin{pmatrix} 1/2 & 0 \\ 0 & -1/2 \end{pmatrix} \tag{14-41}$$

and (14-21) gives

$$S_+ = \hbar \begin{pmatrix} 0 & 1 \\ 0 & 0 \end{pmatrix} \qquad S_- = \hbar \begin{pmatrix} 0 & 0 \\ 1 & 0 \end{pmatrix} \tag{14-42}$$

We may write this representation as

$$\mathbf{S} = \tfrac{1}{2}\hbar\boldsymbol{\sigma} \tag{14-43}$$

where

$$\sigma_x = \begin{pmatrix} 0 & 1 \\ 1 & 0 \end{pmatrix} \qquad \sigma_y = \begin{pmatrix} 0 & -i \\ i & 0 \end{pmatrix} \qquad \sigma_z = \begin{pmatrix} 1 & 0 \\ 0 & -1 \end{pmatrix} \tag{14-44}$$

are the *Pauli matrices*. They satisfy the commutation relations

$$[\sigma_x, \sigma_y] = 2i\sigma_z \tag{14-45}$$

and so on, as they must, to satisfy (14-40), and they also satisfy

$$\sigma_x{}^2 = \sigma_y{}^2 = \sigma_z{}^2 = \begin{pmatrix} 1 & 0 \\ 0 & 1 \end{pmatrix} \equiv \mathbf{1} \tag{14-46}$$

and

$$\sigma_x\sigma_y = -\sigma_y\sigma_x$$

$$\sigma_z\sigma_x = -\sigma_x\sigma_z$$

$$\sigma_y\sigma_z = -\sigma_z\sigma_y \qquad (14\text{-}47)$$

which are relations peculiar to the spin $\frac{1}{2}$ representations and do not hold for the $l = 1$ matrices, for example.

The eigenstates of S_z will be represented by a two component column vector, which we call *spinor*. To find these eigenspinors, we solve

$$S_z\begin{pmatrix} u \\ v \end{pmatrix} = \pm\tfrac{1}{2}\hbar\begin{pmatrix} u \\ v \end{pmatrix} \qquad (14\text{-}48)$$

that is,

$$\begin{pmatrix} 1 & 0 \\ 0 & -1 \end{pmatrix}\begin{pmatrix} u \\ v \end{pmatrix} = \pm\begin{pmatrix} u \\ v \end{pmatrix}$$

or

$$\begin{pmatrix} u \\ -v \end{pmatrix} = \pm\begin{pmatrix} u \\ v \end{pmatrix}$$

The plus eigensolution has $v = 0$, and the minus eigensolution has $u = 0$. We thus write

$$\chi_+ = \begin{pmatrix} 1 \\ 0 \end{pmatrix} \qquad \chi_- = \begin{pmatrix} 0 \\ 1 \end{pmatrix} \qquad (14\text{-}49)$$

for the eigenspinors corresponding to spin up $[S_z = +(1/2)\hbar]$ and spin down $[S_z = -(1/2)\hbar]$, respectively.

An arbitrary spinor can be expanded in this complete set

$$\begin{pmatrix} \alpha_+ \\ \alpha_- \end{pmatrix} = \alpha_+\begin{pmatrix} 1 \\ 0 \end{pmatrix} + \alpha_-\begin{pmatrix} 0 \\ 1 \end{pmatrix} \qquad (14\text{-}50)$$

and the expansion postulate yields the interpretation that $|\alpha_+|^2$ and $|\alpha_-|^2$, when properly normalized, so that

$$|\alpha_+|^2 + |\alpha_-|^2 = 1 \qquad (14\text{-}51)$$

yield the probabilities that a measurement of S_z on the state $\begin{pmatrix} \alpha^+ \\ \alpha_- \end{pmatrix}$ yields $+(1/2)\,\hbar$ and $-(1/2)\,\hbar$, respectively.

It is not necessary to keep S_z diagonal. If we look for the eigenstates of the operator $S_x \cos \phi + S_y \sin \phi$, we must solve

$$(S_x \cos \phi + S_y \sin \phi) \begin{pmatrix} u \\ v \end{pmatrix} = \tfrac{1}{2}\hbar\lambda \begin{pmatrix} u \\ v \end{pmatrix} \qquad (14\text{-}52)$$

that is,

$$\begin{pmatrix} 0 & \cos \phi - i \sin \phi \\ \cos \phi + i \sin \phi & 0 \end{pmatrix} \begin{pmatrix} u \\ v \end{pmatrix} = \lambda \begin{pmatrix} u \\ v \end{pmatrix}$$

This implies that

$$v\, e^{-i\phi} = \lambda u$$
$$u\, e^{i\phi} = \lambda v \qquad (14\text{-}53)$$

Hence

$$\lambda = \pm 1 \qquad (14\text{-}54)$$

The eigenvectors corresponding to $\lambda = +1$ and $\lambda = -1$ are

$$\frac{1}{\sqrt{2}} \begin{pmatrix} e^{-i\phi/2} \\ e^{i\phi/2} \end{pmatrix} \qquad \frac{1}{\sqrt{2}} \begin{pmatrix} e^{-i\phi/2} \\ -e^{i\phi/2} \end{pmatrix} \qquad (14\text{-}55)$$

respectively. It is interesting to observe that if we change ϕ to $\phi + 2\pi$ *the solutions change sign.* This is characteristic of odd half-integer spin wave functions (fermion states); although this does not violate quantum mechanics, since -1 is just a phase factor, it does mean that no classical macroscopic wave packet can be constructed that has odd half-integral angular momentum.

Given an arbitrary state α, the expectation value of $\mathbf{S}$ may be calculated. We have

$$\langle \alpha | \mathbf{S} | \alpha \rangle = \sum_i \sum_j \langle \alpha | i \rangle \langle i | \mathbf{S} | j \rangle \langle j | \alpha \rangle$$

or, equivalently,

$$(\alpha_+^*, \alpha_-^*)\, \mathbf{S} \begin{pmatrix} \alpha_+ \\ \alpha_- \end{pmatrix}$$

Thus

$$\langle S_x \rangle = (\alpha_+^*, \alpha_-^*)\, \tfrac{1}{2}\hbar \begin{pmatrix} 0 & 1 \\ 1 & 0 \end{pmatrix} \begin{pmatrix} \alpha_+ \\ \alpha_- \end{pmatrix}$$

$$= \tfrac{1}{2}\hbar(\alpha_+^*, \alpha_-^*)\begin{pmatrix} \alpha_- \\ \\ \alpha_+ \end{pmatrix} = \tfrac{1}{2}\hbar(\alpha_+^*\alpha_- + \alpha_-^*\alpha_+)$$

$$\langle S_y \rangle = \tfrac{1}{2}\hbar(\alpha_+^*, \alpha_-^*)\begin{pmatrix} 0 & -i \\ i & 0 \end{pmatrix}\begin{pmatrix} \alpha_+ \\ \alpha_- \end{pmatrix}$$

$$= \tfrac{1}{2}\hbar(\alpha_+^*, \alpha_-^*)\begin{pmatrix} -i\alpha_- \\ \\ i\alpha_+ \end{pmatrix} = -\frac{i\hbar}{2}(\alpha_+^*\alpha_- - \alpha_-^*\alpha_+)$$

$$\langle S_z \rangle = \tfrac{1}{2}\hbar(\alpha_+^*, \alpha_-^*)\begin{pmatrix} \alpha_+ \\ \\ -\alpha_- \end{pmatrix} = \tfrac{1}{2}\hbar(|\alpha_+|^2 - |\alpha_-|^2)$$

$$(14\text{-}56)$$

Note that all of these are real, as expected for hermitian operators.

We shall see later that the spin of an electron appears in the Hamiltonian for the hydrogen atom, for example, coupled to the orbital angular momentum. When an electron is localized at a crystal lattice site, for example, it is often possible to treat the spin as the only degree of freedom that the electron possesses. The electron will have an intrinsic magnetic dipole moment by virtue of its spin, and that magnetic moment[1] is

$$\mathbf{M} = -\frac{eg}{2mc}\mathbf{S} \tag{14-57}$$

where g, the gyromagnetic ratio, is very close to 2,

$$g = 2\left(1 + \frac{\alpha}{2\pi} + \dots\right) = 2.0023192 \tag{14-58}$$

and m is the electron mass.

For such a localized electron, the Hamiltonian in the presence of an external magnetic field B is just the potential energy

$$H = -\mathbf{M}\cdot\mathbf{B} = \frac{eg\hbar}{4mc}\mathbf{\sigma}\cdot\mathbf{B} \tag{14-59}$$

The Schrödinger equation for the state $\psi(t) = \begin{bmatrix} \alpha_+(t) \\ \alpha_-(t) \end{bmatrix}$ is

[1] A "classical" electron moving in a circle with angular momentum $\mathbf{L}$ will form a current loop whose magnetic moment is $\mathbf{M} = -e\mathbf{L}/2mc$. Since the spin is a purely quantum-mechanical variable, one can argue (14-57) only by analogy. For its justification one needs the relativistic Dirac equation from which the value $g = 2$ also emerges. The corrections to $g = 2$ come from quantum electrodynamics. The nonclassical aspects of spin were pointed out by its discoverers, S. Goudsmit and G. Uhlenbeck (1925).

$$i\hbar \frac{d\psi(t)}{dt} = \frac{eg\hbar}{4mc} \, \mathbf{d} \cdot \mathbf{B}\psi(t) \tag{14-60}$$

If **B** is taken to define the z-axis, and if we write

$$\psi(t) = \begin{bmatrix} \alpha_+(t) \\ \alpha_-(t) \end{bmatrix} = e^{-i\omega t} \begin{bmatrix} \alpha_+ \\ \alpha_- \end{bmatrix} \tag{14-61}$$

then the equation becomes

$$\hbar\omega \begin{pmatrix} \alpha_+ \\ \alpha_- \end{pmatrix} = \frac{eg\hbar B}{4mc} \begin{pmatrix} 1 & 0 \\ 0 & -1 \end{pmatrix} \begin{pmatrix} \alpha_+ \\ \alpha_- \end{pmatrix} \tag{14-62}$$

The solutions correspond to different frequencies ω. We have, for $\omega = egB/4mc$, $\begin{pmatrix} \alpha_+ \\ \alpha_- \end{pmatrix} = \begin{pmatrix} 1 \\ 0 \end{pmatrix}$, and for $\omega = -(egB/4mc)$, $\begin{pmatrix} \alpha_+ \\ \alpha_- \end{pmatrix} = \begin{pmatrix} 0 \\ 1 \end{pmatrix}$. Thus, if the initial state is

$$\psi(0) = \begin{pmatrix} a \\ b \end{pmatrix} \tag{14-63}$$

then the state at a later time will be

$$\psi(t) = \begin{pmatrix} a\,e^{-i\omega t} \\ b\,e^{i\omega t} \end{pmatrix} \qquad \omega = \frac{geB}{4mc} \tag{14-64}$$

Suppose that at $t = 0$ the spin is an eigenstate of S_x with eigenvalue $+(1/2)\,\hbar$, that is, it "points in the x-direction." This means that

$$\tfrac{1}{2}\hbar \begin{pmatrix} 0 & 1 \\ 1 & 0 \end{pmatrix} \begin{pmatrix} a \\ b \end{pmatrix} = \tfrac{1}{2}\hbar \begin{pmatrix} a \\ b \end{pmatrix}$$

that is, $\begin{pmatrix} a \\ b \end{pmatrix} = \dfrac{1}{\sqrt{2}} \begin{pmatrix} 1 \\ 1 \end{pmatrix}$. Then, at a later time

$$\langle S_x \rangle = \tfrac{1}{2}\hbar \, \frac{1}{\sqrt{2}} \, (e^{i\omega t}, e^{-i\omega t}) \begin{pmatrix} 0 & 1 \\ 1 & 0 \end{pmatrix} \frac{1}{\sqrt{2}} \begin{pmatrix} e^{-i\omega t} \\ e^{i\omega t} \end{pmatrix}$$

$$= \frac{\hbar}{4} (e^{i\omega t}, e^{-i\omega t}) \begin{pmatrix} e^{i\omega t} \\ e^{-i\omega t} \end{pmatrix} = \frac{\hbar}{2} \cos 2\omega t \tag{14-65}$$

Similarly

$$\langle S_y \rangle = \tfrac{1}{2}\hbar \frac{1}{\sqrt{2}} (e^{i\omega t}, e^{-i\omega t}) \begin{pmatrix} 0 & -i \\ i & 0 \end{pmatrix} \frac{1}{\sqrt{2}} \begin{pmatrix} e^{-i\omega t} \\ e^{i\omega t} \end{pmatrix}$$

$$= \frac{\hbar}{4} (-i\, e^{2i\omega t} + i\, e^{-2i\omega t})$$

$$= \frac{\hbar}{2} \sin 2\omega t \tag{14-66}$$

Thus the spin precesses about the z-axis, the direction of B, with frequency

$$2\omega = \frac{egB}{2mc} \cong \frac{eB}{mc} \tag{14-67}$$

In a solid the gyromagnetic factor g of an electron is affected by the nature of the forces acting in the solid. A knowledge of g provides very useful constraints on what these forces could be, and it is therefore important to be able to measure g. This can be done by the *paramagnetic resonance method*, which we now describe.

Consider an electron, whose only degrees of freedom are the spin states, under the influence of a large magnetic field B_0 pointing in the z-direction, and constant in time, and a small oscillating field $B_1 \cos \omega t$, pointing in the x-direction. The Schrödinger equation now reads

$$i\hbar \frac{d}{dt} \begin{bmatrix} a(t) \\ b(t) \end{bmatrix} = \frac{eg\hbar}{4mc} \begin{pmatrix} B_0 & B_1 \cos \omega t \\ B_1 \cos \omega t & -B_0 \end{pmatrix} \begin{bmatrix} a(t) \\ b(t) \end{bmatrix} \tag{14-68}$$

or, with

$$\omega_0 = \frac{egB_0}{4mc} \qquad \omega_1 = \frac{egB_1}{4mc} \tag{14-69}$$

$$i\, \frac{da(t)}{dt} = \omega_0 a(t) + \omega_1 \cos \omega t\, b(t)$$

$$i\, \frac{db(t)}{dt} = \omega_1 \cos \omega t\, a(t) - \omega_0 b(t) \tag{14-70}$$

Let

$$A(t) = a(t)\, e^{i\omega_0 t}$$

$$B(t) = b(t)\, e^{-i\omega_0 t} \tag{14-71}$$

These satisfy the equations

$$i\, \frac{dA(t)}{dt} = \omega_1 \cos \omega t\, B(t)\, e^{2i\omega_0 t}$$

$$\approx \tfrac{1}{2}\,\omega_1\,e^{i(2\omega_0-\omega)t}\,B(t)$$

$$i\,\frac{dB(t)}{dt} = \omega_1\cos\omega t\,A(t)\,e^{-2i\omega_0 t}$$

$$\approx \tfrac{1}{2}\,\omega_1\,e^{-i(2\omega_0-\omega)t}\,A(t) \tag{14-72}$$

In obtaining these, we made an approximation. We wrote

$$\cos\omega t\,e^{2i\omega_0 t} = \tfrac{1}{2}\left[e^{i(2\omega_0+\omega)t} + e^{i(2\omega_0-\omega)t}\right]$$

$$\approx \tfrac{1}{2}\,e^{i(2\omega_0-\omega)t}$$

Since we will be interested in values of $\omega = 2\omega_0$, and since both are large, the term that has been dropped oscillates very rapidly, and we may expect that its contribution averages to zero. A more detailed treatment supports this observation. We may eliminate $B(t)$:

$$B(t) = \frac{2i}{\omega_1}\,\frac{dA(t)}{dt}\,e^{-i(2\omega_0-\omega)t} \tag{14-73}$$

and use this to obtain a second order differential equation for $A(t)$:

$$\frac{d^2A(t)}{dt^2} - i(2\omega_0 - \omega)\,\frac{dA(t)}{dt} + \frac{\omega_1^2}{4}\,A(t) = 0 \tag{14-74}$$

A trial solution is

$$A(t) = A(0)\,e^{i\lambda t} \tag{14-75}$$

When this is inserted into (14-74), the roots of the equation

$$-\lambda^2 + (2\omega_0 - \omega)\,\lambda + \frac{\omega_1^2}{4} = 0$$

that is,

$$\lambda_\pm = \frac{2\omega_0 - \omega \pm \sqrt{(2\omega_0 - \omega)^2 + \omega_1^2}}{2} \tag{14-76}$$

determine λ.

The most general solution is

$$A(t) = A_+\,e^{i\lambda_+ t} + A_-\,e^{i\lambda_- t} \tag{14-77}$$

and hence

$$B(t) = -\,\frac{2}{\omega_1}\,e^{-i(2\omega_0-\omega)t}\,(\lambda_+ A_+\,e^{i\lambda_+ t} + \lambda_- A_-\,e^{i\lambda_- t}) \tag{14-78}$$

This finally yields

$$a(t) = e^{-i\omega_0 t}\,(A_+\,e^{i\lambda_+ t} + A_-\,e^{i\lambda_- t})$$

$$b(t) = -\,\frac{2}{\omega_1}\,e^{-i(\omega_0-\omega)t}\,(\lambda_+ A_+\,e^{i\lambda_+ t} + \lambda_- A_-\,e^{i\lambda_- t}) \tag{14-79}$$

If at $t = 0$ the electron spin points in the positive z-direction, then $a(0) = 1$ and $b(0) = 0$, that is,

$$A_+ + A_- = 1$$

$$\lambda_+ A_+ + \lambda_- A_- = 0$$

so that

$$A_+ = \frac{\lambda_-}{\lambda_- - \lambda_+}$$

$$A_- = -\frac{\lambda_+}{\lambda_- - \lambda_+} \qquad (14\text{-}80)$$

The probability that at some later time t the spin points in the negative z-direction is $|b(t)|^2$:

$$|b(t)|^2 = \frac{4}{\omega_1^2} \left| \frac{\lambda_+ \lambda_-}{\lambda_- - \lambda_+} e^{i\lambda_+ t} - \frac{\lambda_+ \lambda_-}{\lambda_- - \lambda_+} e^{i\lambda_- t} \right|^2$$

$$= \frac{\omega_1^2}{(2\omega_0 - \omega)^2 + \omega_1^2} \left| 1 - e^{-i(\lambda_+ - \lambda_-)t} \right|^2$$

$$= \frac{\omega_1^2}{(2\omega_0 - \omega)^2 + \omega_1^2} \frac{1 - \cos\sqrt{(2\omega_0 - \omega)^2 + \omega_1^2}\, t}{2} \qquad (14\text{-}81)$$

This quantity is small, since $\omega_1 \ll \omega,\ \omega_0$. When the frequency of the field B_1 is "tuned" to match $2\omega_0$, then the probability becomes

$$|b(t)|^2 \rightarrow \frac{1 - \cos\omega_1 t}{2} \qquad (14\text{-}82)$$

that is, it approaches unity. Since the energy of the "up" state is different from that of the "down" state, such an energy difference, absorbed from the external field, signals the resonance frequency, so that ω_0, and hence g can be measured with great precision.

Problems

1. If the ground state vector for the harmonic oscillator is given by

$$u_0 = \begin{pmatrix} 1 \\ 0 \\ 0 \\ \cdot \\ \cdot \\ \cdot \end{pmatrix}$$

use (14-2) and (14-10) to calculate u_1, u_2, u_3. What is the general pattern? Satisfy yourself that

$$\langle u_m | u_n \rangle = \delta_{mn}$$

2. Given a vector

$$\psi = \frac{1}{\sqrt{6}} \begin{pmatrix} 1 \\ 2 \\ 1 \\ 0 \\ \cdot \\ \cdot \\ \cdot \end{pmatrix}$$

calculate with the harmonic oscillator operators (14-9), (14-10), (14-11) the quantities

(a) $\langle H \rangle$.

(b) $\langle x^2 \rangle$, $\langle x \rangle$, $\langle p^2 \rangle$, $\langle p \rangle$.

(c) Use this to calculate $\Delta p \, \Delta x$.

[*Note.* The expression for p and x in terms of A and $A^\dagger$ are to be found in (7-4).]

3. Calculate the top left 4 x 4 corner of the matrix representation of x^4 for the harmonic oscillator.

4. Use (14-20) and (14-21) to calculate the matrix representation of L_x, L_y, and L_z for angular momentum 3/2, Check that the commutation relations

$$[L_x, L_y] = i\hbar \, L_z$$

and so on are satisfied.

5. You are given the Hamiltonian

$$H = \frac{1}{2I_1} L_x^2 + \frac{1}{2I_2} L_y^2 + \frac{1}{2I_3} L_z^2$$

Find the eigenvalues of H (a) when the angular momentum of the system is 1; (b) when the angular momentum of the system is 2.

(*Note.* The matrix representations of L_x, L_y, L_z for angular momentum 2 are obtainable from

$$L_z = \hbar \begin{pmatrix} 2 & 0 & 0 & 0 & 0 \\ 0 & 1 & 0 & 0 & 0 \\ 0 & 0 & 0 & 0 & 0 \\ 0 & 0 & 0 & -1 & 0 \\ 0 & 0 & 0 & 0 & -2 \end{pmatrix}$$

$$L_+ = \hbar \begin{pmatrix} 0 & 2 & 0 & 0 & 0 \\ 0 & 0 & \sqrt{6} & 0 & 0 \\ 0 & 0 & 0 & \sqrt{6} & 0 \\ 0 & 0 & 0 & 0 & 2 \\ 0 & 0 & 0 & 0 & 0 \end{pmatrix} \qquad L_- = (L_+)^\dagger$$

6. Calculate the eigenvalues of the matrix

$$H = \begin{pmatrix} 8 & 4 & 6 \\ 4 & 14 & 4 \\ 6 & 4 & 8 \end{pmatrix}$$

What are the eigenvectors?

7. Consider an angular momentum 1 system, represented by the state vector

$$u = \frac{1}{\sqrt{26}} \begin{pmatrix} 1 \\ 4 \\ -3 \end{pmatrix}$$

What is the probability that a measurement of L_x yields the value 0?

8. Consider a system of angular momentum 1: What are the eigenfunctions and eigenvalues of the operator $L_x L_y + L_y L_x$?

9. Consider a system of spin 1/2. What are the eigenvalues and eigenvectors of the operator $S_x + S_y$? Suppose a measurement of this operator is made, and the system is found to be in the state corresponding to the larger eigenvalue. What is the probability that a measurement of S_z yields $\hbar/2$?

10. The equation for the rate of change of an operator in the Heisenberg picture is given by Eq. (7-47). Consider the operators $S_x(t)$, . . . What are the equations of motion of these operators, if the Hamiltonian is given by

$$H = \frac{eg}{2mc} \mathbf{S}(t) \cdot \mathbf{B}$$

and the commutation relations are $[S_x(t), S_y(t)] = i\hbar S_z(t)$, and so on. If $\mathbf{B} = (0,0,B)$, solve for $\mathbf{S}(t)$ in terms of $\mathbf{S}(0)$.

11. A spin 1/2 object is in an eigenstate of S_x with eigenvalue $+\hbar/2$ at time $t = 0$. At that time it is placed in a magnetic field $\mathbf{B} = (0,0,B)$ in which it is allowed to precess for a time T. At that instant the magnetic field is very rapidly rotated in the y-direction, so that its components are $(0,B,0)$. After another time interval T a measurement of S_x is carried out. What is the probability that the value $\hbar/2$ will be found?

12. Work out the behavior of a spin 1 particle in an external magnetic field. Choose $\mathbf{B} = (0,0,B)$ and take the initial state to be an eigenstate of

$$\mathbf{S \cdot n} = S_x \sin \theta \cos \phi + S_y \sin \theta \sin \phi + S_z \cos \theta$$

with eigenvalues $\hbar$, 0, $-\hbar$ in succession.

[*Hint.* Use the matrix representations given by (14-22) to (14-24).]

References

The material on spin is standard, and discussions may be found in all of the books listed at the end of this volume.

chapter 15

The Addition of Angular Momenta

Suppose we have two electrons, whose spins are described by the operators S_1 and S_2. Each of these sets of operators satisfies the standard angular momentum commutation relations

$$[S_{1x}, S_{1y}] = i\hbar S_{1z}$$

and so on,

$$[S_{2x}, S_{2y}] = i\hbar S_{2z} \tag{15-1}$$

and so on, but the two sets of operators commute with each other, since the degrees of freedom associated with different particles are independent, that is,

$$[S_1, S_2] = 0 \tag{15-2}$$

Let us now define the total spin S by

$$S = S_1 + S_2 \tag{15-3}$$

The commutation relations obeyed by the components of S are

$$\begin{aligned}
[S_x, S_y] &= [S_{1x} + S_{2x}, S_{1y} + S_{2y}] \\
&= [S_{1x}, S_{1y}] + [S_{2x}, S_{2y}] \\
&= i\hbar(S_{1z} + S_{2z}) = i\hbar S_z \tag{15-4}
\end{aligned}$$

and so on. We are therefore justified in calling S the total *spin*. We may now determine the eigenvalues and eigenfunctions of S^2 and S_z.

The two-spin system actually has four states. If we denote the spinor of the first electron by $\chi_\pm^{(1)}$, so that

$$S_1{}^2 \chi_\pm^{(1)} = \tfrac{1}{2}(\tfrac{1}{2} + 1)\, \hbar^2 \chi_\pm^{(1)}$$
$$S_{1z} \chi_\pm^{(1)} = \pm\tfrac{1}{2}\hbar \chi_\pm^{(1)} \tag{15-5}$$

and similarly for the spinor $\chi_\pm^{(2)}$ of the second electron, then the four states are

$$\chi_+^{(1)}\chi_+^{(2)}, \; \chi_+^{(1)}\chi_-^{(2)}, \; \chi_-^{(1)}\chi_+^{(2)}, \; \chi_-^{(1)}\chi_-^{(2)} \tag{15-6}$$

The eigenvalues of S_z for the four states are

$$S_z\chi_\pm^{(1)}\chi_\pm^{(2)} = (S_{1z} + S_{2z})\, \chi_\pm^{(1)}\chi_\pm^{(2)}$$
$$= (S_{1z}\chi_\pm^{(1)})\, \chi_\pm^{(2)} + \chi_\pm^{(1)}(S_{2z}\chi_\pm^{(2)})$$

that is,

$$S_z\chi_+^{(1)}\chi_+^{(2)} = \hbar\chi_+^{(1)}\chi_+^{(2)}$$
$$S_z\chi_+^{(1)}\chi_-^{(2)} = S_z\chi_-^{(1)}\chi_+^{(2)} = 0$$
$$S_z\chi_-^{(1)}\chi_-^{(2)} = -\hbar\chi_-^{(1)}\chi_-^{(2)} \tag{15-7}$$

There are two states with m-value 0. One might expect that one linear combination of them will form an $S = 1$ state, to form a triplet with the $m = 1$ and $m = -1$ states, and the orthogonal combination will form a singlet $S = 0$ state. To check this expectation, let us construct the lowering operator

$$S_- = S_{1-} + S_{2-} \tag{15-8}$$

and apply this to the $m = 1$ state. This should give us the $m = 0$ state that belongs to the $S = 1$ triplet, aside from a coefficient in front. Indeed, using the fact that

$$S_-^{(i)}\chi_+^{(i)} = \hbar\chi_-^{(i)} \tag{15-9}$$

which can be established by noting that

$$\frac{1}{2}\hbar\left[\begin{pmatrix} 0 & 1 \\ 1 & 0 \end{pmatrix} - i\begin{pmatrix} 0 & -i \\ i & 0 \end{pmatrix}\right]\begin{pmatrix} 1 \\ 0 \end{pmatrix} = \hbar\begin{pmatrix} 0 \\ 1 \end{pmatrix} \tag{15-10}$$

we get

$$S_-\chi_+^{(1)}\chi_+^{(2)} = (S_{1-}\chi_+^{(1)})\, \chi_+^{(2)} + \chi_+^{(1)}S_{2-}\chi_+^{(2)}$$
$$= \hbar\chi_-^{(1)}\chi_+^{(2)} + \hbar\chi_+^{(1)}\chi_-^{(2)}$$
$$= \sqrt{2}\hbar\, \frac{\chi_+^{(1)}\chi_-^{(2)} + \chi_-^{(1)}\chi_+^{(2)}}{\sqrt{2}} \tag{15-11}$$

The linear combination has been normalized, and the compensating factor in front, $\sqrt{2}\hbar$, agrees with what one would expect from (10-36) and (10-48) with $l = m = 1$. If we now apply S_- to this linear combination, and note that

$$S_-^{(i)}\chi_-^{(i)} = 0 \tag{15-12}$$

we get

$$S_- \frac{\chi_+^{(1)}\chi_-^{(2)} + \chi_-^{(1)}\chi_+^{(2)}}{\sqrt{2}} = \frac{\hbar}{\sqrt{2}} (\chi_-^{(1)}\chi_-^{(2)} + \chi_-^{(1)}\chi_-^{(2)})$$

$$= \sqrt{2}\hbar\chi_-^{(1)}\chi_-^{(2)} \qquad (15\text{-}13)$$

as we should, for an angular momentum state $S = 1$.

The remaining state, constructed to be orthogonal to (15-11) and properly normalized, is

$$\frac{1}{\sqrt{2}} (\chi_+^{(1)}\chi_-^{(2)} - \chi_-^{(1)}\chi_+^{(2)}) \qquad (15\text{-}14)$$

and because it has no partners, we conjecture that it is an $S = 0$ state. In order to check this, we compute $\mathbf{S}^2$ for the two states

$$X_\pm = \frac{1}{\sqrt{2}} (\chi_+^{(1)}\chi_-^{(2)} \pm \chi_-^{(1)}\chi_+^{(2)}) \qquad (15\text{-}15)$$

We have

$$\mathbf{S}^2 = (\mathbf{S}_1 + \mathbf{S}_2)^2 = \mathbf{S}_1{}^2 + \mathbf{S}_2{}^2 + 2\mathbf{S}\cdot\mathbf{S}_2$$

$$= \mathbf{S}_1{}^2 + \mathbf{S}_2{}^2 + 2S_{1z}S_{2z} + S_{1+}S_{2-} + S_{1-}S_{2+} \qquad (15\text{-}16)$$

First of all,

$$\mathbf{S}_1{}^2 X_\pm = \frac{1}{\sqrt{2}} (\chi_-^{(2)}\mathbf{S}_1{}^2\chi_+^{(1)} \pm \chi_+^{(2)}\mathbf{S}_1{}^2\chi_-^{(1)})$$

$$= \tfrac{3}{4}\hbar^2 X_\pm \qquad (15\text{-}17)$$

and similarly

$$\mathbf{S}_2{}^2 X_\pm = \tfrac{3}{4}\hbar^2 X_\pm \qquad (15\text{-}18)$$

Next, we calculate

$$2S_{1z}S_{2z}X_\pm = 2(\tfrac{1}{2}\hbar)(-\tfrac{1}{2}\hbar) X_\pm = -\tfrac{1}{2}\hbar^2 X_\pm \qquad (15\text{-}19)$$

Finally

$$(S_{1+}S_{2-} + S_{1-}S_{2+}) X_\pm = \frac{1}{\sqrt{2}} (S_{1+}\chi_+^{(1)}S_{2-}\chi_-^{(2)} + S_{1-}\chi_+^{(1)}S_{2+}\chi_-^{(2)}$$

$$\pm S_{1+}\chi_-^{(1)}S_{2-}\chi_+^{(2)} \pm S_{1-}\chi_-^{(1)}S_{2+}\chi_+^{(2)})$$

which, with the help of (15-9) and (15-12) yields

$$(S_{1+}S_{2-} + S_{1-}S_{2+}) X_\pm = \pm\hbar^2 X_\pm \qquad (15\text{-}20)$$

Thus

$$\mathbf{S}^2 X_\pm = \hbar^2(\tfrac{3}{4} + \tfrac{3}{4} - \tfrac{1}{2} \pm 1)\, X_\pm = \begin{pmatrix} 2 \\ \\ 0 \end{pmatrix} \hbar^2 X_\pm$$

$$= \hbar^2 S(S+1)\, X_\pm \tag{15-21}$$

with $S = 1$ and 0 corresponding to the $\pm$ states.

What we have shown is that the totality of the four states of two spin 1/2 particles may be recombined into a triplet and into a singlet total spin state. For free spins, the two descriptions are completely equivalent. If, however, we have a physical system in which the forces depend on the spin, the eigenfunctions of the individual spins are no longer simultaneous eigenfunctions of H and, say, $\mathbf{S}_1^2$, S_{1z}, $\mathbf{S}_2^2$, S_{2z}, but they may be simultaneous eigenfunctions of H, $\mathbf{S}^2$, S_z, $\mathbf{S}_1^2$, and $\mathbf{S}_2^2$. This is most easily seen in an example.

If we have a potential between two electrons that depends on the spin, so that

$$V(r) = V_1(r) + \frac{1}{\hbar^2}\, \mathbf{S}_1 \cdot \mathbf{S}_2 V_2(r) \tag{15-22}$$

we can easily see that S_{1z} and S_{2z} do not commute with the second term, so that the eigenstates of H containing this potential cannot just be simple products of eigenstates of S_{1z} and S_{2z}. If we observe, however, that

$$\mathbf{S}_1 \cdot \mathbf{S}_2 = \tfrac{1}{2}(\mathbf{S}^2 - \mathbf{S}_1^2 - \mathbf{S}_2^2) \tag{15-23}$$

so that this term can be replaced by the eigenvalue, when acting on an eigenfunction of $\mathbf{S}^2$, $\mathbf{S}_1^2$, and $\mathbf{S}_2^2$, then

$$V(r) = V_1(r) + \frac{1}{2}\, V_2(r) \left[S(S+1) - \frac{3}{2} \right]$$

$$= V_1(r) + \frac{1}{4} \begin{pmatrix} 1 \\ \\ -3 \end{pmatrix} V_2(r) \begin{cases} S = 1 \\ S = 0 \end{cases} \tag{15-24}$$

Such a spin-dependent potential is actually observed in the neutron-proton system. The bound state is an $S = 1$ state—this is the deuteron—but there is also an unbound $S = 0$ state, which is only possible if $V_2(r) \neq 0$.

Much more important for future applications is the combination of a spin with an orbital angular momentum. Since $\mathbf{L}$ depends on spatial coordinates and $\mathbf{S}$ does not, they commute

$$[\mathbf{L}, \mathbf{S}] = 0 \tag{15-25}$$

It is therefore evident that the components of the total angular momentum $\mathbf{J}$, defined by

$$\mathbf{J} = \mathbf{L} + \mathbf{S} \tag{15-26}$$

will satisfy the angular momentum commutation relations. We can now ask for linear combinations of the Y_{lm} and $\chi_{\pm}$ that form eigenstates of

$$J_z = L_z + S_z \tag{15-27}$$

and

$$\mathbf{J}^2 = (\mathbf{L}^2 + 2\mathbf{L}\cdot\mathbf{S} + \mathbf{S}^2)$$
$$= \mathbf{L}^2 + \mathbf{S}^2 + 2L_zS_z + L_+S_- + L_-S_+ \tag{15-28}$$

Let us consider the linear combination

$$\psi_{j,m+1/2} = \alpha Y_{lm}\chi_+ + \beta Y_{l,m+1}\chi_- \tag{15-29}$$

It is, by construction, an eigenfunction of J_z with eigenvalue $(m + \frac{1}{2})\hbar$. We now determine α and β such that it is also an eigenfunction of J^2. We shall make use of the fact that

$$L_+Y_{lm} = [l(l + 1) - m(m + 1)]^{1/2}\hbar Y_{l,m+1}$$
$$= [(l + m + 1)(l - m)]^{1/2}\hbar Y_{l,m+1}$$
$$L_-Y_{lm} = [(l - m + 1)(l + m)]^{1/2}\hbar Y_{l,m-1}$$
$$S_+\chi_+ = S_-\chi_- = 0 \qquad S_\pm\chi_\mp = \hbar\chi_\pm \tag{15-30}$$

Then

$$\mathbf{J}^2\psi_{j,m+1/2} = \alpha\hbar^2\left\{l(l + 1)\ Y_{lm}\chi_+ + \tfrac{3}{4}Y_{lm}\chi_+ + 2m(\tfrac{1}{2})\ Y_{lm}\chi_+ \right.$$
$$+ [(l - m)(l + m + 1)]^{1/2}\ Y_{l,m+1}\chi_-\} + \beta\hbar^2\{l(l + 1)\ Y_{l,m+1}\chi_-$$
$$+ \tfrac{3}{4}Y_{l,m+1}\chi_- + 2(m + 1)(-\tfrac{1}{2})\ Y_{l,m+1}\chi_-$$
$$+ [(l - m)(l + m + 1)]^{1/2}\ Y_{lm}\chi_+\} \tag{15-31}$$

This will be of the form

$$\hbar^2 j(j + 1)\ \psi_{j,m+1/2} = \hbar^2 j(j + 1)(\alpha Y_{lm}\chi_+ + \beta Y_{l,m+1}\chi_-) \tag{15-32}$$

provided that

$$\alpha[l(l + 1) + \tfrac{3}{4} + m] + \beta[(l - m)(l + m + 1)]^{1/2} = j(j + 1)\ \alpha$$
$$\beta[l(l + 1) + \tfrac{3}{4} - m - 1] + \alpha[(l - m)(l + m + 1)]^{1/2} = j(j + 1)\ \beta$$
$$\tag{15-33}$$

This requires that

$$(l - m)(l + m + 1) = [j(j + 1) - l(l + 1) - \tfrac{3}{4} - m]$$
$$\times [j(j + 1) - l(l + 1) - \tfrac{3}{4} + m + 1]$$

which evidently has two solutions,

$$j(j + 1) - l(l + 1) - \tfrac{3}{4} = \begin{cases} -l - 1 \\ \\ l \end{cases} \qquad (15\text{-}34)$$

that is,

$$j = \begin{cases} l - \tfrac{1}{2} \\ \\ l + \tfrac{1}{2} \end{cases} \qquad (15\text{-}35)$$

For $j = l + 1/2$, we get, after a little algebra

$$\alpha = \sqrt{\frac{l + m + 1}{2l + 1}} \qquad \beta = \sqrt{\frac{l - m}{2l + 1}} \qquad (15\text{-}36)$$

(Actually we just get the ratio; these are already normalized forms). Thus

$$\psi_{l+1/2,m+1/2} = \sqrt{\frac{l + m + 1}{2l + 1}} Y_{lm}\chi_+ + \sqrt{\frac{l - m}{2l + 1}} Y_{l,m+1}\chi_- \qquad (15\text{-}37)$$

We can guess that the $j = l - 1/2$ solution must have the form

$$\psi_{l-1/2,m+1/2} = \sqrt{\frac{l - m}{2l + 1}} Y_{lm}\chi_+ - \sqrt{\frac{l + m + 1}{2l + 1}} Y_{l,m+1}\chi_- \qquad (15\text{-}38)$$

in order to be orthogonal to the $j = l + 1/2$ solution.

These two examples illustrate the general features that are involved in the addition of angular momenta: If we have the eigenstates $Y_{l_1 m_1}^{(1)}$ of $\mathbf{L}_1^2$ and L_{1z}, and the eigenstates $Y_{l_2 m_2}^{(2)}$ of $\mathbf{L}_2^2$ and L_{2z}, then we can form $(2l_1 + 1)(2l_2 + 1)$ product wave functions

$$Y_{l_1 m_1}^{(1)} Y_{l_2 m_2}^{(2)} \begin{Bmatrix} -l_1 \le m_1 \le l_1 \\ \\ -l_2 \le m_2 \le l_2 \end{Bmatrix} \qquad (15\text{-}39)$$

These may be classified by the eigenvalue of

$$J_z = L_{1z} + L_{2z} \qquad (15\text{-}40)$$

which is $m_1 + m_2$, and which ranges from a maximum value of $l_1 + l_2$ down to $-l_1 - l_2$. As in the simple cases discussed above, different linear combinations of functions with the same m value will belong to different values of j. In the table below we list the possible combinations for the special example of $l_1 = 4$, $l_2 = 2$. We shall use the simple abbreviation (m_1, m_2) for $Y_{l_1 m_1}^{(1)} Y_{l_2 m_2}^{(2)}$.

m-value	m_1, m_2 combinations	number
6	(4,2)	1
5	(4,1) (3,2)	2
4	(4,0) (3,1) (2,2)	3
3	(4,−1) (3,0) (2,1) (1,2)	4
2	(4,−2) (3,−1) (2,0) (1,1) (0,2)	5
1	(3,−2) (2,−1) (1,0) (0,1) (−1,2)	5
0	(2,−2) (1,−1) (0,0) (−1,1) (−2,2)	5
−1	(1,−2) (0,−1) (−1,0) (−2,1) (−3,2)	5
−2	(0,−2) (−1,−1) (−2,0) (−3,1) (−4,2)	5
−3	(−1,−2) (−2,−1) (−3,0) (−4,1)	4
−4	(−2,−2) (−3,−1) (−4,0)	3
−5	(−3,−2) (−4,−1)	2
−6	(−4,−2)	1

There are a total of 45 combinations, consistent with $(2l_1 + 1)(2l_2 + 1)$.

The highest state has total angular momentum $l_1 + l_2$ as can easily be checked by applying J^2 to $Y^{(1)}_{l_1 l_1} Y^{(2)}_{l_2 l_2}$:

$$\mathbf{J}^2 Y^{(1)}_{l_1 l_1} Y^{(2)}_{l_2 l_2} = (\mathbf{L}_1^2 + \mathbf{L}_2^2 + 2L_{1z}L_{2z} + L_{1+}L_{2-} + L_{1-}L_{2+}) Y^{(1)}_{l_1 l_1} Y^{(2)}_{l_2 l_2}$$

$$= \hbar^2[l_1(l_1 + 1) + l_2(l_2 + 1) + 2l_1 l_2] Y^{(1)}_{l_1 l_1} Y^{(2)}_{l_2 l_2}$$

$$= \hbar^2(l_1 + l_2)(l_1 + l_2 + 1) Y^{(1)}_{l_1 l_1} Y^{(2)}_{l_2 l_2} \qquad (15\text{-}41)$$

This is $j = 6$ in the example discussed in the table. Successive applications of

$$J_- = L_{1-} + L_{2-} \qquad (15\text{-}42)$$

will pick out one linear combination from each row in the table. These will form the 13 states that belong to $j = 6$. When this is done, there remains a single state with $m = 5$, two with $m = 4, \ldots$, one with $m = -5$. It is extremely plausible, and can, in fact, be checked, that the $m = 5$ state belongs to $j = 5$. Again successive applications of J_- pick out another linear combination from each row in the table, forming 11 states that belong to $j = 5$. Repetition of this procedure shows that we get, after this, sets that belong to $j = 4, j = 3$, and finally $j = 2$. The multiplicities add up to 45:

$$13 + 11 + 9 + 7 + 5 = 45$$

We shall not work out the details of this decomposition, as it is beyond the scope of this book. We merely state the results.

(a) The products $Y^{(1)}_{l_1 m_1} Y^{(2)}_{l_2 m_2}$ can be decomposed into eigenstates of $\mathbf{J}^2$, with eigenvalues $j(j + 1)\hbar^2$, where j can take on the values

$$j = l_1 + l_2, l_1 + l_2 - 1, \ldots |l_1 - l_2| \qquad (15\text{-}43)$$

(b) It is possible to generalize (15-37) and (15-38) to give the Clebsch-Gordan series

$$\psi_{jm} = \sum_{m1} C(jm;\, l_1 m_1 l_2 m_2)\, Y_{l_1 m_1}^{(1)} Y_{l_2 m_2}^{(2)} \qquad (15\text{-}44)$$

The coefficients $C(jm;\, l_1 m_1 l_2 m_2)$ are known as Wigner coefficients, and have been tabulated for many values of the arguments. We shall use only the coefficients for $l_2 = 1/2$, which we have calculated explicitly.

We can verify that the multiplicities check in (15-43): if we sum the number of states we get $(l_1 \geq l_2)$

$$[2(l_1 + l_2) + 1] + [2(l_1 + l_2 - 1) + 1] + \ldots + [2(l_1 - l_2) + 1]$$

$$= \sum_{n=0}^{2l_2} [2(l_1 - l_2 + n) + 1]$$

$$= (2l_2 + 1)(2l_1 + 1) \qquad (15\text{-}45)$$

A final comment is in order. We noted, when discussing identical particles, that a system of two electrons (or more generally, two fermions) must be in a state that is antisymmetric under the interchange of the two particles. This interchange involves not only the exchange of the spatial coordinates, but also of the spin labels. For a system of two identical spin $1/2$ particles, the $S = 1$ triplet of states

$$\frac{1}{\sqrt{2}} \begin{matrix} \chi_+^{(1)} \chi_+^{(2)} \\ (\chi_+^{(1)} \chi_-^{(2)} + \chi_-^{(1)} \chi_+^{(2)}) \\ \chi_-^{(1)} \chi_-^{(2)} \end{matrix} \qquad (15\text{-}46)$$

is symmetric under spin label interchange, while the $S = 0$ (singlet)

$$\frac{1}{\sqrt{2}} (\chi_+^{(1)} \chi_-^{(2)} - \chi_-^{(1)} \chi_+^{(2)}) \qquad (15\text{-}47)$$

is antisymmetric. Thus for a triplet state, the spatial wave function must be antisymmetric, and for a singlet state, it must be symmetric. The spatial wave function of a two-particle system in their center of mass system is of the general form

$$u(\mathbf{r}) = R_{nlm}(r)\, Y_{lm}(\theta,\phi) \qquad (15\text{-}48)$$

An interchange of the coordinates of the two particles is equivalent to the change

$$\begin{aligned} r &\to r \\ \theta &\to \pi - \theta \\ \phi &\to \phi + \pi \end{aligned} \qquad (15\text{-}49)$$

Thus the radial function remains unchanged. However under this transformation

$$Y_{lm}(\theta,\phi) \rightarrow Y_{lm}(\pi - \theta, \phi + \pi)$$
$$= (-1)^l \, Y_{lm}(\theta,\phi) \qquad (15\text{-}50)$$

Thus triplet states must have odd orbital angular momentum l, and singlet states must have even orbital angular momentum. We shall see an application of this when we discuss the states of helium.

An interesting application of these remarks occurs in elementary particle physics. One of the first highly unstable elementary particles to be discovered was the π meson predicted by Yukawa. This particle, which plays an important role in nuclear forces, comes in three charge states π^+, π^0, π^-. It was found to have spin 0, and the question arose whether the wave function of a pion—as this meson came to be called—was even or odd under reflection, assuming that the known particles, the proton and the neutron, had positive intrinsic parity. The following experiment was suggested.

Consider the capture of a π^- by a deuteron. A slow pion in liquid deuterium loses energy by a variety of mechanisms, till it finally ends up in the lowest Bohr orbit about the (pn) nucleus, and is then captured through the action of the nuclear forces. In the nuclear reaction

$$\pi^- + d \rightarrow n + n$$

the angular momentum is 1; the pion has zero spin, the orbital angular momentum is zero in the lowest Bohr state, so that the only contribution is the angular momentum of the deuteron, which is 1. The two neutrons must therefore be in an angular momentum 1 state. If the total spin of the two neutrons is 0, then the orbital angular momentum must be 1. If the total spin of the two-neutron state is 1, then orbital angular momentum 0, 1, and 2 is possible, since adding two angular momenta of one unit each can yield 0, 1, and 2, and adding one unit to two units of angular momentum can yield 3, 2, and 1. However a singlet state of two identical fermions must have even angular momentum, and is thus excluded. A triplet state must have odd orbital angular momentum, and this is possible if the orbital angular momentum is 1. Such a state, however, has odd parity by (15-50), and hence the pion must have odd parity. In terms of the spectroscopic notation, which we shall use, where a state is labeled according to

$$^{2S+1}L_J \qquad (15\text{-}51)$$

the two neutron states, from the total class of states $^1S_0$, $^1P_1$, $^1D_2$, $^1F_3$, $\ldots$, $^3S_1$, $^3P_2$, $^3P_1$, $^3P_0$, $^3D_3$, $^3D_2$, $^3D_1$, $^3F_4$, $^3F_3$, $^3F_2$, $\ldots$, are restricted to $^1S_0$, $^1D_2$, $\ldots$, $^3P_{2,1,0}$, $^3F_{4,3,2}$, $\ldots$ by the Fermi-Dirac statistics argument, and of these there is only one state, the $^3P_1$ state, that has angular momentum 1.

Problems

1. Work out the generalization of (15-37) and (15-38) to the addition of orbital angular momentum L to spin 1.

(a) Find the eigenstates of S^2 and S_z, where

$$S_z = \hbar \begin{pmatrix} 1 & 0 & 0 \\ 0 & 0 & 0 \\ 0 & 0 & -1 \end{pmatrix}$$

(b) If these eigenstates are labeled ξ_{+1}, ξ_0, and ξ_{-1}, find the action of S_+ and S_- on these states.

(c) Calculate the effect of

$$J^2 = L^2 + S^2 + 2L_zS_z + L_+S_- + L_-S_+$$

on combinations like

$$\Psi_{jm+1} = \alpha Y_{lm}\xi_1 + \beta Y_{l,m+1}\xi_0 + \gamma Y_{l,m+2}\xi_{-1}$$

(d) Determine the relations between α, β, and γ obtained from

$$J^2\Psi_{j,m} = \hbar^2 j\,(j+1)\,\Psi_{j,m}$$

2. Find the analog of (15-46) for two spin 1 particles, which can combine to form spin 2, 1, and 0 states. Use the notation $\xi_{+1}^{(i)}$, $\xi_0^{(i)}$, $\xi_{-1}^{(i)}$ for the one-particle spin vectors.

3. A deuteron has spin 1. What are the possible spin and total angular momentum states of two deuterons in an arbitrary angular momentum state L? Do not forget the Pauli principle.

4. A particle of spin 1 moves in a central potential of the form

$$V(r) = V_1(r) + S \cdot L V_2(r) + (S \cdot L)^2 V_3(r)$$

What are the values of $V(r)$ in the states $J = L + 1$, L, and $L - 1$?

5. Consider the discussion of the determination of the parity of the π^-. Suppose the π^- had spin 1, but was still captured in an $L = 0$ orbital state in the reaction

$$\pi^- + d \rightarrow 2n$$

What are the possible two-neutron states? Which states are allowed if the π^- had negative parity?

6. Suppose the π^- has spin 0 and negative parity, but is captured in the reaction

$$\pi^- + d \rightarrow 2n$$

from the P orbit. Show that the two neutrons must be in a singlet state.

7. The Hamiltonian of a spin system is given by

$$H = A + \frac{B\mathbf{S}_1 \cdot \mathbf{S}_2}{\hbar^2} + \frac{C(S_{1z} + S_{2z})}{\hbar}$$

Find the eigenvalues and eigenfunctions of the system of two particles, (a) when both particles have spin 1/2; (b) when one of the particles has spin 1/2 and the other has spin 1. Assume in (a) that the two particles are identical.

8. Consider two spin 1/2 particles, whose spins are described by the Pauli operators $\boldsymbol{\sigma}_1$ and $\boldsymbol{\sigma}_2$. Let $\hat{\mathbf{e}}$ be the unit vector connecting the two particles and define the operator

$$S_{12} = 3(\boldsymbol{\sigma}_1 \cdot \hat{\mathbf{e}})(\boldsymbol{\sigma}_2 \cdot \hat{\mathbf{e}}) - \boldsymbol{\sigma}_1 \cdot \boldsymbol{\sigma}_2$$

Show that if the two particles are in a $S = 0$ state (singlet) then

$$S_{12} X_{\text{singlet}} = 0$$

Show that for a triplet state

$$(S_{12} - 2)(S_{12} + 4)\, X_{\text{triplet}} = 0$$

References

The material discussed here is also discussed in one way or another in every textbook on quantum mechanics. Many details can be found in

M. E. Rose, *Elementary Theory of Angular Momentum*, John Wiley and Sons, Inc., 1957.

chapter 16

Time Independent
Perturbation Theory

There are few potentials $V(r)$ for which the Schrödinger equation is exactly solvable, and we have already discussed most of them. We must therefore develop approximation techniques to obtain the eigenvalues and eigenfunctions for potentials that do not lead to exactly soluble equations. In this chapter we discuss perturbation theory. We assume that we have found the eigenvalues and the complete set of eigenfunctions for a Hamiltonian H_0,

$$H_0 \phi_n = E_n{}^0 \phi_n \qquad (16\text{-}1)$$

and we ask for the eigenvalues and eigenfunctions for the Hamiltonian

$$H = H_0 + \lambda H_1 \qquad (16\text{-}2)$$

that is, for the solutions of

$$(H_0 + \lambda H_1)\, \psi_n = E_n \psi_n \qquad (16\text{-}3)$$

We will express the desired quantities as power series in λ. The question of convergence of the series will not be discussed. Frequently one can show that the series cannot be convergent, and yet the first few terms, when λ is small, do properly describe the physical system. We will assume that as $\lambda \to 0$, $E_n \to E_n{}^0$ and $\psi_n \to \phi_n$.

Since the ϕ_i form a complete set, we may expand ψ_n in a series involving all the ϕ_i. We write

$$\psi_n = N(\lambda) \left\{ \phi_n + \sum_{k \neq n} C_{nk}(\lambda)\, \phi_k \right\} \qquad (16\text{-}4)$$

The factor $N(\lambda)$ is there to allow us to normalize the ψ_n. We have the freedom to choose the phase of ψ_n, and we choose it such that the coefficient of ϕ_n in the expansion is real and positive. Since we require that $\psi_n \to \phi_n$ as $\lambda \to 0$, we have

$$N(0) = 1$$

$$C_{nk}(0) = 0 \tag{16-5}$$

More generally, we have

$$C_{nk}(\lambda) = \lambda C_{nk}^{(1)} + \lambda^2 C_{nk}^{(2)} + \ldots \tag{16-6}$$

and

$$E_n = E_n{}^0 + \lambda E_n^{(1)} + \lambda^2 E_n^{(2)} + \ldots \tag{16-7}$$

The Schrödinger equation then reads

$$(H_0 + \lambda H_1)\left\{ \phi_n + \sum_{k \neq n} \lambda C_{nk}^{(1)}\phi_k + \sum_{k \neq n} \lambda^2 C_{nk}^{(2)}\phi_k + \ldots \right\}$$

$$= (E_n{}^0 + \lambda E_n^{(1)} + \lambda^2 E_n^{(2)} + \ldots)\left\{ \phi_n + \sum_{k \neq n} \lambda C_{nk}^{(1)}\phi_k + \sum_{k \neq n} \lambda^2 C_{nk}^{(2)}\phi_k + \ldots \right\} \tag{16-8}$$

Note that the normalization factor $N(\lambda)$ does not appear in this linear equation. Identifying powers of λ yields a series of equations. The first one is

$$H_0 \sum_{k \neq n} C_{nk}^{(1)}\phi_k + H_1\phi_n = E_n{}^0 \sum_{k \neq n} C_{nk}^{(1)}\phi_k + E_n^{(1)}\phi_n \tag{16-9}$$

Using $H_0\phi_k = E_k{}^0\phi_k$ we obtain

$$E_n^{(1)}\phi_n = H_1\phi_n + \sum_{k \neq n} (E_k{}^0 - E_n{}^0) C_{nk}^{(1)}\phi_k \tag{16-10}$$

If we now take a scalar product with ϕ_n, and use the orthonormality condition

$$\langle \phi_k | \phi_l \rangle = \delta_{kl} \tag{16-11}$$

we obtain

$$\lambda E_n^{(1)} = \langle \phi_n | \lambda H_1 | \phi_n \rangle \tag{16-12}$$

This is a *very important formula*. It states that the first order energy shift for a given state is just the expectation value of the perturbing potential in that state. If the change in the potential is of a definite sign, then the energy shift will have the same sign. The explicit form of

$$\lambda E_n^{(1)} = \int d^3r \, \phi_n^*(\mathbf{r}) \, \lambda H_1(\mathbf{r}) \, \phi_n(\mathbf{r}) \tag{16-13}$$

shows that for the shift to be significant, both the potential change and the probability density $|\phi_n(\mathbf{r})|^2$ must be large.

If we take the scalar product of (16-10) with ϕ_m, for $m \neq n$, then

$$\langle \phi_m | H_1 | \phi_n \rangle + (E_m{}^0 - E_n{}^0)\, C_{nm}^{(1)} = 0$$

that is,

$$\lambda C_{nk}^{(1)} = \frac{\langle \phi_k | \lambda H_1 | \phi_n \rangle}{E_n{}^0 - E_k{}^0} \qquad (16\text{-}14)$$

The numerator is the matrix element of H_1 in the basis of states in which H_0 is diagonal. This formula is used in the next equation, which comes from the identification of terms proportional to λ^2:

$$H_0 \sum_{k \neq n} C_{nk}^{(2)} \phi_k + H_1 \sum_{k \neq n} C_{nk}^{(1)} \phi_k$$
$$= E_n{}^0 \sum_{k \neq n} C_{nk}^{(2)} \phi_k + E_n^{(1)} \sum_{k \neq n} C_{nk}^{(1)} \phi_k + E_n^{(2)} \phi_n \qquad (16\text{-}15)$$

Taking the scalar product with ϕ_n yields

$$E_n^{(2)} = \sum_{k \neq n} \langle \phi_n | H_1 | \phi_k \rangle\, C_{nk}^{(1)} = \sum_{k \neq n} \frac{\langle \phi_n | H_1 | \phi_k \rangle \langle \phi_k | H_1 | \phi_n \rangle}{E_n{}^0 - E_k{}^0}$$
$$= \sum_{k \neq n} \frac{|\langle \phi_k | H_1 | \phi_n \rangle|^2}{E_n{}^0 - E_k{}^0} \qquad (16\text{-}16)$$

The last line follows from the hermiticity of H_1:

$$\langle \phi_n | H_1 | \phi_k \rangle = \langle \phi_k | H_1 | \phi_n \rangle^* \qquad (16\text{-}17)$$

This, too, is a very important formula, especially since the first order shift frequently vanishes on grounds of symmetry. We may interpret the formula as follows: the second order energy shift is the sum of terms, whose strength is given by the square of the matrix element connecting the given state ϕ_n to all other states by the perturbing potential, weighted by the reciprocal of the energy difference between the states. We can draw several conclusions from the formula.

(a) If ϕ_n is the *ground state*, that is, the state of lowest energy, then the denominator in the sum is always negative, and hence (16-16) is always negative.

(b) All other things being equal, that is, if all the matrix elements of H_1 are of roughly the same order of magnitude (which is the kind of guess one would make without more specific knowledge), then nearby levels have a bigger effect on the second order energy shift than distant ones have.

(c) If an important level "k"—important in the sense of lying nearby, or of $\langle \phi_k | H_1 | \phi_n \rangle$ being large—lies above the given level "n," then the second order shift is downwards; if it lies below, the shift is upward. We speak of this as a tendency of levels to repel each other.

An expression for $C_{nk}^{(2)}$ may be obtained from (16-15) by taking the scalar product with ϕ_m, $m \neq n$, but we shall not require this formula. Also $N(\lambda)$ can be determined from

$$\langle \psi_n | \psi_n \rangle = N^2(\lambda) \left\{ 1 + \lambda^2 \sum_{k \neq n} |C_{nk}^{(1)}|^2 + \dots \right\}$$

$$= 1 \qquad (16\text{-}18)$$

It is therefore 1 to first order in λ. Hence, to first order in λ, we may write

$$\psi_n = \phi_n + \sum_{k \neq n} \frac{\langle \phi_k | \lambda H_1 | \phi_n \rangle}{E_n^0 - E_k^0} \phi_k \qquad (16\text{-}19)$$

a formula that is sometimes useful.

The above development needs modification when there is degeneracy, since, on the face of it, the denominator involving energy differences could vanish. The difficulty is associated with the fact that, instead of a unique ϕ_n, there is a finite set of $\phi_n^{(i)}$, all of which have the same energy E_n^0. This set can be made orthonormal with respect to the label "i," because, as we have seen in Chapter 4 this label can be associated with the eigenvalues of some other, simultaneously commuting, hermitian operators. We thus choose the set of $\phi_n^{(i)}$ such that

$$\langle \phi_m^{(j)} | \phi_n^{(i)} \rangle = \delta_{mn} \delta_{ij} \qquad (16\text{-}20)$$

The natural way to take the degeneracy into account is to replace (16-4) by an expression that involves linear combinations of the degenerate eigenfunctions of H_0:

$$\psi_n = N(\lambda) \left\{ \sum_i \alpha_i \phi_n^{(i)} + \lambda \sum_{k \neq n} C_{nk}^{(1)} \sum_i \beta_i \phi_k^{(i)} + \dots \right\} \qquad (16\text{-}21)$$

The coefficients α_i, β_i, . . . will have to be determined. When the above is substituted into the Schrödinger equation (16-3), we get, to first order in λ,

$$H_0 \sum_{k \neq n} C_{nk}^{(1)} \sum_i \beta_i \phi_k^{(i)} + H_1 \sum_i \alpha_i \phi_n^{(i)}$$

$$= E_n^{(1)} \sum_i \alpha_i \phi_n^{(i)} + E_n^0 \sum_{k \neq n} C_{nk}^{(1)} \sum_i \beta_i \phi_k^{(i)} \qquad (16\text{-}22)$$

Taking the scalar product with $\phi_n^{(j)}$ gives the first order shift equation

$$\sum_i \alpha_i \langle \phi_n^{(j)} | \lambda H_1 | \phi_n^{(i)} \rangle = \lambda E_n^{(1)} \alpha_j \qquad (16\text{-}23)$$

This is a finite-dimensional eigenvalue problem. For example, if there is a two-fold degeneracy, and if we use the notation

$$\langle \phi_n^{(j)} | H_1 | \phi_n^{(i)} \rangle = h_{ji} \qquad (16\text{-}24)$$

this equation reads

$$h_{11}\alpha_1 + h_{12}\alpha_2 = E_n^{(1)} \alpha_1$$

$$h_{21}\alpha_1 + h_{22}\alpha_2 = E_n^{(1)} \alpha_2 \tag{16-25}$$

Both the eigenvalues—there will, in general, be two possible values of $E_n^{(1)}$—and the α_i, can be determined from this equation, if we add the condition that

$$\sum_i |\alpha_i|^2 = 1 \tag{16-26}$$

We do not bother with the determination of the β_i, since we shall only use degenerate perturbation theory for the first order energy eigenvalues in our applications. If it so happens that $h_{ij} = 0$ for $i \neq j$, that is, that the matrix h_{ij} is diagonal, then the first order shifts are just the diagonal elements of this matrix. This will happen when the perturbation H_1 commutes with the operator whose eigenvalues the "i" labels represent. For example, in the hydrogen atom, there is a degeneracy associated with the eigenvalues of L_z, that is, all m-values have the same energy. If it happens that

$$[H_1, L_z] = 0 \tag{16-27}$$

and if we choose our $\phi_n^{(i)}$ to be eigenfunctions of L_z, then h_{ij} will be diagonal. To see this, note that with

$$L_z\phi_n^{(i)} = \hbar m^{(i)}\phi_n^{(i)} \tag{16-28}$$

$$\langle \phi_n^{(j)}|[H_1,L_z]|\phi_n^{(i)}\rangle = \langle \phi_n^{(i)}|H_1L_z - L_zH_1|\phi_n^{(i)}\rangle$$

$$= \hbar(m^{(i)} - m^{(j)}) h_{ij}$$

$$= 0 \tag{16-29}$$

that is, (16-27) implies

$$h_{ij} = 0 \quad \text{for} \quad m^{(i)} \neq m^{(j)} \tag{16-30}$$

Some of these features will be illustrated in the example below; others will appear later in our discussion of the real hydrogen atom.

To illustrate the application of perturbation theory to a real problem, we will consider the effect of an external electric field on the energy levels of a hydrogenlike atom. This is the *Stark Effect*. The unperturbed hamiltonian is

$$H_0 = \frac{\mathbf{p}^2}{2\mu} - \frac{Ze^2}{r} \tag{16-31}$$

whose eigenfunctions we denote by $\phi_{nlm}(\mathbf{r})$. The perturbing potential is

$$\lambda H_1 = e\,\mathcal{E} \cdot \mathbf{r} = e\,\mathcal{E}z \tag{16-32}$$

where $\mathcal{E}$ is the electric field. The quantity $e\mathcal{E}$ will play the role of the parameter λ.

The energy shift of the ground state, which is nondegenerate, is given by the expression

$$E_{100}^{(1)} = e\,\mathcal{E}\langle\phi_{100}|z|\phi_{100}\rangle = e\,\mathcal{E}\int d^3r\,|\phi_{100}(r)|^2\,z \qquad (16\text{-}33)$$

This integral vanishes, since the square of the wave function is always an even function under parity, and the perturbing potential is an odd function under reflections. Thus for the ground state there is no energy shift that is linear in the electric field $\mathcal{E}$. Classically, a system that has an electric dipole moment $\mathbf{d}$, will experience an energy shift of magnitude $-\mathbf{d}\cdot\mathcal{E}$. Thus the atom, in its ground state, has no permanent dipole moment. The argument given above may be generalized: *systems in nondegenerate states cannot have permanent dipole moments.* The statement of nondegeneracy is important: it is only then that the states are also eigenstates of the parity operator, and then $|\phi(\mathbf{r})|^2$ is even, and the expectation value of z vanishes.

Many molecules do have permanent dipole moments, and it is often said that this is because the ground states are degenerate. The expectation value of z in a state like $\alpha\psi_+ + \beta\psi_-$, where the subscripts indicate the parity, certainly does not vanish, and a state like the above will be degenerate with its space-inverted state $\alpha\psi_+ - \beta\psi_-$ if the two states ψ_+, ψ_- have the same energy. This explanation is not quite correct. The reason is that the lowest-lying states are never quite degenerate. Consider, for example, a molecule like ammonia, NH_3. Its structure is tetrahedral, with the three H nuclei forming an equilateral triangle. The N can be at a position (determined by the condition that the energy is minimum) either "above" or "below" the triangle. The even and odd linear combinations of these two states do not have quite the same energy, though the energy difference is very tiny (-10^{-4} eV), because of the large barrier between the "above" and "below" locations.[1] Thus, strictly speaking, the ground state is nondegenerate. However, if $\mathcal{E}d$, where

$$d = e\int \psi_{\text{above}}^*\,z\psi_{\text{above}} = -e\int \psi_{\text{below}}^*\,z\psi_{\text{below}} \qquad (16\text{-}34)$$

is much larger than the tiny splitting, then the energy shift will be linear in the electric field, and the molecule will behave as if it had an electric dipole moment (cf. Eq. 16-64).

Let us look at the second-order term. It reads

$$E_{100}^{(2)} = e^2\mathcal{E}^2 \sum_{nlm} \frac{|\langle\phi_{nlm}|z|\phi_{100}\rangle|^2}{E_1{}^0 - E_n{}^0} \qquad (16\text{-}35)$$

The matrix element in this expression is

$$\langle\phi_{nlm}|z|\phi_{100}\rangle = \int d^3r\,R_{nlm}(r)\,Y_{lm}^*\theta,\phi)\,r\cos\theta\,R_{100}(r)\,Y_{00}(\theta,\phi) \qquad (16\text{-}36)$$

[1] See our simple model of a molecule, and the discussion on p. 98.

where we have replaced z (which appears with $\mathcal{E}$ pointing in the z-direction) by the more convenient $r \cos \theta$. The angular part of the integration can be carried out, since

$$Y_{00} = \frac{1}{\sqrt{4\pi}}$$

$$\cos \theta = \sqrt{\frac{4\pi}{3}} \, Y_{10} \qquad (16\text{-}37)$$

It is therefore

$$\int d\Omega \, Y_{lm}^{*}(\theta,\phi) \, \frac{1}{\sqrt{3}} \, Y_{10}(\theta,\phi) = \frac{1}{\sqrt{3}} \, \delta_{l1}\delta_{m0} \qquad (16\text{-}38)$$

by the orthonormality of the Y_{lm}.

The fact that the m-value must be the same for the two states is our first example of a *selection rule*, which can be stated in the form

$$\Delta m = 0 \qquad (16\text{-}39)$$

It follows from the fact that

$$[L_z, z] = 0 \qquad (16\text{-}40)$$

that is, that the perturbation commutes with L_z. One thus has to evaluate the radial integral to obtain the answer:

$$R = \int_0^\infty r^2 dr \, R_{n10}(r) \, r \, R_{100}(r) \qquad (16\text{-}41)$$

This can be done,[2] and the result is

$$|\langle \phi_{n10} | z | \phi_{100} \rangle|^2 = \frac{1}{3} \frac{2^8 n^7 (n-1)^{2n-5}}{(n+1)^{2n+5}} \, a_0^2 \qquad (16\text{-}42)$$

which we write as $f(n) \, a_0^2$. For the second order shift, this gives

$$E_{100}^{(2)} = -e^2 \mathcal{E}^2 a_0^2 \sum_{n=2}^{\infty} \frac{f(n)}{\frac{1}{2}\mu c^2 \alpha^2 (1 - 1/n^2)}$$

$$= -\frac{2e^2 \mathcal{E}^2 a_0^2}{\mu c^2 \alpha^2} \sum_{n=2}^{\infty} \frac{n^2 f(n)}{n^2 - 1}$$

$$= -2a_0^3 \mathcal{E}^2 \sum_{n=2}^{\infty} \frac{n^2 f(n)}{n^2 - 1} \qquad (16\text{-}43)$$

---

[2] See H. A. Bethe and E. E. Salpeter, *Quantum Mechanics of One- and Two-Electron Atoms*, Academic Press, New York, 1957, p. 262.

On the face of it, the sum can only be evaluated term by term.[3] The $n = 2$ term contributes 0.74 to the series, and the $n = 3$ term contributes 0.10. The convergence is not spectacularly rapid, and the sum actually adds up to 1.125. The first term in the series does, however, give us an estimate of the order of magnitude of the effect. The dependence on a_0^3 is, of course, automatic, on purely dimensional grounds. The factor $\mathcal{E}^2$ must be multiplied by something that is a (length)[3], and the only natural length is the Bohr radius. If we speak of a hydrogenlike atom, rather than hydrogen, we must make the substitution $a_0 \rightarrow a_0/Z$.

If we differentiate the energy shift with respect to the electric field, we get an expression for the dipole moment

$$d = -\frac{\partial E_{100}^{(2)}}{\partial \mathcal{E}} = 4\mathcal{E}a_0^3 \sum_n \frac{n^2 f(n)}{n^2 - 1} \tag{16-44}$$

This is proportional to the electric field strength, that is, the dipole moment is induced. The *polarizability*, defined by

$$P = \frac{d}{\mathcal{E}} \tag{16-45}$$

can thus be calculated.

In making estimates of sums of the sort that occur in (16-35), one may sometimes find useful upper bounds. For example

$$\sum_{nlm} \frac{\langle \phi_{100}|z|\phi_{nlm}\rangle\langle \phi_{nlm}|z|\phi_{100}\rangle}{E_{100}^0 - E_n^0}$$

$$< \frac{1}{|E_1^0 - E_2^0|} \sum_{nlm} \langle \phi_{100}|z|\phi_{nlm}\rangle\langle \phi_{nlm}|z|\phi_{100}\rangle \tag{16-46}$$

However, because of the completeness of the states, we may replace

$$\sum_{nlm} |\phi_{nlm}\rangle\langle \phi_{nlm}| \rightarrow 1 \tag{16-47}$$

as argued in (14-17), so that

$$\sum_{nlm} \langle \phi_{100}|z|\phi_{nlm}\rangle\langle \phi_{nlm}|z|\phi_{100}\rangle = \langle \phi_{100}|z^2|\phi_{100}\rangle \tag{16-48}$$

This, however, is easy to evaluate. Since the ground state wave function is spherically symmetric, we have

$$\langle z^2 \rangle = \langle x^2 \rangle = \langle y^2 \rangle = \frac{1}{3}\langle \phi_{100}|r^2|\phi_{100}\rangle = a_0^2 \tag{16-49}$$

---

[3] Actually the second order shift can be evaluated in closed form. See, for example, S. Borowitz, *Fundamentals of Quantum Mechanics*, W. A. Benjamin, New York, 1968, pp. 328–330.

where the last step follows from (12-31). From this we find that

$$\sum_{n=2}^{\infty} f(n) = 1 \qquad (16\text{-}50)$$

and therefore

$$\sum_{n=2}^{\infty} \frac{n^2 f(n)}{n^2 - 1} < \frac{4}{3} \sum_{n=2}^{\infty} f(n) = \frac{4}{3} \qquad (16\text{-}51)$$

The relation

$$\sum_{nlm} |\langle \phi_{nlm} | z | \phi_{100} \rangle|^2 = \langle \phi_{100} | z^2 | \phi_{100} \rangle \qquad (16\text{-}52)$$

is called a *sum rule*, and is an example of relations that are useful in making estimates.

To illustrate degenerate perturbation theory, we calculate the first order (linear in $\mathcal{E}$) Stark effect for the $n = 2$ states of the hydrogen atom. For the unperturbed system there are really four $n = 2$ states that have the same energy. These are

$$\phi_{200} = (2a_0)^{-3/2} \, 2 \left(1 - \frac{r}{2a_0} \right) e^{-r/2a_0} \, Y_{00}$$

$$\phi_{211} = (2a_0)^{-3/2} \, 3^{-1/2} \left(\frac{r}{a_0} \right) e^{-r/2a_0} \, Y_{11}$$

$$\phi_{210} = (2a_0)^{-3/2} \, 3^{-1/2} \left(\frac{r}{a_0} \right) e^{-r/2a_0} \, Y_{10}$$

$$\phi_{2,1,-1} = (2a_0)^{-3/2} \, 3^{-1/2} \left(\frac{r}{a_0} \right) e^{-r/2a_0} \, Y_{1,-1} \qquad (16\text{-}53)$$

The $l = 0$ state has even parity, and the $l = 1$ states have odd parity. We want to solve an equation like (16-23) and, on the face of it, four equations are involved. If we note, however, that (a) the perturbing potential (that is, z) commutes with L_z so that it only connects states with the same m-value, and (b) parity forces us to consider only terms in which the perturbing potential must connect $l = 1$ to $l = 0$ terms, that is,

$$\langle \phi_{2,1,\pm1} | z | \phi_{2,1,\pm1} \rangle = 0 \qquad (16\text{-}54)$$

then the matrix in (16-23) is only a 2 × 2 matrix. The equation reads

$$e\mathcal{E} \begin{pmatrix} \langle \phi_{200} | z | \phi_{200} \rangle & \langle \phi_{200} | z | \phi_{210} \rangle \\ \langle \phi_{210} | z | \phi_{200} \rangle & \langle \phi_{210} | z | \phi_{210} \rangle \end{pmatrix} \begin{pmatrix} \alpha_1 \\ \alpha_2 \end{pmatrix} = E^{(1)} \begin{pmatrix} \alpha_1 \\ \alpha_2 \end{pmatrix} \qquad (16\text{-}55)$$

The diagonal elements are zero, because of parity, and the off-diagonal elements are equal, since they are complex conjugates of each other, and each may be chosen to be real. We have

$$\langle \phi_{200} | z | \phi_{210} \rangle = \int_0^\infty r^2 dr (2a_0)^{-3} e^{-r/a_0} \frac{2r}{\sqrt{3}a_0} \left(1 - \frac{r}{2a_0}\right) r$$

$$\cdot \int d\Omega Y_{00}^* (\sqrt{4\pi/3}\ Y_{10})\ Y_{10} \qquad (16\text{-}56)$$

$$= -3a_0$$

and hence (16-55) becomes

$$\begin{pmatrix} -E^{(1)} & -3e\mathcal{E}a_0 \\ -3e\mathcal{E}a_0 & -E^{(1)} \end{pmatrix} \begin{pmatrix} \alpha_1 \\ \alpha_2 \end{pmatrix} = 0 \qquad (16\text{-}57)$$

The eigenvalues of this are

$$E^{(1)} = \pm 3e\mathcal{E}a_0 \qquad (16\text{-}58)$$

and the corresponding eigenstates, when properly normalized are

$$\frac{1}{\sqrt{2}} \begin{pmatrix} 1 \\ -1 \end{pmatrix} \quad \text{and} \quad \frac{1}{\sqrt{2}} \begin{pmatrix} 1 \\ 1 \end{pmatrix},$$

respectively. Thus the linear Stark effect for the $n = 2$ states yields a splitting of degenerate levels as shown in Fig. 16-1.

There are some general comments that can be abstracted from the calculations just concluded.

(a) The states in the presence of the electric field are no longer eigenstates of $\mathbf{L}^2$, since in the above case, for example, we found that the states that diagonalize the perturbation were equal mixtures of $l = 0$ and $l = 1$, though they are still eigenstates of L_z. The reason, is that the perturbation changes the Hamiltonian, so that it no longer commutes with $\mathbf{L}^2$. This can be worked out in detail, but it is really evident that the external field specifies a preferred direction, so that the physical system is no longer invariant under arbitrary rotations. It is still invariant under rotations about the preferred axis, here the z-direction, and hence L_z is still a good constant of the motion.

(b) Quite generally, whenever there is a perturbation that does not conserve some quantity (for example, $\mathbf{L}^2$ here), then the states that "diagonalize" the new Hamiltonian in any approximation, are superpositions of states with different values of the previously conserved quantum numbers, and thus degenerate levels will be split.

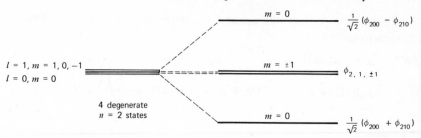

Fig. 16-1. Pattern of Stark splitting of hydrogen atom in $n = 2$ state. The four-fold degeneracy is partly lifted by the perturbation. The $m = \pm 1$ states remain degenerate and are not shifted in the Stark effect.

(c) We may summarize the procedure in degenerate perturbation theory in matrix language as follows. If H_0 is diagonal, but H_1 is not, then, since H_0 and H_1 do not commute, it is not possible to diagonalize H_1 by itself, without "un-diagonalizing" H_0. One must work with

$$H = H_0 + H_1$$

as a whole. If we work with a subset of degenerate states, all of which are eigenstates of H_0 *with the same eigenvalue*, then, as far as these states are concerned, H_0 is not merely diagonal, but it is proportional to the unit matrix. Since H_1 (and everything else) commutes with the unit matrix, one may diagonalize H_1 by itself, without affecting H_0.

The hydrogenlike atoms considered here were somewhat idealized. As we will see in Chapter 17, there are small relativistic and spin-orbit coupling effects that actually remove some of the degeneracies. Does this mean that we never really need to use degenerate perturbation theory? Actually, even if, say, ϕ_{200} and ϕ_{210} do not have exactly the same energy, it may still be sensible to take some linear combination of them in the perturbation expansion. If we have, for example

$$H_0\,\phi_{200} = (E_2{}^0 - \Delta)\,\phi_{200}$$

$$H_0\,\phi_{210} = (E_2{}^0 + \Delta)\,\phi_{210} \tag{16-59}$$

with Δ small, then the Schrödinger equation, with the linear combinations, reads

$$(H_0 + \lambda H_1)\left(\alpha_1\phi_{200} + \alpha_2\phi_{210} + \lambda\sum_{n \neq 2} C_n\phi_n\right)$$

$$= E\left(\alpha_1\phi_{200} + \alpha_2\phi_{210} + \lambda\sum_{n \neq 2} C_n\phi_n\right) \tag{16-60}$$

Taking the scalar product with ϕ_{200} and ϕ_{210}, respectively, leads to the following equation to order λ:

$$\begin{pmatrix} E_2^0 - \Delta - \langle\phi_{200}|\lambda H_1|\phi_{200}\rangle & \langle\phi_{200}|\lambda H_1|\phi_{210}\rangle \\ \langle\phi_{210}|\lambda H_1|\phi_{200}\rangle & E_2^0 + \Delta - \langle\phi_{210}|\lambda H_1|\phi_{210}\rangle \end{pmatrix} \begin{pmatrix} \alpha_1 \\ \alpha_2 \end{pmatrix} = E \begin{pmatrix} \alpha_1 \\ \alpha_2 \end{pmatrix}$$

$$(16\text{-}61)$$

If we write

$$\langle\phi_{200}|\lambda H_1|\phi_{210}\rangle = \langle\phi_{210}|\lambda H_1|\phi_{200}\rangle = a\lambda \qquad (16\text{-}62)$$

we must find the eigenvalues of the matrix

$$\begin{pmatrix} E_2^0 - \Delta & \lambda a \\ \lambda a & E_2^0 + \Delta \end{pmatrix} \qquad (16\text{-}63)$$

and these are

$$E = E_2^0 \pm \sqrt{a^2\lambda^2 + \Delta^2} \qquad (16\text{-}64)$$

(In the above we have set $\langle\phi_{200}|H_1|\phi_{200}\rangle = \langle\phi_{210}|H_1|\phi_{210}\rangle = 0$.) We see that when $\Delta \gg a\lambda$, we get a "quadratic" effect only. This corresponds to no degeneracy. When $\Delta \ll a\lambda$ we get the result of the form (16-58). In the intermediate region, the above, more careful treatment is necessary. Furthermore, when the new linear combinations are used, then in second order perturbation theory there no longer appear very tiny energy differences in the denominators. We do not discuss this in detail, but this is not difficult to establish.

As a final comment we point out two apparently contradictory facts. (1) The predictions of perturbation theory concerning the Stark effect are borne out very well by experiment, and (2) the perturbation series evidently diverges, since the perturbing potential $e\mathcal{E}z$ grows without bound as z becomes very large, no matter how small $e\mathcal{E}$ is. The question arises whether one has any right to believe in the accuracy of the first few terms of a mathematically divergent series, since it is well known that a mathematically divergent series can be rearranged to give entirely different expansions. The answer lies in the physics and not in the mathematics of the problem. The reason for the divergence can be seen in Fig. 16-2, which gives a rough picture of the total potential for x, y fixed. It appears that there is a barrier created for the bound electron. This barrier is ultimately penetrable, even though for small $e\mathcal{E}$ it is very broad. What the mathematical divergence of the series is responding to is the possibility that the electron in the ground state, for example, has a finite (although very, very small) probability of being sufficiently far away from the nucleus, where the external electric field is stronger than the Coulomb field, and the electron is carried away by the electric field. Thus the new "shifted" energy levels of the hydrogen atom are no longer stationary states, but rather metastable states. If the field is weak,

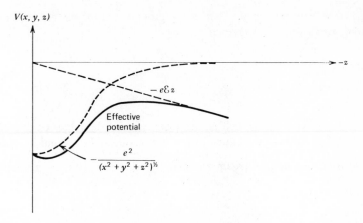

Fig. 16-2. Schematic picture of potential energy as a function of z with x and y held fixed. The dotted line represents the Coulomb potential, the dashed line the potential energy due to the external field, and the solid line the total potential.

however, they may be stable on a time scale of the age of the universe,[4] and hence the observations agree perfectly with what the first few terms of the perturbation series predict.

Problems

1. Consider the hydrogen atom, and assume that the proton, instead of being a point-source of the Coulomb field, is a uniformly charged sphere of radius R, so that the Coulomb potential is now modified to

$$V(r) = -\frac{3e^2}{2R^3}\left(R^2 - \frac{1}{3}r^2\right) \qquad r < R(\ll a_0)$$

$$= -\frac{e^2}{r} \qquad\qquad r > R$$

Calculate the energy shift for the $n = 1$, $l = 0$ state, and for the $n = 2$ states, caused by this modification, using the wave functions given in (12-25).

[4] Actually a simple barrier penetration calculation of the type carried out in Chapter 5 shows that the time scale is more like 10^{1000} lifetimes of the universe, for fairly reasonable fields!

2. Calculate the energy shift in the ground state of the one-dimensional harmonic oscillator, when the perturbation

$$V = \lambda x^4$$

is added to

$$H = \frac{p^2}{2m} + \tfrac{1}{2}m\omega^2 x^2$$

3. Consider a square well in one dimension. If the edges of the well are rounded off as shown in the figure, what is the change in the ground state energy? Choose your rounding-off parametrization such that $\int_{-\infty}^{\infty} V(x)\,dx$ remains unchanged.

4. The bottom of an infinite well is changed to have the shape

$$V(x) = \epsilon \sin \frac{\pi x}{b} \qquad 0 \le x \le b$$

Calculate the energy shifts for all the excited states to first order in ϵ. Note that the well originally had $V(x) = 0$ for $0 \le x \le b$, with $V = \infty$ elsewhere.

5. Prove the sum rule (Thomas-Reiche-Kuhn sum rule)

$$\sum_n (E_n - E_a)|\langle n|x|a\rangle|^2 = \frac{\hbar^2}{2m}$$

[*Hint.* (a) Write the commutation relation $[p,x] = \hbar/i$ in the form

$$\sum_n \left\{\langle a|p|n\rangle\langle n|x|a\rangle - \langle a|x|n\rangle\langle n|p|a\rangle\right\} = \frac{\hbar}{i}\langle a|a\rangle = \frac{\hbar}{i}$$

(b) Use the fact that

$$\langle a|p|n\rangle = \left\langle a\left|m\frac{dx}{dt}\right|n\right\rangle = m\frac{i}{\hbar}\langle a|[H,x]|n\rangle$$

in working out the problem.]

6. Check the above sum rule for the one-dimensional harmonic oscillator, with "a" taken in the ground state.

7. Work out the first order Stark effect in the $n = 3$ state of the hydrogen atom. Do not bother to work out all the integrals.

8. Consider an electron in a state n in a hydrogen atom. The atom is placed

in an external electric field $\mathcal{E}$. Estimate the lifetime of the atom, or, equivalently, the transmission coefficient through the barrier made up of the Colulomb attraction to the nucleus. It is enough to consider a one-dimensional model of the problem.

9. Consider a two-dimensional harmonic oscillator described by the Hamiltonian

$$H = \frac{1}{2m}(p_x{}^2 + p_y{}^2) + \tfrac{1}{2}m\omega^2(x^2 + y^2)$$

Generalize the approach of Chapter 7 to obtain solutions of this problem in terms of raising operators acting on the ground state. Calculate the energy shifts due to the perturbation

$$V = 2\lambda xy$$

in the ground state, and in the degenerate first excited states, using first order perturbation theory. Can you interpret your result very simply? Solve the problem exactly, and compare it with a second order perturbation calculation.

[*Hints.* (a) Examine the symmetries of the unperturbed Hamiltonian. (b) Decompose the motion into center of mass motion and internal motion.]

References

There are many examples of the application of first-order perturbation theory in the textbook literature, and the references listed at the end of this book may serve as a source of further examples. For a discussion of the exact calculation of the Stark effect see

S. Borowitz, *Fundamentals of Quantum Mechanics*, W. enjamin, Inc., 1967.

chapter 17

The Real Hydrogen Atom

The discussion of hydrogenlike atoms in Chapter 12 was based on the Hamiltonian

$$H_0 = \frac{\mathbf{p}^2}{2\mu} - \frac{Ze^2}{r} \tag{17-1}$$

In a more realistic treatment, several corrections must be taken into account. First of all, the expression for the kinetic energy of the electron is altered when relativistic corrections are taken into account. In the original electron-proton Hamiltonian we replace

$$\frac{\mathbf{p}_e^2}{2m} + \frac{\mathbf{p}_P^2}{2M} = \mathbf{p}^2 \left(\frac{1}{2m} + \frac{1}{2M} \right) = \frac{\mathbf{p}^2}{2\mu}$$

(in the center of mass frame) by

$$(p_e^2 c^2 + m^2 c^4)^{1/2} + \frac{\mathbf{p}_P^2}{2M} \cong mc^2 + \frac{\mathbf{p}^2}{2m} - \frac{1}{8} \frac{(\mathbf{p}^2)^2}{m^3 c^2} + \frac{\mathbf{p}^2}{2M}$$

$$= mc^2 + \frac{\mathbf{p}^2}{2\mu} - \frac{1}{8} \frac{(\mathbf{p}^2)^2}{m^3 c^2} \tag{17-2}$$

The electron rest mass term is irrelevant. The nonrelativistic term still involves the reduced mass, but there is now a correction term,

$$H_1 = - \frac{1}{8} \frac{(\mathbf{p}^2)^2}{m^3 c^2} \tag{17-3}$$

that should be added to the Hamiltonian H_0. We may estimate the magnitude of the correction:

$$\frac{\langle H_1 \rangle}{\langle H_0 \rangle} \approx \frac{\langle p^2 \rangle}{m^2 c^2} \approx \frac{(mcZ\alpha)^2}{m^2 c^2} \approx (Z\alpha)^2 \tag{17-4}$$

271

For hydrogen this is of the order of 10^{-5}, smaller than the reduced mass effects.

The existence of the electron spin gives rise to another correction that is of the same order of magnitude. It may be qualitatively understood as follows: if the electron were at rest relative to the proton (we are discussing this on a classical level), it would only see an electric field due to the proton charge. This is the Coulomb potential term that appears in H_0. Because the electron is moving, there are additional effects. In the electron rest frame, the proton is moving, so that there is a current present, and the electron "sees" a magnetic field. If the relative motion were rectilinear, the magnetic field, as seen by the electron, would be $\mathbf{v} \times \mathbf{E}/c$. This magnetic field interacts with the spin of the electron, or more precisely, with the magnetic moment of the electron. We might expect an interaction of the form

$$-\mathbf{M} \cdot \mathbf{B} = \frac{e}{mc} \mathbf{S} \cdot \mathbf{B}$$

$$= \frac{e}{mc^2} \mathbf{S} \cdot \mathbf{v} \times \mathbf{E} = -\frac{e}{m^2 c^2} \mathbf{S} \cdot \mathbf{p} \times \nabla\phi(r)$$

$$= -\frac{e}{m^2 c^2} \mathbf{S} \cdot \mathbf{p} \times \mathbf{r} \frac{1}{r} \frac{d\phi}{dr}$$

$$= \frac{1}{m^2 c^2} \mathbf{S} \cdot \mathbf{r} \times \mathbf{p} \frac{1}{r} \frac{d}{dr} e\phi(r) \tag{17-5}$$

where $\phi(r)$ is the potential due to the nuclear charge. Actually this is not correct. It turns out that relativistic effects associated with the fact that the electron does not move in a straight line (the Thomas precession effect) reduce the above by a factor of 2. Thus the correct perturbation is

$$H_2 = \frac{1}{2m^2 c^2} \mathbf{S} \cdot \mathbf{L} \frac{1}{r} \frac{d[e\phi(r)]}{dr} \tag{17-6}$$

Let us now use first order perturbation theory to calculate the effects of H_1 and H_2 on the spectrum of hydrogenlike atoms. We may rewrite H_1 in the form

$$H_1 = -\frac{1}{8} \frac{(\mathbf{p}^2)^2}{m^3 c^2} = -\frac{1}{2mc^2} \left(\frac{\mathbf{p}^2}{2m} \right)^2$$

$$= -\frac{1}{2mc^2} \left(H_0 + \frac{Ze^2}{r} \right) \left(H_0 + \frac{Ze^2}{r} \right) \tag{17-7}$$

if we neglect reduced mass effects in H_1.

Hence

$$\langle \phi_{nlm} | H_1 | \phi_{nlm} \rangle = -\frac{1}{2mc^2} \left\langle \phi_{nlm} \left| \left(H_0 + \frac{Ze^2}{r} \right) \left(H_0 + \frac{Ze^2}{r} \right) \right| \phi_{nlm} \right\rangle$$

$$= -\frac{1}{2mc^2}\left[E_n{}^2 + 2E_nZe^2\left\langle\frac{1}{r}\right\rangle_{nl} + (Ze^2)^2\left\langle\frac{1}{r^2}\right\rangle_{nl}\right]$$

$$= -\frac{1}{2mc^2}\left\{\left[\frac{mc^2(Z\alpha)^2}{2n^2}\right]^2 - 2Ze^2\frac{mc^2(Z\alpha)^2}{2n^2}\left(\frac{Z}{a_0n^2}\right)\right.$$

$$\left. + (Ze^2)^2\frac{Z^2}{a_0{}^2n^3(l+1/2)}\right\}$$

$$= -\frac{1}{2}mc^2(Z\alpha)^2\left[\frac{(Z\alpha)^2}{n^3(l+1/2)} - \frac{3(Z\alpha)^2}{4n^4}\right] \tag{17-8}$$

In calculating the above, we have used expressions for

$$\left\langle\frac{1}{r}\right\rangle_{nl} \equiv \left\langle\phi_{nlm}\left|\frac{1}{r}\right|\phi_{nlm}\right\rangle \quad \text{and} \quad \left\langle\frac{1}{r^2}\right\rangle_{nl} \equiv \left\langle\phi_{nlm}\left|\frac{1}{r^2}\right|\phi_{nlm}\right\rangle$$

from (12-31). The spin of the electron does not enter into this energy shift, since H_1 does not depend on the spin. H_2 does depend on the spin, and for our unperturbed wave functions we must take two-component wave functions, since what we want to calculate is the expectation value of

$$\frac{1}{2m^2c^2}\mathbf{S}\cdot\mathbf{L}\frac{1}{r}\frac{ed\phi(r)}{dr} = \frac{Ze^2}{2m^2c^2}\mathbf{S}\cdot\mathbf{L}\frac{1}{r^3} \tag{17-9}$$

Here, again, we have an example of degenerate perturbation theory. For a given n and l, there are $2(2l+1)$ degenerate eigenstates of H_0, with the additional factor of 2 coming from the two spin states. Thus the calculation of the energy shift involves a diagonalization of a submatrix, as in Eq. 16-23. We can save ourselves a great deal of labor by noting that

$$\mathbf{S} + \mathbf{L} = \mathbf{J} \tag{17-10}$$

implies that

$$\mathbf{S}^2 + 2\mathbf{S}\cdot\mathbf{L} + \mathbf{L}^2 = \mathbf{J}^2$$

that is,

$$\mathbf{S}\cdot\mathbf{L} = \tfrac{1}{2}(\mathbf{J}^2 - \mathbf{L}^2 - \mathbf{S}^2) \tag{17-11}$$

Thus if we combine the degenerate eigenfunctions into linear combinations that are eigenfunctions of $\mathbf{J}^2$ (they already are eigenfunctions of $J_z = L_z + S_z$), then these linear combinations will diagonalize H_2. The appropriate linear combinations were obtained in Chapter 15, Eq. 15-37 and 15-38. With these linear combinations we have

$$\mathbf{S}\cdot\mathbf{L}\,\psi_{\substack{j=l+(1/2)\\m_j=m+(1/2)}} = \frac{1}{2}\left(\mathbf{J}^2 - \mathbf{L}^2 - \mathbf{S}^2\right)\psi_{\substack{j=l+(1/2)\\m_j=m+(1/2)}}$$

$$= \frac{1}{2} \hbar^2 \left[\left(l + \frac{1}{2} \right) \left(l + \frac{3}{2} \right) - l(l+1) - \frac{3}{4} \right] \psi_{\substack{j=l+(1/2) \\ m_j = m+(1/2)}}$$

$$= \frac{1}{2} \hbar^2 l \, \psi_{\substack{j=l+(1/2) \\ m_j=m+(1/2)}} \tag{17-12}$$

and

$$\mathbf{S \cdot L} \, \psi_{\substack{j=l-(1/2) \\ m_j=m+(1/2)}} = \frac{1}{2} \hbar^2 \left[\left(l - \frac{1}{2} \right) \left(l + \frac{1}{2} \right) - l(l+1) - \frac{3}{4} \right] \psi_{\substack{j=l-(1/2) \\ m_j=m+(1/2)}}$$

$$= -\frac{1}{2} \hbar^2 (l+1) \, \psi_{\substack{j=l-(1/2) \\ m_j=m+(1/2)}} \tag{17-13}$$

For a given l value there are $[2(l + 1/2) + 1] + [2(l - 1/2) + 1]$ states. What has happened is that the degenerate states have merely been rearranged, but the two groups that they have been split into behave differently under the action of H_2. If we call the linear combinations ϕ_{jm_jl} then

$$\langle \phi_{jm_jl} | H_2 | \phi_{jm_jl} \rangle = \frac{Ze^2}{2m^2c^2} \frac{\hbar^2}{2} \left\{ \begin{matrix} l \\ -l-1 \end{matrix} \right\}$$

$$\times \int_0^\infty dr \, r^2 [R_{nl}(r)]^2 \frac{1}{r^3} \tag{17-14}$$

for $j = l \pm 1/2$, respectively.

With the help of

$$\left\langle \frac{1}{r^3} \right\rangle_{nl} = \frac{Z^3}{a_0^3} \frac{1}{n^3 l(l + 1/2)(l + 1)} \tag{17-15}$$

we get the energy shift

$$\Delta E = \frac{1}{4} mc^2 (Z\alpha)^4 \frac{\left\{ \begin{matrix} l \\ -l-1 \end{matrix} \right\}}{n^3 l(l + 1/2)(l + 1)} \tag{17-16}$$

We must, of course combine the effects of H_1 and H_2.
When this is done, we obtain after some algebra

$$\Delta E = -1/2 mc^2 (Z\alpha)^4 \frac{1}{n^3} \left[\frac{1}{j + 1/2} - \frac{3}{4n} \right] \tag{17-17}$$

for both values of $l = j \pm 1/2$. It is necessary to work with the relativistic Dirac equation to show that the result is also correct when $l = 0$, even though the product in (17-14) is not well defined.

The splitting is depicted graphically in Fig. 17-1. A very interesting result is that the corrections add up in a manner that leaves the $^2P_{1/2}$ and the $^2S_{1/2}$ states

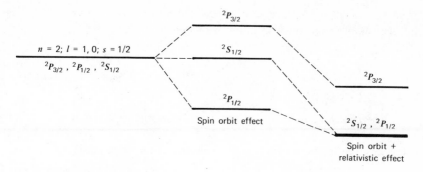

Fig. 17-1. Splitting of the $n = 2$ levels by (1) the spin-orbit coupling (which leaves the S state unaffected) and (2) the relativistic effect. The final degeneracy of the $^2S_{1/2}$ and $^2P_{1/2}$ states is actually lifted by quantum electrodynamic effects. The upward shift of the $^2S_{1/2}$ state is called the Lamb shift.

degenerate. A more careful discussion, using the relativistic Dirac equation, does not alter this result. In 1947, a very delicate microwave absorption experiment carried out by Lamb and Retherford showed that there was, indeed, a tiny splitting of the two levels. The magnitude of the splitting, of order $mc^2(Z\alpha)^4 \, \alpha \log \alpha$ could be explained by the additional interaction of the electron with its own electromagnetic field, that is, as a self-energy effect. These matters are outside of the scope of this book.

Let us now turn to the discussion of the behavior of hydrogenlike atoms in an external magnetic field, that is, to the *anomalous Zeeman effect*. There is, of course, nothing anomalous about the effect; it is just that the Zeeman effect that could be explained classically was exhibited only by atoms in states in which the total electronic spin was zero. For the other states, for which there was no classical explanation (since that involves spin), the Zeeman splitting pattern was different, and therefore "anomalous."

For the unperturbed Hamiltonian we take the usual H_0 together with the spin orbit term. The reason for doing this is that the external perturbation may be small compared with the effect of what we called H_2. Thus

$$H_0 = \frac{p^2}{2\mu} - \frac{Ze^2}{r} + \frac{1}{2m^2c^2} \frac{Ze^2}{r^3} \mathbf{L} \cdot \mathbf{S} \qquad (17\text{-}18)$$

The perturbation now reads

$$H_1 = \frac{e}{2mc} (\mathbf{L} + 2\mathbf{S}) \cdot \mathbf{B} \qquad (17\text{-}19)$$

The first term is, in effect, the interaction of the magnetic dipole moment arising from the circulating charge, and the second term is the contribution of the intrinsic dipole moment of an object with spin

$$\mathbf{M} = - \frac{eg}{2mc} \mathbf{S} \tag{17-20}$$

with $g = 2$.

The choice of H_0 dictates that we calculate the expectation value of the perturbation in eigenstates of $\mathbf{J}^2$ and J_z (15-37) and (15-38). If we choose the z-axis as given by the direction of $\mathbf{B}$, then we need to calculate

$$\langle \phi_{jm_jl} | \frac{eB}{2mc} (L_z + 2S_z) | \phi_{jm_jl} \rangle = \langle \phi_{jm_jl} | \frac{eB}{2mc} (J_z + S_z) | \phi_{jm_jl} \rangle$$

$$= \frac{eB}{2mc} (\hbar m_j + \langle \phi_{jm_jl} | S_z | \phi_{jm_jl} \rangle) \tag{17-21}$$

To calculate the last matrix element, we carry out the calculation explicitly, using the eigenfunctions given in (15-37) and (15-38). Thus for $j = l + 1/2$ we have

$$\left\langle \sqrt{\frac{l+m+1}{2l+1}} Y_{lm}\chi_+ + \sqrt{\frac{l-m}{2l+1}} Y_{l,m+1}\chi_- \,\middle|\, S_z \,\middle|\, \sqrt{\frac{l+m+1}{2l+1}} Y_{lm}\chi_+ \right.$$

$$\left. + \sqrt{\frac{l-m}{2l+1}} Y_{l,m+1}\chi_- \right\rangle = \frac{\hbar}{2} \left(\frac{l+m+1}{2l+1} - \frac{l-m}{2l+1} \right)$$

$$= \frac{\hbar}{2} \frac{2m+1}{2l+1} = \frac{\hbar m_j}{2l+1} \tag{17-22}$$

and for $j = -1/2$, we have

$$\left\langle \sqrt{\frac{l-m}{2l+1}} Y_{ml}\chi_+ - \sqrt{\frac{l+m+1}{2l+1}} Y_{l,m+1}\chi_- \,\middle|\, S_z \,\middle|\, \sqrt{\frac{l-m}{2l+1}} Y_{lm}\chi_+ \right.$$

$$\left. - \sqrt{\frac{l+m+1}{2l+1}} Y_{l,m+1}\chi_- \right\rangle = \frac{\hbar}{2} \left(\frac{l-m}{2l+1} - \frac{l+m+1}{2l+1} \right)$$

$$= - \frac{\hbar}{2} \frac{2m+1}{2l+1} = - \frac{\hbar m_j}{2l+1} \tag{17-23}$$

In both cases we used the fact that $m_j = m + 1/2$ in the above. Inserting the above into (17-21) yields

$$\Delta E = \frac{e\hbar B}{2mc} m_j \left(1 \pm \frac{1}{2l+1} \right) \qquad j = l \pm \tfrac{1}{2} \tag{17-24}$$

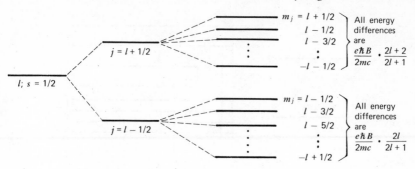

Fig. 17-2. General representation of anomalous Zeeman effect.

The splitting is depicted in Fig. 17-2. The selection rule[1] for the transitions is still

$$\Delta m_j = \pm 1, 0 \qquad (17\text{-}25)$$

but since the splitting between the lines is not the same for every multiplet, we do not get just the three lines that we obtained for the normal Zeeman effect in Chapter 13. For example, for $n = 2$, the $^2P_{3/2}$ state splits into four lines, with the splitting two times as large as that of the two states in the $^2P_{1/2}$ lines (Fig. 17-3). If the external field is very strong, so that the spin-orbit coupling can be neglected, we may use the ordinary hydrogenic wave functions simply multiplied by spinors, that is, eigenstates of $\mathbf{L}^2$, L_z, $\mathbf{S}^2$, and S_z. If we call the eigenvalues of L_z and S_z, m_l and m_s, respectively, then the expectation value of H_1 in (17-19), with $\mathbf{B}$ pointing in the z-direction, is

$$\langle H_1 \rangle = \frac{e\hbar B}{2mc}(m_l + 2m_s) \qquad (17\text{-}26)$$

Thus the $n = 2$, $l = 1$ states are split into five levels, corresponding to the values of $m_l = 1, 0, -1$; $m_s = 1/2, -1/2$.

In addition to the *fine structure* of the levels caused by the spin-orbit coupling, there is a very tiny *hyperfine splitting*, which is really a permanent Zeeman effect due to the magnetic field generated by the magnetic dipole moment of the nucleus. If the spin of the nucleus is $\mathbf{I}$, then the magnetic dipole moment operator is

$$\mathbf{M} = \frac{Zeg_N}{2M_Nc}\mathbf{I} \qquad (17\text{-}27)$$

where Ze is the charge of the nucleus, M_N its mass, and g_N its gyromagnetic

---

[1] The derivation of this selection rule (and others) will be discussed in Chapter 22.

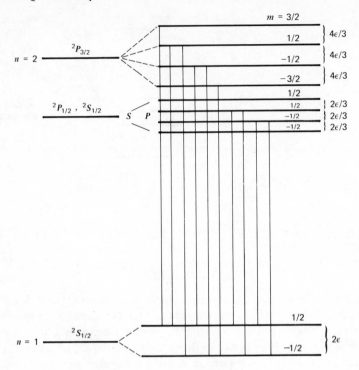

Fig. 17-3. Zeeman effect in hydrogen, ϵ represents the energy $e\hbar B/2mc$. The transitions for which $\Delta l = 1$, $\Delta m = 1, 0, -1$ are drawn in the figure. The location of the unperturbed states is given by Fig. 17-1.

ratio. The vector potential due to a point dipole is, from electromagnetic theory

$$\mathbf{A(r)} = -\frac{1}{4\pi}(\mathbf{M} \times \boldsymbol{\nabla})\frac{1}{r} \tag{17-28}$$

so that the magnetic field is

$$\mathbf{B} = \boldsymbol{\nabla} \times \mathbf{A} = -\frac{\mathbf{M}}{4\pi}\nabla^2 \frac{1}{r} + \frac{1}{4\pi}\boldsymbol{\nabla}(\mathbf{M}\cdot\boldsymbol{\nabla})\frac{1}{r} \tag{17-29}$$

Thus the perturbation is

$$H_1 = -\mathbf{M}_e \cdot \mathbf{B}$$

$$= \frac{e}{mc}\mathbf{S}\cdot\mathbf{B}$$

$$= \frac{Ze^2 g_N}{2mM_N c^2}\frac{1}{4\pi}\mathbf{S}\cdot\left[-\mathbf{I}\nabla^2\frac{1}{r} + \boldsymbol{\nabla}(\mathbf{I}\cdot\boldsymbol{\nabla})\frac{1}{r}\right] \tag{17-30}$$

The expectation value of the term on the right can be calculated very straight-forwardly. First, we note that the magnitude of the splitting is

$$\frac{Ze^2 g_N}{8\pi m M_N c^2}\, \hbar^2 \left(\frac{Z\alpha m c}{\hbar}\right)^3 \simeq \frac{g_N}{8\pi}\, (Z\alpha)^4 \left(\frac{m}{M_N}\right) mc^2 \tag{17-31}$$

that is, it is a factor of m/M_N smaller than the typical spin-orbit splittings. The calculation of the expectation value of (17-30) in the state characterized by $l = 0$, for example, the ground state, is simplified. We have

$$\int d^3 r \phi^*(r)\left((\mathbf{S}\cdot\boldsymbol{\nabla})(\mathbf{I}\cdot\boldsymbol{\nabla})\,\frac{1}{r}\right)\phi(r) = S_i I_k \int d^3 r |\phi(r)|^2 \frac{\partial}{\partial x_i}\frac{\partial}{\partial x_k}\frac{1}{r}$$

Because of the spherical symmetry of all the terms in the integrand except for the derivatives, the angular integration will vanish unless $i = k$. All the $i = k$ contributions will be equal for the same reason, so that the above yields

$$\frac{1}{3}\, S_i I_i \int d^3 r |\phi|(r)|^2 \nabla^2 \frac{1}{r}$$

Thus, when inserted between $l = 0$ states (and only then), we may write

$$(\mathbf{S}\cdot\boldsymbol{\nabla})(\mathbf{I}\cdot\boldsymbol{\nabla})\,\frac{1}{r} = \frac{1}{3}\,\mathbf{S}\cdot\mathbf{I}\nabla^2\,\frac{1}{r} \tag{17-32}$$

Thus what is needed is

$$\langle H_1 \rangle = -\frac{2\cdot Ze^2 g_N}{24\pi m M_N c^2}\,(\mathbf{S}\cdot\mathbf{I})\left\langle \nabla^2\,\frac{1}{r}\right\rangle \tag{17-33}$$

We use the fact that[2]

$$\nabla^2\,\frac{1}{4\pi r} = -\,\delta(\mathbf{r}) \tag{17-34}$$

to obtain

$$\langle H_1 \rangle = \frac{1}{3}\, g_N(Z\alpha)\,\frac{m}{M_N}\, mc^2 \left(\frac{\mathbf{S}\cdot\mathbf{I}}{\hbar^2}\right)\left(\frac{\hbar}{mc}\right)^3 \int d^3 r \phi^*_{n00}(\mathbf{r})\,\delta(\mathbf{r})\,\phi_{n00}(\mathbf{r})$$

$$= \frac{g_N}{3}\,\frac{m}{M_N}\,(Z\alpha)\, mc^2\,\frac{\mathbf{S}\cdot\mathbf{I}}{\hbar^2}\left(\frac{\hbar}{mc}\right)^3 |R_{n0}(0)|^2 \tag{17-35}$$

[2] Only the radial part of ∇^2 is relevant. To show this, we prove that $(1/r^2)(d/dr)$ $[r^2(d/dr)]\,(1/r) = 0$ for $r \neq 0$, and that $\nabla^2(1/r)$ integrated over a small sphere of radius ϵ, gives a result -4π independent of ϵ.

When the value of the radial function at the origin is inserted into the above, then[3]

$$|R_{n0}(0)|^2 = \frac{4}{n^3}\left(\frac{Z\alpha mc}{\hbar}\right)^3 \tag{17-36}$$

leads to the result

$$\langle H_1 \rangle = \frac{4}{3}\, g_N\, \frac{m}{M_N}\, (Z\alpha)^4\, mc^2\, \frac{1}{n^3}\left(\frac{\mathbf{S}\cdot\mathbf{I}}{\hbar^2}\right) \tag{17-37}$$

If we take the total spin of the electron and nucleus to be $\mathbf{F}$,

$$\mathbf{F} = \mathbf{S} + \mathbf{I} \tag{17-38}$$

then

$$\frac{\mathbf{S}\cdot\mathbf{I}}{\hbar^2} = \frac{\mathbf{F}^2 - \mathbf{S}^2 - \mathbf{I}^2}{2\hbar^2} = \frac{[F(F+1) - 3/4 - I(I+1)]}{2}$$

$$= \frac{1}{2}\begin{cases} I & F = I + \frac{1}{2} \\ -I - 1 & F = I - \frac{1}{2} \end{cases} \tag{17-39}$$

For hydrogen, $g_N = g_P \cong 5.56$, and the energy difference between the excited state, characterized by $F = 1$ and the ground state of $F = 0$ is

$$\Delta E = \frac{4}{3}\,(5.56)\frac{1}{1840}\frac{1}{(137)^4}\,(mc^2)$$

The wavelength of the radiation corresponding to the transition between the $F = 1$ and $F = 0$ states is

$$\lambda \simeq 21.4 \text{ cm} \tag{17-40}$$

and the frequency[4] is

$$\nu = \frac{c}{\lambda} \simeq 1420 \text{ megacycles} \tag{17-41}$$

The radiation arising from this transition plays an important role in astronomy. In a gas of neutral atoms, the $F = 1$ state cannot be excited by ordinary radiation, because of a selection rule that strongly suppresses transitions in which there is no change in orbital angular momentum. Both the $F = 1$ and the $F = 0$ states have zero angular momentum. On the other hand, there are other mechanisms that can cause transitions. The $F = 1$ state can, for example, be excited by

---

[3] See, for example, Bethe and Salpeter, *loc. cit.*

[4] This frequency is one of the most accurately measured quantities in physics. $\nu_{\exp} = 1420405751.800 + 0.028$ cycles (Hz). The number involves the distribution of magnetization in the proton, but there is no theory yet that can deal with a number of this accuracy.

collisions, and the return to the $F = 0$ ground state can be detected. From an analysis of the intensity of the 21 cm radiation received, astronomers have learned a great deal about the density distribution of neutral hydrogen in interstellar space, as well as the motion and the temperature of the gas clouds containing the hydrogen. The average number of neutral hydrogen atoms appears to be about 1 cm^{-3} in the galactic plane near the sun, and the temperature is of the order of 100° K.

Problems

1. What effect does the addition of a constant to the Hamiltonian have on the wave function?

2. If the general form of a spin-orbit coupling for a particle of mass m and spin $\mathbf{S}$ moving in a potential $V(r)$ is

$$H_{SO} = \frac{1}{2m^2c^2} \mathbf{S} \cdot \mathbf{L} \frac{1}{r} \frac{dV(r)}{dr}$$

what is the effect of that coupling on the spectrum of a three-dimensional harmonic oscillator?

3. Consider the $n = 2$ states in the real hydrogen atom. What is the spectrum in the absence of a magnetic field? How is that spectrum changed when the atom is placed in a magnetic field of 25,000 gauss?

4. Show that

$$\nabla^2 \frac{1}{r} = -4\pi\delta(\mathbf{r})$$

Use the procedure outlined in the footnote to Eq. 17-34.

5. Consider a gas of hydrogen atoms at low temperature and density. At what temperature will the $F = 1$ and the $F = 0$ states be equally occupied? (*Note.* The Boltzmann factor

$$ge^{-E/kT}$$

gives the relative probability of occupation of a given state with degeneracy g when the system is in equilibrium, at temperature T.)

6. Consider a harmonic oscillator in three dimensions. If the relativistic expression for the kinetic energy is used, what is the shift in the ground state energy?

7. The deuteron consists of a proton (charge $+e$) and a neutron (charge 0)

in a state of total spin 1 and total angular momentum $J = 1$. The g-factors for the proton and neutron are

$$g_P = 2(2.7896)$$
$$g_N = 2(-1.9103)$$

(a) What are the possible orbital angular momentum states for this system? If it is known that the state is primarily $^3S_1$, what admixture is allowed given that parity is conserved?

(b) Write an expression for the interaction of the deuteron with an external magnetic field and calculate the Zeeman splitting. Show that if the interaction with the magnetic field is written in the form

$$V = -\mu_{\text{eff}} \cdot \mathbf{B}$$

then the effective magnetic moment of the deuteron is the sum of the proton and neutron magnetic moments, and any deviation from that result is due to an admixture of non-S state to the wave function.

8. Consider positronium, a hydrogenlike atom consisting of an electron and a positron (same mass, opposite charge). Calculate (a) the ground state energy, and that for the $n = 2$ states; (b) the relativistic kinetic energy effect and the spin-orbit coupling; (c) the hyperfine splitting of the ground state. Compare your results with those for the hydrogen atom and explain major differences.

References

The most detailed discussion of the physics of hydrogenlike atoms may be found in

H. A. Bethe and E. E. Salpeter, *Quantum Mechanics of One- and Two-Electron Atoms*, Springer Verlag, 1957.

The Thomas precession is discussed in

R. M. Eisberg, *Fundaments of Modern Physics*, Wiley, New York (1961)

chapter 18

The Helium Atom

The helium atom consists of a nucleus of charge $Z = 2$ and two electrons, which we label 1 and 2. Each electron is attracted to the nucleus, and the two electrons repel each other. We assume, and this will turn out to be correct, that no forces, other than the electromagnetic ones (Coulomb to a very good approximation), are necessary to describe the dynamics of the helium atom with the help of quantum mechanics.

If the nucleus is placed at the origin, and if the electron coordinates are labeled $\mathbf{r}_1$ and $\mathbf{r}_2$, then the Hamiltonian for the atom is (Fig. 18-1)

$$ H = \frac{1}{2m}\mathbf{p}_1{}^2 + \frac{1}{2m}\mathbf{p}_2{}^2 - \frac{Ze^2}{r_1} - \frac{Ze^2}{r_2} + \frac{e^2}{|\mathbf{r}_1 - \mathbf{r}_2|} \tag{18-1} $$

Here m is the electron mass. We shall ignore the small effects connected with the motion of the nucleus,[1] relativistic effects, spin-orbin effects, and the effect of the current caused by the motion of one electron, upon the other electron. The above Hamiltonian may be written as

$$ H = H^{(1)} + H^{(2)} + V \tag{18-2} $$

with

$$ H^{(i)} = \frac{1}{2m}\mathbf{p}_i{}^2 - \frac{Ze^2}{r_i} \tag{18-3} $$

and

$$ V = \frac{e^2}{|\mathbf{r}_1 - \mathbf{r}_2|} \tag{18-4} $$

We shall work with the nuclear charge Z and set $Z = 2$ later. Our work on the hydrogen atom provides us with a complete set of eigenfunctions for $H^{(1)}$ and

---

[1] The reduced mass effect takes a somewhat different form since one is trying to convert a three-particle problem into an effective two-particle problem. This is worked out in D. Park, *Introduction to the Quantum Theory*, McGraw-Hill Co. (1964).

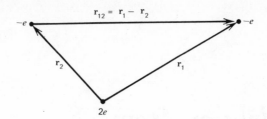

Fig. 18-1. Coordinates used in the formulation of the helium Hamiltonian.

$H^{(2)}$. Thus, if we were to ignore V in the total Hamiltonian, we would have a solution to the eigenvalue problem for the two-electron system. The eigenfunctions would be

$$u(\mathbf{r}_1,\mathbf{r}_2) = \phi_{n_1 l_1 m_1}(\mathbf{r}_1)\,\phi_{n_2 l_2 m_2}(\mathbf{r}_2) \tag{18-5}$$

for the equation

$$[H^{(1)} + H^{(2)}]\,u(\mathbf{r}_1,\mathbf{r}_2) = Eu(\mathbf{r}_1,\mathbf{r}_2) \tag{18-6}$$

and the energy would be given by (Fig. 18.2a)

$$E = E_{n_1} + E_{n_2} \tag{18-7}$$

where $E_n = -(mc^2/2)(Z\alpha)^2/n^2$. Thus in the idealized model in which the two electrons ignore each other, the lowest energy is

$$E = -2E_1 = -mc^2(2\alpha)^2 = -108.8 \text{ eV} \tag{18-8}$$

Note that this is $2 \times Z^2 = 8$ times the hydrogen energy of -13.6 eV.

The first excited state is one in which one electron is in its ground state, $n = 1$, and the second electron is raised to the first excited $n = 2$ state. Then

$$E = E_1 + E_2 = -68.0 \text{ eV} \tag{18-9}$$

The ionization energy, that is, the energy required to remove one electron from the ground state to infinity is

$$E_{\text{ioniz}} = (E_1 + E_\infty) - 2E_1 = 54.4 \text{ eV} \tag{18-10}$$

and, interestingly enough, the onset of the continuum lies *lower* than the excited state for which both electrons are in the $n = 2$ state. The energy of the latter state is

$$E = 2E_2 = -27.2 \text{ eV} \tag{18-11}$$

and it brings up a new phenomenon: the existence of a discrete state in the continuum for the Hamiltonian $H^{(1)} + H^{(2)}$. We shall briefly discuss the implications of this at the end of the chapter.

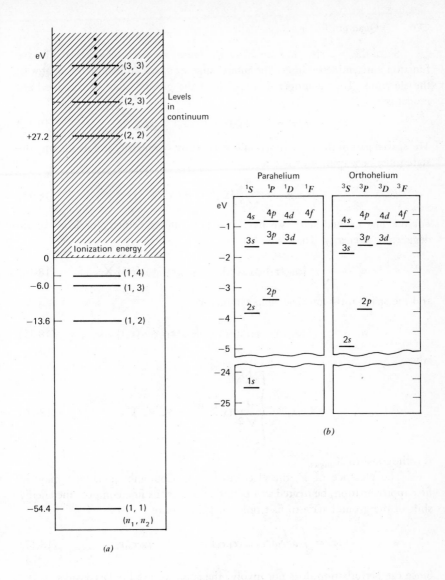

Fig. 18-2. (*a*) The spectrum of helium as it would look in the absence of the electron–electron interaction. The zero energy point is chosen at the ionization energy. (*b*) The actual spectrum of helium for the singlet (parahelium) and triplet (orthohelium) states. The level labeling has a suppressed (1*s*), so that the level (2*p*) is approximately described by the (1*s*)(2*p*) orbital.

Since the two electrons are *identical fermions* we must make the total wave function antisymmetric under the interchange of space and spin coordinates of the electrons. Thus a proper description of the ground state of this idealized model is

$$u_0(\mathbf{r}_1,\mathbf{r}_2) = \phi_{100}(\mathbf{r}_1)\, \phi_{100}(\mathbf{r}_2)\, X_{\text{singlet}} \tag{18-12}$$

The spatial part of the wave function is necessarily symmetric, and that is why the state must be a spin singlet state

$$X_{\text{singlet}} = \frac{1}{\sqrt{2}}\, (\chi_+^{(1)}\chi_-^{(2)} - \chi_-^{(1)}\chi_+^{(2)}) \tag{18-13}$$

For the first excited state, we have two possibilities, which, for $V = 0$, are degenerate in energy. These are

$$u_1^{(s)} = \frac{1}{\sqrt{2}}\, [\phi_{100}(\mathbf{r}_1)\, \phi_{2lm}(\mathbf{r}_2) + \phi_{2lm}(\mathbf{r}_1)\, \phi_{100}(\mathbf{r}_2)]\, X_{\text{singlet}} \tag{18-14}$$

and the space-antisymmetric, spin symmetric

$$u_1^{(t)} = \frac{1}{\sqrt{2}}\, [\phi_{100}(\mathbf{r}_1)\, \phi_{2lm}(\mathbf{r}_2) - \phi_{2lm}(\mathbf{r}_1)\, \phi_{100}(\mathbf{r}_2)]\, X_{\text{triplet}} \tag{18-15}$$

where

$$X_{\text{triplet}} = \begin{cases} \chi_+^{(1)}\chi_+^{(2)} \\ \dfrac{1}{\sqrt{2}}\, (\chi_+^{(1)}\chi_-^{(2)} + \chi_-^{(1)}\chi_+^{(2)}) \\ \chi_-^{(1)}\chi_-^{(2)} \end{cases} \tag{18-16}$$

is orthogonal to X_{singlet}.

The presence of V, the electron-electron Coulomb interaction may, in first approximation, be treated as a perturbation. Let us first compute the energy shift of the ground state to first order in V. We have

$$\Delta E = \int d^3r_1 d^3r_2\, u_0^*(\mathbf{r}_1,\mathbf{r}_2)\, \frac{e^2}{|\mathbf{r}_1 - \mathbf{r}_2|}\, u_0(\mathbf{r}_1,\mathbf{r}_2) \tag{18-17}$$

Since the perturbation does not involve the spin, we need only consider

$$\Delta E = \int d^3r_1 d^3r_2\, |\phi_{100}(\mathbf{r}_1)|^2\, \frac{e^2}{|\mathbf{r}_1 - \mathbf{r}_2|}\, |\phi_{100}(\mathbf{r}_2)|^2 \tag{18-18}$$

The integral has a simple physical interpretation. Since $|\phi_{100}(\mathbf{r}_1)|^2$ is the probability density of finding electron 1 at $\mathbf{r}_1$, we may interpret $e|\phi_{100}(\mathbf{r}_1)|^2$ as the charge density due to electron 1. Hence

$$U(\mathbf{r}_2) = \int d^3r_1\, \frac{e|\phi_{100}(\mathbf{r}_1)|^2}{|\mathbf{r}_1 - \mathbf{r}_2|} \tag{18-19}$$

is the potential at $\mathbf{r}_2$ due to the charge distribution of electron 1, and

$$\Delta E = \int d^3 r_2 \, e \, |\phi_{100}(\mathbf{r}_2)|^2 \, U(\mathbf{r}_2) \tag{18-20}$$

is therefore the electrostatic energy of interaction of electron 2 with that potential. The integral can be carried out. With $\phi_{100} = (2/\sqrt{4\pi})(Z/a_0)^{3/2} \, e^{-Zr/a_0}$ we have

$$\Delta E = \left[\frac{1}{\pi} \, (Z/a_0)^3\right]^2 e^2 \int_0^\infty r_1{}^2 \, dr_1 \, e^{-2Zr_1/a_0} \int_0^\infty r_2{}^2 \, dr_2 \, e^{-2Zr_2/a_0}$$

$$\int d\Omega_1 \int d\Omega_2 \frac{1}{|\mathbf{r}_1 - \mathbf{r}_2|} \tag{18-21}$$

In writing this, we used the separation

$$\int d^3 r = \int_0^\infty r^2 \, dr \, d\Omega$$

and isolated the only term that depends on the angles between $\mathbf{r}_1$ and $\mathbf{r}_2$. We have

$$\frac{1}{|\mathbf{r}_1 - \mathbf{r}_2|} = \frac{1}{(r_1{}^2 + r_2{}^2 - 2r_1 r_2 \cos\theta)^{1/2}} \tag{18-22}$$

where θ is the angle between $\mathbf{r}_1$ and $\mathbf{r}_2$. We may proceed in one of two ways.

(a) Most directly, we choose the direction of $\mathbf{r}_1$ as z-axis for the $d\Omega_2$ integration, and get

$$\int d\Omega_2 \frac{1}{|\mathbf{r}_1 - \mathbf{r}_2|} = \int_0^{2\pi} d\phi \int_{-1}^{1} d(\cos\theta) \frac{1}{(r_1{}^2 + r_2{}^2 - 2r_1 r_2 \cos\theta)^{1/2}}$$

$$= -2\pi \frac{1}{2r_1 r_2} \left[(r_1{}^2 + r_2{}^2 - 2r_1 r_2 \cos\theta)^{1/2}\right]_{\cos\theta = -1}^{\cos\theta = +1}$$

$$= \frac{\pi}{r_1 r_2} (r_1 + r_2 - |r_1 - r_2|) \tag{18-23}$$

The integration over $d\Omega_1$ is trivial, since nothing depends on that angle, so that

$$\int d\Omega_1 = 4\pi \tag{18-24}$$

and we are left with

$$4e^2 \left(\frac{Z}{a_0}\right)^6 \int_0^\infty r_1 \, dr_1 \, e^{-2Zr_1/a_0} \int_0^\infty r_2 \, dr_2 \, e^{-2Zr_2/a_0}$$

$$\times (r_1 + r_2 - |r_1 - r_2|) \tag{18-25}$$

(b) A very useful expansion, necessary when there is additional angular dependence in the numerator, is the following. For $r_1 > r_2$,

$$(r_1^2 + r_2^2 - 2r_1r_2 \cos \theta)^{-1/2} = r_1^{-1} \left(1 + \frac{r_2^2}{r_1^2} - 2\frac{r_2}{r_1} \cos \theta \right)^{-1/2}$$

$$= \frac{1}{r_1} \sum_{L=0}^{\infty} \left(\frac{r_2}{r_1} \right)^L P_L(\cos \theta) \qquad (18\text{-}26)$$

with the roles of r_1 and r_2 reversed when $r_2 > r_1$. Thus

$$\int d\Omega_1 \int d\Omega_2 \frac{1}{|\mathbf{r}_1 - \mathbf{r}_2|} = \int d\Omega_1 \int d\Omega_2 \sum_{L=0}^{\infty} \frac{r_<^L}{r_>^{L+1}} P_L(\cos \theta) \qquad (18\text{-}27)$$

where $r_>$ ($r_<$) is the larger (smaller) of r_1 and r_2. We can now proceed as before, using the fact that

$$\frac{1}{2} \int_{-1}^{1} d(\cos \theta) \, P_L(\cos \theta) = \delta_{L0} \qquad (18\text{-}28)$$

as a special case of

$$\frac{1}{2} \int_{-1}^{1} d(\cos \theta) \, P_L(\cos \theta) \, P_{L'}(\cos \theta) = \frac{\delta_{LL'}}{2L + 1} \qquad (18\text{-}29)$$

In any case, (18-25) becomes

$$\Delta E = 4e^2(Z/a_0)^6 \int_0^{\infty} r_1 \, dr_1 \, e^{-2Zr_1/a_0} \left\{ 2 \int_0^{r_1} r_2^2 \, dr_2 \, e^{-2Zr_2/a_0} \right.$$

$$\left. + 2r_1 \int_{r_1}^{\infty} r_2 \, dr_2 \, e^{-2Zr_2/a_0} \right\} \qquad (18\text{-}30)$$

The integrals are straightforward, and yield the answer

$$\Delta E = \frac{5}{8} \frac{Ze^2}{a_0} = \frac{5}{4} Z \left(\frac{1}{2} mc^2\alpha^2 \right) \qquad (18\text{-}31)$$

This is a positive contribution, since it arises from a repulsive force, and its magnitude, for $Z = 2$ is 34 eV. When this is added to the zero order result of -108.8 eV we obtain, to first order

$$E \simeq -74.8 \text{ eV} \qquad (18\text{-}32)$$

When this is compared with

$$E_{\text{exp}} = -78.975 \text{ eV} \qquad (18\text{-}33)$$

a sizable discrepancy is seen. Physically, we may attribute this discrepancy to the fact that in our calculation we took no account of "screening", that is, the effect that the presence of one electron tends to decrease the net charge "seen" by the other electron. Very roughly, if one argues that, for example, electron 1 is half the time "between" electron 2 and the nucleus, then half the time electron 2 sees a charge Z and half the time it sees a charge $Z - 1$, that is, effectively, in the expression

$$E + \Delta E = -\frac{1}{2} mc^2\alpha^2 \left(2Z^2 - \frac{5}{4} Z \right) \tag{18-34}$$

$(Z - 1/2)$ should be substituted for Z. This does improve agreement, but the crude argument advanced is not sufficient justification for the choice of 50% for the probability of effective screening. We will return to this subject later in this chapter, when we discuss the Rayleigh-Ritz variational principle for the ground state energy.

We next consider the first excited state of helium. It will be sufficient to calculate the energy shift with the singlet and triplet $m = 0$ states listed in (18-14) and (18-15), since the shift is caused by a perturbation that commutes with L_z. For such a perturbation, the shift must be independent of the m-value. Again, because of the spin-independence of the perturbing potential, V, we have

$$\Delta E_1^{(s,t)} = \frac{1}{2} e^2 \int d^3r_1 \int d^3r_2 \left[\phi_{100}(\mathbf{r}_1) \, \phi_{210}(\mathbf{r}_2) \pm \phi_{210}(\mathbf{r}_1) \, \phi_{100}(\mathbf{r}_2) \right]^*$$

$$\times \frac{1}{|\mathbf{r}_1 - \mathbf{r}_2|} \left[\phi_{100}(\mathbf{r}_1) \, \phi_{210}(\mathbf{r}_2) \pm \phi_{210}(\mathbf{r}_1) \, \phi_{100}(\mathbf{r}_2) \right]$$

$$= e^2 \int d^3r_1 \int d^3r_2 |\phi_{100}(\mathbf{r}_1)|^2 |\phi_{210}(\mathbf{r}_2)|^2 \frac{1}{|\mathbf{r}_1 - \mathbf{r}_2|}$$

$$\pm e^2 \int d^3r_1 \int d^3r_2 \, \phi_{100}^*(\mathbf{r}_1) \, \phi_{210}^*(\mathbf{r}_2) \frac{1}{|\mathbf{r}_1 - \mathbf{r}_2|} \, \phi_{210}(\mathbf{r}_1) \, \phi_{100}(\mathbf{r}_2) \tag{18-35}$$

In obtaining this simplified form, we made use of the symmetry of V under $\mathbf{r}_1 \leftrightarrow \mathbf{r}_2$.

The energy shift is seen to consist of two terms: the first has the familiar form of an electrostatic interaction between two "electron clouds" distributed according to the wave functions of the two electrons. This term is just a simple generalization of the term that we found for the ground state energy shift. The second term has no classical interpretation. Its origin lies in the Pauli principle, and its sign depends on whether the state has spin 0 or 1. Thus, because of this *exchange* contribution, the singlet and triplet terms are no longer degenerate. Although we considered $n = 2$ here, we have quite generally

$$\Delta E_{n,l}^{(t)} = J_{nl} - K_{nl}$$

$$\Delta E_{n,l}^{(s)} = J_{nl} + K_{nl} \tag{18-36}$$

The integrals can be evaluated in closed form [it is here that (18-27) becomes useful], but we shall not do this here. The integral J_{nl} is manifestly positive, and it turns out that this is also the case for K_{nl}. For $l = n - 1$ this is obvious: the wave functions appearing in (18-35) have no nodes in that case. That the triplet state should have a lower energy than the singlet state, that is, that

$$J_{nl} - K_{nl} < J_{nl} + K_{nl}$$

that is,

$$K_{nl} > 0 \qquad (18\text{-}37)$$

can be argued on qualitative grounds. For the triplet state the spatial wave function is antisymmetric, so that the electrons are somewhat constrained to stay away from each other. This tends to reduce the screening effect, so that each electron "sees" more of the nuclear charge, and it also tends to make the repulsion between the electrons less effective than for the spatially symmetric singlet state. An interesting aspect of this result is that, although the perturbing potential $e^2/|\mathbf{r}_1 - \mathbf{r}_2|$ does not depend on the spins of the electrons, the symmetry of the wave function does make the potential act as if it were spin-dependent. We may write (18-36) in a form that exhibits this. Let the spins of the two electrons be $\mathbf{s}_1$ and $\mathbf{s}_2$. Then the total spin $\mathbf{S} = \mathbf{s}_1 + \mathbf{s}_2$, and

$$\mathbf{S}^2 = \mathbf{s}_1{}^2 + \mathbf{s}_2{}^2 + 2\,\mathbf{s}_1 \cdot \mathbf{s}_2 \qquad (18\text{-}38)$$

If we act with this on triplet and singlet states (18-16) and (18-13) that are also eigenstates of $\mathbf{s}_1{}^2$ and $\mathbf{s}_2{}^2$, we get

$$S(S + 1)\hbar^2 = \frac{3}{4}\hbar^2 + \frac{3}{4}\hbar^2 + 2\mathbf{s}_1 \cdot \mathbf{s}_2$$

that is,

$$2\mathbf{s}_1 \cdot \mathbf{s}_2/\hbar^2 = S(S + 1) - \frac{3}{2} = \begin{cases} \dfrac{1}{2} & \text{triplet} \\[2ex] -\dfrac{3}{2} & \text{singlet} \end{cases} \qquad (18\text{-}39)$$

We may thus write, in terms of the $\boldsymbol{\sigma}$'s related to the spins by $\mathbf{s}_i = (1/2)\,\hbar\boldsymbol{\sigma}_i$,

$$\Delta E_{n,l} = J_{n,l} - \frac{1}{2}(1 + \boldsymbol{\sigma}_1 \cdot \boldsymbol{\sigma}_2) K_{nl} \qquad (18\text{-}40)$$

We shall see this phenomenon again when we discuss the H_2 molecule. Usually spin-dependent forces between atoms are quite weak. As illustrated in the example of spin-orbit coupling, the spin-dependent forces tend to arise from relativistic corrections to the static forces. In the spin-orbit example, these forces

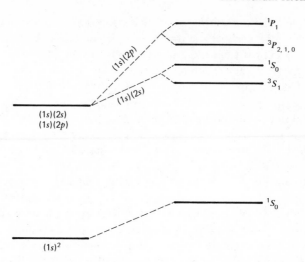

Fig. 18-3.   Schematic sketch of splitting of the first excited states of helium.

are down by a factor of α^2, which is just (v/c).[2] Such forces could not be strong enough to keep the electron spins aligned in a ferromagnet, except at unrealistically low temperatures.[2] The spin dependence due to exchange is much stronger than that: the force is of the same order of magnitude as the electrostatic force, and, as first observed by Heisenberg, it is responsible for the phenomenon of ferromagnetism.

The spectrum of the first few excited states of helium is shown in Fig. 18-3. The notation used for the unperturbed states is that of *orbitals*, that is, the quantum numbers of the unperturbed electrons. Thus both electrons in the ground state are in $n = 1$, $l = 0$ states, and we write this as $(1s, 1s)$, or more briefly $(1s)^2$. It should be understood that when we write $(1s)(2p)$, as for the first excited state, this does not mean that one electron is in one state, and the other electron in the other, since we must write totally antisymmetric wave functions for the electrons. Another way of labeling the state is by the $^{2S+1}L_J$ notation, which we use for the perturbed states in the figure. We see that the singlet states lie above the triplet states in a given multiplet. This follows from the symmetry (cf. our argument that $K_{nl} > 0$) and is a special example of one of *Hund's Rules: Other things being equal, the states of highest spin will have the lowest energy.*

If we excite helium from the ground state by shining ultraviolet light on it, we find that the *selection rule* $\Delta L = 1$, which we will derive later, implies an ex-

---

[2] A useful numerical relation is that in $E = kT$ a temperature of 300° K corresponds to an energy E of 1/40 eV.

citation to the P states. Furthermore, there is a selection rule $\Delta S = 0$, that is, only transitions singlet $\rightarrow$ singlet and triplet $\rightarrow$ triplet are probable.[3] Hence the state most strongly excited from the ground state is the $^1P_1$ state. The other levels may also become occupied through other mechanisms, for example, collisional excitation. Once occupied, the radiative transitions to the ground state are very improbable. The ^3P state, which may be populated when atoms in the $^1P_1$ state undergo collisions with other atoms in the gas, can only decay to the $^3S_1$ state, and that state is *metastable*, since it cannot decay to the ground state easily. The fact that there are no transitions, to good approximation, between triplet states and singlet states, led, at one time, to the belief that there existed two kinds of helium, ortho-helium (triplet) and para-helium (singlet).

The spectrum of helium that we saw in Fig. 18-2b shows that the excited states $(1s)(nl)$ have energies that do not differ very much from those of the hydrogen atom levels. Thus the binding energy of one electron in the atom is 24.6 eV (total binding energy minus binding energy of singly ionized helium = $79.0 - 54.4 = 24.6$ eV), whereas the energy that would be liberated if one electron were to be removed from the $2s$ state is of the order of $4 - 5$ eV, which is comparable to the energy 3.4 eV ($= 13.6/n^2$ eV) for hydrogen. The reason for this effect is that the "outer" electron sees only a unit positive charge, since the "inner" electron in the $(1s)$ orbital tends to shield the nucleus, leaving a net effective charge $\approx Z - 1$. This is not the case for the ground state, since both electrons have access to the nucleus. Thus the ground state lies quite a bit deeper than the hydrogen ground state.

In our discussion of the first order calculation of the ground state energy, we noted that there was a discrepancy of about 4 eV from the experimental value. Rather than attempt an estimate of the second order result, which would be very tedious, we turn to an entirely different method of calculating the ground state energy—the *Ritz variational method*.

Consider a Hamiltonian H, and an arbitrary square integrable function Ψ, which we choose to be normalized to unity, so that

$$\langle \Psi | \Psi \rangle = 1 \qquad (18\text{-}41)$$

This function Ψ can be expanded in a complete set of eigenstates of H, denoted by ψ_n,

$$H\psi_n = E_n\psi_n \qquad (18\text{-}42)$$

The expression reads

$$\Psi = \sum_n C_n\psi_n \qquad (18\text{-}43)$$

---

[3] Selection rules will be discussed in Chapter 22.

Now

$$\langle \Psi | H | \Psi \rangle = \sum_n \sum_m C_n^* \langle \psi_n | H | \psi_m \rangle \, C_m$$

$$= \sum_n \sum_m C_n^* C_m E_m \langle \psi_n | \psi_m \rangle$$

$$= \sum_n | C_n |^2 E_n$$

$$\geq E_0 \sum_n | C_n |^2 \qquad (18\text{-}44)$$

Since (18-41) implies that

$$\sum_n | C_n |^2 = 1 \qquad (18\text{-}45)$$

we obtain the result that

$$E_0 \leq \langle \Psi | H | \Psi \rangle \qquad (18\text{-}46)$$

We may use this result to calculate an upper bound on E_0. This can be done by choosing a Ψ that depends on a number of parameters ($\alpha_1, \alpha_2, \ldots$), calculating $\langle \Psi | H | \Psi \rangle$, and minimizing this with respect to the parameters.

We illustrate the utility of this procedure by calculating the ground state energy of helium with a Ψ chosen to be a product of hydrogenlike wave functions in the (1s) orbitals, but corresponding to an arbitrary charge Z^*. We take

$$\Psi(\mathbf{r}_1, \mathbf{r}_2) = \psi_{100}(\mathbf{r}_1) \, \psi_{100}(\mathbf{r}_2) \qquad (18\text{-}47)$$

where

$$\left(\frac{\mathbf{p}^2}{2m} - \frac{Z^* e^2}{r} \right) \psi_{100}(\mathbf{r}) = \epsilon \psi_{100}(\mathbf{r}) \qquad (18\text{-}48)$$

with $\epsilon = -(1/2) \, mc^2 (Z^*\alpha)^2$. We now need

$$\int d^3 r_1 \int d^3 r_2 \, \psi_{100}^*(\mathbf{r}_1) \, \psi_{100}^*(\mathbf{r}_2) \left(\frac{\mathbf{p}_1^2}{2m} + \frac{\mathbf{p}_2^2}{2m} - \frac{Z e^2}{r_1} - \frac{Z e^2}{r_2} \right.$$

$$\left. + \frac{e^2}{|\mathbf{r}_1 - \mathbf{r}_2|} \right) \psi_{100}(\mathbf{r}_1) \, \psi_{100}(\mathbf{r}_2) \qquad (18\text{-}49)$$

We have

$$\int d^3 r_1 \int d^3 r_2 \, \psi_{100}^*(\mathbf{r}_1) \, \psi_{100}^*(\mathbf{r}_2) \left(\frac{\mathbf{p}_1^2}{2m} - \frac{Z e^2}{r_1} \right) \psi_{100}(\mathbf{r}_1) \, \psi_{100}(\mathbf{r}_2)$$

$$= \int d^3 r_1 \, \psi_{100}^*(\mathbf{r}_1) \left(\frac{\mathbf{p}_1^2}{2m} - \frac{Z^* e^2}{r_1} + \frac{(Z^* - Z) \, e^2}{r_1} \right) \psi_{100}(\mathbf{r}_1)$$

$$= \epsilon + (Z^* - Z) e^2 \int d^3 r_1 |\psi_{100}(\mathbf{r}_1)|^2 \frac{1}{r_1}$$

$$= \epsilon + (Z^* - Z) e^2 \frac{Z^*}{a_0}$$

$$= \epsilon + Z^*(Z^* - Z) mc^2\alpha^2 \tag{18-50}$$

An identical factor comes from the Hamiltonian for electron 2, and the expectation value of the electron-electron repulsion has already been calculated in (18-31), except that we must substitute Z^* for Z there. Adding up the terms, we get

$$\langle \Psi | H | \Psi \rangle = -\frac{1}{2} mc^2\alpha^2 \left(2Z^{*2} + 4Z^*(Z - Z^*) - \frac{5}{4} Z^* \right)$$

$$= -\frac{1}{2} mc^2\alpha^2 \left(4ZZ^* - 2Z^{*2} - \frac{5}{4} Z^* \right) \tag{18-51}$$

Minimizing this with respect to Z^* yields

$$Z^* = Z - \frac{5}{16} \tag{18-52}$$

which is an improvement on the guess we made earlier $(Z - 1/2)$. We thus obtain

$$E_0 \le -\frac{1}{2} mc^2\alpha^2 \left[2 \left(Z - \frac{5}{16} \right)^2 \right] = -77.38 \text{ eV} \tag{18-53}$$

when we substitute $Z = 2$. This is much better than the first order perturbation result.

The variational calculation can be done with more complicated trial wave functions. Pekeris[4] used a 1075 term wave function and minimized $\langle \Psi | H | \Psi \rangle$ on a computer. The resulting bound agrees, within experimental errors, with what is measured. It is, of course, true that such a complicated wave function does not have a form that is as easily interpretable as (18-47), with its partial screening effects. It does, however, provide strong support for the correctness of quantum mechanics, and for the assumption that only electromagnetic forces are required to explain the structure of atoms.

In conclusion, we briefly return to our observation that there exist eigenvalues of $H^{(1)} + H^{(2)}$ that lie above the ionization threshold and that are nevertheless discrete. The states labeled by the orbitals $(2s)^2$ or $(2s)(2p)$, for example, lie well above the ionization energy. This has some dramatic physical consequences. Consider, for example, the $(2s)(2p)$ state. If the electrons form a spin

---

[4] This is discussed in Bethe and Jackiw, *loc. cit.*

singlet state, then this will be a $^1P_1$ state, and it can be excited from the ground state by the absorption of radiation, since the selection rules $\Delta l = 1$ and $\Delta S = 0$ are not being violated. This state, once excited, need not decay back to the ground state ($^1S_0$) or to another state allowed by the selection rules (a $^1D_2$ state, say), because it can go into another *channel*: it can decay into an electron and singly ionized helium, He^+, with the electron energy determined by energy conservation. This process is described as *autoionization*.

The $(2s)(2p)$ state in the continuum will show up very clearly in the scattering of electrons by He^+ ions. When the electron energy is such that the *compound state* can be formed, a very dramatic peak will occur in the scattering rate. Similarly, in the absorption of radiation by helium, in the vicinity of the energy of the compound state ($e^- - He^+$), a sharp peak is seen in the absorption (Fig. 18-4). There is absorption at other energies, too, since the process

$$\text{radiation} + He \rightarrow e^- + He^+$$

can occur, but the absorption at energies away from the compound state energy will vary very smoothly with energy. We can describe the state in still another way by calling it a *resonant state*. Since it decays into its constituents $e^- + He^+$,

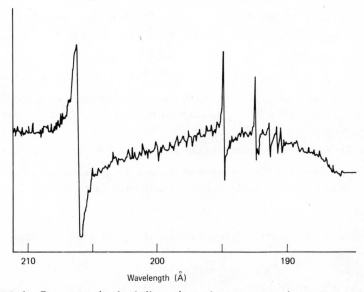

Wavelength (Å)

Fig. 18-4. Resonance in the helium absorption spectrum above the continuum threshold; the first peak occurs at the energy corresponding to the location of the $(2s)(2p)$ level. (From R. P. Madden and K. Codling, *Phys. Rev. Letters*, **10**, *516* (1963), by permission.)

it does not exist forever. Hence, by the uncertainty relation, $\Delta E \gtrsim \hbar/\Delta t$, it appears that its energy is not precisely defined, which seems to contradict the fact that the $(2s)(2p)$ state does have a well-defined energy. It turns out that if the coupling of the discrete state to the continuum state is taken into account, the state ceases to be discrete, and its energy may lie anywhere in a narrow range about the energy as calculated without the coupling. We shall return to this topic in Chapter 23 and in Special Topics section 4, "Lifetimes, Line Widths, and Resonances."

Problems

1. Consider the helium atom in the approximation in which the electron-electron interaction is neglected. What is the lowest orthohelium (spin 1) state? What is its degeneracy in the above approximation? Write down the expression of the splitting due to electron-electron repulsion in first order perturbation theory, and estimate its magnitude.

2. Calculate the energy shift $\Delta E_{2,l}^{(t)}$ ($l = 0, 1$).

3. Consider the lowest state of orthohelium. What is its magnetic moment, that is, calculate the interaction with an external magnetic field.

4. Consider

$$"E" = \langle \Psi | H | \Psi \rangle$$

with an arbitrary trial wave function Ψ. Show that if Ψ differs from the correct ground state wave function ψ_0 by terms of order ϵ, then "E" differs from the ground state energy by terms of order ϵ^2.

(Note. Do not forget the normalization condition $\langle \Psi | \Psi \rangle = 1$.)

5. Use the variational principle to estimate the ground state energy of the three-dimensional harmonic oscillator, using the trial wave function

$$\Psi = N e^{-\alpha r}$$

6. Consider a one-dimensional cut-off harmonic oscillator of the form

$$V(x) = \frac{1}{2} m\omega^2(x^2 - a^2) \qquad |x| < a$$

$$= 0 \qquad |x| > a$$

Use the variational principle to calculate the best upper bound to the ground state energy using the exponential form $N e^{-\beta|x|}$.

7. Consider the binding of a proton and a neutron (both with $mc^2 = 938$ MeV, approx) by means of a potential

$$V(r) = V_0 \frac{e^{-r/r_0}}{r/r_0}$$

with the system in an $L = 0$ state. The range of the potential is given by r_0. Use the following procedure to calculate the depth of the potential required to give the binding energy E_B. (a) Calculate an approximate value of the binding energy using the variational principle. (b) In the expression that connects the approximate value with r_0 and the depth of the potential, insert the experimental value of E_B. Do your numerical evaluation using $r_0 = 2.8 \times 10^{-13}$ cm and $E_B = -2.23$ MeV. (Do not forget the reduced mass.)

8. Consider a finite-dimensional matrix H_{ij}. Show that the condition for minimizing

$$\langle \Psi | H | \Psi \rangle = \sum_{i,j=1}^{n} a_i^* H_{ij} a_j$$

subject to the condition

$$\langle \Psi | \Psi \rangle = \sum_{i=1}^{n} a_i^* a_i = 1$$

yields the eigenvalues of the matrix H.

(*Hint.* Use the method of Lagrange multipliers.)

9. Use the variational principle to show that a one-dimensional attractive potential will always have a bound state.

(*Hint.* Evaluate $\langle \Psi | H | \Psi \rangle$ with a convenient trial function, for example, $Ne^{-\beta^2 x^2}$ and show that the above can always be made negative.)

10. Use the data of Fig. 18-4 to compute the location of the $(2s)(2p)$ level above the ground state of helium and compute the velocity of the electron emitted in autoionization, if the He^+ ion is in its lowest state at the end. What will it be if the He^+ ion is in its first excited state?

11. Consider a wave function $\psi(\alpha_1, \alpha_2, \ldots \alpha_n)$ for which only the dependence on some parameters is exhibited. The wave function is normalized

$$\langle \psi(\alpha_1, \alpha_2, \ldots \alpha_n) | \psi(\alpha_1, \alpha_2, \ldots \alpha_n) \rangle = 1$$

and the dependence on the parameters is so chosen that

$$\mathcal{E} = \langle \psi(\alpha_1, \ldots) | H | \psi(\alpha_1, \ldots) \rangle$$

is a minimum. Show that the parameters are determined by the set of equations

$$\left\langle \psi(\alpha_1, \ldots) | H \left| \frac{\partial \psi}{\partial \alpha_i} \right. \right\rangle - \mu \left\langle \psi(\alpha_1, \ldots) \left| \frac{\partial \psi}{\partial \alpha_i} \right. \right\rangle = 0 \qquad i = 1, 2, \ldots n$$

where μ is a Lagrange multiplier. Let H depend on a parameter λ (e.g., the nuclear charge or some distance, say the internuclear distance in a molecule). Then the α_i will depend on that parameter. Prove that

$$\frac{d\mathcal{E}}{d\lambda} = \left\langle \psi(\alpha_1, \ldots) \Big| \frac{\partial H}{\partial \lambda} \Big| \psi(\alpha_1, \ldots) \right\rangle$$

This is known as the Feynman-Hellmann theorem and is very useful in molecular physics calculations.

12. Use the variational principle to estimate the ground state energy for the anharmonic oscillator

$$H = \frac{p^2}{2m} + \lambda x^4$$

Compare your result with the exact result

$$E_0 = 1.060\lambda^{1/3}\left(\frac{\hbar^2}{2m}\right)^{2/3}$$

References

A very nice discussion of the spectrum of helium may be found in

H. A. Bethe and R. W. Jackiw, *Intermediate Quantum Mechanics*, W. A. Benjamin, Inc., 1968.

chapter 19

The Structure of Atoms

The energy eigenvalue problem for an atom with Z electrons has the form

$$\left(\sum_{i=1}^{Z} \frac{\mathbf{p}_i^2}{2m} - \frac{Ze^2}{r_i} + \sum_{i>j} \frac{e^2}{|\mathbf{r}_i - \mathbf{r}_j|} \right) \psi(\mathbf{r}_1, \mathbf{r}_2, \ldots, \mathbf{r}_Z) = E\psi(\mathbf{r}_1, \mathbf{r}_2, \ldots, \mathbf{r}_Z) \quad (19\text{-}1)$$

and is a partial differential equation in $3Z$ dimensions. For light atoms it is possible to solve such an equation on a computer, but such solutions are only meaningful to the expert. We shall base our discussion of atomic structure on a different approach. As in the example of helium ($Z = 2$), it is both practical and enlightening to treat the problem as one involving Z independent electrons in a single potential, and to consider the electron-electron interaction later. Perturbation theory turned out to be adequate for $Z = 2$, but as the number of electrons increases, the shielding effects, not taken into account by first order perturbation theory, become more and more important. The variational principle discussed at the end of Chapter 18 had the virtue of maintaining the single-particle picture, while at the same time yielding single particle functions that incorporate the screening corrections.

To apply the variational principle, let us assume that the trial wave function is of the form

$$\psi(\mathbf{r}_1, \mathbf{r}_2, \ldots, \mathbf{r}_Z) = \phi_1(\mathbf{r}_1)\, \phi_2(\mathbf{r}_2) \ldots \phi_Z(\mathbf{r}_Z) \quad (19\text{-}2)$$

Each of the functions is normalized to unity. If we calculate the expectation value of H in this state, we obtain

$$\langle H \rangle = \sum_{i=1}^{Z} \int d^3r_i\, \phi_i^*(\mathbf{r}_i) \left(-\frac{\hbar^2}{2m} \nabla_i^2 - \frac{Ze^2}{r_i} \right) \phi_i(\mathbf{r}_i)$$

$$+ e^2 \sum_{i>j} \sum_j \iint d^3\mathbf{r}_i\, d^3\mathbf{r}_j\, \frac{|\phi_i(\mathbf{r}_i)|^2 |\phi_j(\mathbf{r}_j)|^2}{|\mathbf{r}_i - \mathbf{r}_j|} \quad (19\text{-}3)$$

The procedure of the variational principle is to pick the $\phi_i(\mathbf{r}_i)$ such that $\langle H \rangle$ is a minimum. If we were to choose the $\phi_i(\mathbf{r}_i)$ to be hydrogenlike wave functions, with a different Z_i for each electron (and with each electron in a different quantum state to satisfy the Pauli exclusion principle), we would get a set of equations analogous to (18-51) and (18-52). A more general approach is that due to Hartree. If the $\phi_i(\mathbf{r}_i)$ were the single particle wave functions that minimized $\langle H \rangle$, then an alteration in these functions by an infinitesimal amount

$$\phi_i(\mathbf{r}_i) \rightarrow \phi_i(\mathbf{r}_i) + \lambda f_i(\mathbf{r}_i) \tag{19-4}$$

should only change $\langle H \rangle$ by a term of order λ^2. The alterations must be such that

$$\int d^3 r_i |\phi_i(\mathbf{r}_i) + \lambda f_i(\mathbf{r}_i)|^2 = 1 \tag{19-5}$$

that is, to first order in λ,

$$\int d^3 r_i [\phi_i^*(\mathbf{r}_i) f_i(\mathbf{r}_i) + \phi_i(\mathbf{r}_i) f_i^*(\mathbf{r}_i)] = 0 \tag{19-6}$$

Let us compute the terms linear in λ that arise when (19-4) is substituted into (19-3). Term by term, we have

$$\sum_i \int d^3 r_i \left[\phi_i^*(\mathbf{r}_i) \left(-\frac{\hbar^2}{2m} \mathbf{\nabla}_i^2 \right) \lambda f_i(\mathbf{r}_i) + \lambda f_i^*(\mathbf{r}_i) \left(-\frac{\hbar^2}{2m} \mathbf{\nabla}_i^2 \right) \phi_i(\mathbf{r}_i) \right]$$
$$= \lambda \sum_i \int d^3 r_i \left\{ f_i(\mathbf{r}_i) \left[-\frac{\hbar^2}{2m} \mathbf{\nabla}_i^2 \phi_i^*(\mathbf{r}_i) \right] + f_i^*(\mathbf{r}_i) \left[-\frac{\hbar^2}{2m} \mathbf{\nabla}_i^2 \phi_i(\mathbf{r}_i) \right] \right\} \tag{19-7}$$

To obtain this we have integrated by parts two times, and used the fact that $f_i(\mathbf{r}_i)$ must vanish at infinity in order to be an acceptable variation of a square integrable function. Next we have

$$-\lambda \sum_i \int d^3 r_i \left[f_i^*(\mathbf{r}_i) \frac{Ze^2}{r_i} \phi_i(\mathbf{r}_i) + \phi_i^*(\mathbf{r}_i) \frac{Ze^2}{r_i} f_i(\mathbf{r}_i) \right] \tag{19-8}$$

and finally

$$\lambda e^2 \sum_{i>j} \sum_j \int d^3 r_i \int d^3 r_j \frac{1}{|\mathbf{r}_i - \mathbf{r}_j|} \{ [f_i^*(\mathbf{r}_i) \phi_i(\mathbf{r}_i) + f_i(\mathbf{r}_i) \phi_i^*(\mathbf{r}_i)] |\phi_j(\mathbf{r}_j)|^2$$
$$+ [f_j^*(\mathbf{r}_j) \phi_j(\mathbf{r}_j) + f_j(\mathbf{r}_j) \phi_j^*(\mathbf{r}_j)] |\phi_i(\mathbf{r}_i)|^2 \} \tag{19-9}$$

We cannot just set the sum of these three terms equal to zero because the $f_i(\mathbf{r}_i)$ are constrained by (19-6). The proper way to account for the constraint is by the use of Lagrange multipliers, that is, we multiply each of the constraining relations (19-6) by a constant (the "multiplier") and add the sum to our three terms. The total can then be set equal to zero, since the constraints on the $f_i(\mathbf{r}_i)$ are now taken care of. With a certain amount of notational foresight we label the multipliers $-\epsilon_i$, and thus get

$$\sum_i \int d^3 \mathbf{r}_i \left\{ f_i^*(\mathbf{r}_i) \left[-\frac{\hbar^2}{2m} \boldsymbol{\nabla}_i^2 \phi_i(\mathbf{r}_i) \right] - f_i^*(\mathbf{r}_i) \frac{Ze^2}{r_i} \phi_i(\mathbf{r}_i) \right\}$$

$$+ e^2 \sum_{i \neq j} \sum_j \iint d^3 \mathbf{r}_i \, d^3 \mathbf{r}_j \, f_i^*(\mathbf{r}_i) \frac{|\phi_j(\mathbf{r}_j)|^2}{|\mathbf{r}_i - \mathbf{r}_j|} \phi_i(\mathbf{r}_i)$$

$$- \epsilon_i \int d^3 r_i \, f_i^*(\mathbf{r}_i) \, \phi_i(\mathbf{r}_i) + (\text{complex conjugate term}) = 0 \qquad (19\text{-}10)$$

In deriving the second line, first we converted the double sum $\sum_{i>j} \sum_j$ into $(1/2) \sum_{i \neq j} \sum_j$, which is unrestricted except for the requirement that $i \neq j$, and then used the fact that the integrand in (19-9) is symmetric in i and j. Now $f_i(\mathbf{r}_i)$ is completely unrestricted, so that we may treat $f_i(\mathbf{r}_i)$ and $f_i^*(\mathbf{r}_i)$ as completely independent (each one has a real and an imaginary part). Furthermore, other than being square integrable, they are completely arbitrary, so that for (19-10) to hold, the coefficients of $f_i(\mathbf{r}_i)$ and $f_i^*(\mathbf{r}_i)$ must separately vanish *at each point* $\mathbf{r}_i$, since we are allowed to make local variations in the functions $f_i(\mathbf{r}_i)$ and $f_i^*(\mathbf{r}_i)$. We are thus led to the condition that

$$\left[-\frac{\hbar^2}{2m} \boldsymbol{\nabla}_i^2 - \frac{Ze^2}{r_i} + e^2 \sum_{j \neq i} \int d^3 \mathbf{r}_j \frac{|\phi_j(\mathbf{r}_j)|^2}{|\mathbf{r}_i - \mathbf{r}_j|} \right] \phi_i(\mathbf{r}_i) = \epsilon_i \phi_i(\mathbf{r}_i) \qquad (19\text{-}11)$$

and the complex conjugate relation.

This equation has a straightforward interpretation: it is an energy eigenvalue equation for electron "i" located at $\mathbf{r}_i$, moving in a potential

$$V_i(\mathbf{r}_i) = -\frac{Ze^2}{r_i} + e^2 \sum_{j \neq i} \int d^3 \mathbf{r}_j \frac{|\phi_j(\mathbf{r}_j)|^2}{|\mathbf{r}_i - \mathbf{r}_j|} \qquad (19\text{-}12)$$

that consists of an attractive Coulomb potential due to a nucleus of charge Z, and a repulsive contribution due to the charge density of all the other electrons. We do not, of course, know the charge densities

$$\rho_j(\mathbf{r}_j) = e|\phi_j(\mathbf{r}_j)|^2 \qquad (19\text{-}13)$$

of all the other electrons, so that we must search for a *self-consistent* set of $\phi_i(\mathbf{r}_i)$, in the sense that their insertion in the potential leads to eigenfunctions that reproduce themselves. The equation (19-11) is a rather complicated integral equation, but it is at least an equation in three dimensions (we can replace the variable $\mathbf{r}_i$ by $\mathbf{r}$), and that makes numerical work much easier. An even greater simplification occurs when $V_i(\mathbf{r})$ is replaced by its angular average

$$V_i(r) = \int \frac{d\Omega}{4\pi} V_i(\mathbf{r}) \qquad (19\text{-}14)$$

for then the self-consistent potential becomes central, and the self-consistent solutions can be decomposed into angular and radial functions, that is, they will be functions that can be labeled by n_i, l_i, m_i, σ_i, with the last label referring to the spin state ($S_{iz} = \pm 1/2$).

The trial wave function (19-2) does not take into account the exclusion principle. The latter plays an important role, since if all the electrons could be in the same quantum state, the energy would be minimum with all the electrons in the $n = 1$, $l = 0$ "orbital." Atoms do not have such a simple structure. To take the exclusion principle into account, we add to the *Ansatz* represented by (19-2) the rule: *every electron must be in a different state*, if the spin states are included in the labeling. A more sophisticated way of doing this automatically is to replace (19-2) by a trial wave function that is a *Slater determinant* [cf. (8-60)]. The resulting equations differ from (19-11) by the addition of an exchange term. The new Hartree-Fock equations have eigenvalues that turn out to differ by 10–20% from those obtained using Hartree equations (with the rule stated above), and since it is a little easier to talk about the physics of atomic structure in terms of the Hartree picture, we will not discuss the Hartree-Fock equations.

The potential (19-14) no longer has the $1/r$ form, and thus the degeneracy of all states with a given n and $l \leq n - 1$ is no longer present. We may expect, however, that for low Z at least, the splitting for different l values for a given n will be smaller than the splitting between different n-values, so that electrons placed in the orbitals $1s$, $2s$, $2p$, $3s$, $3p$, $3d$, $4s$, $4p$, $4d$, $4f$, . . . will be successively less strongly bound.[1] Screening effects will accentuate this: whereas s orbitals do overlap the small r region significantly, and thus feel the full nuclear attraction, the p-, d-, . . . orbitals are forced out by the centrifugal barrier, and feel less than the full attraction. This effect is so strong that the energy of the $3d$ electrons is very close to that of the $4s$ electrons, so that the anticipated ordering is sometimes disturbed. The same is true for the $4d$ and $5s$ electrons, the $4f$ and $6s$ electrons and so on. The dominance of the l-dependence over the n-dependence becomes more important' as we go to larger Z values, as we shall see in our discussion of the *periodic table*.

The number of electrons that can be placed in orbitals with a given (n,l) is $2(2l + 1)$, since there are two spin states for given m-value. When all these $2(2l + 1)$ states are filled, we speak of the *closing of a shell*. The charge density for a closed shell has the form

$$e \sum_{m=-l}^{l} |R_{nl}(r)|^2 |Y_{lm}(\theta,\phi)|^2 \tag{19-15}$$

---

[1] The notation is the same as that used for hydrogen. A more sensible notation, used by nuclear shell-structure physicists, is to replace the n by $n - l$, which is just an index representing the ordering of a given l-state. Thus instead of starting with $3d$ states, for example, it might be more sensible to have the lowest d state called the $1d$ state, and so on. We shall nevertheless continue to use the conventional notation, even though the n-value does not have much to do with the ordering of levels for large Z atoms.

and this is spherically symmetric because of the property of spherical harmonics that

$$\sum_{m=-l}^{l} |Y_{lm}(\theta,\phi)|^2 = \frac{2l+1}{4\pi} \tag{19-16}$$

Let us now discuss the building up of atoms by the addition of more and more electrons to the appropriate nucleus, whose only role, in our approximation, is to provide the charge Z.

Hydrogen. Here there is only one electron, and the ground state configuration is $(1s)$. The spectroscopic description of the electronic state is $^2S_{1/2}$, and the binding energy, as is well known, is 13.6 eV.

Helium. Here $Z = 2$, and, as we saw in Chapter 18, the ground state configuration is $(1s)^2$, which is a shorthand notation for $(1s)(1s)$. The state, in the (L,S) description is a $^1S_0$ state, and the total binding energy is 79 eV. After one electron is removed, the remaining electron is in a $(1s)$ orbit about a $Z = 2$ charge, so that its binding energy is $13.6Z^2 = 54.4$ eV. Thus the energy required to remove the least bound electron, the *ionization energy* is the difference, that is, 24.6 eV (see Fig. 18-2b). It is also interesting to estimate the energy of the first excited state, which is $(1s)(2s)$: this is $13.6\,Z^2 + (13.6/n^2)(Z-1)^2$ because of the shielding, that is, approximately 58 eV. Thus it takes approximately $79 - 58 \cong 20$ eV to excite the helium atom.[2] Because the electrons form a closed shell it is chemically inert, a property shared by all atoms whose electrons form closed shells.

Lithium. Here $Z = 3$, and the exclusion principle forbids a $(1s)^3$ configuration. The lowest lying accessible configuration is the $(1s)^2(2s)$. Since we are adding a single electron to a closed shell $(^1S_0)$, the spectroscopic description of the state is $^2S_{1/2}$ as for hydrogen. If screening were perfect, the additional electron would only "see" a $Z = 1$, and since $n = 2$, we would have an energy of $13.6/4 = 3.4$ eV. The screening is not perfect; in fact, since the orbital of the extra electron is $(2s)$, there is a reasonable overlap of the wave function at $r = 0$, and hence the effective Z is larger than 1. The experimental energy, 5.4 eV shows that $Z^* = 1.3$.

Beryllium. With $Z = 4$, the natural place for the fourth electron to go is into the second space in the $2s$ orbital, so that the configuration is $(1s)^2(2s)^2$, and we again have a closed shell, with a $^1S_0$ spectroscopic state description. As far as the energy is concerned, the situation is very much like that of helium. If the

---

[2] This is a crude estimate that ignores the electron-electron repulsion and exchange effects. The difference between the 20 eV and the 24.6 eV is the 4–5 eV that will be released when the excited atom decays to its ground state. (See Fig. 18-2b.)

screening were perfect, the only difference would be that the last electron is in an $n = 2$ state, giving a binding energy of $24.6/n^2 = 6.2$ eV. Screening is not perfect, and the experimental value is 9.3 eV. Although a shell is closed, the excitation of an electron to a $2p$ state will require relatively little energy. Thus in the presence of another element a rearrangement of electrons may yield enough energy to break up the closed shell. Hence beryllium is not as inert as helium. In general this type of shell is not quite as stable as the shell in which, for a given n, all the possible l-states are filled.

Boron. After the closing of the second shell, the fifth electron can either be put in the $3s$ or in the $2p$ orbital. The latter is lower in energy, and it is the $2p$ shell that begins to fill up, starting with boron. The configuration is $(1s)^2(2s)^2(2p)$, and the state is $^2P_{1/2}$. The last deserves a comment: if we add spin $1/2$ to orbital angular momentum 1, we may have $J = 3/2$ or $1/2$. These are split by the spin-orbit interaction

$$\frac{1}{2m^2c^2} \mathbf{L} \cdot \mathbf{S} \frac{1}{r} \frac{dV(r)}{dr} = \frac{1}{4m^2c^2} [J(J+1) - L(L+1) - S(S+1)] \frac{1}{r} \frac{dV(r)}{dr}$$

$$(19\text{-}17)$$

and the form of this leads to the higher J value having a higher energy, since the expectation value of $(1/r)[dV(r)/dr]$, even though no longer equal to the value given in (17-16), is still positive. This conclusion may not hold when there are more electrons in an unfilled shell. The ionization energy might be expected to be somewhat smaller than that of beryllium, since the $2p$ state energy is somewhat higher than that of the $2s$ orbital, because of the centrifugal barrier. The experimental value is 8.3 eV.

Carbon. Here $Z = 6$, and the $2p$ shell continues to be filled. The configuration is $(1s)^2(2s)^2(2p)^2$. The total spin may be 0 or 1 and the total orbital angular momentum may be 2, 1, or 0 (since we are adding two orbital angular momenta 1). Since the wave function must be antisymmetric for the two electrons outside the closed shells, a singlet state must have even L, and a triplet odd L, so that there are only the possibilities $^1S_0$, $^3P_{2,1,0}$ and $^1D_2$. We now invoke *Hund's rule*, referred to in our discussion of helium: "The state of highest spin has the lowest energy." Thus we must have a ^3P state. The result, that the state is $^3P_0$, follows from Hund's second rule, that has been abstracted from spinorbit calculations:

If the incomplete shell is not more than half filled, then the lowest level has $J = |L - S|$, its minimum value. If the shell is more than half filled, then the maximum J value, $J = L + S$ has the least energy.

Since it takes six electrons to fill the $2p$ shell, we obtain $J = 0$. As far as the ionization energy is concerned, we have increased Z by one. Since the second $2p$

electron can stay "out of the way" of the first one, by being in a different m-state, the repulsion between the electrons will be of less importance, and we expect a somewhat larger binding. The experimental value is 11.3 eV.

Nitrogen. The $Z = 7$ atom has the configuration $(1s)^2(2s)^2(2p)^3$, or, $(2p)^3$, if, for brevity, we omit the closed shells from our description. By Hund's rule, the spin of the ground state is the maximum value $S = 3/2$. This is a symmetric spin state (this is most evident in the $S_z = 3/2$ state, for which all the spins must be parallel), and hence the three filled $3p$ orbitals must each be in a different m-state. Of the total L values of 3, 2, 1, 0 that can be obtained by vectorially adding three unit orbital angular momenta, the $L = 3$ state is clearly excluded. One must look at the detailed construction of the states to find out that the totally antisymmetric state is the $L = 0$ state, so that the ground state is $^4S_{3/2}$. The ionization potential might be expected to be a little larger than that for carbon, since Z is again increased by one, and the third electron can be put in the third p orbital without significantly overlapping the other two electrons in the $2p$ shell, that is, by reducing somewhat the effect of the electron-electron repulsion. The experimental value is 14.5 eV.

Oxygen. ($Z = 8$) The configuration may be abbreviated by $(2p)$,[4] and the shell is more than half full, and it appears that the determination of the electronic state is very complicated indeed. We can, however, look at the shell in another way: we know that when the shell is filled, that is, when the configuration is $(2p)^6$ ($Z = 10$), then the total state has $L = S = 0$. We may thus think of oxygen as having a closed $2p$ shell with two *holes* in it. These holes are just like "anti-electrons" (though they are not positrons!) and we can look at possible two-hole configurations. These will be the same as two-electron configurations, since holes also have spin $1/2$. Thus, as with carbon, the possible states consistent with the antisymmetry of the two-fermion (two-hole) wave function are ^1S, ^3P, ^1D, and the four electrons that, when added to these give $L = S = 0$, must be in similar states. The highest spin is $S = 1$, and by Hund's second rule, the angular momentum, for a more than half-filled shell, must be the maximum $J = 2$. Thus the state is $^3P_2$. When the fourth electron is added to the $2p$ shell, it must be put into an orbital with an m-value that is already occupied, so that the overlap between two of the electrons is larger than before. Hence it is not surprising that the ionization energy drops to the value of 13.6 eV.

Fluorine. Here $Z = 9$, and the configuration is $(2p)^5$, that is, we have one hole in a p orbital. The state must be $^2P_{3/2}$ since the maximum of $J = 1/2$ or $3/2$ must be chosen. The monotonic increase in the ionization energy resumes, with the value of 17.4 eV.

Neon. With $Z = 10$, the $2p$ shell is closed, the ground state is a $^1S_0$ state, and the ionization energy, continuing the monotonic trend is 21.6 eV.

At this point, the addition of another electron requires putting it in an orbit with a higher n value ($n = 3$), and thus neon marks the end of a *period* in the periodic table, as did helium. In neon, as in helium, the first available state into which an electron can be excited has a higher n-value, so that it takes quite a lot of energy to perturb the atom. Neon shares with helium the property of being an *inert gas*.

The next period again has eight elements in it. First the ($3s$) shell is filled, with sodium ($Z = 11$) and magnesium ($Z = 12$) and then the $3p$ shell, which includes, in order, aluminum ($Z = 13$), silicon ($Z = 14$), phosphorus ($Z = 15$), sulphur ($Z = 16$), chlorine ($Z = 17$) and, closing the shell, argon ($Z = 18$). These elements are chemically very much like the series: lithium, . . . , neon, and the spectroscopic description of the ground states are the same. The only difference is that, since $n = 3$, the ionization energies are somewhat smaller, as can be seen from the periodic table at the end of the chapter.

It might appear a little strange that the period ends with argon, since the ($3d$) shell, accommodating ten elements, remains to be filled. The fact is, that the self-consistent potential is not of the $1/r$ form, and the intrashell splitting here is sufficiently large that the ($4s$) state lies lower than the ($3d$) state, though not by much. Hence a competition develops, and in the next period we have ($4s$), ($4s$)^2, ($4s$)^2($3d$), ($4s$)^2($3d$)^2, ($4s$)^2($3d$)^3, ($4s$)($3d$)^5, ($4s$)^2($3d$)^5, ($4s$)^2($3d$)^6, ($4s$)^2($3d$)^7, ($4s$)^2($3d$)^8, ($4s$)($3d$)^{10}, ($4s$)^2($3d$)^{10} and then the $4p$ shell gets filled until the period ends with krypton ($Z = 36$). The chemical properties of elements at the beginning and end of this period are similar to those of elements at the beginning and end of other periods. Thus potassium, with the single ($4s$) electron, is an alkali metal, like sodium with its single ($3s$) electron outside a closed shell. Bromine, with the configuration ($4s$)^2($3d$)^{10}($4p$)^5, has a single hole in a p-shell and thus is chemically like chlorine and fluorine. The series of elements in which the ($3d$) states are being filled all have rather similar chemical properties. The reason for this again has to do with the details of the self-consistent potential. It turns out that the radii of these orbits[3] are somewhat smaller than those of the ($4s$) electrons, so that when the ($4s$)^2 shell is filled, these electrons tend to shield the ($3d$) electrons, no matter how many there are, from outside influences. The same effect occurs when the ($4f$) shell is being filled, just after the ($6s$) shell has been filled. The elements here are called the *rare earths*.

Limitations of space prevent us from a more detailed discussion of the periodic table. A few additional comments are, however, in order.

(a) There is nothing in atomic structure that limits the number of elements. The reason that atoms with $Z \gtrsim 100$ do not occur naturally is that heavy *nuclei* undergo spontaneous fission. If new, superheavy (meta)stable nuclei are ever discovered, there will presumably exist corresponding atoms, and it is expected

---

[3] It is understood that this is just a way of talking about the peaking tendencies of the charge distribution.

Z	Element	Configuration	Term[1]	Ionization Potential eV
1	H	$(1s)$	$^2S_{1/2}$	13.6
2	He	$(1s)^2$	$^1S_0$	24.6
3	Li	$(He) (2s)$	$^2S_{1/2}$	5.4
4	Be	$(He)(2s)^2$	$^1S_0$	9.3
5	B	$(He)(2s)^2(2p)$	$^2P_{1/2}$	8.3
6	C	$(He)(2s)^2(2p)^2$	$^3P_0$	11.3
7	N	$(He)(2s)^2(2p)^3$	$^4S_{3/2}$	14.5
8	O	$(He)(2s)^2(2p)^4$	$^3P_2$	13.6
9	F	$(He)(2s)^2(2p)^5$	$^2P_{3/2}$	17.4
10	Ne	$(He)(2s)^2(2p)^6$	$^1S_0$	21.6
11	Na	$(Ne)(3s)$	$^2S_{1/2}$	5.1
12	Mg	$(Ne)(3s)^2$	$^1S_0$	7.6
13	Al	$(Ne)(3s)^2(3p)$	$^2P_{1/2}$	6.0
14	Si	$(Ne)(3s)^2(3p)^2$	$^3P_0$	8.1
15	P	$(Ne)(3s)^2(3p)^3$	$^4S_{3/2}$	11.0
16	S	$(Ne)(3s)^2(3p)^4$	$^3P_2$	10.4
17	Cl	$(Ne)(3s)^2(3p)^5$	$^2P_{3/2}$	13.0
18	Ar	$(Ne)(3s)^2(3p)^6$	$^1S_0$	15.8
19	K	$(Ar)(4s)$	$^2S_{1/2}$	4.3
20	Ca	$(Ar)(4s)^2$	$^1S_0$	6.1
21	Sc	$(Ar)(4s)^2(3d)$	$^2D_{3/2}$	6.5
22	Ti	$(Ar)(4s)^2(3d)^2$	$^3F_2$	6.8
23	V	$(Ar)(4s)^2(3d)^3$	$^4F_{3/2}$	6.7
24	Cr	$(Ar)(4s)(3d)^5$	$^7S_3$	6.7
25	Mn	$(Ar)(4s)^2(3d)^5$	$^6S_{3/2}$	7.4
26	Fe	$(Ar)(4s)^2(3d)^6$	$^5D_4$	7.9
27	Co	$(Ar)(4s)^2(3d)^7$	$^4F_{9/2}$	7.8
28	Ni	$(Ar)(4s)^2(3d)^8$	$^3F_4$	7.6
29	Cu	$(Ar)(4s)(3d)^{10}$	$^2S_{1/2}$	7.7
30	Zn	$(Ar)(4s)^2(3d)^{10}$	$^1S_0$	9.4
31	Ga	$(Ar)(4s)^2(3d)^{10}(4p)$	$^2P_{1/2}$	6.0
32	Ge	$(Ar)(4s)^2(3d)^{10}(4p)^2$	$^3P_0$	8.1
33	As	$(Ar)(4s)^2(3d)^{10}(4p)^3$	$^4S_{3/2}$	10.0
34	Se	$(Ar)(4s)^2(3d)^{10}(4p)^4$	$^3P_2$	9.8
35	Br	$(Ar)(4s)^2(3d)^{10}(4p)^5$	$^2P_{3/2}$	11.8
36	Kr	$(Ar)(4s)^2(3d)^{10}(4p)^6$	$^1S_0$	14.0
37	Rb	$(Kr)(5s)$	$^2S_{1/2}$	4.2
38	Sr	$(Kr)(5s)^2$	$^1S_0$	5.7
39	Y	$(Kr)(5s)^2(4d)$	$^2D_{3/2}$	6.6
40	Zr	$(Kr)(5s)^2(4d)^2$	$^3F_2$	7.0

(Continued)

Z	Element	Configuration	Term[1]	Ionization Potential eV
41	Nb	$(Kr)(5s)(4d)^4$	$^6D_{1/2}$	6.8
42	Mo	$(Kr)(5s)(4d)^5$	$^7S_3$	7.2
43	Tc	$(Kr)(5s)^2(4d)^5$	$^6S_{5/2}$	Not known
44	Ru	$(Kr)(5s)(4d)^7$	$^5F_5$	7.5
45	Rh	$(Kr)(5s)(4d)^8$	$^4F_{9/2}$	7.7
46	Pd	$(Kr)(4d)^{10}$	$^1S_0$	8.3
47	Ag	$(Kr)(5s)(4d)^{10}$	$^2S_{1/2}$	7.6
48	Cd	$(Kr)(5s)^2(4d)^{10}$	$^1S_0$	9.0
49	In	$(Kr)(5s)^2(4d)^{10}(5p)$	$^2P_{1/2}$	5.8
50	Sn	$(Kr)(5s)^2(4d)^{10}(5p)^2$	$^3P_0$	7.3
51	Sb	$(Kr)(5s)^2(4d)^{10}(5p)^3$	$^4S_{3/2}$	8.6
52	Te	$(Kr)(5s)^2(4d)^{10}(5p)^4$	$^3P_2$	9.0
53	I	$(Kr)(5s)^2(4d)^{10}(5p)^5$	$^2P_{3/2}$	10.4
54	Xe	$(Kr)(5s)^2(4d)^{10}(5p)^6$	$^1S_0$	12.1
55	Cs	$(Xe)(6s)$	$^2S_{1/2}$	3.9
56	Ba	$(Xe)(6s)^2$	$^1S_0$	5.2
57	La	$(Xe)(6s)^2(5d)$	$^2D_{3/2}$	5.6
58	Ce	$(Xe)(6s)^2(4f)(5d)$	$^3H_5$	6.9
59	Pr	$(Xe)(6s)^2(4f)^3$	$^4I_{9/2}$	5.8
60	Nd	$(Xe)(6s)^2(4f)^4$	$^5I_4$	6.3
61	Pm	$(Xe)(6s)^2(4f)^5$	$^6H_{5/2}$	Not known
62	Sm	$(Xe)(6s)^2(4f)^6$	$^7F_0$	5.6
63	Eu	$(Xe)(6s)^2(4f)^7$	$^8S_{7/2}$	5.7
64	Gd	$(Xe)(6s)^2(4f)^7(5d)$	$^9D_2$	6.2
65	Tb	$(Xe)(6s)^2(4f)^9$	$^6H_{15/2}$	6.7
66	Dy	$(Xe)(6s)^2(4f)^{10}$	$^5I_8$	6.8
67	He	$(Xe)(6s)^2(4f)^{11}$	$^4I_{15/2}$	Not known
68	Er	$(Xe)(6s)^2(4f)^{12}$	$^3H_6$	Not known
69	Tm	$(Xe)(6s)^2(4f)^{13}$	$^2F_{7/2}$	Not known
70	Yb	$(Xe)(6s)^2(4f)^{14}$	$^1S_0$	6.2
71	Lu	$(Xe)(6s)^2(4f)^{14}(5d)$	$^2D_{3/2}$	5.0
72	Hf	$(Xe)(6s)^2(4f)^{14}(5d)^2$	$^3F_2$	5.5
73	Ta	$(Xe)(6s)^2(4f)^{14}(5d)^3$	$^4F_{3/2}$	7.9
74	W	$(Xe)(6s)^2(4f)^{14}(5d)^4$	$^5D_0$	8.0
75	Re	$(Xe)(6s)^2(4f)^{14}(5d)^5$	$^6S_{5/2}$	7.9
76	Os	$(Xe)(6s)^2(4f)^{14}(5d)^6$	$^5D_4$	8.7
77	Ir	$(Xe)(6s)^2(4f)^{14}(5d)^7$	$^4F_{9/2}$	9.2
78	Pt	$(Xe)(6s)(4f)^{14}(5d)^9$	$^3D_3$	9.0
79	Au	$(Xe)(6s)(4f)^{14}(5d)^{10}$	$^2S_{1/2}$	9.2
80	Hg	$(Xe)(6s)^2(4f)^{14}(5d)^{10}$	$^1S_0$	10.4
81	Tl	$(Xe)(6s)^2(4f)^{14}(5d)^{10}(6p)$	$^2P_{1/2}$	6.1

Elements 57–71 are bracketed as Lanthanides (Rare Earths).

(*Continued*)

PERIODIC TABLE—(*continued*)

Z	Element	Configuration	Term[1]	Ionization Potential eV
82	Pb	$(Xe)(6s)^2(4f)^{14}(5d)^{10}(6p)^2$	$^3P_0$	7.4
83	Bi	$(Xe)(6s)^2(4f)^{14}(5d)^{10}(6p)^3$	$^4S_{3/2}$	7.3
84	Po	$(Xe)(6s)^2(4f)^{14}(5d)^{10}(6p)^4$	$^3P_2$	8.4
85	At	$(Xe)(6s)^2(4f)^{14}(5d)^{10}(6p)^5$	$^2P_{3/2}$	Not known
86	Rn	$(Xe)(6s)^2(4f)^{14}(5d)^{10}(6p)^6$	$^1S_0$	10.7
87	Fr	$(Rn)(7s)$		Not known
88	Ra	$(Rn)(7s)^2$	$^1S_0$	5.3
89	Ac	$(Rn)(7s)^2(6d)$	$^2D_{3/2}$	6.9
90	Th	$(Rn)(7s)^2(6d)^2$	$^3F_2$	
91	Pa	$(Rn)(7s)^2(5f)^2(6d)$	$^4K_{11/2}$	
92	U	$(Rn)(7s)^2(5f)^3(6d)$	$^5L_6$	
93	Np	$(Rn)(7s)^2(5f)^4(6d)$	$^6L_{11/2}$	
94	Pu	$(Rn)(7s)^2(5f)^6$	$^7F_0$	
95	Am	$(Rn)(7s)^2(5f)^7$	$^8S_{7/2}$	
96	Cm	$(Rn)(7s)^2(5f)^7(6d)$	$^9D_2$	
97	Bk	$(Rn)(7s)^2(5f)^9$	$^6H_{15/2}$	
98	Cf	$(Rn)(7s)^2(5f)^{10}$	$^5I_8$	
99	Es	$(Rn)(7s)^2(5f)^{11}$	$^4I_{15/2}$	
100	Fm	$(Rn)(7s)^2(5f)^{12}$	$^3H_6$	
101	Md	$(Rn)(7s)^2(5f)^{13}$	$^2F_{7/2}$	
102	No	$(Rn)(7s)^2(5f)^{14}$	$^1S_0$	

(Elements 89–102 are bracketed together as **Actinides**.)

[1] Term designation is equivalent to spectroscopic description.

that their structure will conform to the prediction of the building-up approach outlined in this chapter.

(b) We went to a great deal of trouble to specify the S, L, and J quantum numbers of the ground states of the various elements. The reason for doing this is that in spectroscopy, the quantum numbers are of particular interest because of the selection rules

$$\Delta S = 0$$

$$\Delta L = \pm 1$$

$$\Delta J = 0, \pm 1 \quad (\text{no } 0 - 0) \tag{19-18}$$

that will be derived later, and that may then be used to determine the quantum numbers of the excited states. The spectroscopy of atoms, once we get beyond hydrogen and helium, is very complicated. Consider, as a relatively simple example, the first few states of carbon, which are formed from different con-figurations of the two electrons that lie outside the closed shell in the $(2p)^2$

orbitals. As already pointed out, the possible states are $^1S_0$, $^3P_{2,1,0}$ and $^1D_2$. The $^3P_0$ state lies lowest, but the other states are still there. The first excited states may be described by the orbitals $(2p)(3s)$. Here $S = 0$ or 1, but $L = 1$ only. Since the n-values are different, the exclusion principle does not restrict the states in any way, and all of the states $^1P_1$, $^3P_{2,1,0}$ are possible, while the excited states that arise from the orbitals $(2p)(3p)$ can have $S = 0$, 1 and $L = 2$, 1, 0, leading to all the states $^1D_2$, $^1P_1$, $^1S_0$, $^3D_{3,2,1}$, $^3P_{2,1,0}$, and $^3S_1$. Even with the restrictions provided by the selection rules, there are numerous transitions. Needless to say, the ordering of these levels represents a delicate balance between various competing effects, and the prediction of the more complex spectra is very difficult.[4] That task is not really of interest to us, since the main point that we want to make is that quantum mechanics provides a qualitative, and sometimes quantitative, detailed explanation of the chemical properties of atoms and of their spectra, without assuming an interaction other than the electromagnetic interaction between charged particles. We shall have occasion to return to the topic of spectra.

Problems

1. List the spectroscopic states (in the form $^{2S+1}L_J$) that can arise from combining

$$S = 1/2, L = 3$$
$$S = 2, L = 1$$
$$S_1 = 1/2, S_2 = 1, L = 4$$
$$S_1 = 1, S_2 = 1, L = 3$$
$$S_1 = 1/2, S_2 = 1/2, L = 2$$

Which states are excluded, among the two-spin questions, if the particles are identical?

2. Consider the following states

$$^1D, \ ^2P, \ ^4F, \ ^3G, \ ^2D, \ ^3H$$

What are the possible J values associated with each?

3. Consider the states ^1D, ^3P, ^3S, ^5G, ^5P, ^5S. Given that in each one the state consists of two identical particles in their largest possible spin state, which of the states are disallowed by the spin-statistics theorem?

[4] The spectroscopic labeling that we have used has its limitations. Behind its validity lies the physical picture that the total orbital and spin angular momenta are separately conserved, except for the perturbation caused by spin-orbit coupling. This coupling is small for the light elements, but for large Z it cannot be treated as a perturbation. There it is better to couple the l and s for each electron to form a j, and then consider the j-j coupling. Detailed consideration of this is beyond the scope of this book.

4. Use Hund's rules to find the spectroscopic description of the ground states of the following atoms:

$$N(Z = 7), K(Z = 19), Sc(Z = 21), Co(Z = 27).$$

Figure out the electronic configurations as far as you are able to.

5. List the possible spectroscopic states that can arise in the following electronic configurations: $(1s)^2$, $(2p)^2$, $(2p)^3$, $(2p)^4$, $(3d)^2$, $(2p)^5(3s)$, $(3d)^4$. Take into account the exclusion principle.

(*Note.* Remember that if all the spins or orbital angular momenta are pointing in the same direction, then the state is symmetric; also remember about holes in closed shells.)

6. Plot the ionization potentials given in the periodic table against Z. Observe the peaks indicating the shell structure of the atoms.

References

An excellent introductory treatment of atomic structure may be found in

G. Herzberg, *Atomic Spectra and Atomic Structure*, Dover Publishers (1944).

An advanced treatment that is definitive is

I. I. Sobel'man, *Introduction to the Theory of Atomic Spectra*, Pergamon Press, New York (1972). This is a very advanced book.

chapter 20

Molecules

Just as atoms are aggregates of electrons and a single nucleus, so molecules are aggregates of electrons and several nuclei. Molecules in their lowest energy state are stable, that is, it takes a certain amount of energy to dissociate them into their components. Since dissociation of molecules into atoms is the most common occurrence when enough energy is transferred to the system, we may call molecules bound states of atoms, although we shall see that this description hides much of what makes up the structure of molecules. The purpose of this chapter, and the next, is to show that quantum mechanics is successful in describing the properties and behavior of molecules.

The simplest molecules are those that involve two nuclei, the diatomic molecules. Even they are more complex systems than atoms because, after the center of mass is fixed in space, the nuclei are still free to move. This leads to an increase in the number of degrees of freedom. Thus for the simplest of all molecules, the H_2^+ molecule, consisting of two protons and one electron, there are still six degrees of freedom left, three for the electron and three for the relative motion of the two protons. As with atoms, a frontal attack on the problem of the dynamics of molecules, that is, a numerical solution of the Schrödinger equation in many dimensions is possible. For our purposes cruder but more physical approaches will be more enlightening.

Insight into the dynamics of molecules can be obtained from use of the fact that nuclei are a great deal more massive than electrons ($M/m_e \gg 10^3$) and thus their motion is a great deal slower. One may view the motion of the electrons as if the nuclei were fixed in space. The motion of the nuclei, on the other hand, is in an average field due to the electrons. For a given set of nuclear coordinates, there will be a Hamiltonian for the electrons. The lowest eigenvalue of that Hamiltonian will depend on these coordinates, and its minimum value will determine the positions of the nuclei. This picture must be modified a little for nuclei that are not infinitely massive, since they can also move. Their motion depends on the electrons, but they only "see" an average charge distribution due to the rapid motion of the electrons, and to first approximation, they move in a harmonic potential about the locations determined by the minimum in the energy of the electrons.

We can describe the situation mathematically as follows. The Schrödinger equation describing the nuclei and the electrons has the form

$$[T_R + T_r + V(r,R)]\,\Psi(r,R) = E\Psi(r,R) \tag{20-1}$$

where T_R is the sum of the kinetic energies of the nuclei, T_r is the sum of the kinetic energies of the electrons, and $V(r,R)$ is the potential energy, which consists of the Coulomb attraction of the electrons to the nuclei, the electron-electron repulsions, and the repulsion among the nuclei. Consider first the Hamiltonian describing the electronic motion for a set of fixed $\{R\}$

$$H_0 = T_r + V(r,R) \tag{20-2}$$

The eigenvalue problem can, in principle, be solved

$$[T_r + V(r,R)]\,u_n(r,R) = \epsilon_n(R)\,u_n(r,R) \tag{20-3}$$

Both the eigenvalues and the eigenfunctions depend on the values of R, which here play the role of fixed parameters. Since the $u_n(r,R)$ form a complete set, we may expand $\Psi(r,R)$ in terms of them

$$\Psi(r,R) = \sum_m \phi_m(R)\,u_m(r,R) \tag{20-4}$$

To determine the coefficients $\phi_m(R)$ we insert this into (20-1), and obtain, using (20-3), the equation

$$T_R \sum_m \phi_m(R)\,u_m(r,R) + \sum_m \epsilon_m(R)\,\phi_m(R)\,u_m(r,R) = E \sum_m \phi_m(R)\,u_m(r,R)$$

$$\tag{20-5}$$

The first term will consist of terms of the type

$$\left(-\frac{\hbar^2}{2M_1}\nabla_{R_1}^2 + \dots\right) \sum_m \phi_m(R)\,u_m(r,R) = \sum_m [T_R\phi_m(R)]\,u_m(r,R)$$

$$-\frac{\hbar^2}{2M_1}\sum_m \nabla_{R_1}\phi_m(R)\cdot\nabla_{R_1} u_m(r,R) - \dots$$

$$-\frac{\hbar^2}{2M_1}\sum_m \phi_m(R)\,\nabla_{R_1}^2 u_m(r,R) - \dots \tag{20-6}$$

As a first approximation we neglect the second and third terms on the right side of this equation, that is, we assume that the eigenfunctions $u_m(r,R)$ are slowly varying functions of the nuclear coordinates R, at least in the region of the solutions for the minimum electronic energy, R_0 defined by

$$\nabla_{R_i}\,\epsilon_n(R)\,|_{R_i=R_{i0}} = 0 \tag{20-7}$$

This approximation

$$T_R \sum_m \phi_m(R) \, u_m(r,R) \approx \sum_m [T_R \phi_m(R)] \, u_m(r,R) \qquad (20\text{-}8)$$

is the first step in a sequence of approximations by which variations with respect to R can be taken into account. The procedure was developed by Born and Oppenheimer, and the first approximation is generally a good one. If we now take the scalar product with $u_n(r,R)$, and use orthonormality

$$\int \prod_i d^3r_i \, u_n^*(r,R) \, u_m(r,R) = \delta_{mn} \qquad (20\text{-}9)$$

then (20-5), together with (20-8), reduces to

$$T_R \phi_m(R) + \epsilon_m(R) \, \phi_m(R) = E\phi_m(R) \qquad (20\text{-}10)$$

This is just the Schrödinger equation for the nuclear motion in a potential $\epsilon_m(R)$, the electronic energy. For R close to the minimum points R_0, we may expand

$$\epsilon_m(R) \cong \epsilon_m(R_0) + \frac{1}{2} (R - R_0)^2 \left(\frac{\partial^2 \epsilon_m}{\partial R^2} \right)_0 + \dots \qquad (20\text{-}11)$$

If only the first two terms are important the nuclei move in harmonic oscillator wells. Thus the nuclei will undergo vibrational motion.[1] They will also undergo rotational motion, since T_R involves angular coordinates. We may estimate the magnitudes of the various energies.

Let us assume that the size of the molecule is of order a. Then, by the uncertainty principle, the electronic energy is of order

$$\epsilon \simeq \frac{1}{2m} \left(\frac{\hbar}{a} \right)^2 \qquad (20\text{-}12)$$

The frequency of the vibrational motion of the nucleus is, according to (20-11), given by the formula

$$M\omega^2 \cong \frac{\partial^2 \epsilon(R)}{\partial R^2} \qquad (20\text{-}13)$$

Given the potential energy in (20-12), a dimensional argument gives us

$$\frac{\partial^2 \epsilon}{\partial R^2} \cong \frac{\hbar^2}{ma^4}$$

that is,

$$\omega \cong \left(\frac{m}{M} \right)^{1/2} \frac{\hbar}{ma^2} \qquad (20\text{-}14)$$

[1] The motion will be somewhat more complicated if we do not limit ourselves to a two-term expansion of (20-11) but it will still have the qualitative properties that we are discussing.

Thus the ratio of the vibrational energy of the nuclei to the electronic energy is

$$\frac{E_{\text{vib}}}{\epsilon} \simeq \frac{\hbar\omega}{\hbar^2/ma^2} \simeq \left(\frac{m}{M}\right)^{1/2} \tag{20-15}$$

The molecule can also rotate about the center of mass. Typically

$$E_{\text{rot}} \cong \frac{J(J+1)\,\hbar^2}{2I} \cong \frac{\hbar^2}{2Ma^2} \cong \frac{m}{M}\,\epsilon \tag{20-16}$$

Thus, as far as *molecular structure* is concerned, one may neglect the rotational and vibrational degrees of freedom. Nevertheless rotational and vibrational energy levels will exist, and in molecular spectroscopy there will be:

(a) *Electronic transitions.* If the dimensions of the molecule are of the order of 1 Å, that is,

$$a \simeq \frac{2\hbar}{mc\alpha} \tag{20-17}$$

then

$$h\nu = 2\pi\hbar\,\frac{c}{\lambda} \simeq \frac{\hbar^2}{2ma^2} \simeq \frac{mc^2\alpha^2}{8}$$

that is,

$$\lambda \simeq \frac{16\pi}{\alpha}\,\frac{\hbar}{mc\alpha} \simeq 16\pi \times 137 \times 0.5 \text{ Å}$$

$$\simeq 3500 \text{ Å} \tag{20-18}$$

Thus the radiation emitted in electronic transitions lies in the ultraviolet.

(b) *Vibrational transitions.* These are transitions between different levels in the approximate harmonic oscillator well. The typical energies will be of the order of $(m/M)^{1/2}\,\epsilon$, that is, the wavelengths will be of the order of $(M/m)^{1/2} \sim 50$ times larger. The range of wavelengths, $\sim 2 - 3 \times 10^{-3}$ cm, lies in the infrared region.

(c) The rotational spectra will be characterized by wavelengths $M/m \sim 10^3 - 10^4$ times larger than the electronic optical wavelengths, and $\lambda \sim 0.1 - 1$ cm is typical of the microwave region.

To see what form $\epsilon_0(R)$ can take, let us turn to the simplest molecule of all, the H_2^+ ion. After separating out the center of mass of the two nuclei (we ignore the electron in doing this), we are left with the energy eigenvalue equation

$$\left(-\frac{\hbar^2}{2M}\,\boldsymbol{\nabla}_R^{\,2} - \frac{\hbar^2}{2m}\,\boldsymbol{\nabla}_r^{\,2} - \frac{e^2}{|\mathbf{r} - \mathbf{R}/2|} - \frac{e^2}{|\mathbf{r} + \mathbf{R}/2|} + \frac{e^2}{R} - E\right)\Psi(\mathbf{r},\mathbf{R}) = 0 \tag{20-19}$$

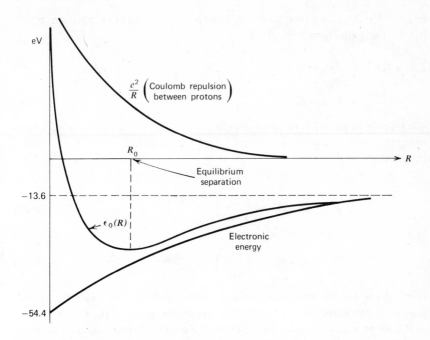

Fig. 20-1. Contributions to "nuclear potential." The Coulomb repulsion and the electronic energy combine to give a curve with a minimum at R_0.

The first term represents the kinetic energy of the protons with

$$M = \frac{M_P^2}{M_P + M_P} = \frac{M_P}{2} \tag{20-20}$$

the reduced mass of the two-proton system. The second is the kinetic energy of the electron. The next two terms represent the attraction between the electron and the two protons located at $\mathbf{R}/2$ and $-\mathbf{R}/2$, and the last term represents the repulsion between the two protons separated by a distance $R = |\mathbf{R}|$. The qualitative features of the solution with $\mathbf{R}$ held fixed, and the proton kinetic energy term absent are shown in Fig. 20-1. For R very large, the electron will be bound to one of the protons, and the energy of the system is -13.6 eV, the energy of a single hydrogen atom. When $R \rightarrow 0$, and we leave out the proton-proton repulsion, the electron will be bound to a $Z = 2$ nucleus, and the binding energy will be $-13.6\,Z^2 = -54.4$ eV. The electronic energy, as a function of R, interpolates smoothly between these points. When the energy or repulsion e^2/R is added to this, the curve $\epsilon_0(R)$ results. This curve has a minimum for the H_2^+ molecule. A minimum does not always exist, so that some atoms do not form molecules, as we shall soon see. The electronic eigenvalue equation, which has the form anticipated in (20-3),

$$H_0 u_0(\mathbf{r},\mathbf{R}) = \left(\frac{p_e^2}{2m} - \frac{e^2}{|\mathbf{r} - \mathbf{R}/2|} - \frac{e^2}{|\mathbf{r} + \mathbf{R}/2|} + \frac{e^2}{R} \right) u_0(\mathbf{r},\mathbf{R})$$

$$= \epsilon_0(R) \, u_0(\mathbf{r},\mathbf{R}) \qquad (20\text{-}21)$$

can actually be solved in elliptical coordinates, but we will get more insight from using the *variational principle*, with trial wave functions that reflect some physical intuition about the system.

A reasonable trial wave function is a linear combination of

$$\psi_1(\mathbf{r},\mathbf{R}) = \left(\frac{1}{\pi a_0^3} \right)^{1/2} e^{-|\mathbf{r} - \mathbf{R}/2|/a_0} \qquad (20\text{-}22)$$

and

$$\psi_2(\mathbf{r},\mathbf{R}) = \left(\frac{1}{\pi a_0^3} \right)^{1/2} e^{-|\mathbf{r} + \mathbf{R}/2|/a_0} \qquad (20\text{-}23)$$

representing the electron bound to one or the other proton. Since the Hamiltonian is symmetric about reflections in the origin ($\mathbf{p}_e \rightarrow -\mathbf{p}_e$, $\mathbf{r} \rightarrow -\mathbf{r}$, $\mathbf{R} \rightarrow -\mathbf{R}$), we may take as trial wave functions even and odd combinations of these[2]

$$\psi_g(\mathbf{r},\mathbf{R}) = C_+(R)[\psi_1(\mathbf{r},\mathbf{R}) + \psi_2(\mathbf{r},\mathbf{R})]$$

$$\psi_u(\mathbf{r},\mathbf{R}) = C_-(R)[\psi_1(\mathbf{r},\mathbf{R}) - \psi_2(\mathbf{r},\mathbf{R})] \qquad (20\text{-}24)$$

The normalization factors are given by

$$\frac{1}{C_\pm^2} = \langle \psi_1 \pm \psi_2 | \psi_1 \pm \psi_2 \rangle$$

$$= 2 \pm 2 \int d^3r \, \psi_1(\mathbf{r},\mathbf{R}) \, \psi_2(\mathbf{r},\mathbf{R}) \qquad (20\text{-}25)$$

The integral appearing above is called the *overlap integral*, and it can be calculated. The calculation of

$$S(R) = \int d^3\mathbf{r} \, \psi_1(\mathbf{r},\mathbf{R}) \, \psi_2(\mathbf{r},\mathbf{R})$$

$$= \frac{1}{\pi a_0^3} \int d^3r \, e^{-|\mathbf{r} - \mathbf{R}/2|/a_0} \, e^{-|\mathbf{r} + \mathbf{R}/2|/a_0}$$

$$= \frac{1}{\pi a_0^3} \int d^3r' \, e^{-|\mathbf{r}' - \mathbf{R}|/a_0} \, e^{-r'/a_0} \qquad (20\text{-}26)$$

---

[2] The labeling is historical: "g" stands for "gerade," which means even in German, and "u" for "ungerade," odd.

is straightforward, though tedious. The result is

$$S(R) = \left(1 + \frac{R}{a_0} + \frac{R^2}{3a_0{}^2}\right) e^{-R/a_0} \tag{20-27}$$

The expectation value of H_0 in the two states is

$$\langle H \rangle_{g,u} = \frac{1}{2[1 \pm S(R)]} \langle \psi_1 \pm \psi_2 | H_0 | \psi_1 \pm \psi_2 \rangle$$

$$= \frac{1}{2[1 \pm S(R)]} \{ \langle \psi_1 | H_0 | \psi_1 \rangle + \langle \psi_2 | H_0 | \psi_2 \rangle \pm \langle \psi_1 | H_0 | \psi_2 \rangle \pm \langle \psi_2 | H_0 | \psi_1 \rangle \}$$

$$= \frac{\langle \psi_1 | H_0 | \psi_1 \rangle \pm \langle \psi_1 | H_0 | \psi_2 \rangle}{1 \pm S(R)} \tag{20-28}$$

where use has been made of the symmetry under $\mathbf{R} \to -\mathbf{R}$. The two terms in the numerator can be calculated:

$$\langle \psi_1 | H_0 | \psi_1 \rangle = \int d^3r \psi_1^*(\mathbf{r},\mathbf{R}) \left(\frac{p_e{}^2}{2m} - \frac{e^2}{|\mathbf{r} - \mathbf{R}/2|} - \frac{e^2}{|\mathbf{r} + \mathbf{R}/2|} + \frac{e^2}{R} \right)$$
$$\times \psi_1(\mathbf{r},\mathbf{R})$$

$$= E_1 + \frac{e^2}{R} - e^2 \int d^3r \frac{|\psi_1(\mathbf{r},\mathbf{R})|^2}{|\mathbf{r} + \mathbf{R}/2|} \tag{20-29}$$

The first term is just the energy of a single hydrogen atom $E_1 = -13.6$ eV; the second term is the proton-proton repulsion, and the third term is the electrostatic potential energy due to the electron charge distribution about one proton being attracted to the other proton. The last integral can be evaluated, so that finally

$$\langle \psi_1 | H_0 | \psi_1 \rangle = E_1 + \frac{e^2}{R} \left(1 + \frac{R}{a_0} \right) e^{-2R/a_0} \tag{20-30}$$

Similarly

$$\langle \psi_1 | H_0 | \psi_2 \rangle = \int d^3r_1 \, \psi_1^*(\mathbf{r},\mathbf{R}) \left(E_1 + \frac{e^2}{R} - \frac{e^2}{|\mathbf{r} + \mathbf{R}/2|} \right) \psi_2(\mathbf{r},\mathbf{R})$$

$$= \left(E_1 + \frac{e^2}{R} \right) S(R) - e^2 \int d^3r \frac{\psi_1^*(\mathbf{r},\mathbf{R}) \, \psi_2(\mathbf{r},\mathbf{R})}{|\mathbf{r} + \mathbf{R}/2|} \tag{20-31}$$

Here the last term is the exchange integral, which can also be evaluated, yielding

$$e^2 \int d^3r \frac{\psi_1^*(\mathbf{r},\mathbf{R}) \, \psi_2(\mathbf{r},\mathbf{R})}{|\mathbf{r} + \mathbf{R}/2|} = \frac{e^2}{a_0} \left(1 + \frac{R}{a_0} \right) e^{-R/a_0} \tag{20-32}$$

When all of this is put together, the resulting energies can be calculated as functions of R. Figure 20-2 shows the calculated energies.

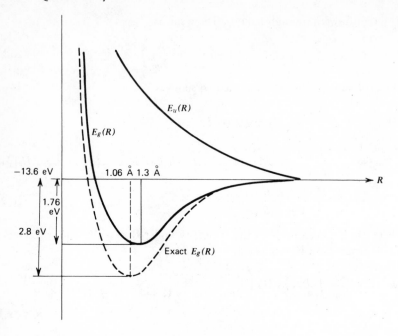

Fig. 20-2. Results of variational calculation for H_2^+.

The exact solution, which according to the variational principle must lie below the curves obtained, differs little from the minimum. In our approximation, we see that the even solution yields binding, while the odd one does not.[3] The difference between the even and the odd solutions is that in the former, the electron has a high probability of being located between the two protons, where the attractive contribution is maximized; for the odd solution, which has a node midway between the protons, the electron tends to be excluded from that region.

The experimental separation between the protons is 1.06 Å, and the binding energy is -2.8 eV. The calculations outlined above lead to a separation of 1.3 Å and a binding energy of -1.76 eV. Thus our wave function is not as compact as it should be. The reason is that when R is small, the wave function should approach that of a He^+ ion, which (20-24) does not. One could improve the calculation by introducing an effective charge for the proton and minimizing $\langle H_0 \rangle_g$ with respect to that parameter in addition to R, as in our illustration involving the helium atom. Since we are more interested in a qualitative understanding of the problem than in improving the variational calculation, we do not pursue this idea.

---

[3] One might worry that the true $\epsilon_u(R)$, lying below the $\langle H \rangle_u = E_u(R)$ curve, still dips down and gives a weaker bound state. Detailed calculations show that it does not.

The orbitals that we have considered do not depend on the azimuthal angle about the axis of the molecule. Since the Hamiltonian is invariant under rotations about the axis, we may classify the solutions by the angular momentum component along the axis. If we choose $\mathbf{R}$ to define the z-axis, our eigenstates will be simultaneous eigenstates of L_z. The solutions will, in general, have the dependence $e^{im\phi}$ with $m = 0, \pm1, \pm2, \ldots$. These are labeled $\sigma, \pi, \delta, \ldots$ in analogy with $S, P, D, \ldots$. There is also the labeling "g" and "u" which is applicable to all diatomic molecules for which the atoms are the same (*homonuclear* molecules). Thus in our example the ground state could be labeled $1s\sigma_g$, and the antisymmetric state could be labeled $1s\sigma_u{}^*$, the asterisk indicating that the state is unbound. Excited states of the $H_2{}^+$ molecule may be formed with higher orbitals.

We will not deal with the rotational and vibrational degrees of freedom of the molecule except to note their roles in the two topics that we discuss next. First we will be concerned with the effect of the Pauli Exclusion Principle on homonuclear molecules. Consider, for example, the H_2 molecule for which the two nuclei are identical and each has spin $1/2$. Thus the total wave function must be antisymmetric under the interchange of the two nuclei. The two protons in this example may be in the antisymmetric spin singlet ($S = 0$) state, in which case the rotational state must be described by a symmetric function, so that the angular momentum is even. If the two protons are in the symmetric spin triplet ($S = 1$) state, the angular momentum of rotation must be odd. In a gas, collisions among the H_2 molecules will randomize the distribution of spin states, and assuming that they have equal probability, the number of molecules in a given spin state will be proportional to the degeneracy $(2S + 1)$. Thus there will be three times as many odd L molecules as there are even L H_2 molecules in the gas This will manifest itself in the intensity of the spectral lines associated with the transitions between rotational levels. More generally, if each nucleus has spin I. then the spin states $2I, 2I - 2, 2I - 4, \ldots$ and the spin states $2I - 1, 2I - 3, \ldots$, will have opposite symmetry. If, for example, I is an integer, then the first series of spin states will be associated with even orbital angular momentum, since the nuclei are bosons in this case. Their total number is

$$\sum_{k=0}^{I} [2(2I - 2k) + 1] = (4I + 1)(I + 1) - \frac{4I(I + 1)}{2}$$

$$= (2I + 1)(I + 1) \qquad (20\text{-}33)$$

whereas the remaining

$$(2I + 1)^2 - (2I + 1)(I + 1) = (2I + 1)\,I \qquad (20\text{-}34)$$

states will be associated with odd orbital angular momentum. Thus for integral I, the ratio of even L to odd L intensities for a given L is $(I + 1)/I$. For fermions that ratio is inverted.

To a good approximation, the energies of the rotational states are

$$E_L = \frac{\hbar^2 L(L+1)}{2\mathfrak{I}} \tag{20-35}$$

where $\mathfrak{I}$ is the moment of inertia of the homonuclear molecule under consideration. Transitions between adjacent L values (to conform with the selection rule
$\Delta L = \pm 1$, still to be derived) yield radiation with frequencies

$$\omega(L+1 \to L) = \frac{\hbar}{2\mathfrak{I}}[(L+1)(L+2) - L(L+1)]$$

$$= \frac{\hbar}{\mathfrak{I}}(L+1) \tag{20-36}$$

From the study of rotational spectra one can identify the rotational levels and
find their L values. The intensities then give a way of discriminating between
even and odd spin. Historically, a study of the rotational spectrum of the N_2
molecule led to the conclusion that its spin was even. This could not be understood on the basis of a nuclear model in which the nitrogen nucleus consisted
of fourteen protons and seven electrons; such a nucleus would have odd half-
integral spin. The discovery of the neutron, and the realization that the nitrogen
nucleus consisted of seven protons and seven neutrons removed the difficulty.

The existence of the hierarchy of excitation energies, rotational, vibrational, and electronic manifests itself in the form of the specific heat at constant
volume as a function of temperature. We take from statistical mechanics the
following facts.

(a) The specific heat at constant volume is given by

$$C_V = N_0 \frac{\partial}{\partial T} \bar{E}(T) \tag{20-37}$$

where $\bar{E}(T)$ is the average energy of a molecule in equilibrium at temperature T,
and N_0 is Avogadro's number.

(b) The average energy can be calculated from the Boltzmann distribution

$$\bar{E}(T) = \int dE\, E g(E)\, e^{-E/kT} \Big/ \int dE g(E)\, e^{-E/kT}$$

$$= -\frac{\partial}{\partial(1/kT)} \int dE g(E)\, e^{-E/kT} \Big/ \int dE g(E)\, e^{-E/kT}$$

$$= kT^2 \frac{\partial}{\partial T} \log \int dE g(E)\, e^{-E/kT} \tag{20-38}$$

where $g(E)$ is the degeneracy of states with energy E.

(c) The average energy can be written as a sum of contributions from independent degrees of freedom, so that we can write

$$\bar{E}(T) = \bar{E}_{\text{trans}}(T) + \bar{E}_{\text{rot}}(T) + \bar{E}_{\text{vib}}(T) + \ldots \qquad (20\text{-}39)$$

For the translational contribution, where E is the kinetic energy, we have[4]

$$\int dE g(E)\, e^{-E/kT} = \int \frac{d^3\mathbf{p}}{(2\pi\hbar)^3}\, e^{-p^2/2MkT} \qquad (20\text{-}40)$$

so that

$$\int dE g(E)\, e^{-E/kT} = C'T^{3/2} \qquad (20\text{-}41)$$

as can be seen from dimensional considerations. Thus (20-38) yields

$$C_V = N_0 \frac{\partial}{\partial T}\left(kT^2 \frac{\partial}{\partial T}\log C'T^{3/2}\right)$$

$$= \frac{3}{2}N_0 k$$

$$= \frac{3}{2}R \qquad (20\text{-}42)$$

where $R = N_0 k = 1.98$ calorie/mole K°. This is just the Dulong-Petit result.

For the rotational contribution we have

$$\int dE g(E)\, e^{-E/kT} \rightarrow \sum g_s(2L+1)\, e^{-\hbar^2 L(L+1)/2\mathcal{J}kT} \qquad (20\text{-}43)$$

where g_s is the spin multiplicity corresponding to the given L, and $(2L+1)$ is the usual degeneracy corresponding to a given value of L. For the H_2 molecule we have the special situation mentioned before: the existence of para- and ortho-hydrogen, for which the nuclei are in the spin states $S = 0$ and $S = 1$ respectively.

For para-hydrogen, L is restricted to the even values, $L = 0, 2, 4, \ldots$ and $g_s = 1$; for ortho-hydrogen, L is odd, $L = 1, 3, 5, \ldots$ and $g_s = 3$. At low temperatures—and here "low" depends on the moment of inertia, so that for H_2 the relevant number is[5]

$$\frac{\hbar^2}{2\mathcal{J}k} = 84.8° K \qquad (20\text{-}44)$$

---

[4] For a discussion of the degeneracy of states applicable to free motion, see Chapter 22, Section C.

[5] The moment of inertia can be determined from the spacing in rotational spectra.

—the $L = 0$ state will be primarily occupied, that is, the gas will consist of para-hydrogen.[6] At room temperature the difference between even and odd L's becomes insignificant, and the ratio is determined by the ratio of the g_s, that is, it is $3:1$ ortho- to para-hydrogen.

In the rest of the discussion we shall, for brevity, ignore the complication of the two forms of molecules. At temperatutes where the rotational degrees of freedom become excited we have

$$\overline{E}_{\rm rot} = \frac{\hbar^2/2\Im \sum\limits_L (2L + 1) L(L + 1) \ e^{-\hbar^2 L(L+1)/2\Im kT}}{\sum\limits_L (2L + 1) \ e^{-\hbar^2 L(L+1)/2\Im kT}} \tag{20-45}$$

This can be evaluated numerically. At high temperatures, the level-spacing is so small, compared to kT, that we can replace the sum by an integral, and use (20-38). We have

$$\int dE g(E) \ e^{-E/kT} \sim \int dl \ 2l \ e^{-\hbar^2 l^2/2\Im kT} \sim C''T \tag{20-46}$$

so that, using (20-38) and (20-37) we get, for large T,

$$(C_V)_{\rm rot} = kN_0 \frac{\partial}{\partial T} \left(T^2 \frac{\partial}{\partial T} \log C''T \right) = R \tag{20-47}$$

At higher temperatures the vibrational states can become excited. The harmonic oscillator potential in which the nucleus moves need not be symmetric. If the line connecting the two nuclei is taken in the z-direction, it is plausible that the potential walls will be steeper in the x- and y-direction than in the z-direction and thus in the expression for the energy

$$E = \hbar\omega_x(n_x + \tfrac{1}{2}) + \hbar\omega_y(n_y + \tfrac{1}{2}) + \hbar\omega_z(n_z + \tfrac{1}{2}) \tag{20-48}$$

the first excitation will be from the ground state $E_0 = \tfrac{1}{2}\hbar\omega_x + \tfrac{1}{2}\hbar\omega_y + \tfrac{1}{2}\hbar\omega_z$ to $E_1 = E_0 + \hbar\omega_z$. Thus

$$\overline{E} = \frac{E_0 \ e^{-E_0/kT} + E_1 \ e^{-E_1/kT}}{e^{-E_0/kT} + e^{-E_1/kT}}$$

$$= \frac{E_0 + E_1 \ e^{-\hbar\omega_z/kT}}{1 + e^{-\hbar\omega_z/kT}}$$

$$\cong E_0 \left(1 + \frac{E_1}{E_0} e^{-\hbar\omega_z/kT} \right) \left(1 - e^{-\hbar\omega_z/kT} \right)$$

$$\cong E_0 + \hbar\omega_z \ e^{-\hbar\omega_z/kT} \tag{20-49}$$

[6] Actually transitions between the ortho- and para-states are very slow, so that cooling the gas is not enough to make para-hydrogen. In practice one uses a catalyst.

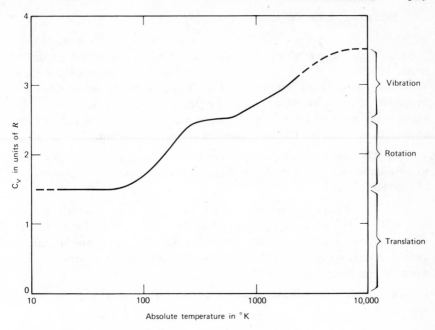

Fig. 20-3. Specific heat of H gas as a function of temperature.

Hence

$$(C_V)_{\text{vib}} = \frac{\partial \overline{E}}{\partial T} N_0 = N_0 k \left(\frac{\hbar\omega_z}{kT}\right)^2 e^{-\hbar\omega_z/kT} \qquad (20\text{-}50)$$

For H_2 the vibrational effects do not set in until 600°K (Fig. 20-3). On the other hand, for Cl_2, $\hbar\omega_z$ has quite a low value, and at room temperature, the contribution of the first vibrational level is $\sim 0.5\ R$. In general ω_x, ω_y are quite a bit larger than ω_z for diatomic molecules, and therefore the high energy contribution of the vibrational excitations computed from

$$\int g(E)\ e^{-E/kT}\ dE = \sum_{n_z} e^{-\hbar\omega_z(n_z+1/2)/kT}$$

$$\simeq \int dn\ e^{-\hbar\omega_z(n+1/2)/kT} \simeq C'''T \qquad (20\text{-}51)$$

is R as for the rotational levels. This is shown in Fig. 20-3 showing C_V for H_2. The electronic levels only contribute at extremely high energies.

Problems

1. In HCl a number of absorption lines with wave numbers (in cm^{-1}) 83.03, 103.73, 124.30, 145.03, 165.51, 185.86 have been observed. Are these vibrational or rotational transitions? If the former, what is the characteristic frequency? If the latter, what J values do they correspond to, and what is the moment of inertia of HCl? In that case, estimate the separation between the nuclei. (In radiation the quantum numbers change by one unit.)

2. What is the ratio of the number of HCl molecules in a state with $J = 10$ to the number in a state with $J = 0$, if the gas of molecules is at a temperature of 300° K?

3. The frequency of vibration of the CO molecule in its lowest state is $v_0 = 2 \times 10^{13}$ Hz. What is the wavelength of the radiation emitted in the lowest vibrational excitation? What is the probability that the first vibrational state is excited, relative to the probability that CO is in its vibrational ground state, when the temperature is 300° K?

4. Consider the vibrational and rotational energy of a molecule in the approximation

$$E_J(R) = \frac{1}{2} m\omega^2(R - R_0)^2 + \frac{J(J+1)\,\hbar^2}{2mR^2}$$

Find the position where the energy is a minimum. If the moment of inertia of the molecule is calculated using the new internuclear separation, show that the rotational energy can be written in the form

$$E_J = AJ(J+1) + B[J(J+1)]^2 + \ldots$$

Determine the coefficients A and B (the latter is the effect of centrifugal distortion).

References

Several elementary books that treat the matters in this chapter are listed at the end of Chapter 21.

chapter 21

Molecular Structure

In this chapter we discuss, albeit only qualitatively, how the electronic structure of molecules determines their shapes and other properties. We begin with the H_2 *molecule*, and discuss it in some detail, because there are two electrons (in contrast to the H_2^+ molecule) and the exclusion principle and electron spin considerations make their first appearance. Both here, and in the rest of the chapter, the nuclear motion will be neglected.

The nuclei (protons) will be labeled A, B and the two electrons "1" and "2" (Fig. 21-1). The Hamiltonian has the form

$$H = H_1 + H_2 + \frac{e^2}{r_{12}} + \frac{e^2}{R_{AB}} \tag{21-1}$$

where

$$H_i = \frac{\mathbf{p}_i^2}{2m} - \frac{e^2}{r_{Ai}} - \frac{e^2}{r_{Bi}} \qquad (i = 1, 2) \tag{21-2}$$

depends only on the coordinates of the electron i relative to the nuclei. We will again compute an upper bound to $E(R_{AB})$ by constructing the expectation value of H with a trial wave function. Since

$$\tilde{H}_i = H_i + \frac{e^2}{R_{AB}} \tag{21-3}$$

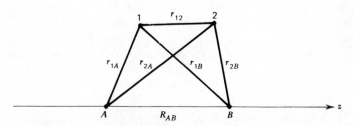

Fig. 21-1. Coordinate labels in the discussion of the H_2 molecule.

are just Hamiltonians for the H_2^+ molecule (Eq. 20-21) it is suggestive to take as our trial wave function a product of two $1s\sigma_g$ functions (Eq. 20-24) for the H_2^+ molecule:

$$\psi_g(\mathbf{r}_1, \mathbf{r}_2) = \frac{1}{2[1 + S(R_{AB})]} [\psi_A(\mathbf{r}_1) + \psi_B(\mathbf{r}_1)] [\psi_A(\mathbf{r}_2) + \psi_B(\mathbf{r}_2)] \, \mathrm{X}_{\text{singlet}} \quad (21\text{-}4)$$

The electron spin state is a singlet, since the spatial part of the wave function is taken to be symmetric. In this trial wave function, *each electron is associated with both protons*, that is, the trial wave function is said to be a product of *molecular orbitals*. The description in terms of molecular orbitals is sometimes called the *MO* method.

The calculation of $\langle \psi_g | H | \psi_g \rangle$ yields

$$\left\langle \psi_g \left| \left(\tilde{H}_1 - \frac{e^2}{R_{AB}} \right) + \left(\tilde{H}_2 - \frac{e^2}{R_{AB}} \right) + \frac{e^2}{r_{12}} + \frac{e^2}{R_{AB}} \right| \psi_g \right\rangle$$

$$= \epsilon(R_{AB}) + \epsilon(R_{AB}) + \left\langle \psi_g \left| \frac{e^2}{r_{12}} \right| \psi_g \right\rangle - \frac{e^2}{R_{AB}}$$

$$= 2\epsilon(R_{AB}) - \frac{e^2}{R_{AB}} + \left\langle \psi_g \left| \frac{e^2}{r_{12}} \right| \psi_g \right\rangle \quad (21\text{-}5)$$

where $\epsilon(R_{AB})$ is the energy of the H_2^+ molecule calculated in Chapter 20. The first order electron-electron repulsion contribution can also be calculated, and when the total energy so computed is minimized with respect to the separation R_{AB}, it is found that the binding energy and internuclear separation are given by

$$E_b = -2.68 \text{ eV}$$
$$R = 0.85 \text{ Å} \quad (21\text{-}6)$$

The experimental values are

$$E_b = -4.75 \text{ eV}$$
$$R = 0.74 \text{ Å} \quad (21\text{-}7)$$

Evidently the approximation is not a very good one. We noted in our discussion of the H_2^+ molecule that the trial wave functions (the MO's) are inaccurate for small proton-proton separations, and the fact that the MO's are too spread out in space shows up in the numbers above. The trial wave function also has some undesirable features for large R_{AB}. The product in (21-4) may be rewritten in the form

$$[\psi_A(\mathbf{r}_1) + \psi_B(\mathbf{r}_1)][(\psi_A(\mathbf{r}_2) + \psi_B(\mathbf{r}_2)]$$

$$= [\psi_A(\mathbf{r}_1) \, \psi_A(\mathbf{r}_2) + \psi_B(\mathbf{r}_1) \, \psi_B(\mathbf{r}_2)] + [\psi_A(\mathbf{r}_1) \, \psi_B(\mathbf{r}_2) + \psi_A(\mathbf{r}_2) \, \psi_B(\mathbf{r}_1)]$$

$$(21\text{-}8)$$

The first term is called an "ionic" term, since it describes both electrons bound to one proton or the other. The second term, the "covalent" term, is a description in terms of linear combinations of atomic orbitals (LCAO). Our trial wave function thus implies, since the two terms enter with equal weight, that for large R_{AB} the molecule is as likely to dissociate into the ions H^+ and H^-, as it is into two hydrogen atoms, and this is patently false.

The last difficulty can be avoided with the use of the *Valence Bond* (also called Heitler-London) method, in which linear combinations of atomic orbitals are used. The singlet wave function used as a trial wave function in the variational principle is taken to be

$$\psi(\mathbf{r}_1, \mathbf{r}_2) = \left\{ \frac{1}{2[1 + S^2(R_{AB})]} \right\}^{1/2} [\psi_A(\mathbf{r}_1)\,\psi_B(\mathbf{r}_2) + \psi_A(\mathbf{r}_2)\,\psi_B(\mathbf{r}_1)]\, X_{\text{singlet}} \quad (21\text{-}9)$$

where, as before, the $\psi_A(\mathbf{r}_i)$ are hydrogenic wave functions for the i-th electron about proton A. We could, in principle, add a triplet term to our variational trial wave function. However, a triplet wave function must be spatially anti-symmetric and has low probability for the electrons being located in the region between the protons. We saw in our discussion of the H_2^+ molecule that just this configuration led to the lowest energy. Although it is not immediately obvious that the attraction is still largest in this configuration when there are *two* electrons that repel each other in the system, it is in fact so. The results of a variational calculation with the VB trial wave function is

$$E_b = -3.14 \text{ eV}$$

$$R = 0.87 \text{ Å} \quad (21\text{-}10)$$

This is not a significant improvement over the MO results, for the simple reason that the inadequacy of the trial wave functions for small R_{AB} carries more weight. There should be no question about the quantitative successes of quantum mechanics in molecular physics. More sophisticated trial wave functions have to be used; for example, a 50-term trial wave function yields complete agreement with observations for the H_2 molecule, but it does not, as the MO and VB functions do, give us something of a qualitative feeling of what goes on between the atoms. In what follows, we will explore the relevance of these approaches to a qualitative understanding of some aspects of chemistry.

The expectation value of H for the H_2 molecule in the VB approach has the following schematic form

$$\langle \psi | H | \psi \rangle = \frac{1}{2(1 + S^2)} \langle \psi_{A1}\psi_{B2} + \psi_{A2}\psi_{B1} | H | \psi_{A1}\psi_{B2} + \psi_{A2}\psi_{B1} \rangle$$

$$= \frac{1}{1 + S^2} \left\langle \psi_{A1}\psi_{B2} \left| \left(T_1 + T_2 - \frac{e^2}{r_{A1}} - \frac{e^2}{r_{A2}} - \frac{e^2}{r_{B1}} - \frac{e^2}{r_{B2}} \right. \right. \right.$$

$$\left. + \frac{e^2}{r_{12}} + \frac{e^2}{R_{AB}}\right) \middle| \psi_{A1}\psi_{B2} + \psi_{A2}\psi_{B1}\right\rangle$$

where T_i is the kinetic energy of the i-th electron, and since

$$\left(T_1 - \frac{e^2}{r_{A1}}\right)\psi_{A1} = E_1\psi_{A1}$$

and so forth, this can be simplified to

$$\frac{1}{1+S^2}\left(\left\langle\psi_{A1}\psi_{B2}\middle| 2E_1 - \frac{e^2}{r_{B1}} - \frac{e^2}{r_{A2}} + \frac{e^2}{r_{12}} + \frac{e^2}{R_{AB}}\middle|\psi_{A1}\psi_{B2}\right\rangle\right.$$

$$\left. + \left\langle\psi_{A1}\psi_{B2}\middle| 2E_1 - \frac{e^2}{r_{B2}} - \frac{e^2}{r_{A1}} + \frac{e^2}{r_{12}} + \frac{e^2}{R_{AB}}\middle|\psi_{A2}\psi_{B1}\right\rangle\right)$$

$$= \frac{1}{1+S^2}\left\{\left(2E_1 + \frac{e^2}{R_{AB}}\right)(1+S^2) - 2e^2\left\langle\psi_{A1}\middle|\frac{1}{r_{B1}}\middle|\psi_{A1}\right\rangle\right.$$

$$- 2e^2 S\left\langle\psi_{A1}\middle|\frac{1}{r_{A1}}\middle|\psi_{B1}\right\rangle + e^2\iint\frac{|\psi_{A1}|^2|\psi_{B2}|^2}{r_{12}}$$

$$\left. + e^2\iint\frac{\psi_{A1}^*\psi_{B1}\psi_{B2}^*\psi_{A2}}{r_{12}}\right\}$$

$$(21\text{-}11)$$

In obtaining this, liberal use has been made of symmetry. The terms that can make this expression more negative are

$$\left\langle\psi_{A1}\middle|\frac{1}{r_{B1}}\middle|\psi_{A1}\right\rangle \quad \text{and} \quad \frac{S}{1+S^2}\left\langle\psi_{A1}\middle|\frac{1}{r_{A1}}\middle|\psi_{B1}\right\rangle$$

The former is just the attraction of the electron cloud about one proton to the other proton; the second is the overlap of the two electrons (weighted with $1/r_{A1}$). If this can be large, there will be binding. The two electrons can only overlap significantly, however, if their spins are antiparallel; this is a consequence of the exclusion principle. The region of overlap is between the two nuclei, and there the attraction to the nuclei generally overcomes the electrostatic repulsion between the electrons.

In the MO picture, too, it is an overlap term—the last term in (20-31)— that is crucial to bonding, and again, bonding occurs because the electron charge distribution is large between the nuclei. Thus, although here the orbitals belong to the whole molecule rather than to individual atoms, the physical reason for bonding is the same.

We will discuss some molecules in terms of these two approaches to the description of the electronic charge distribution. An important simplification occurs because we really do not need to take all electrons into account. In the construction of orbitals, be it valence or molecular, only the outermost electrons,

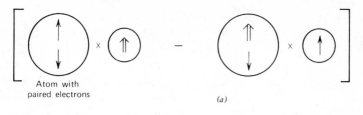

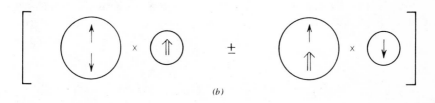

Fig. 21-2. Illustration of why paired electrons do not give rise to bonding. (*a*) If parallel electrons exchange, wave function is spatially antisymmetric. (*b*) If antiparallel electrons exchange, one term in the wave function has electrons in the same spin state, which may require promotion to a higher energy orbital.

not in closed shells, that is, the so-called *valence electrons* have a chance to contribute to the bonding. The inner electrons, being closer to the nucleus, are less affected by the presence of another atom in the vicinity.[1] Furthermore, not all valence electrons contribute equally: if two electrons are in a spin 0 state—we call them *paired electrons*—they will *not give rise to bonding*. To see why this is so, consider what happens when an atom with a single valence electron is brought near an atom with two paired electrons. There are two cases to be considered (Fig. 21-2).

(a) If the two electrons that are parallel exchange (i.e., are put into a form such as (21-9) with a ± sign between the terms) then they must be in a triplet state, and hence the spatial wave function of this pair must be antisymmetric. This reduces the overlap, and it turns out that the exchange integral gives a repulsive contribution to the energy.

(b) When the two electrons that are antiparallel exchange, then one atom finds itself some of the time with two electrons in the same spin state. The

---

[1] It may happen in atoms that even the valence electrons are rather close to the nucleus. This is the case for the rare earths. A consequence of the fact that the outer electrons in $5d$ and $4f$ shells lie close in is that the rare earths are chemically less active than the transition metals ($Z \simeq 20 - 30$).

original atomic state will frequently no longer be a possible one, and one of the electrons will have to be promoted into another atomic orbital. Sometimes this may cost very little energy, but usually this is not the case, and again bonding is not achieved. *Chemical activity depends on the presence of unpaired outer electrons.* An example of this is the nonexistence of the H-He molecule. In He we have two electrons in the $1s$ state; promotion of one of them into a $2s$ state costs a lot of energy. It is for this reason that the atoms for which the outer shells are closed are *inert*. Not all unpaired electrons are of equal significance. As noted before, the unpaired d- and f-electrons in the transition elements tend to be close to the nucleus, and hence inactive. Thus, mainly s- and p-electrons in the outer shells contribute to chemical activity. The pairing effect is also responsible for what is called the "saturation of chemical binding forces": once two unpaired electrons

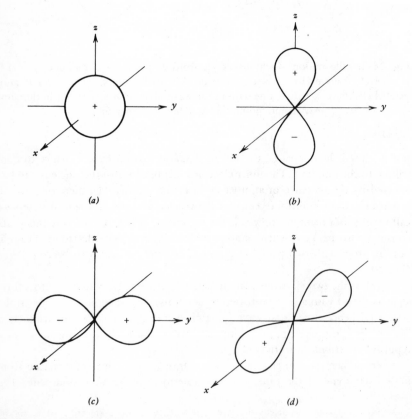

Fig. 21-3. Pictorial representation of shapes of (*a*) the s-orbital, (*b*) the $p_z(Y_{10})$, (*c*) the $p_y(Y_{11} - Y_{1,-1})$, and (*d*) the $p_x(Y_{11} + Y_{1,-1})$ orbitals. The signs refer to the signs of the wave function in the given region.

from different atoms form a singlet state (and cause bonding), they become paired; an electron from a third atom must find an unpaired electron elsewhere, that is, participate in a different bond. Another consequence is that molecules have spin 0 in most cases.

Let us next go through a process analogous to the building up of the electronic shells in atoms. In Fig. 21-3 we show pictures of atomic orbitals, in particular the $s(Y_{00})$ orbital and the p-orbitals. For the latter the linear combinations $p_x(Y_{11} + Y_{1,-1})$ and $p_y(Y_{11} - Y_{1,-1})$ are plotted in addition to $p_z(Y_{10})$. The corresponding d-orbitals, d_{xy}, d_{xz}, d_{yz}, d_{zz}, and $d_{xx} - d_{yy}$ are not shown, because the d-electrons will play no role in our discussion. Figure 21-4 represents what happens when atomic orbitals are brought together and exchange occurs. Thus, two 1s atomic orbitals may combine into a spatially symmetric MO (hence with spin 0) or into a spatially antisymmetric MO, which is antibonding since the wave function between the nuclei is small. Similarly, the formation of bonding and antibonding MO's with p-orbitals is illustrated in the figure. Note that (a) the parity "g" or "u" can be read off from the figures, since these indicate the signs of the wave functions; the distributions that change sign upon reflection in the $x = y$ plane, here represented by a vertical line, are odd; (b) since the p_x- and p_y-orbitals have $m_l = \pm 1$, the molecular orbital formed from them is a π-orbital. It should be stressed that in the figure we are *not* just bringing two charge distributions together, but are trying to suggest the probability amplitude that results when wave functions are combined, that is, the MO's such as $\psi_{1s}(\mathbf{r}_A) \pm \psi_{1s}(\mathbf{r}_{Bz})$ and $\psi_{2p_y}(\mathbf{r}_A) \pm \psi_{2p_y}(\mathbf{r}_B)$ for R_{AB} large and for R_{AB} small.

We can use the MO's to discuss the properties of a few diatomic homonuclear molecules:

H$_2$. This molecule was discussed in some detail. We merely repeat that the two electrons can go into a $1s\sigma_g$ MO, and since this orbital has a lower energy than the separated 1s atomic orbitals, there is stability.

He$_2$. Of the four electrons, only two can go into a bonding $1s\sigma_g$ orbital; the other two must form an antibonding $1s\sigma_u{}^*$ orbital. The net energy is greater than that of the separated He atoms, so that no molecule is formed. In terms of the Valence Bond picture, both atoms have paired electrons and the conclusion is the same. In general, electrons in bonding orbitals and in antibonding orbitals tend to cancel each other out. Since there are two electrons involved in a full bond, we may speak of a *bond number*, given by

$$\binom{\text{Bond}}{\text{number}} = \frac{1}{2}\left[\binom{\text{Electrons in}}{\text{bonding orbitals}} - \binom{\text{Electrons in}}{\text{antibonding orbitals}}\right]$$

This number vanishes for He$_2$.

Li$_2$. The atomic structure of Li is $(1s)^2(2s)$. Thus, the 2s electrons are unpaired, and they can form $2s\sigma_g$ bonding orbitals. We thus expect the molecule

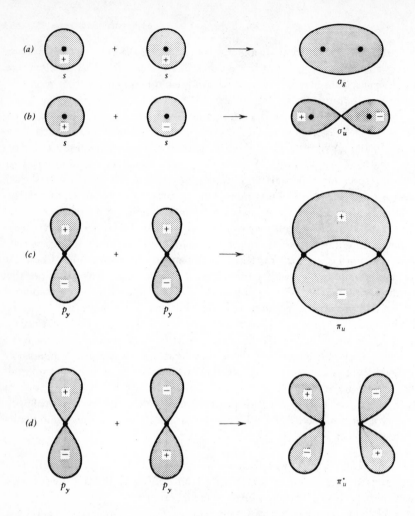

Fig. 21-4. Molecular orbitals resulting when two atomic orbitals are brought together. (*a*) Two *s* orbitals combine to form the spatially symmetric *MO* σ_g that gives rise to bonding; (*b*) Two *s* orbitals combine to form the spatially antisymmetric antibonding *MO* $\sigma_u{}^*$. (*c*) and (*d*) show bonding and antibonding with p_y atomic orbitals; (*e*) and (*f*) show bonding and antibonding with p_z orbitals. The *z*-axis is along the line connecting the nuclei, which are represented by black dots.

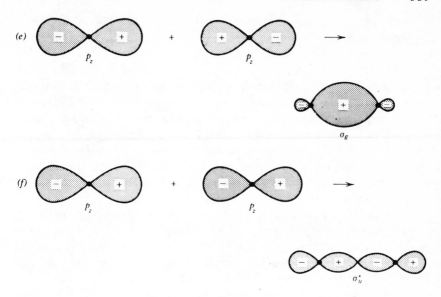

Fig. 21-.4. continued

to exist, but because of the $n = 2$ value of the orbital, we would expect the binding to be significantly smaller than for the H_2 molecule.

Be$_2$. Here the atomic structure is $(1s)^2(2s)^2$; there are no unpaired electrons, and hence we expect no molecule to exist. This is indeed so.

B$_2$. The atomic structure indicates that there is an unpaired $2p$ electron in each atom. It can be in any one of the states $2p_x$, $2p_y$, and $2p_z$. They may combine either into a $2p\pi_u$ or into a $2p\sigma_g$ MO. The former has a lower energy, so that here the ground state is a triplet. This is in agreement with Hund's Rule: The state with highest multiplicity has the lowest energy.

The reason why $2p\sigma_g$ has a higher energy is that there exist $2s\sigma_g$ orbitals. Whenever there are states that have the same quantum numbers, "mixing" occurs, and states that are almost degenerate tend to repel each other. The state that is largely $2p\sigma_g$ is pushed up. We begin to see the appearance of complications similar to the ones that appeared in our discussion of atomic structure!

C$_2$. The atomic structure is $(1s)^2(2s)^2(2p)^2$, that is, each atom has two unpaired electrons. Since each electron can be in any one of three p states, two bonding MO's can be formed. The MO description turns out to be $(2p\sigma_g)(2p\pi_u)$.

N_2. Here the situation is very similar to that of C_2 except that three bonding MO's can be formed. The MO description turns out to be $(2p\sigma_g)(2p\pi_u)^2$.

O_2. Here things get a little more interesting, because the atomic structure is $(1s)^2(2s)^2(2p)^4$, that is, there are four valence electrons. In terms of molecular orbitals, three bonds, as in N_2, can be formed, but this leaves two electrons that cannot possibly form a bonding orbital. What is the least harmful antibonding orbital? The two electrons should avoid each other as much as possible, and this can be done by means of a triplet state, with the electrons in orthogonal orbitals, for example, one in a p_x, the other in a p_y state, with the two spatially antisymmetrized. In this case the spin of O_2 is 1, an exception to the strong tendency toward zero spin that was mentioned earlier.

In the Valence Bond picture, two of the four valence electrons in oxygen must be paired, so that two bonds will exist orthogonal to each other, as p_x is to p_z, for example. One may see the effect of this directionality in a molecule like H_2O. Each H uses up one bond, and we would expect the shape of the molecule to be an **L** with 90° between the equal length arms. Actually, the two hydrogen nuclei repel each other, and one might expect the angle to be a little larger than 90°. Experimentally it is around 105°! It is the directionality of the p-orbitals that explains the shape of simple molecules.

Lest the reader feel that all of chemistry can be understood with the material at hand, we will point out just a few of the many complications that show the subtle sources of the incredible variety in the structure of matter. For example, a carbon atom has two valence electrons, and one might expect a CH_2 molecule, with a shape similar to the H_2O molecule, to exist. Actually, C turns out to be tetravalent (four bonds) rather than divalent, so that it is CH_4 that is actually formed. The reason is that although the ground state of C is $(1s)^2(2s)^2(2p)^2$, the excited state $(1s)^2(2s)(2p)^3$ differs very little from it in energy. This state, however, has four valence electrons, and the molecular bonding with four bonds is sufficiently stronger than that with two, to compensate for the electronic excitation energy. More precisely, the near degeneracy of the $2s$ and $2p$ states in the atom allows the formation of linear combinations that allow larger overlaps. Figure 21-5 shows that a linear combination of an s- and a p-orbital gives a lopsided wave function allowing an increased overlap. This "mixing" is actually quite common. We may give a more detailed description of the water molecule by working with *hybrid* orbitals that involve s- and p-orbitals. If we ignore the difference between the $2s$ and $2p$ electrons in oxygen, then we must really deal with the more general states

$$\chi^{(i)} = \alpha^{(i)}\phi_{2s} + \beta_x^{(i)}\phi_{2p_y} + \beta_y^{(i)}\phi_{2p_y} + \beta_z^{(i)}\phi_{2pz} \tag{21-12}$$

with the coefficients constrained by

$$\langle \chi^{(i)} | \chi^{(j)} \rangle = \delta_{ij} \tag{21-13}$$

Fig. 21-5. Combination of s and p_z orbitals leading to unsymmetric wave function.

rather than with ϕ_{2s}, ϕ_{2p_x}, ϕ_{2p_y}, and ϕ_{2p_z}, and molecular orbitals or valence orbitals constructed for use as trial wave functions in the variational principle should be made up out of the χ's. The minimization of the energy will determine the coefficients $\alpha^{(i)}$ and $\beta^{(i)}$. For the H_2O molecule, it turns out that the four orthonormal combinations are

$$\chi^{(1)} = \frac{1}{\sqrt{4}}\,(\phi_{2s} + \phi_{2p_x} + \phi_{2p_y} + \phi_{2p_z})$$

$$\chi^{(2)} = \frac{1}{\sqrt{4}}\,(\phi_{2s} + \phi_{2p_x} - \phi_{2p_y} - \phi_{2p_z})$$

$$\chi^{(3)} = \frac{1}{\sqrt{4}}\,(\phi_{2s} - \phi_{2p_x} + \phi_{2p_y} - \phi_{2p_z})$$

$$\chi^{(4)} = \frac{1}{\sqrt{4}}\,(\phi_{2s} - \phi_{2p_x} - \phi_{2p_y} + \phi_{2p_z}) \tag{21-14}$$

Consider now the 10 electrons in H_2O. Two $1s$ electrons remain strongly bound to the oxygen atom. (One could, for consistency, describe them in terms of $1s\sigma_g$ and $1s\sigma_u$ MO's, analogous to the H_2^+ molecule, but this changes nothing.) The remaining eight electrons go, two each, into the hybridized orbitals $\chi^{(1)}, \ldots \chi^{(4)}$. The geometrical shape of the molecule can be determined from the form of these hybrid orbitals. If we try to draw what $\chi^{(1)}$ looks like, we see that the last three terms are all negative in the octant $(x, y, z < 0)$ and there they effectively cancel the s term. Roughly speaking, $\chi^{(1)}$ looks like a fat cigar pointing from the origin to the point $(1, 1, 1)$. Similarly $\chi^{(2)}$ points from the origin to the point $(1, -1, -1)$, and so on. The shape is a tetrahedron, and some simple geometry shows that the angle between the bonds, θ satisfies $\sin \theta/2 = \sqrt{2/3}$ so that $\theta \cong 109°$. Without hybridization we found the angle to be 90°, and the truth lies somewhere in between. In more accurate calculations, (21-14) is somewhat modified, so that only three of the four orbitals are so pronouncedly "p-like" and the remaining orbital has less directionality, that is, is more "s-like".

This more detailed picture allows us to understand some of the properties of water. If we think of the water molecule as oxygen, with four tetrahedrally oriented arms that are charged, with protons, from the H's attached to two of them by Coulomb forces, we first of all see that the water molecule may be expected to have a large dipole moment coming from the two negatively charged arms pointing away from the proton. We recall (cf. Chapter 16) that a ground state can only have an electric dipole moment if it is degenerate. We easily see that since the ϕ_{2p} change sign under reflection and the ϕ_{2s} do not, the $\chi^{(i)}$ are not eigenstates of parity. The reflected orbitals, for example,

$$\chi^{(1)'} = \frac{1}{\sqrt{4}} \left(\phi_{2s} - \phi_{2p_x} - \phi_{2p_y} - \phi_{2p_z} \right) \qquad (21\text{-}15)$$

have exactly the same energy and the reflected shape. The ground state is there-fore degenerate, and a dipole moment can exist.

When water molecules get close to each other in a liquid, the negatively charged arms of one may come close to the proton of another. The electrostatic attraction between the two will lower the energy, and there will be a tendency for the two molecules to bind. The bond is fairly weak (0.2 eV) and is called a *hydrogen bond*. Each water molecule can bond four other ones at once, and one thus expects to find large clusters of water molecules in the liquid, effectively molecules of the form $H_{2n}O_n$. This leads to a strong temperature dependence of the viscosity of water. In cold water, the large clusters easily tangle together; heating the water breaks them up and reduces the size of the clusters and hence the viscosity.

Many other molecules also form hydrogen bonds. This is how the process of dissolving works; water molecules form hydrogen bonds with the substance, and the molecules of the substance would rather stick to the water molecules than to each other; the substance dissolves. Oils do not form good hydrogen bonds and thus do not dissolve in water.

Not all hybrid orbitals are of the form (21-14). CH_4 *is* tetrahedral, and in fact the bond angle is 109.6°, but in C_2H_4 the molecule has a planar structure. It turns out that for this molecule, the carbon orbitals are hybridized as follows:

$$\chi^{(1)} = \phi_{2p_z}$$

$$\chi^{(2)} = \frac{1}{\sqrt{3}} \phi_{2s} + \sqrt{\frac{2}{3}} \phi_{2p_x}$$

$$\chi^{(3)} = \frac{1}{\sqrt{3}} \phi_{2s} - \frac{1}{\sqrt{6}} \phi_{2p_x} + \frac{1}{\sqrt{2}} \phi_{2p_y}$$

$$\chi^{(4)} = \frac{1}{\sqrt{3}} \phi_{2s} - \frac{1}{\sqrt{6}} \phi_{2p_x} - \frac{1}{\sqrt{2}} \phi_{2p_y} \qquad (21\text{-}16)$$

The first one points along the z-axis, and the last three are oriented at 120° intervals in the x-y plane. The four outer electrons in carbon [specifically, $(2s)^2(2p)^2$] go into these orbitals. Given two carbon atoms, the two electrons in

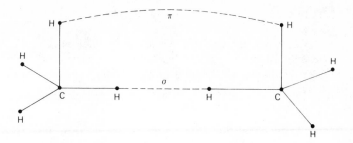

Fig. 21-6. Bonds in the C_2H_4 molecule.

the $\chi^{(3)}$ and $\chi^{(4)}$ orbitals bind to the hydrogens, while those in the $\chi^{(1)}$ and $\chi^{(2)}$ states form bonds to the other carbon (Fig. 21-6). The electrons in the $\chi^{(2)}$ state form a σ bond between the carbons, while the electrons in the $\chi^{(1)}$ state, being at right angles to the bonding axis, form a π bond. It is these bonds that force the two p_z orbitals to be parallel, and thus make the rest of the structure planar.

As a final comment, we note that in the so-called "aromatic" compounds, for example, benzene, we cannot speak of such well-localized orbitals. The carbons are hybridized as in C_2H_4 and they form a planar structure, with the p_z out of the plane (Fig. 21-7). The σ-orbitals form a core, but the π-orbitals can be paired according to (12) (34) (56) or (23) (45) (61), both of which have the same energy. As usual, linear combinations of these degenerate possibilities lower the energy, so that the stable state does not have localized orbitals. This additional exchange effect strongly affects the physical properties of the compounds, but a discussion of these would carry us too far afield. For more information, the reader should turn to books on quantum chemistry.

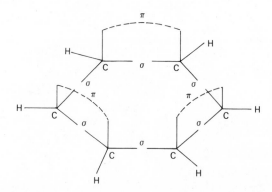

Fig. 21-7. Schematic picture of Benzene molecule (C_6H_6).

References

This chapter owes much to the brief discussion of molecules in G. Baym, *Lectures on Quantum Mechanics*, W. A. Benjamin, Inc., New York, 1969. The interested reader should consult the following books for more information:

M. Karplus and R. N. Porter, *Atoms and Molecules*, W. A. Benjamin, Inc., New York, 1970.

M. W. Hanna, *Quantum Mechanics in Chemistry*, W. A. Benjamin, Inc., New York, 1969.

U. Fano and L. Fano, *Physics of Atoms and Molecules*, Chicago University Press, Chicago, 1972.

G. W. King, *Spectroscopy and Molecular Structure*, Holt, Rinehart, and Winston, Inc., New York, 1964.

There are, of course, hundreds of books on quantum chemistry, molecular structure, and molecular spectroscopy, and the sampling listed above is just the one familiar to the author. For a better reference list, the reader should consult any physical chemist.

chapter 22

The Radiation of Atoms

In the study of spectra, that is, the study of transitions between atomic levels accompanied by the emission or absorption of radiation, one is interested in the interaction between atoms and the electromagnetic field. Since the radiation field oscillates, it is time dependent. It is therefore necessary to study the effect of time-dependent perturbations.

A. Time-Dependent Perturbation Theory

The problem is, given the complete set of solutions to

$$H_0 \phi_n = E_n^0 \phi_n \tag{22-1}$$

to solve for $\psi(t)$, which obeys the equation

$$i\hbar \frac{\partial \psi(t)}{\partial t} = [H_0 + \lambda V(t)] \, \psi(t) \tag{22-2}$$

The standard procedure is to expand $\psi(t)$ in a complete set of states:

$$\psi(t) = \sum_n c_n(t) \, e^{-iE_n^0 t/\hbar} \, \phi_n \tag{22-3}$$

The time-dependence associated with the ϕ_n is explicitly inserted in the expansion, so that if $V(t) = 0$, the $c_n(t)$ would be constants. The expansion coefficients $c_n(t)$ satisfy a set of equations that may be obtained by substituting (22-3) into the time-dependent Schrödinger equation (22-2). We get

$$\sum_n \left[i\hbar \frac{dc_n(t)}{dt} + E_n^0 c_n(t) \right] e^{-iE_n^0 t/\hbar} \, \phi_n = H\psi(t)$$

$$= \sum_n \left[E_n^0 + \lambda V(t) \right] c_n(t) \, e^{-iE_n^0 t/\hbar} \, \phi_n$$

that is,

$$i\hbar \sum_n \frac{dc_n(t)}{dt} e^{-iE_n^0 t/\hbar} \phi_n = \lambda \sum_n V(t) c_n(t) e^{-iE_n^0 t/\hbar} \phi_n \qquad (22\text{-}4)$$

Taking the scalar product with ϕ_m and using the orthonormality of the ϕ_m,

$$\langle \phi_m | \phi_n \rangle = \delta_{mn} \qquad (22\text{-}5)$$

yields, after the factor $e^{-iE_m^0 t/\hbar}$ is divided out, the set of equations

$$i\hbar \frac{dc_m(t)}{dt} = \lambda \sum_n c_n(t) e^{i(E_m^0 - E_n^0)t/\hbar} \langle \phi_m | V(t) | \phi_n \rangle \qquad (22\text{-}6)$$

We shall solve these to first order in the parameter λ. As an initial condition at $t = 0$ we take the system to be in a particular state ϕ_k, so that $\psi(0) = \phi_k$, that is,

$$c_{nk}(0) = \delta_{nk} \qquad (22\text{-}7)$$

Since departures from these values at later times will depend on λ, we may, for a first-order calculation, substitute the above into the right side of (22-6). This yields the differential equation (for $m \neq k$)

$$i\hbar \frac{dc_m(t)}{dt} = \lambda e^{i(E_m^0 - E_k^0)t/\hbar} \langle \phi_m | V(t) | \phi_k \rangle \qquad (22\text{-}8)$$

which is easily solved

$$c_m(t) = \frac{\lambda}{i\hbar} \int_0^t dt' \, e^{i(E_m^0 - E_k^0)t'/\hbar} \langle \phi_m | V(t') | \phi_k \rangle \qquad (22\text{-}9)$$

The probability that at a later time t, the state $\psi(t)$, is an eigenstate of H_0 with energy E_n^0, that is, that it is ϕ_n, is, according to the expansion postulate

$$P_n(t) = |\langle \phi_n | \psi(t) \rangle|^2 = |c_n(t)|^2 \qquad (22\text{-}10)$$

This general result can only be made more specific if $V(t)$ is known. The perturbation will be specified next.[1]

B. The Electromagnetic Interaction

The Hamiltonian describing the interaction of an electron in a static potential $V(r)$ with an electromagnetic field described by the vector potential $\mathbf{A}(\mathbf{r},t)$ is given by

$$H = \frac{[\mathbf{p} + (e/c)\,\mathbf{A}(\mathbf{r},t)]^2}{2m} + V(r) \qquad (22\text{-}11)$$

[1] Actually, more can be said if it is known that either $V(t)$ varies very slowly or changes very rapidly compared to the typical frequencies (e.g., $E_n^0/\hbar$) in the system. So-called "adiabatic" and "sudden" approximations are discussed in more advanced texts.

as we saw in Chapter 13. Thus, if we write

$$H_0 = \frac{\mathbf{p}^2}{2m} + V(r) \qquad (22\text{-}12)$$

we find that

$$\lambda V(t) = \frac{e}{mc}\, \mathbf{A}(\mathbf{r},t)\cdot\mathbf{p} \qquad (22\text{-}13)$$

In obtaining the last expression we have specified the gauge so that

$$\boldsymbol{\nabla}\cdot\mathbf{A}(\mathbf{r},t) = 0 \qquad (22\text{-}14)$$

Under these circumstances, $\mathbf{p}\cdot\mathbf{A} = \mathbf{A}\cdot\mathbf{p}$, and we have dropped the term quadratic in $\mathbf{A}(\mathbf{r},t)$. If we treat e, the electron charge as the parameter of smallness λ, then the $\mathbf{A}^2$ term is a second-order term. We will see that the $\mathbf{A}^2$ term will contribute to the scattering of light by an atom and to the transition with the emission of two photons, but it will not contribute to the transition accompanied by the emission (or absorption) of a single photon. The probability for a transition involving two photons involves a factor $(e^2)^2$, whereas the one-photon transition probability is proportional to e^2. Recalling that the appropriate dimensionless number involving e^2 is $\alpha = e^2/\hbar c \cong 1/137$, we are justified in concentrating on transitions that are accompanied by the emission of a single photon.

To give a real justification of the association of each $\mathbf{A}(\mathbf{r},t)$ with the emission or absorption of a single photon—so that higher powers of $\mathbf{A}(\mathbf{r},t)$ imply the presence of more photons—one must treat the electromagnetic field quantum mechanically, that is, treat the fields at each point $\mathbf{r}$ as operators. This is fundamentally not terribly complicated, but it is outside the scope of this book. The reader will have to take the following assertions on faith.

If we write

$$\mathbf{A}(\mathbf{r},t) = \mathbf{A}_0^*(\mathbf{r})\, e^{i\omega t} + \mathbf{A}_0(\mathbf{r})\, e^{-i\omega t} \qquad (22\text{-}15)$$

then in the *emission of a photon*, only the first term, with the time dependence $e^{i\omega t}$, is to be included in $\lambda V(t)$, whereas in the *absorption of a photon*, only the second term, with the time dependence $e^{-i\omega t}$, appears. This is a consequence of the general association of $\mathbf{A}_0^*(\mathbf{r})$ with the creation of a photon and $\mathbf{A}_0(\mathbf{r})$ with the annihilation of a photon, and the time dependence is just what one would expect from the harmonic oscillator (7-51). The resemblance to the harmonic oscillator problem is not accidental, since in the quantization of the electromagnetic field, what is done is a normal mode decomposition, according to which one finds that the field is really a collection of simple harmonic oscillators; these are then quantized. The "occupation number" n that labels the harmonic oscillator state vector may be associated with the number of photons, hence $\mathbf{A}_0^*$ raises the photon number by unity and $\mathbf{A}_0$ lowers the photon number by unity.

The more quantitative description of $\mathbf{A}_0^*(\mathbf{r})$ and $\mathbf{A}_0(\mathbf{r})$ need not, fortunately, involve the full machinery of quantum electrodynamics. We may use correspondence principle arguments to find these quantities, and then just state the quantum mechanical modifications. Away from the sources, the electromagnetic field has a very simple spatial behavior. If we look back at (13-12) and substitute (22-15), we find that

$$-\boldsymbol{\nabla}^2\mathbf{A}_0(\mathbf{r}) - \frac{\omega^2}{c^2}\,\mathbf{A}_0(\mathbf{r}) = 0 \tag{22-16}$$

whose solution is

$$\mathbf{A}_0(\mathbf{r}) = \mathbf{A}_0\,e^{\,i\mathbf{k}\cdot\mathbf{r}} \tag{22-17}$$

with

$$\mathbf{k}^2 = \frac{\omega^2}{c^2} \tag{22-18}$$

The choice of gauge (22-14) implies that

$$\mathbf{k}\cdot\mathbf{A}_0 = 0 \tag{22-19}$$

The electric and magnetic fields corresponding to this vector potential are

$$\mathbf{E} = -\frac{1}{c}\frac{\partial\mathbf{A}}{\partial t} = \frac{i\omega}{c}\,\mathbf{A}_0\,e^{i(\mathbf{k}\cdot\mathbf{r}-\omega t)} + \text{complex conjugate}$$

$$\mathbf{B} = \boldsymbol{\nabla}\times\mathbf{A} = i\mathbf{k}\times\mathbf{A}_0\,e^{i(\mathbf{k}\cdot\mathbf{r}-\omega t)} + \text{complex conjugate} \tag{22-20}$$

Now the energy density of the electromagnetic field is given by

$$\frac{1}{8\pi}\,(\mathbf{E}^2 + \mathbf{B}^2) = \frac{1}{8\pi}\left[\,2\,\frac{\omega^2}{c^2}\,\mathbf{A}_0\cdot\mathbf{A}_0^* + 2(\mathbf{k}\times\mathbf{A}_0)\cdot(\mathbf{k}\times\mathbf{A}_0^*) + \text{oscillating terms}\right] \tag{22-21}$$

If we average over time, so that the oscillating terms drop out, and make use of the fact that with (22-19)

$$(\mathbf{k}\times\mathbf{A}_0)\cdot(\mathbf{k}\times\mathbf{A}_0^*) = k^2\mathbf{A}_0\cdot\mathbf{A}_0^* \tag{22-22}$$

and that $k^2 = \omega^2/c^2$, we get

$$\frac{1}{8\pi}\,(\mathbf{E}^2 + \mathbf{B}^2) = \frac{\omega^2}{2\pi c^2}\,\mathbf{A}_0\cdot\mathbf{A}_0^* \tag{22-23}$$

If the system is enclosed in a box of volume, V, then the total energy in the electromagnetic field is

$$\int d^3r\,\frac{1}{8\pi}\,(\mathbf{E}^2 + \mathbf{B}^2) = \frac{\omega^2 V}{2\pi c^2}\,|\mathbf{A}_0|^2 \tag{22-24}$$

If this is to be carried by N photons, each with energy $\hbar\omega$, we have

$$\frac{\omega^2 V}{2\pi c^2} |\mathbf{A}_0|^2 = N\hbar\omega \qquad (22\text{-}25)$$

The direction of $\mathbf{A}_0$ is determined by the polarization of the electric field, and will be denoted by the unit vector $\boldsymbol{\varepsilon}$. It must satisfy

$$\boldsymbol{\varepsilon}\cdot\boldsymbol{\varepsilon} = 1$$
$$\boldsymbol{\varepsilon}\cdot\mathbf{k} = 0 \qquad (22\text{-}26)$$

We therefore obtain

$$\mathbf{A}(\mathbf{r},t) = \left(\frac{2\pi c^2 N\hbar}{\omega V}\right)^{1/2} \boldsymbol{\varepsilon}\, e^{\,i(\mathbf{k}\cdot\mathbf{r}-\omega t)} \qquad (22\text{-}27)$$

The quantum electrodynamic modification is the following: For the absorption of a light quantum by a charged particle from an initial state that already has N photons of frequency ω,

$$A(\mathbf{r},t) = \left(\frac{2\pi c^2 N\hbar}{\omega V}\right)^{1/2} \boldsymbol{\varepsilon}\, e^{\,i(\mathbf{k}\cdot\mathbf{r}-\omega t)} \qquad (22\text{-}28)$$

For the emission of a light quantum by a charged particle into a final state that has $N + 1$ quanta, that is, from an initial state with N quanta of frequency ω,

$$\mathbf{A}(\mathbf{r},t) = \left[\frac{2\pi c^2 (N + 1)\,\hbar}{\omega V}\right]^{1/2} \boldsymbol{\varepsilon}\, e^{\,-i(\mathbf{k}\cdot\mathbf{r}-\omega t)} \qquad (22\text{-}29)$$

Hence for the emission of a single photon of frequency from a state that has no photons, we have, according to (22-13),

$$\lambda V(t) = \frac{e}{mc}\left(\frac{2\pi c^2\hbar}{\omega V}\right)^{1/2} \boldsymbol{\varepsilon}\cdot\mathbf{p}\; e^{\,-i(\mathbf{k}\cdot\mathbf{r}-\omega t)} \qquad (22\text{-}30)$$

Hence

$$c_m(t) = \frac{-ie}{mc\hbar}\left(\frac{2\pi c^2\hbar}{\omega V}\right)^{1/2} \langle\phi_m|e^{-i\mathbf{k}\cdot\mathbf{r}}\,\boldsymbol{\varepsilon}\cdot\mathbf{p}|\phi_k\rangle \int_0^t dt'\, e^{\,i(E_m{}^0 - E_k{}^0 + \hbar\omega)t'/\hbar} \qquad (22\text{-}31)$$

and thus the probability of transition from the initial state k to the state m is given by

$$P_{k\to m}(t) = \frac{2\pi\, e^2}{m^2\hbar\omega V} |\langle\phi_m|e^{-i\mathbf{k}\cdot\mathbf{r}}\,\boldsymbol{\varepsilon}\cdot\mathbf{p}|\phi_k\rangle|^2 \left|\int_0^t dt'\, e^{\,i(E_m{}^0 - E_k{}^0 + \hbar\omega)t'/\hbar}\right|^2 \qquad (22\text{-}32)$$

The time-dependent factor is

$$\left|\int_0^t dt'\, e^{\,i\Delta t'}\right|^2 = \left|\frac{(e^{\,i\Delta t}-1)}{i\,\Delta}\right|^2 = \left|\frac{2}{\Delta}\, e^{\,i\Delta t/2}\sin\frac{\Delta t}{2}\right|^2 = \frac{4}{\Delta^2}\sin^2\frac{t\Delta}{2} \qquad (22\text{-}33)$$

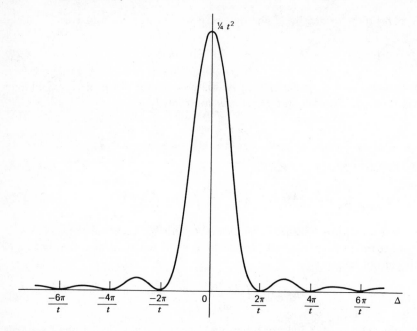

Fig. 22-1.   Plot of the function $1/\Delta^2 \sin^2 t\Delta/2$ versus Δ.

where

$$\Delta = \frac{E_m{}^0 - E_k{}^0 + \hbar\omega}{\hbar} \qquad (22\text{-}34)$$

Figure 22-1 shows the behavior of this function. For large t it becomes strongly peaked at $\Delta = 0$, and away from $\Delta = 0$ it oscillates very rapidly. This is the kind of behavior that we associate with a delta function. In fact, if $f(\Delta)$ is a smooth function of Δ, then, for t large

$$\int_{-\infty}^{\infty} f(\Delta)\, \frac{4}{\Delta^2}\, \sin^2 \frac{t\Delta}{2}\, d\Delta \approx f(0) \int_{-\infty}^{\infty} d\Delta\, \frac{4}{\Delta^2}\, \sin^2 \frac{t\Delta}{2}$$

$$= 2tf(0) \int_{-\infty}^{\infty} dy\, \frac{1}{y^2}\, \sin^2 y = 2\pi t f(0) \qquad (22\text{-}35)$$

that is, for t large

$$\frac{4}{\Delta^2}\, \sin^2 \frac{t\Delta}{2} \rightarrow 2\pi t\, \delta(\Delta) = 2\pi \hbar t \delta(E_m{}^0 + \hbar\omega - E_k{}^0) \qquad (22\text{-}36)$$

Thus the transition probability in (22-32) grows linearly with time, and hence the *transition probability per unit time* is

$$\Gamma_{k \to m} = 2\pi\hbar \, \frac{2\pi e^2}{m^2 \hbar \omega V} \, | \langle \phi_m | e^{-i\mathbf{k} \cdot \mathbf{r}} \, \boldsymbol{\varepsilon} \cdot \mathbf{p} | \phi_k \rangle |^2 \, \delta(E_k^0 - E_m^0 - \hbar\omega) \quad (22\text{-}37)$$

For general purposes, it will be useful to remember that if the time-dependent perturbation is of the form

$$V(t) = V^* \, e^{-i\omega t} \quad (22\text{-}38)$$

then the formula for the transition rate is

$$\Gamma_{k \to m} = \frac{2\pi}{\hbar} \, | \langle \phi_m | V^* | \phi_k \rangle |^2 \, \delta(E_k^0 - E_m^0 - \hbar\omega) \quad (22\text{-}39)$$

As things stand, the reader undoubtedly feels swindled. First of all the manipulations involved in (22-33)–(22-36) certainly are not straightforward. They involve vague notions such as "*t* large," which cannot be taken too seriously, since a transition probability that grows linearly with time must sooner or later exceed unity. Second, they lead to a nonsensical formula, according to which a perfectly reasonable quantity, like the transition rate, is proportional to a delta function. Needless to say, the difficulties are connected, and we will later outline a more satisfactory discussion. At this point we merely note that the fault lies in the use of perturbation theory, and that both (22-37) and (22-39) are correct, if properly used.

For this, we note that $\Gamma_{k \to m}$ is really the *transition probability per unit time for the atom making a transition from the state ϕ_k to the state ϕ_m, accompanied by the emission of a photon of energy $\hbar\omega$.* The delta function, unappealing as it is, does tell us that energy must be conserved, that is,

$$\hbar\omega = E_k^0 - E_m^0 \quad (22\text{-}40)$$

The delta function is actually integrated over, if we take into account that the photon energy $\hbar\omega$ does not uniquely specify the photon state. The photon will in general be detected in some momentum interval $(\mathbf{k}, \mathbf{k} + \Delta\mathbf{k})$ in the vicinity of $|\mathbf{k}| = \omega/c$, and the transition rate that is measured is really

$$R_{k \to m} = \sum_{\Delta\mathbf{k}} \Gamma_{k \to m} \quad (22\text{-}41)$$

summed over all the possible photon states in that interval. Note that the various final states in the interval $\Delta\mathbf{k}$ are in principle distinguishable, so that it is the probabilities that are summed. We will see that the sum (22-41) is well defined, since, in effect, it involves the integral of a delta function and a smooth function. The summation will be treated in the next section.

C. Phase Space

We will now calculate the number of photon states in the momentum interval $(\mathbf{k}, \mathbf{k} + \Delta\mathbf{k})$, that is, the density of photon states. For the purpose at hand, we write the vector potential $\mathbf{A}(\mathbf{r},t)$ in the form

$$\mathbf{A}(\mathbf{r},t) = \frac{1}{\sqrt{V}} \, \mathbf{a} \, e^{i(\mathbf{k}\cdot\mathbf{r}-\omega t)} + \text{complex conjugate} \qquad (22\text{-}42)$$

where V is the volume of the enclosure in which the calculation is done. This "box" is just a convenience to save us the trouble of working with wave packets for the free particles (the photons, here—cf. Chapter 4). Its shape and the conditions at the boundary may be chosen at will, but it must be large. At the end, we will take $V \to \infty$. We will find it convenient to take the box to be a cube of side L, and to impose periodic boundary conditions, that is,

$$\mathbf{A}(x + L, y, z, t) = \mathbf{A}(x, y, z, t) \qquad (22\text{-}43)$$

and so on. This implies, just as in the solution of a particle in a one-dimensional box, that the wave numbers, that is, the momenta, are quantized. The form (22-42) requires that

$$e^{ik_x L} = e^{ik_y L} = e^{ik_z L} = 1 \qquad (22\text{-}44)$$

that is, that the wave numbers be of the form

$$k_x = \frac{2\pi}{L}\, n_x \qquad k_y = \frac{2\pi}{L}\, n_y \qquad k_z = \frac{2\pi}{L}\, n_z \qquad (22\text{-}45)$$

where n_x, n_y, and n_z are integers. We also have

$$\Delta\mathbf{k} = \Delta k_x \, \Delta k_y \, \Delta k_z = \left(\frac{2\pi}{L}\right)^3 \Delta n_x \, \Delta n_y \, \Delta n_z \qquad (22\text{-}46)$$

and

$$\omega = |\mathbf{k}|c = \frac{2\pi c}{L}\, (n_x{}^2 + n_y{}^2 + n_z{}^2)^{1/2} \qquad (22\text{-}47)$$

When we carry out a sum like that in (22-41), we sum over all values of (n_x, n_y, n_z) in the range specified by (22-46) consistent with the constraint of the delta function. Thus

$$R_{k \to m} = \sum_{\Delta\mathbf{k}} \Gamma_{k \to m}$$

$$= \int d^3\mathbf{n} \, \Gamma_{k \to m}$$

$$= \int \frac{L^3}{(2\pi)^3} \, d^3\mathbf{k} \, \Gamma_{k \to m}$$

$$= \int \frac{V d^3\mathbf{p}}{(2\pi\hbar)^3} \, \Gamma_{k \to m} \tag{22-48}$$

In the second line we made use of the fact that as V becomes large the states become very dense, and the sum can be converted into an integral; in the third line (22-46) was used, and in the last line, the relation

$$\mathbf{p} = \hbar\mathbf{k} \tag{22-49}$$

was used. The integration is over the volume in momentum space defined by the experimental arrangement. If we write

$$d^3\mathbf{p} = d\Omega_\mathbf{p} \, p^2 dp = d\Omega_\mathbf{p} \left(\frac{\omega}{c}\right)^2 d\left(\frac{\omega}{c}\right) \hbar^3 \tag{22-50}$$

where $d\Omega_\mathbf{p}$ is the solid angle differential, we find that the energy conserving delta function is integrated over and the result is

$$R_{k \to m} = \int \frac{4\pi^2 e^2}{m^2 \omega V} \, |\langle \phi_m | e^{-i\mathbf{k} \cdot \mathbf{r}} \, \boldsymbol{\varepsilon} \cdot \mathbf{p} | \phi_k \rangle|^2 \, d\Omega_\mathbf{p} \, \frac{V}{(2\pi\hbar)^3}$$

$$\times \hbar^3 \frac{\omega^2}{c^3} \frac{d(\hbar\omega)}{\hbar} \, \delta(E_k{}^0 - E_m{}^0 - \hbar\omega)$$

$$= \int d\Omega_\mathbf{p} \frac{\alpha}{2\pi} \, \omega_{km} \left| \frac{1}{mc} \langle \phi_m | e^{-i\mathbf{k} \cdot \mathbf{r}} \, \boldsymbol{\varepsilon} \cdot \mathbf{p} | \phi_k \rangle \right|^2 \tag{22-51}$$

where

$$\omega_{km} = \frac{E_k{}^0 - E_m{}^0}{\hbar} \tag{22-52}$$

If the experimental apparatus does not discriminate between the polarization states of the photon, the rate calculation must include a sum over those two independent final states. Furthermore, the sum should also include all the final states of the atom. This will be discussed in a later section.

The phase space

$$d^3\mathbf{n} = \frac{V \, d^3\mathbf{p}}{(2\pi\hbar)^3} \tag{22-53}$$

is not restricted to photons. An electron that is free is described by the plane wave function $1/\sqrt{V} \, e^{i\mathbf{p} \cdot \mathbf{r}/\hbar}$ and it will have the same density of states. The only difference is that the relation between energy (which appears in the delta function) and momentum is $E = \mathbf{p}^2/2m$ [or, relativistically, $E = (\mathbf{p}^2 c^2 + m^2 c^4)^{1/2}$] instead of $E = pc$.

If we have several free particles in the final state, the density of states is the product

$$\prod_k \frac{V \, d^3\mathbf{p}_k}{(2\pi\hbar)^3} \tag{22-54}$$

The expression (22-48) combined with (22-39) then generalizes to

$$R_{i \to f} = \frac{2\pi}{\hbar} \int_{\substack{\text{independent} \\ \text{momenta}}} \prod_k \frac{V \, d^3\mathbf{p}_k}{(2\pi\hbar)^3} \, |M_{fi}|^2 \, \delta \left(E_f^0 + \sum_k E_k - E_i^0 \right) \tag{22-55}$$

where M_{fi} is the matrix element of the perturbation between the initial and final states of the unperturbed system. The delta function again expresses energy conservation, that is, the energy carried off by the free particles is equal to the energy change in the system, and *the integration is over independent momenta*. Thus, if a system decays into three particles, there are only two independent momenta, since the third one is determined by momentum conservation. Note, however, that the product of factors in (22-54) is over *all* the particles in the final state, that is, it involves V^n if there are n particles in the final state. Equivalently we could write (22-55) as an integral over *all* momenta, with a delta function that includes a statement of momentum conservation. The reason that such a delta function did not appear in our derivation is we are dealing with atoms that are so much more massive than the photon (precisely $M_{\text{atom}} c^2 \gg \hbar\omega$) that the atomic recoil never entered into the calculation. At any rate, the result

$$R_{i \to f} = \frac{2\pi}{\hbar} \int \prod_k \frac{V \, d^3\mathbf{p}_k}{(2\pi\hbar)^3}$$

$$\times \, |M_{fi}|^2 \, \delta \left(E_i^0 - E_f^0 - \sum E_k \right) \delta \left(\mathbf{p}_i - \mathbf{p}_f - \sum \mathbf{p}_k \right) \tag{22-56}$$

which could also be abbreviated by

$$R_{i \to f} = \frac{2\pi}{\hbar} \, |M_{fi}|^2 \, \rho(E) \tag{22-57}$$

with $\rho(E)$ called the density of states, is a fundamental result, and has been named the *Golden Rule* by Fermi.

Note that the volume of the box always drops out. For n free particles in the final state, there is a V^n from the density of states (phase space) and a $1/\sqrt{V}$ for each free particle in the matrix element, coming from their wave function

$$\prod_k \frac{e^{i\mathbf{p}_k \cdot \mathbf{r}/\hbar}}{\sqrt{V}} \tag{22-58}$$

There are n of these factors, and thus the V dependence of the square of the matrix element just cancels the V^n from the phase space. We will have further occasion to use the Golden Rule, but at this point we turn to the evaluation of the matrix element for the radiative transition.

C. The Matrix Element and Selection Rules

Our next task is to calculate

$$\langle \phi_m | e^{-i\mathbf{k}\cdot\mathbf{r}} \, \boldsymbol{\varepsilon}\cdot\mathbf{p} | \phi_k \rangle \tag{22-59}$$

We begin by estimating its magnitude. For a typical atomic transition

$$\boldsymbol{\varepsilon}\cdot\mathbf{p} \sim |\mathbf{p}| \sim Zmc\alpha \tag{22-60}$$

We also need to estimate the exponent, since it is an oscillating factor and could change the result significantly. With

$$r \sim \frac{\hbar}{mcZ\alpha} \tag{22-61}$$

and

$$|k| \sim \frac{\hbar\omega}{\hbar c} \sim \frac{\frac{1}{2}mc^2(Z\alpha)^2}{\hbar c} \sim \frac{mc}{2\hbar}(Z\alpha)^2 \tag{22-62}$$

we have

$$kr \sim \tfrac{1}{2}Z\alpha \tag{22-63}$$

Hence, for $Z\alpha \ll 1$, the order of magnitude of the matrix element is indeed $Zmc\alpha$, thus

$$R_{k\to m} \sim 2\alpha\omega(Z\alpha)^2 \sim \alpha(Z\alpha)^2 \frac{mc^2(Z\alpha)^2}{\hbar}$$

$$\sim \alpha(Z\alpha)^4 \frac{mc^2}{\hbar} \sim 2 \times 10^{10}\, Z^4\, \text{sec}^{-1} \tag{22-64}$$

It simplifies matters that in the expansion

$$e^{-i\mathbf{k}\cdot\mathbf{r}} = \sum_{n=0}^{\infty} \frac{(-i)^n}{n!}(\mathbf{k}\cdot\mathbf{r})^n \tag{22-65}$$

the successive terms are estimated to decrease as $Z\alpha$. Thus, to order $Z\alpha$,

$$\langle \phi_m | e^{-i\mathbf{k}\cdot\mathbf{r}} \, \boldsymbol{\varepsilon}\cdot\mathbf{p} | \phi_k \rangle \simeq \langle \phi_m | \boldsymbol{\varepsilon}\cdot\mathbf{p} | \phi_k \rangle \tag{22-66}$$

We may write this as

$$
\begin{aligned}
\boldsymbol{\varepsilon} \cdot \langle \phi_m | \mathbf{p} | \phi_k \rangle &= m \boldsymbol{\varepsilon} \cdot \langle \phi_m | d\mathbf{r}/dt | \phi_k \rangle \\
&= \frac{im}{\hbar} \, \boldsymbol{\varepsilon} \cdot \langle \phi_m | [H,\mathbf{r}] | \phi_k \rangle \\
&= im \frac{(E_m{}^0 - E_k{}^0)}{\hbar} \, \boldsymbol{\varepsilon} \cdot \langle \phi_m | \mathbf{r} | \phi_k \rangle \\
&= im\omega \, \boldsymbol{\varepsilon} \cdot \langle \phi_m | \mathbf{r} | \phi_k \rangle
\end{aligned}
\tag{22-67}
$$

Thus we are interested in calculating the matrix element of the operator $\mathbf{r}$, and that is one reason for calling the approximation (22-66) the electric *dipole approximation*.

If the initial state ϕ_k is a hydrogenlike state characterized by the "initial" quantum numbers n_i, l_i, and m_i, and the state ϕ_m, the final state, by the quantum numbers n_f, l_f, and m_f, then what needs to be evaluated is

$$
\begin{aligned}
\langle \phi_m | \boldsymbol{\varepsilon} \cdot \mathbf{r} | \phi_k \rangle &= \int_0^\infty r^2 \, dr \int d\Omega R^*_{n_f l_f}(r) \, Y^*_{l_f m_f}(\theta,\phi) \, \boldsymbol{\varepsilon} \cdot \mathbf{r} R_{n_i l_i}(r) \, Y_{l_i m_i}(\theta,\phi) \\
&= \int_0^\infty r^2 \, dr R^*_{n_f l_f}(r) \, r R_{n_i l_i}(r) \\
&\quad \times \int d\Omega Y^*_{l_f m_f}(\theta,\phi) \, \boldsymbol{\varepsilon} \cdot \hat{\mathbf{r}} Y_{l_i m_i}(\theta,\phi)
\end{aligned}
\tag{22-68}
$$

The radial integral will be discussed for a special case in the next section. Here we concentrate on the angular integral. We have

$$
\boldsymbol{\varepsilon} \cdot \hat{\mathbf{r}} = \epsilon_x \sin \theta \cos \phi + \epsilon_y \sin \theta \sin \phi + \epsilon_z \cos \theta
$$

and making use of

$$
\sqrt{\frac{3}{4\pi}} \, Y_{1,0}(\theta, \phi) = \cos \theta \qquad \sqrt{\frac{3}{8\pi}} \, Y_{1,\pm1}(\theta,\phi) = \mp \sin \theta \, e^{\pm i\phi}
\tag{22-69}
$$

a little algebra yields

$$
\boldsymbol{\varepsilon} \cdot \hat{\mathbf{r}} = \sqrt{\frac{4\pi}{3}} \left(\epsilon_z Y_{1,0} + \frac{-\epsilon_x + i\epsilon_y}{\sqrt{2}} \, Y_{1,1} + \frac{\epsilon_x + i\epsilon_y}{\sqrt{2}} \, Y_{1,-1} \right)
\tag{22-70}
$$

Thus the angular integral in (22-68) involves

$$
\int d\Omega Y^*_{l_f m_f}(\theta,\phi) \, Y_{1,m}(\theta,\phi) \, Y_{l_i m_i}(\theta,\phi)
\tag{22-71}
$$

Let us first consider the azimuthal integration. It yields

$$\int_0^{2\pi} d\phi \, e^{-im_f\phi} \, e^{im\phi} \, e^{im_i\phi} = 2\pi \, \delta_{m=m_f-m_i} \tag{22-72}$$

We thus get the first *selection rule*

$$m_f - m_i = m = 1, 0, -1 \tag{22-73}$$

This was the selection rule that was mentioned in our discussion of the Zeeman Effect. Specifically, if we define the z-axis to lie along the photon momentum direction $\mathbf{k}$, then the condition (22-26) implies that $\epsilon_z = 0$ and hence $m = \pm 1$ only appears, so that

$$m_f - m_i = \pm 1 \tag{22-74}$$

As a special case, we note that if the final state is the ground state, with $l_f = m_f = 0$, then $m = -m_i$. For example, if $m_i = 1$, then $m = -1$ and hence the polarization vector for the radiation is $(\epsilon_x + i\epsilon_y)/\sqrt{2}$. The implication is that if the atom in the initial state is polarized along the z-axis with $m_i = 1$, then in a decay to a state with zero angular momentum, the conservation of the z-component of angular momentum demands that the photon carry this off. The photon must therefore have its spin aligned along the positive z-axis, that is, it must have positive helicity (helicity = +1), or, equivalently, it must be left-circularly polarized. This is just what the term $(\epsilon_x + i\epsilon_y)/\sqrt{2}$ indicates.

The θ integration gives rise to another selection rule. Consider first the special case that $l_f = 0$. Since $Y_{0,0} = 1/\sqrt{4\pi}$, the angular integration (22-71) involves

$$\frac{1}{\sqrt{4\pi}} \int d\Omega Y_{1,m}(\theta,\phi) \, Y_{l_i m_i}(\theta,\phi) = \frac{1}{\sqrt{4\pi}} \delta_{l_i 1} \delta_{m_i,-m} \tag{22-75}$$

which implies that *the initial state must have* $l_i = 1$. In hydrogen. the dominant transitions to the ground state will be $np \rightarrow 1s$.

More generally, when l_i and l_f do not vanish, we still get a selection rule. The derivation, beyond the scope of the mathematical knowledge about special functions assumed in this book, makes use of the *addition theorem* for spherical harmonics, which reads

$$Y_{l_1 m_1}(\theta,\phi) \, Y_{l_2 m_2}(\theta,\phi) = \sum_{L=|l_1-l_2|}^{l_1+l_2} C(L, m_1 + m_2; l_1, l_2, m_1, m_2) \, Y_{L,m_1+m_2}(\theta,\phi) \tag{22-76}$$

The coefficients $C(L, m_1 + m_2; l_1, l_2, m_1, m_2)$ are the same Wigner coefficients that appear in (15-44). The possible angular momenta on the right side are just those that could be obtained from the addition of the angular momenta $\mathbf{l}_1$ and $\mathbf{l}_2$. Substitution into (22-71) yields

$$\int d\Omega Y^*_{l_f m_f}(\theta,\phi) \sum_{L=|l_i-1|}^{l_i+1} C(L,\, m + m_i;\, 1,\, l_i,\, m,\, m_i)\, Y_{L,m+m_i}(\theta,\phi) = 0$$

unless

$$l_f = l_i + 1,\, l_i,\, |l_i - 1| \tag{22-77}$$

This is the general form of the *electric dipole radiation selection rule*

$$\Delta l = 1,\, 0,\, -1 \tag{22-78}$$

with the observation, obvious from (22-75) that there are *no zero-zero transitions*. There is a further constraint that comes from parity conservation. Since $\mathbf{r}$ is odd under reflections, there is an additional selection rule for the electric dipole transitions:

$$\text{\textit{The atomic state must change}} \tag{22-79}$$
$$\text{\textit{parity}}$$

Since parity is given by $(-1)^l$, this implies that the l-value must actually change. Thus, for example, $3p \rightarrow 2p$ transitions are not allowed to order $Z\alpha$.

To the extent that the only perturbation is the coupling

$$\frac{e}{mc}\, \mathbf{p} \cdot \mathbf{A}(\mathbf{r},t) \tag{22-80}$$

there is no spin dependence in it, and hence the spins cannot flip in the transition. This leads to the additional selection rule

$$\Delta S = 0 \tag{22-81}$$

mentioned earlier in connection with the spectrum of helium.

The selection rules stated above are not absolute. The conservation laws of angular momentum and parity (for electromagnetic processes) are absolute, but (22-78) is only approximately true. Transitions between states that involve a change of l larger than 1 cannot take place through the electric dipole mechanism. They can still take place, provided there is a nonvanishing matrix element

$$\langle \phi_f | e^{-i\mathbf{k}\cdot\mathbf{r}}\, \boldsymbol{\varepsilon} \cdot \mathbf{p} | \phi_i \rangle \tag{22-82}$$

For $\Delta l = 2$, the first power of $\mathbf{k}\cdot\mathbf{r}$ will give a nonvanishing contribution. We may write

$$\mathbf{k}\cdot\mathbf{r}\boldsymbol{\varepsilon}\cdot\mathbf{p} = \tfrac{1}{2}(\boldsymbol{\varepsilon}\cdot\mathbf{p}\mathbf{k}\cdot\mathbf{r} + \boldsymbol{\varepsilon}\cdot\mathbf{r}\mathbf{p}\cdot\mathbf{k}) + \tfrac{1}{2}(\boldsymbol{\varepsilon}\cdot\mathbf{p}\mathbf{k}\cdot\mathbf{r} - \boldsymbol{\varepsilon}\cdot\mathbf{r}\mathbf{p}\cdot\mathbf{k})$$
$$= \tfrac{1}{2}(\boldsymbol{\varepsilon}\cdot\mathbf{p}\mathbf{k}\cdot\mathbf{r} + \boldsymbol{\varepsilon}\cdot\mathbf{r}\mathbf{p}\cdot\mathbf{k}) + \tfrac{1}{2}(\mathbf{k}\times\boldsymbol{\varepsilon})\cdot(\mathbf{r}\times\mathbf{p}) \tag{22-83}$$

The first of these terms is called an electric quadrupole term, and the second is clearly related to an $\mathbf{L}\cdot\mathbf{B}$ term, and is called a magnetic dipole term. For these transitions, whose matrix element we estimated to be $Z\alpha$ times smaller than the leading term, we will have $\Delta l = 2$, and, since the operators in (22-83) are even, there will be no parity change between the atomic states. Transitions between $3d \rightarrow 1s$, for example, cannot go via the electric dipole mechanism, but can go via the electric quadrupole mechanism. Actually, it turns out to be much more probable that the $3d$ state decays first into a $2p$ state, and the latter then undergoes the favored $2p \rightarrow 1s$ transition.

The spin selection rule $\Delta S = 0$ too, is not sacred. In addition to the coupling (22-80) there is the coupling discussed in connection with the anomalous Zeeman effect

$$\lambda V(t) = \frac{ge}{2mc} \mathbf{S} \cdot \mathbf{B}(\mathbf{r}, t) \qquad (22\text{-}84)$$

The matrix element for $\Delta S \neq 0$ transition-inducing term can be estimated. We compare it with the electric dipole matrix element

$$\frac{(eg/2mc)\,\hbar|\mathbf{k} \times \boldsymbol{\varepsilon}|}{(2e/mc)|\mathbf{p} \cdot \boldsymbol{\varepsilon}|} \simeq \frac{\hbar|\mathbf{k}|}{|\mathbf{p}|} \simeq \frac{\hbar\omega}{|\mathbf{p}|c} \simeq \frac{mc^2(Z\alpha)^2}{mc^2(Z\alpha)} \simeq Z\alpha \qquad (22\text{-}85)$$

and see that it is suppressed, just like the magnetic dipole matrix element, which it strongly resembles in form. As an example of a situation where the coupling (22-84) plays an important role, we consider the nuclear process of photodisintegration of the deuteron

$$\gamma + d \rightarrow n + p \qquad (22\text{-}86)$$

The deuteron, to a very good approximation is a $^3S_1$ state. An electric dipole transition must involve the final $(n - p)$ system in a ^3P state since $\Delta l = 1$ and $\Delta S = 0$. It turns out, however, that just above threshold for the reaction, the two nucleons are unlikely to be in a relative P-state. In general, particles will be in a relative angular momentum L state with any appreciable probability only if

$$|\mathbf{p}|a \gtrsim \hbar L \qquad (22\text{-}87)$$

where $\mathbf{p}$ is the relative momentum and a are the dimensions of the system. For the deuteron it turns out that for γ's below 10 MeV in energy, the $(n - p)$ system is unlikely to be in a P-state. The additional coupling

$$-\frac{e}{2Mc}(g_p\mathbf{s}_p + g_n\mathbf{s}_n) \cdot \mathbf{B} \qquad (22\text{-}88)$$

can, however, lead to a transition between the $^3S_1$ state, and the unbound $^1S_0$ state. The interaction may be rewritten in the form

$$-\frac{e}{2Mc}[\tfrac{1}{2}(g_p + g_n)(\mathbf{s}_p + \mathbf{s}_n) + \tfrac{1}{2}(g_p - g_n)(\mathbf{s}_p - \mathbf{s}_n)] \cdot \mathbf{B} \qquad (22\text{-}89)$$

The first term is symmetric under the $n \leftrightarrow p$ exchange, and hence cannot contribute to a transition between a symmetric and an antisymmetric spin state. The second term does, however, contribute. The coefficients are actually quite large, since $g_p \cong 5.56$ and $g_n = -3.81$.

There is one selection rule that is sacred, and that is the one forbidding zero-zero transitions (referring to *total* angular momentum $j = 0$) in one-photon processes. A general way of arguing the absoluteness of this selection rule is the following: The matrix element, a scalar quantity, must involve the photon polarization linearly, and must therefore be of the form $\boldsymbol{\varepsilon} \cdot \mathbf{V}$, where $\mathbf{V}$ is some vector that enters into the problem. If the initial and final states are $j = 0$ states, that is, have no directionality associated with them, then the only vector is $\mathbf{k}$, the photon momentum. However $\boldsymbol{\varepsilon} \cdot \mathbf{k} = 0$, so that there is no way of constructing a matrix element. It must therefore not exist.[2]

D. The $2p \rightarrow 1s$ Transition

Let us now specialize to the transition $2p \rightarrow 1s$ in (22-68). We need to evaluate the radial integral

$$\int_0^\infty dr\; r^3 R_{10}^*(r)\; R_{21}(r)$$

$$= \int_0^\infty dr\; r^3 \left[2 \left(\frac{Z}{a_0} \right)^{3/2} e^{-Zr/a_0} \right] \left[\frac{1}{\sqrt{24}} \left(\frac{Z}{a_0} \right)^{5/2} r\, e^{-Zr/2a_0} \right]$$

$$= \frac{1}{\sqrt{6}} \left(\frac{Z}{a_0} \right)^4 \int_0^\infty dr\; r^4\, e^{-3Zr/2a_0}$$

$$= \frac{1}{\sqrt{6}} \left(\frac{Z}{a_0} \right)^4 \left(\frac{2a_0}{3Z} \right)^5 \int_0^\infty dx\; x^4\, e^{-x} = \frac{24}{\sqrt{6}} \left(\frac{2}{3} \right)^5 Z^{-1} a_0 \qquad (22\text{-}90)$$

The angular integral is

$$\int d\Omega\, Y_{0,0}^*\, \boldsymbol{\varepsilon} \cdot \hat{\mathbf{r}}\; Y_{1.m} = \frac{1}{\sqrt{4\pi}} \int d\Omega \sqrt{\frac{4\pi}{3}} \left(\epsilon_z Y_{1,0} + \frac{-\epsilon_x + i\epsilon_y}{\sqrt{2}}\, Y_{1,1} \right.$$

$$\left. + \frac{\epsilon_x + i\epsilon_y}{\sqrt{2}}\, Y_{1,-1} \right)\, Y_{1,m}$$

$$= \frac{1}{\sqrt{3}} \left(\epsilon_z \delta_{m,0} + \frac{-\epsilon_x + i\epsilon_y}{\sqrt{2}}\, \delta_{m,-1} + \frac{\epsilon_x + i\epsilon_y}{\sqrt{2}}\, \delta_{m,1} \right)$$

$$(22\text{-}91)$$

[2] The relation $\boldsymbol{\varepsilon} \cdot \mathbf{k} = 0$ is independent of the choice of gauge, and is a statement about the transversality of the electromagnetic field. Such arguments "by enumeration" are frequently used in elementary particle physics, where the interaction is not really known.

Now the absolute square of the product of (22-90) and (22-91) is

$$96 \left(\frac{2}{3}\right)^{10} \left(\frac{a_0}{Z}\right)^2 \frac{1}{3} [\delta_{m0}\epsilon_z^2 + \tfrac{1}{2} (\delta_{m,1} + \delta_{m,-1})(\epsilon_x^2 + \epsilon_y^2)] \qquad (22\text{-}92)$$

so that the transition rate is for a given m-value of the excited atom,

$$R_{2p \to 1s} = \int d\Omega_\mathbf{p} \left(\frac{\alpha}{2\pi}\right) \frac{\omega}{m^2 c^2} m^2 \omega^2 \frac{2^{15}}{3^{10}} \left(\frac{a_0}{Z}\right)^2$$

$$\times [\delta_{m,0}\epsilon_z^2 + \tfrac{1}{2} (\delta_{m,1} + \delta_{m,-1})(\epsilon_x^2 + \epsilon_y^2)] \qquad (22\text{-}93)$$

where

$$\omega = \frac{1}{\hbar}\left[\frac{1}{2} mc^2 (Z\alpha)^2 \left(1 - \frac{1}{4}\right)\right]$$

$$= \frac{3}{8} \frac{mc^2}{\hbar} (Z\alpha)^2 \qquad (22\text{-}94)$$

is the frequency of the radiation emitted in the transition.

The angular integration in (22-93) is over the photon directions, and this is not trivial, since $\boldsymbol{\varepsilon}$ is constrained to be perpendicular to the photon momentum direction. The integration is very simple if the initial p-state is unaligned, that is, it occurs in the three possible m-states ($m = 1, 0, -1$) with equal probability. The rate is then

$$R_{2p \to 1s} = \frac{1}{3} \sum_{m=-1}^{1} R_{2p \to 1s}(m) \qquad (22\text{-}95)$$

Since

$$\sum_{m=-1}^{1} [\delta_{m0}\epsilon_z^2 + \tfrac{1}{2} (\delta_{m1} + \delta_{m,-1})(\epsilon_x^2 + \epsilon_y^2)] = \epsilon_x^2 + \epsilon_y^2 + \epsilon_z^2 = 1 \qquad (22\text{-}96)$$

the integrand becomes independent of the photon direction. This result should also be multiplied by a factor of 2. The reason is that there are two possible polarization states for the photon, and we are detecting both of them. A more careful way of writing (22-51) would have been

$$\int d\Omega_\mathbf{p} \frac{\alpha}{2\pi} \frac{\omega_{km}}{m^2 c^2} \sum_{\lambda=1}^{2} |\langle \phi_m | e^{-i\mathbf{k}\cdot\mathbf{r}} \, \boldsymbol{\varepsilon}^{(\lambda)} \cdot \mathbf{p} | \phi_k \rangle|^2 \qquad (22\text{-}97)$$

with λ denoting the polarizations. The two polarization states are orthogonal, so that we have

$$\boldsymbol{\varepsilon}^{(\lambda)} \cdot \boldsymbol{\varepsilon}^{(\lambda')} = \delta_{\lambda\lambda'} \qquad (22\text{-}98)$$

358   Quantum Physics

When all of this is put together, we get

$$R_{2p \to 1s} = 2 \cdot 4\pi \, \frac{\alpha}{2\pi} \, \frac{1}{c^2} \left(\frac{3}{8} \frac{mc^2}{\hbar} Z^2 \alpha^2 \right)^3 \frac{2^{15}}{3^{10}} \left(\frac{\hbar}{mcZ\alpha} \right)^2 \frac{1}{3}$$

$$= \frac{2^8}{3^8} \frac{mc^2}{\hbar} \alpha(Z\alpha)^4 \cong 0.6 \times 10^9 \, Z^4 \, \text{sec}^{-1} \qquad (22\text{-}99)$$

This differs by a factor of about 25 from the estimate made in (22-64). Thus detailed factors in the matrix elements are important and guesses cannot replace a calculation. Nevertheless, dimensional considerations and a proper counting of powers of α do give us an order of magnitude guidance to how large physical quantities in atomic physics are.

The expression for the rate

$$R_{fi} = \frac{d\Omega_\mathbf{p}}{2\pi} \frac{e^2}{\hbar c} \frac{\omega^3}{c^2} \sum_{\lambda=1}^{2} |\langle f|\mathbf{r}|i\rangle \cdot \boldsymbol{\varepsilon}^{(\lambda)}|^2 \qquad (22\text{-}100)$$

may be translated into a formula for the intensity of radiation by multiplying it by the energy of the light quantum $\hbar\omega$. Thus

$$I_{fi} = d\Omega_\mathbf{p} \frac{e^2}{2\pi c^3} \omega^4 \sum_{\lambda=1}^{2} |\langle f|\mathbf{r}|i\rangle \cdot \boldsymbol{\varepsilon}^{(\lambda)}|^2 \qquad (22\text{-}101)$$

This, however, is just the *classical* formula for the intensity of light emitted by an oscillating dipole, of dipole moment

$$\mathbf{d} = e\langle f|\mathbf{r}|i\rangle e^{-i\omega t} \qquad (22\text{-}102)$$

providing another illustration of the correspondence principle.

F. Spin and Intensity Rules

The inclusion of spin does not change things very much. It is true that the initial states and the final states can each be in an "up" or a "down" spin state, but since the interaction in atomic transitions is spin independent, only "up" → "up" and "down" → "down" transitions are allowed. Hence the transition rates will not only be independent of m_l (as we saw in the last section) but also of m_s, and hence, m_j. With the inclusion of spin-orbit coupling, there will be small (on the scale of the $2p - 1s$ energy difference) level splittings. For example, the $n = 1$ and $n = 2$ level structure is changed, as shown in Fig. 22-2. The spectral line corresponding to the transition $2p \to 1s$ is split into two lines, $2^2P_{3/2} \to 1^2S_{1/2}$ and $2^2P_{1/2} \to 1^2S_{1/2}$. For the split states, the radial integral and the phase space are almost unchanged, and hence *the ratio of the intensity of the two lines can be determined from the angular parts of the integral alone, that is, purely from angular momentum considerations.*

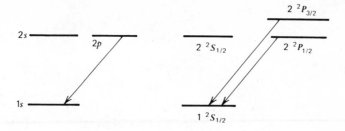

Fig. 22-2.   The splitting of the $2p - 1s$ spectral line by spin-orbit coupling.

The table below lists the wave functions for the states in question.

J	m_j	odd parity $l = 1$	even parity $l = 0$
3/2	3/2	$Y_{11}\chi_+$	—
3/2	1/2	$\sqrt{2/3}\,Y_{10}\chi_+ + \sqrt{1/3}\,Y_{11}\chi_-$	—
3/2	−1/2	$\sqrt{1/3}\,Y_{1,-1}\chi_+ + \sqrt{2/3}\,Y_{10}\chi_-$	—
3/2	−3/2	$Y_{1,-1}\chi_-$	—
1/2	1/2	$\sqrt{1/3}\,Y_{10}\chi_+ - \sqrt{2/3}\,Y_{11}\chi_-$	$Y_{00}\chi_+$
1/2	−1/2	$\sqrt{2/3}\,Y_{1,-1}\chi_+ - \sqrt{1/3}\,Y_{10}\chi_-$	$Y_{00}\chi_-$

In the squares of the matrix elements, the radial parts are common to all of them. Thus, in considering the rates for $P_{3/2} \to S_{1/2}$ we must add the squares of the transition matrix elements for $m_j = 3/2 \to m_j = 1/2$, $m_j = 3/2 \to m_j = -1/2, \ldots m_j = -3/2 \to m_j = -1/2$, while the rate for $P_{1/2} \to S_{1/2}$ involves the sum of the squares of the matrix elements for $m_j = 1/2 \to m_j = 1/2, \ldots$ $m_j = -1/2 \to m_j = -1/2$. This can be done directly by techniques that are quite sophisticated and beyond the scope of this book. One can, however, work out these quantities in detail, using the fact that the spin wave functions are orthonormal.

$P_{3/2} \to S_{1/2}$

$m_j = 3/2 \to m_j = 1/2$	$|\langle Y_{11}|\mathbf{r}\cdot\boldsymbol{\varepsilon}|Y_{00}\rangle|^2 = C$	
$3/2 \to -1/2$	0 since $\chi_+^*\chi_- = 0$	
$1/2 \to 1/2$	$|\langle\sqrt{2/3}\,Y_{10}|\mathbf{r}\cdot\boldsymbol{\varepsilon}|Y_{00}\rangle|^2 = 0$	$(\Delta m = 0)$
$1/2 \to -1/2$	$|\langle\sqrt{1/3}\,Y_{11}|\mathbf{r}\cdot\boldsymbol{\varepsilon}|Y_{00}\rangle|^2 = C/3$	
$-1/2 \to 1/2$	$|\langle\sqrt{1/3}\,Y_{1,-1}|\mathbf{r}\cdot\boldsymbol{\varepsilon}|Y_{00}\rangle|^2 = C/3$	
$-1/2 \to -1/2$	$|\langle\sqrt{2/3}\,Y_{10}|\mathbf{r}\cdot\boldsymbol{\varepsilon}|Y_{00}\rangle|^2 = 0$	$(\Delta m = 0)$
$-3/2 \to 1/2$	0	
$-3/2 \to -1/2$	$|\langle Y_{1,-1}|\mathbf{r}\cdot\boldsymbol{\varepsilon}|Y_{00}\rangle|^2 = C$	

If we sum the terms we get

$$\sum R = \frac{8C}{3} \qquad (22\text{-}103)$$

Similarly

$$\underline{P_{1/2} \to S_{1/2}}$$

$m_j = 1/2 \to m_j = 1/2$	$\lvert \langle \sqrt{1/3}\, Y_{10} \lvert \boldsymbol{\varepsilon} \cdot \mathbf{r} \rvert Y_{00} \rangle \rvert^2 = 0$
$1/2 \to -1/2$	$\lvert \langle -\sqrt{2/3}\, Y_{11} \lvert \boldsymbol{\varepsilon} \cdot \mathbf{r} \rvert Y_{00} \rangle \rvert^2 = 2C/3$
$-1/2 \to 1/2$	$\lvert \langle \sqrt{2/3}\, Y_{1,-1} \lvert \boldsymbol{\varepsilon} \cdot \mathbf{r} \rvert Y_{00} \rangle \rvert^2 = 2C/3$
$-1/2 \to -1/2$	$\lvert \langle -\sqrt{1/3}\, Y_{10} \lvert \boldsymbol{\varepsilon} \cdot \mathbf{r} \rvert Y_{00} \rangle \rvert^2 = 0$

Again

$$\sum R = \frac{4C}{3} \qquad (22\text{-}104)$$

Thus the ratio of the intensities is

$$\frac{R(P_{3/2} \to S_{1/2})}{R(P_{1/2} \to S_{1/2})} = \frac{8C/3}{4C/3} = 2 \qquad (22\text{-}105)$$

The reason for *summing* over all the initial states is that when the atom is excited, all the p-levels are equally occupied, since their energy difference is so tiny compared to the $2p - 1s$ energy difference. We also sum over all the final states if we perform an experiment that does not discriminate between them, as is the case for a spectroscopic measurement. In our calculation of the $2p \to 1s$ transition rate, we *averaged* over the initial m-states. There we were concerned with the problem of asking: "If we have N atoms in the $2p$ states, how many will decay per second?" The averaging came about because of the fact that under most circumstances, when N atoms are excited, about $N/3$ go into each one of the $m = 1, 0, -1$ states. Here, the fact that there are more levels in the $P_{3/2}$ state than there are in the $P_{1/2}$ state is relevant. There will be altogether six levels, (four with $j = 3/2$ and two with $j = 1/2$) and there will be on the average $N/6$ atoms in each of the states. The fact that there are more atoms in the $j = 3/2$ subset of levels just means that more decay, and that therefore the intensity will be larger.

Problems

1. A hydrogen atom is placed in an electric field $\mathbf{E}(t)$ that is uniform and has the time dependence

$$\begin{aligned} \mathbf{E}(t) &= \phantom{\mathbf{E}_0 e^{-\gamma t}}0 & t &< 0 \\ &= \mathbf{E}_0\, e^{-\gamma t} & t &> 0 \end{aligned}$$

What is the probability that as $t \to \infty$, the hydrogen atom, if initially in the ground state, makes a transition to the $2p$ state?

2. Repeat the above calculation with the time dependence of the electric field given by

$$\mathbf{E}(t) = \mathbf{E}_0 \, e^{-\alpha^2 t^2}$$

and with the condition that the hydrogen atom be in its ground state at $t = -\infty$. [*Hint.* As a first step, modify Eq. 22-9 appropriately.] Discuss your result when the time-variation of the electric field is extremely slow.

3. Consider a harmonic oscillator described by

$$H = \frac{1}{2m} \, p_x{}^2 + \tfrac{1}{2} m \omega^2(t) \, x^2$$

where

$$\omega(t) = \omega_0 + \delta\omega \cos ft$$

and $\delta\omega \ll \omega_0$.

Calculate the probability that a transition occurs from the ground state, as a function of time, given that the system is in the ground state at $t = 0$. Use perturbation theory. Use the fact that for $n \neq 0$,

$$\langle n \, | \, x^2 \, | \, 0 \rangle = \hbar/2\sqrt{2}m\omega \quad \text{for} \quad n = 2$$

$$= 0 \qquad \text{otherwise.}$$

Can you derive this formula using the material from Chapter 7?

4. Suppose a particle of rest mass M decays into two particles of rest mass m_1 and m_2, respectively. Use the relativistic relation between energy and momentum to compute the density of states ρ that appears in (22-57). [*Hint.* There is only one independent momentum, say $\mathbf{p}$, and what is needed is

$$\int \frac{d^3\mathbf{p}}{(2\pi\hbar)^6} \, \delta \left(E_{\text{initial}} - \sum_{\substack{\text{final} \\ \text{states}}} E \right)$$

5. Consider the above calculation when the decay is of the form

$$A \to B + C + D$$

with particles C and D massless.
[*Hint.* There are now two independent momenta.]

6. In this problem the *adiabatic theorem* is to be illustrated. The theorem states that if the Hamiltonian is changed very slowly from H_0 to H, then a system in a given eigenstate of H_0 goes over into the corresponding eigenstate of H,

but does not make any transitions. To be specific, consider the ground state, so that

$$H_0\phi_0 = E_0\phi_0$$

Let $V(t) = f(t)V$ where $f(t)$ is a slowly varying function, as shown in the graph. If the ground state of $H = H_0 + V$ is w_0, the theorem states that

$$|\langle w_0|\psi(t)\rangle| \to 1$$

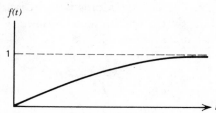

The steps to be carried out are the following:

(a) Show that

$$\frac{1}{i\hbar}\int_0^t dt'\, e^{i(E_m{}^0 - E_0{}^0)t'/\hbar} f(t') \to \frac{e^{i(E_m{}^0 - E_0{}^0)t/\hbar}}{E_m{}^0 - E_0{}^0}$$

for times t such that $f(t) = 1$. Use the fact that

$$\frac{df(t')}{dt'} \ll \frac{E_m{}^0 - E_0{}^0}{\hbar} f(t')$$

Either construct an example of a function $f(t)$ or use integration by parts, that is, write

$$e^{i\omega t'} = \frac{1}{i\omega}\frac{d}{dt'}\, e^{i\omega t'}$$

in the above.

(b) Calculate $\psi(t)$ using (22-3) and (22-9). Compare this with the formula (16-19) which here reads

$$w_0 = \phi_0 + \sum_{m\neq 0}\frac{\langle\phi_m|V|\phi_0\rangle}{E_0{}^0 - E_m{}^0}\phi_m$$

and thus show that

$$|\langle w_0|\psi(t)\rangle| \to 1$$

7. Work out the $2p \to 1s$ transition rate for the three-dimensional oscillator, following the steps carried out in this chapter.

8. Nuclei sometimes decay from excited states to the ground state by *internal conversion*, a process in which one of the $1s$ electrons is emitted instead of a photon. Let the initial and final nuclear wave functions be

$$\phi_I(\mathbf{r}_1, \mathbf{r}_2, \ldots, \mathbf{r}_A) \quad \text{and} \quad \phi_F(\mathbf{r}_1, \mathbf{r}_2, \ldots, \mathbf{r}_A)$$

where $\mathbf{r}_i$ ($i = 1, 2, \ldots Z$) describe the protons. The perturbation giving rise to the transition is just the nucleus-electron interaction

$$V = -\sum_{i=1}^{Z} \frac{e^2}{|\mathbf{r} - \mathbf{r}_i|}$$

where $\mathbf{r}$ is the electron coordinate. Thus the matrix element is given by

$$-\int d^3\mathbf{r} \int d^3\mathbf{r}_1 \ldots d^3\mathbf{r}_A \phi_F^* \frac{e^{-i\mathbf{p}\cdot\mathbf{r}/\hbar}}{\sqrt{V}} \sum_{i=1}^{Z} \frac{e^2}{|\mathbf{r} - \mathbf{r}_i|} \phi_I \psi_{100}(\mathbf{r})$$

(a) What is the magnitude of $\mathbf{p}$, the free electron momentum?

(b) Calculate the rate for the process for a dipole transition in terms of

$$\mathbf{d} = \sum \int d^3\mathbf{r}_1 \ldots d^3\mathbf{r}_A \phi_F^* \mathbf{r}_i \phi_I$$

by making use of the expansion

$$\frac{1}{|\mathbf{r} - \mathbf{r}_i|} \simeq \frac{1}{r} + \frac{\mathbf{r}\cdot\mathbf{r}_i}{r^3}$$

Hint. The integral can be evaluated using

$$\int d^3\mathbf{r}\, e^{-i\mathbf{p}\cdot\mathbf{r}/\hbar}\, \mathbf{r}\cdot\mathbf{r}_i \psi_{100}(\mathbf{r}) = i\hbar\mathbf{r}_i\cdot\nabla_{\mathbf{p}} \int d^3\mathbf{r}\, e^{-i\mathbf{p}\cdot\mathbf{r}/\hbar}\, \psi_{100}(\mathbf{r})$$

See the discussion of the photoelectric effect in Chapter 25.

References

The addition theorem that leads to the more general derivation of selection rules is discussed in all of the more advanced textbooks listed at the end of this volume, and also in

M. E. Rose, *Elementary Theory of Angular Momentum*, John Wiley & Sons, 1957.

The radial integrals for the more general case are discussed in

H. A. Bethe and R. W. Jackiw, *Intermediate Quantum Mechanics*, W. A. Benjamin, Inc., 1968.

H. A. Bethe and E. E. Salpeter, *Quantum Mechanics of One- and Two-Electron Atoms*, Springer Verlag, 1957.

E. U. Condon and G. H. Shortley, *The Theory of Atomic Spectra*, Cambridge University Press, Cambridge, 1959.

chapter 23

Selected Topics in Radiative Transitions

A. Lifetime and Line Width

The number $R(i \to f)$ that we learned to calculate in Chapter 22 represents the probability for the transition $i \to f$, divided by the time during which the perturbation has acted. This time must be long compared to $\hbar/(E_m{}^0 - E_k{}^0 + \hbar\omega)$ in order that the transition probability be proportional to t, but it clearly cannot be too long. If we ask for the probability that the initial state remain intact, we get

$$P_i(t) = 1 - \left[\sum_{f \neq i} R(i \to f) \right] t \qquad (23\text{-}1)$$

where the sum is over all final states that are accessible. This ·clearly has no meaning for long enough times, since probabilities are positive. It turns out that if the calculation of the time development of the system is done more carefully,[1] then it can be shown that the right side of (23-1) just represents an approximation (to lowest order in the perturbation) to the correct expression—again only true for long times—that

$$P_i(t) = \exp\left[-t \sum_{f \neq i} R(i \to f) \right] \qquad (23\text{-}2)$$

One may thus speak of a *lifetime* of the initial state

$$\tau = \frac{1}{R} = \frac{1}{\displaystyle\sum_{f \neq i} R(i \to f)} \qquad (23\text{-}3)$$

[1] This is done in the Special Topics section, "Lifetimes, Line Widths, and Resonances."

The total transition rate R is the sum of partial transition rates into the possible *channels f*. In the example that was discussed in detail, the $2p \rightarrow 1s$ transition in hydrogenlike atoms, no other channels are available, so that the lifetime of the $2p$ state is

$$\tau = 1.6 \times 10^{-9} \, Z^{-4} \text{ sec} \tag{23-4}$$

This is in excellent agreement with experiment. Let us compare this (we take $Z = 1$) with the time it takes the electron to "go once around the nucleus." The velocity is αc, and the distance is of the order of 3×10^{-8} cm, so that the characteristic time is of the order of 1.4×10^{-16} sec. In terms of this time, the $2p$ state is very long lived.

Since the $2p$ state has a finite lifetime, it should, by the uncertainty principle, have an uncertainty in the energy, of magnitude

$$\Delta E \sim \frac{\hbar}{\tau} \tag{23-5}$$

The way in which this manifests itself is that the intensity of the line, as a function of frequency, is not completely sharp at the value $\omega_0 = (E_{2p} - E_{1s})/\hbar$ but it has a distribution of the form

$$I(\omega) \propto \frac{R/2}{(\omega - \omega_0)^2 + R^2/4} \tag{23-6}$$

Note that in the limit that $R \rightarrow 0$, that is, in the limit that perturbation theory is strictly applicable, we get, as a consequence of the formula

$$\operatorname*{Lim}_{\epsilon \to 0} \frac{\epsilon}{(\omega - \omega_0)^2 + \epsilon^2} = \pi \, \delta(\omega - \omega_0) \tag{23-7}$$

the line shape represented by the energy-conservation delta function. The width of the line (23-6) is R, and this is a measure of the uncertainty in the energy.

This line shape, sometimes called the Lorentzian line shape, is not what is generally observed, since there are other effects that broaden it. There is:

(a) *Collision Broadening.* One does not observe a single atom in isolation, but a gas of hot atoms. In the gas there will occur collisions between the atoms. If we define a collision time τ_c as some mean time between collisions, and if $\tau_c < \tau$, then, in effect, the lifetime of the state will be τ_c, and the energy uncertainty $\hbar/\tau_c$.

To get a rough estimate of the collision rate $R_c \, (= 1/\tau_c)$, consider one atom at rest. If its effective area is σ (the collision cross section that will be discussed in Chapter 24), then it will be hit by another atom within 1 sec, if the atom finds itself inside a cylinder of volume $v\sigma$ (Fig. 23-1). If there are n atoms/cm^3, the number of collisions will be

$$R_c = nv\sigma \quad \text{sec}^{-1} \tag{23-8}$$

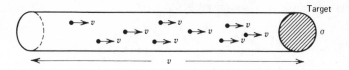

Fig. 23-1. The number of collisions per second for particles moving with a velocity v normal relative to the target.

To see the dependence of this on the pressure and temperature of the gas, we use the kinetic theory relation

$$m\overline{v^2} = 3kT \tag{23-9}$$

and the ideal gas law

$$n = \frac{p}{kT} \tag{23-10}$$

where $k = 1.381 \times 10^{-16}$ erg/deg is Boltzmann's constant. Thus, if for v in (23-8) we take $(\overline{v^2})^{1/2}$, we get

$$R_c = \frac{p\sigma}{kT} \left(\frac{3kT}{m}\right)^{1/2} \tag{23-11}$$

If we now write

$m = 1.6 \times 10^{-24} M$ gm   so that M is the molecular weight

$p = 10^6 p_a$ where p_a is the pressure in atmospheres

$\sigma = \pi \times 10^{-16} D^2$ where D is the atomic or molecular diameter in Angstroms

then

$$R_c = 3.4 \times 10^{10} \frac{p_a D^2}{\sqrt{MT}} \tag{23-12}$$

The collision rate can be decreased by decreasing the pressure, so that in the laboratory (in contrast to stellar surfaces) collision broadening can be controlled.

(b) *Doppler Broadening.* Even at low pressures, the radiating atom is moving quite rapidly (the gas is hot) and its frequency is shifted. If v_x is the velocity of the atom in the direction of the line of sight, then the shift is

$$\Delta\omega = \omega \frac{v_x}{c} \tag{23-13}$$

In terms of the temperature, given by

$$\overline{v_x^2} = \frac{kT}{m} \tag{23-14}$$

we have

$$\frac{\Delta\omega}{\omega} \simeq 0.3 \times 10^{-6} \left(\frac{T}{M}\right)^{1/2} \tag{23-15}$$

where M is again the atomic or molecular weight ($M = 1$ for hydrogen, $M = 4$ for helium, etc.). Thus, the best that we can do is obtain

$$\frac{\Delta\omega}{\omega} \sim 10^{-6} \tag{23-16}$$

whereas for the natural line width, this is $\sim 3 \times 10^{-8}$.

B. The Mössbauer Effect

An atom (or any other quantum system) can act as a very accurate clock, since its transitions are signaled by radiation of a very well-defined frequency. If the only limitation were the natural line width, an accuracy of $1:10^8$ could be achieved in atomic transitions. As noted above, the Doppler broadening reduces this to $1:10^6$. One might think that use of a liquid or solid source would eliminate this, but then broadening caused by the effect of neighboring atoms is just as harmful. One might examine nuclear transitions. A nucleus such as $_{77}\text{Ir}^{191}$ emits a γ-ray of energy ~ 100 keV, with a lifetime of 10^{-10} sec. This corresponds to

$$\frac{\Delta\omega}{\omega} = \frac{\Delta E}{E} = \frac{\hbar/\tau}{E} \simeq \frac{10^{-27}/10^{-10}}{10^5 \times 1.6 \times 10^{-12}} \simeq 0.6 \times 10^{-10} \tag{23-17}$$

There will, unfortunately, be a recoil shift of the line. The γ-ray carries off momentum $\hbar\omega/c$, and the nucleus, to conserve momentum, must recoil with the same momentum. This gives rise to a recoil energy

$$\Delta E = \frac{P_{\text{recoil}}^2}{2M} = \frac{1}{2M}\left(\frac{\hbar\omega}{c}\right)^2 \tag{23-18}$$

and thus a decrease in the energy radiated. The fractional change in frequency is

$$\frac{\Delta E}{\hbar\omega} \simeq \frac{\hbar\omega}{2Mc^2} \simeq \frac{10^{-1}(\text{MeV})}{2 \times 940 \times 191(\text{MeV})} \simeq 3 \times 10^{-7} \tag{23-19}$$

The observation of radiation of this energy cannot be carried out with the conventional, extremely accurate spectroscopic methods, but must utilize a detector that is extremely "well tuned" to the radiation. This is best done by using the same material (e.g., $_{77}Ir^{191}$) as an absorber. The absorption will be very much enhanced at the "resonant" frequency at which the radiation is emitted, but here, too, there will be a recoil shift. The overall shift is thus $\Delta\omega/\omega \simeq 6 \times 10^{-7}$. Thus, the "fine tuning" does not work, since the line is shifted by far more than the width, which is of the order of $10^{-10}\,\omega$. One could try to compensate for the recoil by moving the emitter with the recoil velocity. This is given by

$$\frac{v}{c} = \frac{P_{\text{recoil}}}{Mc} = \frac{\hbar\omega/c}{Mc} = 2\,\frac{\hbar\omega}{2Mc^2} \simeq 6 \times 10^{-7} \qquad (23\text{-}20)$$

that is, $v = 1.7 \times 10^4$ cm/sec. This presents technical difficulties, but it has been achieved with an ultracentrifuge.

A major breakthrough came with the discovery by Mössbauer in 1958 that under certain conditions there is a high probability of *recoilless emission*. The emission is not recoilless, of course, but the recoil is not taken up by the nucleus, but instead by a large part of the crystal that the nucleus is imbedded in. Since the mass of the nucleus is 10^{22} times smaller than that of the crystal, the recoil energy is completely negligible. To get some intuition about what is happening, let us consider the nucleus as moving in a harmonic oscillator well, with characteristic frequency ω_0. The energy levels of the oscillator are

$$E_n = \hbar\omega_0\left(n_x + n_y + n_z + \frac{3}{2}\right) \qquad (23\text{-}21)$$

The harmonic well is just an approximate description of the crystalline forces that are responsible for the properties of the lattice. If the forces that tie the nucleus to its neighbors are strong—if the "springs" are stiff—then ω_0 is large; if the "springs" are soft, then ω_0 is small. In terms of level spacing, a "stiff spring" has widely separated levels, that is, a low density of states, whereas a "soft spring" has a high density of states. Let us now consider the matrix element for a transition from a nuclear state described by $\Psi_i\,(\mathbf{r}_1, \mathbf{r}_2, \ldots \mathbf{r}_N)$ to a nuclea state described by $\Psi_f(\mathbf{r}_1, \mathbf{r}_2, \ldots \mathbf{r}_N)$, and we take the interaction to be

$$-\frac{e}{Mc}\sum_{\text{protons}} \mathbf{p}_k \cdot \mathbf{A}_k(\mathbf{r}_k,t) \qquad (23\text{-}22)$$

The matrix element then is proportional to

$$-\frac{e}{Mc}\int \cdots \int d^3\mathbf{r}_1 \ldots d^3\mathbf{r}_N \Psi_f^*(\mathbf{r}_1 \ldots \mathbf{r}_N) \sum_k \boldsymbol{\varepsilon} \cdot \mathbf{p}_k\, e^{-i\mathbf{k}\cdot\mathbf{r}_k}\, \Psi_i(\mathbf{r}_1, \ldots \mathbf{r}_N)$$

$$(23\text{-}23)$$

If we introduce the center of mass coordinate $\mathbf{R} = (1/N) \sum_i \mathbf{r}_i$ then (a) the interaction term takes the form

$$- \frac{e}{Mc} e^{-i\mathbf{k}\cdot\mathbf{R}} \sum_{\text{protons}} \boldsymbol{\varepsilon} \cdot \mathbf{p}_k \, e^{-i\mathbf{k}\cdot\boldsymbol{\varrho}_k} \qquad (23\text{-}24)$$

where $\boldsymbol{\varrho}_i = \mathbf{r}_i - \mathbf{R}$, and (b) the nuclear wave function decomposes into a product describing the internal motion and the motion of the nuclear center of mass in the harmonic potential

$$\Psi(\mathbf{r}_1, \ldots \mathbf{r}_N) = \psi_{n_x n_y n_z}(\mathbf{R}) \, \phi(\boldsymbol{\varrho}_1, \ldots \boldsymbol{\varrho}_{N-1}) \qquad (23\text{-}25)$$

Thus the matrix element (23-23) becomes

$$- \frac{e}{Mc} \int d^3\mathbf{R}\, \psi_{n_f}^*(\mathbf{R}) \, e^{-i\mathbf{k}\cdot\mathbf{R}} \psi_{n_i}(\mathbf{R})$$

$$\times \int d^3\boldsymbol{\varrho}_1 \ldots d^3\boldsymbol{\varrho}_{N-1} \phi_f^*(\boldsymbol{\varrho}_1 \ldots \boldsymbol{\varrho}_{N-1}) \sum_{\text{protons}} \boldsymbol{\varepsilon} \cdot \mathbf{p}_k \, e^{-i\mathbf{k}\cdot\boldsymbol{\varrho}_k} \phi_i(\boldsymbol{\varrho}_1, \ldots \boldsymbol{\varrho}_{N-1})$$

$$(23\text{-}26)$$

We may write this in the form

$$M = M_{\text{internal}} \int d^3\mathbf{R}\, \psi_{n_f}^*(\mathbf{R}) \, e^{-i\mathbf{k}\cdot\mathbf{R}} \psi_0(\mathbf{R}) \qquad (23\text{-}27)$$

where we have set $n_i = 0$, since the initial state is in the ground state of the lattice. The probability that the radiative transition leaves the nucleus in the lattice ground state is

$$P_0(k) = \frac{|M_{\text{int}}|^2 \left| \int d^3\mathbf{R}\, \psi_0^*(\mathbf{R}) \, e^{-i\mathbf{k}\cdot\mathbf{R}} \psi_0(\mathbf{R}) \right|^2}{|M_{\text{int}}|^2 \sum_{n_f} \left| \int d^3\mathbf{R}\, \psi_{n_f}^*(\mathbf{R}) \, e^{-i\mathbf{k}\cdot\mathbf{R}} \psi_0(\mathbf{R}) \right|^2}$$

$$= \left| \int d^3\mathbf{R}\, \psi_0^*(\mathbf{R}) \, e^{-i\mathbf{k}\cdot\mathbf{R}} \psi_0(\mathbf{R}) \right|^2 \qquad (23\text{-}28)$$

In the last step we replaced the sum in the denominator by unity, using completeness.[2] To calculate this, we use the normalized ground state wave function

---

[2] The formal proof is quickest. We have

$$\sum_{n_f} |\langle n_f | e^{i\mathbf{k}\cdot\mathbf{R}} | 0 \rangle|^2 = \sum_{n_f} \langle 0 | e^{-i\mathbf{k}\cdot\mathbf{R}} | n_f \rangle \, \langle n_f | e^{i\mathbf{k}\cdot\mathbf{R}} | 0 \rangle$$

Using

$$1 = \sum |n_f\rangle\langle n_f|$$

one gets

$$\langle 0 | e^{-i\mathbf{k}\cdot\mathbf{R}} \, e^{i\mathbf{k}\cdot\mathbf{R}} | 0 \rangle = 1$$

of the oscillator. We found in Chapter 7 that the one-dimensional ground state wave function is

$$\psi_0(x) = \left(\frac{m\omega_0}{\pi\hbar}\right)^{1/4} e^{-m\omega_0 x^2/2\hbar}$$

Hence, for three dimensions we have

$$\psi_0(R) = \psi_0(x)\,\psi_0(y)\,\psi_0(z) = \left(\frac{m\omega_0}{\pi\hbar}\right)^{3/4} e^{-m\omega_0 R^2/2\hbar} \tag{23-29}$$

We thus calculate

$$\left| \left(\frac{M_N\omega_0}{\pi\hbar}\right)^{3/2} \int d^3\mathbf{R}\; e^{-M_N\omega_0 R^2/\hbar}\; e^{-i\mathbf{k}\cdot\mathbf{R}} \right|^2$$

where M_N is the mass of the nucleus. We get

$$P_0 = \left(\frac{M_N\omega_0}{\pi\hbar}\right)^3 \left| \int d^3\mathbf{R}\; e^{-(M_N\omega_0/\hbar)[\mathbf{R}+i\mathbf{k}(\hbar/2M_N\omega_0]^2}\; e^{-k^2\hbar/4M_N\omega_0} \right|^2$$

$$= e^{-\hbar^2 k^2/2M_N\hbar\omega_0}$$

$$= \exp\left(-\frac{\text{recoil energy}}{\text{level spacing}}\right) \tag{23-30}$$

since $P_{\text{recoil}} = \hbar k$ and $\hbar\omega_0$ is the level spacing in the lattice. Thus, if the level spacing is large, that is, we have a stiff spring, recoilless emission becomes more probable. The model of the lattice that was used here, that of each nucleus moving in its own harmonic potential, is the Einstein model of a lattice, and the frequency ω_0 is the so-called *Debye frequency*, so that we should really replace ω_0 by ω_D, which is related to the Debye temperature T_D by

$$\hbar\omega_D = kT_D \tag{23-31}$$

A more accurate treatment of the lattice using the Debye model for its description merely changes the exponent by a factor of $3/2$.

It is not quite correct to say that the whole crystal recoils; instead, in a time τ equal to the lifetime of the transition (1.4×10^{-7} sec for Fe^{57}), only a region of the crystal of magnitude

$$L = v_s\tau$$

where v_s is the velocity of propagation of a lattice disturbance, (i.e., the velocity of sound) absorbs the recoil. Now a reasonable estimate of v_s is given by

$$v_s \simeq \frac{a\omega_D}{2\pi}$$

where a is the lattice spacing. Thus

$$\frac{L}{a} \simeq \frac{\omega_D \tau}{2\pi}$$

and with $\omega_D \simeq 10^{13}$ sec^{-1}, the number of nuclei absorbing the recoil, $\sim(L/a)^3$ is still enormous.

The above estimates, combined with the uncertainty relation, may be used to show that it is not possible to determine whether it is a single nucleus that "really" recoils. To measure the recoil energy $\hbar^2 k^2/2M_N$ takes a time of the order of

$$\Delta t \gg \frac{\hbar}{(\hbar^2 k^2/2M_N)}$$

The condition for the Mössbauer effect to occur is that

$$\frac{\hbar^2 k^2}{2M_N} < \hbar\omega_D$$

Hence

$$\Delta t \gg \frac{1}{\omega_D}$$

During that time the disturbance will have travelled a distance

$$d \simeq v_s \, \Delta t \sim \frac{a\omega_D}{2\pi} \Delta t \gg \frac{a}{2\pi}$$

that is, over a distance covering many nuclei.

The question arises of how did we manage to get away from the problem of recoil and momentum conservation by talking about the energy states of the nucleus in the crystal lattice? Where does it say that the crystal absorbed the momentum? The quantum mechanical answer is that, if we want to talk about momentum, we should work in a momentum representation. This, however, is complicated, since it is difficult to describe the crystal forces in that representation. What one must do is to decompose the crystal motion (the crystal is just a lot of oscillators with nearest neighbor "springs") into normal modes and quantize these. The quanta of the lattice motion, analogs of photons, are the *phonons*. Recoilless emission then means a transition in which phonons are not emitted. The resulting formula is very similar to (23-30). Under these circumstances, the recoil broadening is infinitesimal compared to the natural line width. There is still Doppler broadening because of the thermal motion but this can be handled by cooling the emitter and absorber.

Recoilless emitters provide us with a superb clock, and research utilizing the Mössbauer effect has been done in many fields, such as solid-state physics and chemistry. We will mention just one application, the terrestrial measurement of the gravitational red shift. We noted[3] that a photon will have its frequency shifted by

$$\frac{\Delta\omega}{\omega} = \frac{gx}{c^2} \tag{23-32}$$

if it falls through a height x. This can be compensated by a recoil of velocity v, where[4]

$$v^2 = 2gx \tag{23-33}$$

(If the photon and the absorber were to fall freely together, there would be resonant absorption.) If the absorber or the source are allowed to oscillate rapidly—one uses a transducer—and the absorption curve is correlated with the oscillations, it is possible to check the gravitational shift. Since the velocity, for a separation $x = 20$ m, is of the order of ~ 20 m/sec, the experiment is feasible, and was carried out by several groups. Within the errors, the effect is confirmed. For example, for Fe^{57} the predicted shift is $\Delta\omega/\omega = 4.92 \times 10^{-15}$, and the experimental shift found by Pound and Rebka is $(5.13 \pm 0.51) \times 10^{-15}$. A similar experiment in which the energy shift of the γ-ray emitted by Fe^{57} accelerated on a rapidly rotating turntable was measured again yielded results in agreement with the Equivalence Principle.

C. Induced Absorption and Emission

In our discussion of the normalization of the vector potential appropriate to the radiation of an atom in Eqs. 22-28 and 22-29, we saw that the matrix element for emission was proportional to $(N + 1)^{1/2}$, where N was the number of quanta in the initial state and the matrix element for absorption was proportional to $N^{1/2}$. Since this refers to quanta of a particular type, the quantity N should really be labeled by the momentum $\hbar\mathbf{k}$ and the polarization state λ of the photon, that is, N should be replaced by $N_\lambda(\mathbf{k})$. We may use the N-dependence to derive the *Planck Radiation Law*, thus providing a quantum mechanical justification of Planck's approach.

Let us consider a cavity containing radiation. The walls contain atoms that absorb and emit radiation. Since there is a variety of atoms, with a variety of energy levels, there will be a continuous spectrum of frequencies. We will concentrate on a particular frequency, corresponding to transitions between a

---

[3] See the Special Topics section 2 "The Equivalence Principle."

[4] It is one of the subtleties of radiation in a gravitational field that the Doppler shift is the *transverse* shift $v' = v(1 - v^2/c^2)^{1/2}$. Only in this way will the shift in an accelerated frame be the same, whether the absorber is falling or sitting on the edge of a rotating disk, with the emitter in the center.

particular pair of levels in a particular species of atoms, that is, we will describe the atom as having two states of energy, E_1 and E_2, respectively, with $E_1 < E_2$. When equilibrium is established, there are as many photons absorbed as there are photons radiated.[5] The number of photons radiated by the walls is equal to (number of atoms in the upper state "2") $\times$ (transition rate for "2" $\to$ "1"); the number absorbed is equal to (number of atoms in state "1") $\times$ (transition rate for "1" $\to$ "2"), that is,

$$N_2 R_{\text{emission}} = N_1 R_{\text{absorption}} \tag{23-34}$$

We also have

$$R_{\text{emission}} = [N_\lambda(\mathbf{k}) + 1] \, R_{21} \tag{23-35}$$

where R_{21} is the emission rate into a state with one photon. We use (22-57) to write this in the form

$$R_{21} = \frac{1}{2J_2 + 1} \frac{2\pi}{\hbar} \sum |\langle 1 | V^+ | 2 \rangle|^2 \, \rho \tag{23-36}$$

Here ρ stands for the density of photon states; we have the square of the matrix element, and it is summed over the final states of the atom, that is, the $2J_1 + 1$ angular momentum states, and averaged over the initial states. This is exhibited explicitly—the sum is over initial and final states, and is divided by $2J_2 + 1$, the number of angular momentum states for state "2." The reason for averaging over the initial state is that when the state "2" gets excited, then all the states that only differ by the m-value will get excited with equal probability. Only one of the states is excited at a time, and thus the proper counting is done when we sum over all of the $2J_2 + 1$ states and then divide by their number. Note also that we denoted the perturbation by V^+ as the term associated with the time dependence $e^{i\omega t}$ For absorption, we have

$$R_{\text{absorption}} = N_\lambda(\mathbf{k}) \, R_{12} \tag{23-37}$$

where

$$R_{12} = \frac{1}{2J_1 + 1} \frac{2\pi}{\hbar} \sum |\langle 2 | V | 1 \rangle|^2 \, \rho \tag{23-38}$$

The density of states here is the same as in (23-36), since we are dealing with only one frequency. Furthermore,

$$
\begin{aligned}
\sum |\langle 2 | V | 1 \rangle|^2 &= \sum_{m_1} \sum_{m_2} \langle 2 | V | 1 \rangle \langle 2 | V | 1 \rangle^* \\
&= \sum_{m_1} \sum_{m_2} \langle 1 | V^+ | 2 \rangle^* \langle 1 | V^+ | 2 \rangle \\
&= \sum |\langle 1 | V^+ | 2 \rangle|^2
\end{aligned} \tag{23-39}
$$

[5] One must convince oneself that it is permissible to consider equilibrium for one frequency at a time, as we do here. This becomes plausible when we realize that the probability of emission of two photons at a time is small, so that the radiation field in the cavity still obeys linear equations.

This assertion is sometimes called the *Principle of Detailed Balance*. On the face of it, it is an identity, but one could imagine that the perturbation leading to the transition "1" $\rightarrow$ "2" is not the hermitian conjugate of the perturbation that leads to the transition "2" $\rightarrow$ "1," in which case the above derivation would break down. It turns out that the principle holds, provided that the total Hamiltonian is invariant under time reversal.[6] The interaction of charges with the electromagnetic field has this property.

As a consequence of (23-39) we have

$$\frac{R_{\text{emission}}}{R_{\text{absorption}}} = \frac{N_\lambda(\mathbf{k}) + 1}{N_\lambda(\mathbf{k})} \cdot \frac{2J_1 + 1}{2J_2 + 1}$$

$$= \frac{N_\lambda(\mathbf{k}) + 1}{N_\lambda(\mathbf{k})} \cdot \frac{g_1}{g_2} \tag{23-40}$$

where g_i is the conventional notation for the degeneracy of the state "i." On the other hand, we learn from statistical mechanics that at equilibrium, the occupation numbers of the atomic states N_2 and N_1 are related by the Boltzmann factor

$$\frac{N_2}{N_1} = \frac{g_2 \, e^{-E_2/kT}}{g_1 \, e^{-E_1/kT}} = \frac{g_2}{g_1} e^{-\hbar\omega/kT} \tag{23-41}$$

Hence

$$\frac{g_2}{g_1} e^{-\hbar\omega/kT} = \frac{N_2}{N_1} = \frac{R_{\text{absorption}}}{R_{\text{emission}}} = \frac{N_\lambda(\mathbf{k})}{N_\lambda(\mathbf{k}) + 1} \frac{g_2}{g_1}$$

that is,

$$N_\lambda(\mathbf{k}) = \frac{1}{e^{\hbar\omega/kT} - 1} \tag{23-42}$$

The photon energy at the given frequency is given by the product (number of photon states in the interval $d\omega$) $\times$ (number of photons) $\times$ (energy per photon) $\times$ (a factor of 2 to account for the two independent polarization states). Thus

$$dU(\omega) = \frac{V \, d^3\mathbf{p}}{(2\pi\hbar)^3} \cdot \frac{2\hbar\omega}{e^{\hbar\omega/kT} - 1}$$

$$= \frac{V \cdot 4\pi k^2}{(2\pi)^3} \frac{dk}{d\omega} \, d\omega \, \frac{2\hbar\omega}{e^{\hbar\omega/kT} - 1}$$

$$= \frac{8\pi\hbar}{c^3} \left(\frac{\omega}{2\pi}\right)^3 \frac{V}{e^{\hbar\omega/kT} - 1} \, d\omega \tag{23-43}$$

---

[6] Actually, in lowest order perturbation theory, (23-39) always does hold.

To get the energy density, we divide by the volume of the cavity V. If we express this in terms of $\nu = \omega/2\pi$, we get

$$U(\nu) = \frac{8\pi h}{c^3} \frac{\nu^3}{e^{h\nu/kT} - 1} \qquad (23\text{-}44)$$

In the presence of a large number of photons of a given wave length $[N_\lambda(\mathbf{k})$ large] transition rates corresponding to that wavelength will be enormously enhanced. Thus if many atoms can be raised to a given excited state, and the proper environment of the "right" kind of photons is provided, then they will decay in a very short time, thus giving rise to an intense, coherent, and monochromatic pulse of radiation. The laser (Light Amplification by Stimulated Emission) does just that. Under equilibrium conditions it is difficult to obtain a large number of atoms in the excited states from which the transitions are to take place, because the Boltzmann factor $e^{-\hbar\omega/kT}$ is very small, even at high temperatures, so that special techniques must be used to achieve this.

Consider for example, the helium-neon laser. There, advantage is taken of the fact that the 2^1S_0 and 2^3S_1 levels of helium almost coincide with certain sets of levels of neon, the $(2p)^5(5s)$ and $(2p)^5(4s)$ excited states, respectively (Fig. 23-2). The helium levels are easily excited; an electrical discharge in the gas will excite many levels, and they all ultimately decay to these states. The excited helium atoms will collide with unexcited neon atoms in a mixture of the two gases and easily transfer their energy to them. In this way a large number of neon

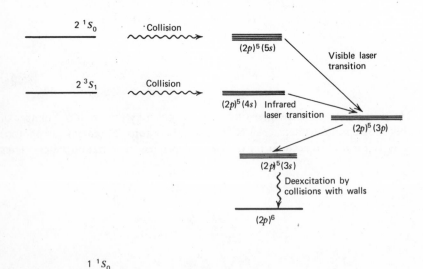

Fig. 23-2. Schematic sketch of relevant energy levels in He–Ne laser.

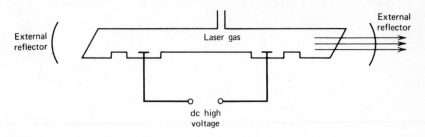

Fig. 23-3.   Schematic sketch of laser.

atoms find themselves in states that would otherwise be sparsely populated. A *population inversion* is created in the neon. These excited states decay to $(2p)^5(4p)$ and $(2p)^5(3p)$ states, emitting photons of a well-defined wavelength. These photons are trapped by mirrors, thus creating the proper environment for the next round of what is now strongly stimulated emission (Fig. 23-3). In this way intense monochromatic and coherent beams of photons are created.

The technological applications of lasers are manifold, and their development provides just one of many examples of the usefulness of quantum theory not only for the understanding of natural phenomena, but also as a source of new, subtle technological tools.

References

The Mössbauer Effect is discussed in detail in

H. Frauenfelder, *The Mössbauer Effect* (A Review with a Collection of Reprints), W. A. Benjamin, Inc., New York, 1962.

A qualitative discussion may be found in

V. F. Weisskopf, "Selected Topics in Theoretical Physics," in *Lectures in Theoretical Physics*, Vol. III, W. E. Brittin, B. W. Downs, J. Downs, Editors, Interscience Publishers, New York, 1961.

chapter 24

Collision Theory

Atomic and molecular structure was largely explored through spectroscopy. When it comes to trying to understand nuclear forces and the laws that govern the interactions of elementary particles, the only technique available is that of scattering a variety of particles by a variety of targets. In some sense, spectroscopy is also a form of "scattering." The atom in the ground state is excited by some projectile (it may be electrons in a discharge tube or collisions with other target particles, as in heating up of the gas), and then an outgoing photon is observed, with the atom going into the ground state again, or possibly another excited state. We do not usually describe these processes as "collision processes" because the atom has very well-defined energy levels, in which it stays for times that are enormously long compared to collision times,[1] so that it is possible to separate the "decay" from the excitation process. In particular, the characteristics of the decay are not sensitive to the particular mode of excitation. In nuclei and also in elementary particles, there exist levels, but frequently the lifetime is not sufficiently long to warrant a separation into excitation and decay, especially since accompanying the "resonant" scattering there is also nonresonant "background" scattering, and the disentangling of the two is sometimes complicated. In this chapter we will therefore discuss the process as a whole.

A. Collision Cross Section

The ideal way to talk about scattering is to formulate equations that describe exactly what happens: an incident particle, described by a wave packet, approaches the target. The wave packet must be spatially large, so that it does not spread appreciably during the experiment, and it must be large compared

---

[1] Recall that the lifetime of a $2p$ hydrogen state is 1.6×10^{-9} sec, which is large compared to the characteristic time $a_0/\alpha c \simeq 2 \times 10^{-17}$ sec.

with the target particle, but small compared with the dimensions of the laboratory, that is, it must not simultaneously overlap the target and detector. The lateral dimensions are, in fact, determined by the beam size in the accelerator. There follows an interaction with the target, and finally we see two wave packets: one continues in the forward direction, describing the unscattered part of the beam, and the other flies off at some angle and describes the scattered particles. The number of particles scattered into a given solid angle per unit time and unit incident flux is defined to be the *differential scattering cross section*. We will not follow this approach directly,[2] but will instead use some of the material developed in Chapter 11 to obtain the differential cross section. We will, however, keep the wave-packet treatment in mind as we interpret our formal results.

In our discussion of the continuum solutions of the Schrödinger equation in Chapter 11 we concluded that: (a) A solution of the Schrödinger equation in the absence of a potential is the plane wave form $e^{i\mathbf{k}\cdot\mathbf{r}}$, which describes a flux

$$\mathbf{j} = \frac{\hbar}{2im}(\psi^*\nabla\psi - \psi\nabla\psi^*) = \frac{\hbar\mathbf{k}}{m} \qquad (24\text{-}1)$$

If we choose $\mathbf{k}$ to define the z-axis, then the large r behavior of this solution may be written (cf. 11-31) in the form of an incoming + an outgoing spherical wave

$$e^{i\mathbf{k}\cdot\mathbf{r}} \Rightarrow \frac{i}{2k}\sum_{l=0}^{\infty}(2l+1)\,i^l\left[\frac{e^{-i(kr-l\pi/2)}}{r} - \frac{e^{i(kr-l\pi/2)}}{r}\right]P_l(\cos\theta) \qquad (24\text{-}2)$$

(b) The conservation of particles forces us to the conclusion that the presence of a radial potential can only alter this to a function, whose asymptotic form is

$$\psi(\mathbf{r}) \Rightarrow \frac{i}{2k}\sum_{l=0}^{\infty}(2l+1)\,i^l\left[\frac{e^{-i(kr-l\pi/2)}}{r} - S_l(k)\frac{e^{i(kr-l\pi/2)}}{r}\right]P_l(\cos\theta) \qquad (24\text{-}3)$$

subject to

$$|S_l(k)| = 1 \qquad (24\text{-}4)$$

The asymptotic form (24-3) may be rewritten, with the help of (24-2), as

$$\psi(\mathbf{r}) \Rightarrow e^{i\mathbf{k}\cdot\mathbf{r}} + \left[\sum_{l=0}^{\infty}(2l+1)\frac{S_l(k)-1}{2ik}P_l(\cos\theta)\right]\frac{e^{ikr}}{r} \qquad (24\text{-}5)$$

corresponding to a plane wave + an outgoing spherical wave.[3] Note that we are working with the effective one-particle Schrödinger equation, so that m is the reduced mass and θ is the center of mass angle between the direction of $\mathbf{k}$ (the z-axis) and the asymptotic point $\mathbf{r}$, where, presumably the counter will be set up.

---

[2] This is done very nicely in R. Hobbie, *American Journal of Physics*, 30, 857 (1962), at the level of mathematics that we use in this book.

[3] See Eq. 11-36, which explains why this is called an *outgoing* spherical wave.

When the target is much more massive than the projectile, there is no distinction between the laboratory angle and the center-of-mass angle. The kinematics are easily worked out using the material of the Special Topics section 1. Note also that we could, of course have set up a solution that has the asymptotic form of a plane wave + an incoming spherical wave since it is the first term in (24-3) that could be modified by a coefficient satisfying (24-4). However the solution that describes the scattering is the one involving the outgoing wave. Let us calculate the flux for the asymptotic solution (24-5).

$$\mathbf{j} = \frac{\hbar}{2im} \left\{ \left[e^{i\mathbf{k}\cdot\mathbf{r}} + f(\theta)\frac{e^{ikr}}{r} \right]^* \mathbf{\nabla} \left[e^{i\mathbf{k}\cdot\mathbf{r}} + f(\theta)\frac{e^{ikr}}{r} \right] - \text{complex conjugate} \right\}$$

(24-6)

where we have defined

$$f(\theta) = \sum_{l=0}^{\infty} (2l+1)\, f_l(k)\, P_l(\cos\theta)$$

(24-7)

with

$$f_l(k) = [S_l(k) - 1]/2ik$$

(24-8)

Calculating the gradient gives

$$\mathbf{j} = \frac{\hbar}{2im} \left\{ \left[e^{-i\mathbf{k}\cdot\mathbf{r}} + f^*(\theta)\frac{e^{-ikr}}{r} \right] \left[i\mathbf{k}\, e^{i\mathbf{k}\cdot\mathbf{r}} + \hat{\imath}_\theta \frac{1}{r}\frac{\partial f(\theta)}{\partial\theta}\frac{e^{ikr}}{r} \right. \right.$$

$$\left. + \hat{\imath}_r f(\theta)\left(ik\frac{e^{ikr}}{r} - \frac{e^{ikr}}{r^2} \right) \right] - \text{complex conjugate} \Bigg\}$$

$$= \frac{\hbar}{2im} \left[i\mathbf{k} + i\mathbf{k}f^*(\theta)\frac{e^{-ikr(1-\cos\theta)}}{r} + ik\hat{\imath}_r f(\theta)\frac{e^{ikr(1-\cos\theta)}}{r} + ik\hat{\imath}_r |f(\theta)|^2\frac{1}{r^2} \right.$$

$$\left. - \hat{\imath}_r f(\theta)\frac{e^{ikr(1-\cos\theta)}}{r^2} + \hat{\imath}_\theta \frac{\partial f(\theta)}{\partial\theta}\frac{e^{ikr(1-\cos\theta)}}{r^2} - \text{complex conjugate} \right\}$$

where we have left out $1/r^3$ terms, and where we have used $\mathbf{k}\cdot\mathbf{r} = kr\cos\theta$, in the exponential factors. Thus the flux is

$$\mathbf{j} = \frac{\hbar\mathbf{k}}{m} + \frac{\hbar k}{m}\,\hat{\imath}_r |f(\theta)|^2\,\frac{1}{r^2}$$

$$+ \frac{\hbar\mathbf{k}}{2m}\frac{1}{r}\left[f^*(\theta)\,e^{-ikr(1-\cos\theta)} + f(\theta)\,e^{ikr(1-\cos\theta)} \right]$$

$$+ \frac{\hbar k}{2m}\frac{\hat{\imath}_r}{r}\left[f^*(\theta)\,e^{-ikr(1-\cos\theta)} + f(\theta)\,e^{ikr(1-\cos\theta)} \right]$$

$$- \frac{\hbar}{2im}\frac{\hat{\imath}_r}{r^2}\left[f(\theta)\,e^{ikr(1-\cos\theta)} - f^*(\theta)\,e^{-ikr(1-\cos\theta)} \right]$$

$$+ \frac{\hbar}{2im}\frac{\hat{\imath}_\theta}{r^2}\left[\frac{\partial f(\theta)}{\partial\theta}\,e^{ikr(1-\cos\theta)} - \frac{\partial f^*(\theta)}{\partial\theta}\,e^{-ikr(1-\cos\theta)} \right]$$

(24-9)

382   Quantum Physics

This rather involved expression simplifies considerably when we consider that $\theta \neq 0$, since one never does a scattering experiment directly in the forward direction,[4] and that in a measurement one always integrates the flux over a small but finite solid angle. Thus in the last four terms of this expression we should replace $e^{ikr(1-\cos\theta)}$ by

$$\int \sin\theta \; d\theta d\phi \; g(\theta,\phi) \; e^{ikr(1-\cos\theta)} \tag{24-10}$$

where $g(\theta,\phi)$ is some sort of smooth, localized acceptance function for the counter. Now, as $r \to \infty$, we have an integral over a product of a smooth function and an extremely rapidly varying one, and this vanishes faster than any power of $1/r$. This is what is known in the mathematical literature as the Riemann-Lesbegue lemma, and the reader can convince himself that this is indeed so by working out an example, with a gaussian acceptance function, say. Thus, only the first two terms remain, so that

$$\mathbf{j} = \frac{\hbar\mathbf{k}}{m} + \frac{\hbar k}{m} \; \hat{\imath}_r \; \frac{|f(\theta)|^2}{r^2} \tag{24-11}$$

In the absence of a potential, only the first term is there: it represents the incident flux. In a wave-packet treatment, $\hbar k/m$ would be multiplied by a function that defines the lateral dimensions of the beam. Thus, if we ask for the *radial flux*, $\hat{\imath}_r \cdot \mathbf{j}$, then that term gives a contribution $\hbar\mathbf{k} \cdot \hat{\imath}_r / m = \hbar k \cos\theta/m$, but only within a finite region of the z-axis (see Fig. 24-1). Since the counter is put outside of that region, this first term does not contribute to the radial flux in the asymptotic region, so that

$$\mathbf{j} \cdot \hat{\imath}_r = \frac{\hbar k}{m} \cdot \frac{|f(\theta)|^2}{r^2} \tag{24-12}$$

Thus the number of particles crossing the area that subtends a solid angle $d\Omega$ at the origin (the target) is

$$\mathbf{j} \cdot \hat{\imath}_r \, dA = \frac{\hbar k}{m} \cdot \frac{|f(\theta)|^2}{r^2} \; r^2 d\Omega \tag{24-13}$$

The differential cross section is this number, divided by the incident flux, $\hbar k/m$, that is,

$$d\sigma = |f(\theta)|^2 \, d\Omega \tag{24-14}$$

If the potential has spin dependence, there may be an azimuthal dependence, so that more generally,

$$\frac{d\sigma}{d\Omega} = |f(\theta,\phi)|^2 \tag{24-15}$$

---

[4] How could one tell scattered from unscattered particles?

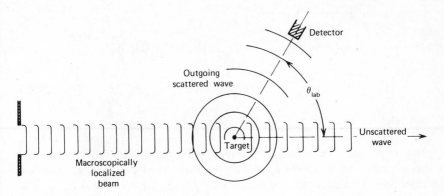

Fig. 24-1. Schematic layout for scattering experiment. The scattering angle is the laboratory angle.

The total cross section is given by

$$\sigma_{\text{tot}}(k) = \int d\Omega \, \frac{d\sigma}{d\Omega} \qquad (24\text{-}16)$$

If we now use $f(\theta)$ as expressed in terms of $S_l(k)$, and express the latter in terms of the phase shift (cf. 11-41) $S_l(k) = e^{2i\delta_l(k)}$, so that

$$f(\theta) = \frac{1}{k} \sum_{l=0}^{\infty} (2l+1) \, e^{i\delta_l(k)} \sin \delta_l(k) \, P_l(\cos\theta) \qquad (24\text{-}17)$$

then

$$\sigma_{\text{tot}} = \int d\Omega \left[\frac{1}{k} \sum_{l} (2l+1) \, e^{i\delta_l(k)} \sin \delta_l(k) \, P_l(\cos\theta) \right]$$

$$\left[\frac{1}{k} \sum_{l'} (2l'+1) \, e^{-i\delta_{l'}(k)} \sin \delta_{l'}(k) \, P_{l'}(\cos\theta) \right]$$

and using

$$\int d\Omega P_l(\cos\theta) \, P_{l'}(\cos\theta) = \frac{4\pi}{2l+1} \, \delta_{ll'} \qquad (24\text{-}18)$$

we get

$$\sigma_{\text{tot}} = \frac{4\pi}{k^2} \sum_{l=0}^{\infty} (2l+1) \, \sin^2\delta_l(k) \qquad (24\text{-}19)$$

It is an interesting fact that

$$\text{Im} f(0) = \frac{1}{k} \sum_{l=0}^{\infty} (2l+1) \, \text{Im}[e^{i\delta_l(k)} \sin \delta_l(k)] \, P_l(1)$$

$$= \frac{1}{k} \sum_{l=0}^{\infty} (2l+1) \sin^2 \delta_l(k) = \frac{k}{4\pi} \sigma_{\text{tot}} \tag{24-20}$$

This relation is known as the *optical theorem* and it is true even when inelastic processes can occur, as they do in nuclear and particle physics scattering processes. It is a very useful relation and in wave language it follows from the fact that the total cross section represents the removal of flux from the incident beam. Such a removal can only occur as a result of destructive interference, and the latter can only occur between the incident wave and the elastically scattered wave in the forward direction. This explains why $f(0)$ appears linearly. A more detailed examination shows why the imaginary part is involved.[5]

The requirement that $|S_l(k)| = 1$ followed from conservation of flux. Actually, in many scattering experiments there is *absorption* of the incident beam; the target may merely get excited, or change its state, or another particle may emerge. Under these circumstances our discussion is unchanged except that

$$S_l(k) = \eta_l(k) \, e^{2i\delta_l(k)} \tag{24-21}$$

is to be used, with

$$0 \le \eta_l(k) \le 1 \tag{24-22}$$

because we are dealing with absorption. The partial wave scattering amplitude is now

$$f_l(k) = \frac{S_l(k) - 1}{2ik} = \frac{\eta_l(k) \, e^{2i\delta_l(k)} - 1}{2ik} = \frac{\eta_l \sin 2\delta_l}{2k} + i \frac{1 - \eta_l \cos 2\delta_l}{2k} \tag{24-23}$$

and the total *elastic* cross section is

$$\sigma_{el} = 4\pi \sum_l (2l+1)|f_l(k)|^2$$

$$= 4\pi \sum_l (2l+1) \frac{1 + \eta_l^2 - 2\eta_l \cos 2\delta_l}{4k^2} \tag{24-24}$$

There is also a cross section for the *inelastic* processes. Since we do not specify what the inelastic processes consist of, we can only talk about the *total inelastic cross section*, which describes the loss of flux. If we look at a particular term in (24-3), the inward radial flux carried by

$$\frac{i}{2k} \frac{e^{-ikr}}{r} P_l(\cos \theta)$$

[5] See L. I. Schiff, *Prog. Theo. Phys.*, (Kyoto), *11*, 288 (1954).

is

$$\left(\frac{\hbar k}{m}\right)\left[\frac{4\pi}{(2k)^2}\right]$$

(cf. Eq. 11-36 and the fact that $Y_{l0} = P_l(\cos\theta)/\sqrt{4\pi}$). The outward radial flux is $(\hbar k/m)(|S_l(k)|^2\, 4\pi/4k^2)$, so that the net flux lost is $(\hbar k/m)(\pi/k^2)[1 - \eta_l^2(k)]$ for each l-value. Hence, dividing by the incident flux, we get

$$\sigma_{\text{inel}} = \frac{\pi}{k^2}\sum_l (2l+1)\,[1 - \eta_l^2(k)] \tag{24-25}$$

Thus the total cross section is

$$\sigma_{\text{tot}} = \sigma_{\text{el}} + \sigma_{\text{inel}}$$

$$= \frac{\pi}{k^2}\sum_l (2l+1)\,(1 + \eta_l^2 - 2\eta_l\cos 2\delta_l + 1 - \eta_l^2)$$

$$= \frac{2\pi}{k^2}\sum_l (2l+1)\,(1 - \eta_l\cos 2\delta_l) \tag{24-26}$$

It also follows from (24-23) that

$$\text{Im}f(0) = \sum_l (2l+1)\,\text{Im}\,f_l(k)$$

$$= \sum_l (2l+1)\,\frac{1 - \eta_l\cos 2\delta_l}{2k} = \frac{k}{4\pi}\sigma_{\text{tot}} \tag{24-27}$$

so that the optical theorem is indeed satisfied.

If $\eta_l(k) = 1$, we have no absorption, and the inelastic cross section vanishes. When $\eta_l(k) = 0$ we have total absorption. Nevertheless there is still elastic scattering in that partial wave. This becomes evident in *scattering by a black disc*. The black disc is described as follows: (a) it has a well-defined edge and (b) it is totally absorbing. Since we will consider scattering for short wavelengths, that is, large k-values, condition (a) specifies that we only consider partial waves $l \lesssim L$, where

$$L = ka \tag{24-28}$$

and a is the radius of the disc. Condition (b) specifies that $\eta_l(k) = 0$ for the relevant values of $l \leq L$. Thus

$$\sigma_{\text{inel}} = \frac{\pi}{k^2}\sum_{l=0}^{L} (2l+1) = \frac{\pi}{k^2}L^2 = \pi a^2 \tag{24-29}$$

and

$$\sigma_{\text{el}} = \frac{\pi}{k^2}\sum_{l=0}^{L} (2l+1) = \pi a^2 \tag{24-30}$$

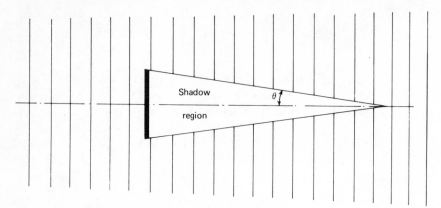

Fig. 24-2.   Black disc scattering and the shadow effect.

so that the total cross section is

$$\sigma_{\text{tot}} = \sigma_{\text{el}} + \sigma_{\text{inel}} = 2\pi a^2 \tag{24-31}$$

The result looks peculiar; on purely classical grounds we might perhaps expect that the total cross section cannot exceed the area presented by the disc; we might also expect to see no elastic scattering when there is total absorption. This is wrong; the absorptive disc takes flux proportional to πa^2 out of the incident beam (Fig. 24-2), and this leads to a shadow behind the disc. Far away, however, the shadow gets filled in—far enough away you cannot "see" the disc—and the only way in which this can happen is through the diffraction of some of the incident wave at the edge of the disc. The amount of incident wave that must be diffracted is the same amount as was taken out of the beam to make the shadow. Thus the elastically scattered flux must also be proportional to πa^2. The elastic scattering that accompanies absorption is called *shadow scattering* for the above reason. It is strongly peaked forward. The angle to which it is confined can be estimated from the uncertainty principle: an uncertainty in the lateral direction of magnitude a will be accompanied by an uncontrolled lateral momentum transfer of magnitude $p_\perp \sim \hbar/a$. This, however, is equal to $p\theta$, so that

$$\theta \sim \frac{\hbar}{ap} \sim \frac{1}{ak} \tag{24-32}$$

This agrees with the optical result $\theta \sim \lambda/a$. These features are observed both in nuclear scattering and in particle scattering at high energies, since the central region of nuclei and of protons is strongly absorptive, and the edges of these objects are moderately sharp. (See Fig. 24-3.)

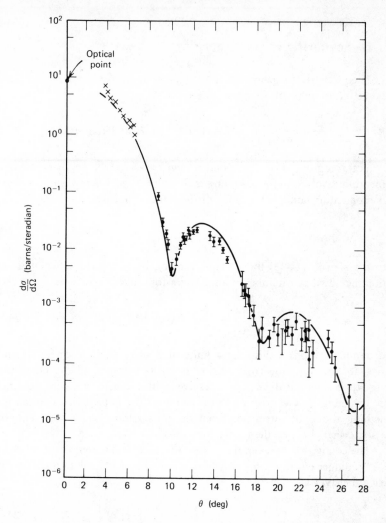

Fig. 24-3. Angular distribution of 1000 MeV (1 BeV) protons scattered by [16]O nuclei. The angular distribution shows the dips that characterize diffraction scattering. The departures from the shape of Frauenhofer scattering in optics is due to the fact that nuclei are not sharp, nor are they totally absorbing. The curve is the result of a theoretical calculation that takes these effects into account. (From H. Palevsky et al., *Phys. Rev. Letters*, *18*, 1200 (1967), by permission.)

B. Scattering at Low Energies

The phase shift expansion (24-17) may be used to express the differential cross section in terms of the phase shifts

$$\frac{d\sigma}{d\Omega} = \frac{1}{k^2} \left| \sum_l (2l + 1)\, e^{i\delta_l(k)} \sin \delta_l(k)\, P_l(\cos \theta) \right|^2 \qquad (24\text{-}33)$$

We expect, on grounds of correspondence with classical theory, that the angular momentum involved in the scattering is bounded by pa where p is the center-of-mass momentum and a is the range of the forces. Thus we expect that

$$l \lesssim \frac{pa}{\hbar} = ka \qquad (24\text{-}34)$$

With the sum in (24-33) limited, one can try, by fitting the differential cross section measured at a number of angles to a form like

$$\frac{d\sigma}{d\Omega} = \sum_{n=0}^{N} A_n (\cos \theta)^n \qquad (24\text{-}35)$$

to determine the phase shifts for a finite number of l-values. There are ambiguities, for example, the cross section is unaltered when all the phase shifts change their sign, but these can be resolved with the help of theory, continuity from low energies, and other tricks of the trade. The hope is that one can learn something about the interaction from the phase shifts, which form empirical data somewhat closer to the theory than the cross sections do.

The connection between the phase shifts $\delta_l(k)$ and the potential $V(r)$ is via the Schrödinger equation; the radial equation will have a solution that asymptotically behaves as

$$R_l(r) \sim \frac{1}{r} \sin\left[kr - \frac{l\pi}{2} + \delta_l(k) \right] \qquad (24\text{-}36)$$

aside from an amplitude factor in front. Thus, given $V(r)$, a straightforward way to calculate $\delta_l(k)$ is to integrate the radial equation numerically to values of r that are far out of the range of the potential, and to examine the asymptotic behavior. This is, in fact, what one does, but this does not give us any insight into the properties of the phase shifts. To learn more about the phase shifts, we consider the square well potential. We found in Chapter 11, Section C that

$$\tan \delta_l(k) = -\frac{C}{B} \qquad (24\text{-}37)$$

where the ratio is obtained by matching the internal to the external radial wave function (11-63)

$$\kappa \, \frac{j_l'(\kappa a)}{j_l(\kappa a)} = k \, \frac{j_l'(ka) + (C/B) \, n_l'(ka)}{j_l(ka) + (C/B) \, n_l(ka)} \qquad (24\text{-}38)$$

in which

$$\kappa^2 = \frac{2m}{\hbar^2} \, (E + V_0) \qquad k^2 = \frac{2mE}{\hbar^2} \qquad (24\text{-}39)$$

the $'$ denotes differentiation, with respect to the argument, and $V_0 > 0$ for an attractive potential. Thus

$$\tan \delta_l(k) = \frac{kj_l'(ka) \, j_l(\kappa a) - \kappa j_l(ka) \, j_l'(\kappa a)}{kn_l'(ka) \, j_l(\kappa a) - \kappa n_l(ka) \, j_l'(\kappa a)} \qquad (24\text{-}40)$$

This is not a particularly transparent expression, but it simplifies in some limiting cases.

(a) Consider the case that

$$ka \ll l \qquad (24\text{-}41)$$

We do not insist that $\kappa a \ll l$. With the help of the formulas (11-25) and (11-26) we get

$$\tan \delta_l(k) \simeq \frac{2l+1}{[1.3.5 \ldots (2l+1)]^2} \, (ka)^{2l+1} \, \frac{lj_l(\kappa a) - \kappa a j_l'(\kappa a)}{(l+1) \, j_l(\kappa a) + \kappa a j_l'(\kappa a)} \qquad (24\text{-}42)$$

after a little algebra. One can show that for large l, this drops faster than e^{-l} even if $ka \gg 1$. The behavior

$$\tan \delta_l(k) \sim k^{2l+1} \qquad (24\text{-}43)$$

for $ka \to 0$ is not restricted to the square well potential, but is true for all reasonably smooth potentials. It is a consequence of the centrifugal barrier, which keeps waves of energy far below the barrier from feeling the effect of the potential.

(b) For certain values of the energy, the denominator in (24-40) will vanish, so that at these energies the phase shift passes through $\pi/2$, or more generally through $(n + 1/2) \, \pi$. When the phase shift is $\pi/2$, then the partial wave cross section

$$\sigma_l(k) = \frac{4\pi(2l+1)}{k^2} \, \sin^2 \delta_l(k) \qquad (24\text{-}44)$$

has the largest possible value. One says that when $\tan \delta_l(k)$ rises rapidly to infinity and continues rising from $-\infty$, we have *resonant scattering*. To justify this terminology, and explain when resonant scattering occurs, let us consider a very deep potential, and also l large, so that

$$\kappa a \gg l \gg ka \qquad (24\text{-}45)$$

We may then use (24-42) for tan $\delta_l(k)$, and this will become infinite when

$$(l + 1) j_l(\kappa a) + \kappa a j_l'(\kappa a) = 0 \qquad (24\text{-}46)$$

Since $\kappa a \gg l$, this condition is approximately equivalent to

$$\frac{(l + 1)}{\kappa a} \cos\left(\kappa a - \frac{l + 1}{2}\pi\right) - \sin\left(\kappa a - \frac{l + 1}{2}\pi\right) = 0$$

that is,

$$\tan\left(\kappa a - \frac{l + 1}{2}\pi\right) \simeq \frac{l + 1}{\kappa a} \qquad (24\text{-}47)$$

Since the right side is very small, the resonance condition is

$$\kappa a - \frac{l + 1}{2}\pi \cong n\pi + \frac{l + 1}{\kappa a} \qquad (24\text{-}48)$$

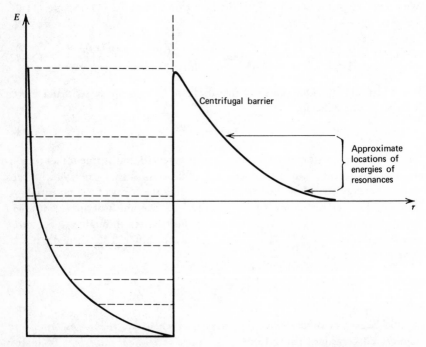

Fig. 24-4. Sketch showing the square well potential with the centrifugal barrier tail. The dashed lines represent the energy levels in an infinite square well of range a, and the approximate locations of the scattering resonance energies are indicated on the right. The lower one will be much sharper than the upper one.

Now this is just the condition (11-50) for the existence of discrete levels in a three-dimensional box, so that resonant scattering occurs when the incident energy is just such as to match an energy level. Since $E > 0$, these levels are not really bound states. As Fig. 24-4 indicates, these are levels that would be bound states if the barrier were infinitely thick. It is not, but a particle being scattered at just the right energy still "knows" that there is a virtual level there.

As (24-42) shows, the phase shift is very tiny for ka small. Nevertheless, as ka changes and goes through the resonance, δ_l rises very rapidly, increasing by π; thus the partial wave cross section (24-44) will exhibit a very sharp peak at the resonant energy. This behavior (Fig. 24-5) is very similar to the cross section for the scattering of electrons by He$^+$ at the energy corresponding to the $(2s)^2$ excited state (Fig. 18-4). In the neighborhood of the resonant energy, the phase shift rises through $\pi/2$ very rapidly. We may represent this behavior by

$$\tan \delta_l \approx \frac{\gamma(ka)^{2l+1}}{E - E_{\text{res}}} \tag{24-49}$$

This leads to the partial wave cross section

$$\sigma_l = \frac{4\pi(2l+1)}{k^2} \frac{\tan^2 \delta_l}{1 + \tan^2 \delta_l} = \frac{4\pi(2l+1)}{k^2} \frac{[\gamma(ka)^{2l+1}]^2}{(E - E_{\text{res}})^2 + [\gamma(ka)^{2l+1}]^2} \tag{24-50}$$

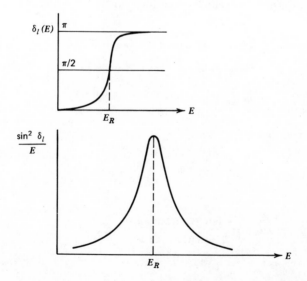

Fig. 24-5. The partial wave cross section corresponding to the phase shift sketched in the upper insert.

This is the well-known *Breit-Wigner formula* for resonant cross sections. Again, the behavior is not a peculiarity of the square well potential, but is characteristic of all potentials that have a shape such that metastable states can simulate bound states above $E = 0$ in it. We just note for completeness that

$$
f_l(k) = \frac{e^{2i\delta_l(k)} - 1}{2ik} = \frac{\dfrac{1 + i \tan \delta_l}{1 - i \tan \delta_l} - 1}{2ik}
$$

$$
= \frac{\tan \delta_l}{k(1 - i \tan \delta_l)} = \frac{\gamma(ka)^{2l+1}/k}{E - E_{\text{res}} - i\gamma(ka)^{2l+1}} \tag{24-51}
$$

If there is nonresonant scattering that is appreciable, then the scattering amplitude is of the form

$$
f_l(k) = f_l^{\text{res}}(k) + f_l^{\text{nonres}}(k) \tag{24-52}
$$

At low energies, the scattering is primarily in S-states, so that we may concentrate on $l = 0$. It is simpler to derive the phase shift directly than to work out (24-40). The solution inside the well that is regular at $r = 0$ is

$$
u(r) = rR(r) = C \sin \kappa r \tag{24-53}
$$

and this is to be matched onto

$$
u(r) = \sin (kr + \delta) \tag{24-54}
$$

the solution outside the well. The continuity of $(1/u)(du/dr)$ at $r = a$ implies that

$$
\kappa \cot \kappa a = k \cot (ka + \delta)
$$

that is,

$$
\tan \delta = \frac{(k/\kappa) \tan \kappa a - \tan ka}{1 + (k/\kappa) \tan \kappa a \tan ka} \tag{24-55}
$$

Note that if we define

$$
\tan qa = \frac{k}{\kappa} \tan \kappa a
$$

then

$$
\tan \delta = \frac{\tan qa - \tan ka}{1 + \tan qa \tan ka} = \tan (qa - ka)
$$

that is,

$$
\delta = \tan^{-1} \left(\frac{k}{\kappa} \tan \kappa a \right) - ka \tag{24-56}
$$

We have, following (24-39),

$$(\kappa a)^2 = (ka)^2 + \frac{2mV_0 a^2}{\hbar^2} \qquad (24\text{-}57)$$

with $V_0 > 0$ for an attractive potential. Thus, at very low energies, using $\tan x \simeq x$ for $x \ll 1$, we get

$$\tan \delta \approx \delta \approx ka \left(\frac{\tan \kappa a}{\kappa a} - 1 \right) \qquad (24\text{-}58)$$

When κa goes through $\pi/2$ (we imagine that we are slowly deepening the potential well), which is just the condition that the well be deep enough for a bound state to develop (cf. Chapter 11, Problem 1), then $\tan \kappa a \to \infty$ and (24-55) shows that

$$\tan \delta = \frac{1}{\tan ka} \to \infty \qquad (24\text{-}59)$$

that is, δ goes through $\pi/2$. In a sense, a bound state at zero energy is like a resonance.

As the well becomes a little deeper, we again have $\tan \delta \sim O~(ka)$, and continuity demands that the branch is such that

$$\delta \approx ka \left(\frac{\tan \kappa a}{\kappa a} - 1 \right) \qquad \text{(no bound state)}$$

$$\delta \approx \pi + ka \left(\frac{\tan \kappa a}{\kappa a} - 1 \right) \qquad \text{(with bound state)} \qquad (24\text{-}60)$$

As the potential becomes still deeper, a second bound state can appear, κa goes through $3\pi/2$, and we have $\delta \approx 2\pi + ka[(\tan \kappa a/\kappa a) - 1]$, and so on. There is a general result known as Levinson's Theorem, which states

$$\delta(0) - \delta(\infty) = N_B \pi \qquad (24\text{-}61)$$

where N_B is the number of bound states, and the above is an example of it. At very low energies the cross section only has the $l = 0$ contribution to it, and it is

$$\sigma \cong \frac{4\pi}{k^2} (ka)^2 \left(\frac{\tan \kappa a}{\kappa a} - 1 \right)^2 = 4\pi a^2 \left(\frac{\tan \kappa a}{\kappa a} - 1 \right)^2 \qquad (24\text{-}62)$$

that is, it is a constant. There will, of course, be a correction of order $(ka)^2$ to this result. If we consider neutron-proton scattering, then we know that the potential must be such as to give the right binding energy of the deuteron. If we let

$$E = - \frac{\hbar^2 \alpha^2}{2m}$$

and

$$\kappa = \sqrt{-\alpha^2 + \frac{2mV_0}{\hbar^2}}$$

(effectively $k^2 = -\alpha^2$ for the bound state problem), then the matching of the wave function outside the potential $u(r) = A e^{-\alpha r}$ to the solution inside $B \sin \kappa r$ at the boundary gives

$$\kappa \cot \kappa a = -\alpha \qquad (24\text{-}63)$$

For $k \ll \kappa$, we have

$$\left(\frac{\tan \kappa a}{\kappa a}\right)_{\text{scatt}} \cong \left(\frac{\tan \kappa a}{\kappa a}\right)_{\text{deuteron}} = -\frac{1}{a\alpha} \qquad (24\text{-}64)$$

Thus

$$\sigma \cong 4\pi a^2 \left(1 + \frac{1}{a\alpha}\right)^2 \cong \frac{4\pi}{\alpha^2} (1 + 2a\alpha) \qquad (24\text{-}65)$$

Thus making the low energy approximation expressed by (24-64) allows us to bypass the problem of determining the potential and *then* calculating the cross section. The approximation only works when the binding energy is small. The quantity $1/\alpha$ is the distance over which the deuteron wave function spills over, and this is always much larger than the range of the potential a for a loosely bound system. It is $1/\alpha$ and not the range of the potential that determines the scattering cross section at low energies.

In the 1930s there was great interest in the form of the neutron-proton potential, since it was hoped that this would give some fundamental clues concerning the nuclear forces in general. Rudimentary experiments at low energies were fitted with a variety of potentials. It became evident after a while that almost any reasonably shaped potential would work, provided that one chose the appropriate depth and range. It was shown in 1947 by Schwinger (and subsequently derived by Bethe in a simpler manner) that at low energies it is always a good approximation to write

$$k \cot \delta = -\frac{1}{A} + \frac{1}{2} r_0 k^2 \qquad (24\text{-}66)$$

where A is called the scattering length, and r_0 is the effective range. The cross section at threshold determines the scattering length

$$\sigma \cong 4\pi A^2 \qquad (24\text{-}67)$$

and the energy dependence determines the effective range. The relation between these parameters and the parameters describing the potential vary with the shape, but a two-parameter fit to the data is always possible. This *effective range formula* shows that if we want to probe the shape of the potential, we must go to higher energies.

The binding energy of the deuteron is 2.23 MeV. Thus, remembering that in our discussion m is the reduced mass, that is, $M_p/2$,

$$\frac{1}{\alpha} = \sqrt{\frac{\hbar^2}{2mE}} = \frac{\hbar c}{\sqrt{M_p c^2 E}} = \frac{\hbar}{M_p c}\sqrt{\frac{M_p c^2}{E}}$$

$$\cong \frac{10^{-27}}{1.6 \times 10^{-24} \times 3 \times 10^{10}}\sqrt{\frac{940}{2.23}} = 4.3 \times 10^{-13} \text{ cm}$$

so that

$$\frac{4\pi}{\alpha^2} \simeq 2.5 \times 10^{-24} \text{ cm}^2 \simeq 2.5 \text{ barns}$$

A more accurate determination leads to the prediction that the cross section at threshold is 4 barns. The measurement, carried out with neutrons at thermal speeds yields 21 barns!

The explanation of this disagreement came with the realization that the spin of the neutron and the proton had not been taken into account. If the potential were spin independent, then all spin states would scatter the same way, that is, it would not matter whether the spins of the particles are "up" or "down." If the potential does depend on the spin, a possible form could be

$$V(r) = V_1(r) + \mathbf{\delta}_p \cdot \mathbf{\delta}_n V_2(r) \tag{24-68}$$

In this case spin is no longer a good quantum number, and the states must be classified by total angular momentum and total spin, that is, with $l = 0$, the four states divide up into a $^3S_1$ triplet of states, and a singlet $^1S_0$. These need not scatter the same way, so that there are really two phase shifts, δ_t for the triplet, and δ_s for the singlet. There are no triplet-singlet transitions, since the total angular momentum J must be the same in the initial and final states. The total cross section is weighted by the number of final states in each case (the cross section involves a *sum* over final states and is independent of the value of the z-component of the angular momentum), so that

$$\sigma = \frac{3}{4}\sigma_t + \frac{1}{4}\sigma_s \tag{24-69}$$

For spin independent forces, $\sigma = \sigma_t = \sigma_s$.

The deuteron is a $^3S_1$ state, so that the four barns are really predicted for σ_t. This implies that

$$\sigma_s = 4\sigma - 3\sigma_t = 72 \text{ barns} \tag{24-70}$$

Since we are at threshold, this implies that

$$|A_s| = \sqrt{\frac{72 \times 10^{-24}}{4\pi}} \cong 2.4 \times 10^{-12} \text{ cm} \qquad (24\text{-}71)$$

The earlier result implied that

$$|A_t| = \sqrt{\frac{4 \times 10^{-24}}{4\pi}} \cong 4.7 \times 10^{-13} \text{ cm} \qquad (24\text{-}72)$$

The question of the signs of A_t and A_s now arises. At threshold we have $k \cot \delta \approx k/\delta \simeq -1/A$ so that $\delta_s = -A_s k$ and $\delta_t = -A_t k$. Thus, the asymptotic wave functions have the form

$$\sin (kr + \delta_{t,s}) \simeq \sin k(r - A_{t,s}) \simeq k(r - A_{t,s}) \qquad (24\text{-}73)$$

The two possible cases are shown in Fig. 24-6. We know that for the triplet state, the wave function turns over just before the edge of the well (since there is a bound state), so that it must correspond to the situation $A_t > 0$.

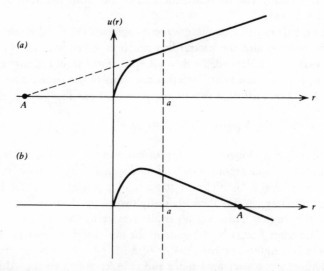

Fig. 24-6. Sketch of the s-wave solution $u(r)$ near threshold. Outside the range radius $r = a$, the wave function has the form $C(r - A)$. [This is not in conflict with (24-73), which is an expansion of $\sin (kr + \delta)$. We could equally well have taken the form of $u(r)$ to be $(C/k) \sin (kr + \delta)$, since the normalization is arbitrary. It is, in fact, the interior wave function and the position of A that determine the slope of the line.] The sign of A depends on whether the interior wave function has or has not turned over [cases (b) and (a), respectively.] Since the wave function must turn over if there is a weakly bound state (so that it can match a slowly falling exponential) and since one does not expect the wave function inside the potential to be very sensitive to variations in E about zero, one expects that for a potential that has a bound state with E_B small, $A > 0$.

If A_s were positive too, one would expect a singlet bound state, with very much weaker binding, since the internal wave function ties onto a much flatter asymptotic form. In fact, the binding energy would be 70 keV. Such a bound state was not found, suggesting that $A_s < 0$.

This choice of sign was actually confirmed by the scattering of neutrons off the H_2 molecule. As we know, the H_2 molecule can exist as ortho-H_2, with the spins in a triplet state, and para-H_2, with the two proton spins in a singlet state. For neutrons at very low energies, such that the wavelength is much larger than the proton-proton separation in the molecule, the scattering amplitude for neutron-H_2 scattering is just the sum of the amplitudes for the individual scatterings. One may show that the amplitude off para-H_2 is different from the amplitude off ortho-H_2 and these separately involve linear combinations of A_s and A_t. The fact that $\sigma_{para} \cong 3.9$ barns, while $\sigma_{ortho} \cong 125$ barns can be explained in this way. The calculation is complicated by a number of effects that must be taken into account, for example, that the effective mass of the proton in a molecule is different from that of a free proton, and that the molecules are not really at rest, but are moving with a distribution appropriate to the (low $\sim 20°K$) temperature. The large discrepancy between the two cross sections is not changed much by these corrections, and it can only be explained if A_s is indeed negative.

C. The Born Approximation

At higher energies many partial waves contribute to the scattering, and it is therefore preferable to avoid the angular momentum decomposition. A procedure that leads to a very useful approximation both when the potential is very weak and when the energy is very high is the Born approximation, in which we consider the scattering process as a transition, just like the transitions studied in Chapter 22. The difference is that here we consider the transitions

$$\text{continuum} \rightarrow \text{continuum}$$

If we work in the center-of-mass system, we have effectively a one-particle problem, and this particle makes a transition from an initial state, described by the eigenfunction

$$\psi_i(\mathbf{r}) = \frac{1}{\sqrt{V}} e^{i\mathbf{p}_i \cdot \mathbf{r}/\hbar} \tag{24-74}$$

to the final state, described by

$$\psi_f(\mathbf{r}) = \frac{1}{\sqrt{V}} e^{i\mathbf{p}_f \cdot \mathbf{r}/\hbar} \tag{24-75}$$

where $\mathbf{p}_i$ and $\mathbf{p}_f$ are the initial and final momenta, respectively. The transition rate, following the Golden Rule (22-55) is given by

$$R_{i \to f} = \frac{2\pi}{\hbar} \int \frac{V \, d^3\mathbf{p}_f}{(2\pi\hbar)^3} \, |M_{fi}|^2 \, \delta\left(\frac{p_f{}^2}{2m} - \frac{p_i{}^2}{2m}\right) \qquad (24\text{-}76)$$

The delta function expresses energy conservation. If the particles that emerge have a different mass from those that enter, or if the target is excited, that delta function takes a somewhat different form. It will, however, always be of the form $\delta[p_f{}^2/2m) - E]$ where E is the energy available for kinetic energy of the final particle. The matrix element M_{fi} is given by

$$M_{fi} = \langle \psi_f | V | \psi_i \rangle = \int d^3\mathbf{r} \, \frac{e^{-i\mathbf{p}_f \cdot \mathbf{r}/\hbar}}{\sqrt{V}} \, V(\mathbf{r}) \, \frac{e^{i\mathbf{p}_i \cdot \mathbf{r}/\hbar}}{\sqrt{V}}$$

$$= \frac{1}{V} \int d^3\mathbf{r} \, e^{-i\Delta \cdot \mathbf{r}} \, V(\mathbf{r}) \qquad (24\text{-}77)$$

$\Delta = \dfrac{1}{\hbar}\,(\mathbf{p}_f - \mathbf{p}_i)$ is the *momentum transfer*. We write the matrix element as

$$M_{fi} = \frac{1}{V} \, \tilde{V}(\Delta) \qquad (24\text{-}78)$$

The integral in (24-76) may be rewritten in the form

$$R_{i \to f} = \frac{2\pi}{\hbar} \int d\Omega \, \frac{V p_f{}^2 dp_f}{(2\pi\hbar)^3} \, \frac{1}{V^2} \, |\tilde{V}(\Delta)|^2 \, \delta\left(\frac{p_f{}^2}{2m} - E\right)$$

$$= \frac{2\pi}{\hbar} \frac{1}{(2\pi\hbar)^3} \frac{1}{V} \int d\Omega p_f m \frac{p_f dp_f}{m} \, \delta\left(\frac{p_f{}^2}{2m} - E\right) |\tilde{V}(\Delta)|^2$$

$$= \frac{1}{4\pi^2\hbar^4} \frac{1}{V} \int d\Omega p_f m |\tilde{V}(\Delta)|^2 \qquad (24\text{-}79)$$

To get the last line, we noted that $p_f dp_f/m = d(p_f{}^2/2m)$ and carried out the delta function integration. Thus, p_f must be evaluated at $p_f = (2mE)^{1/2}$, and we must not forget that m here is the reduced mass in the final state.

This expression has an undesirable dependence on the volume of the quantization box, but this is not really surprising. Our wave functions were normalized to one particle in the box V, so that the number of transitions should certainly go down as V increases. This difficulty arises because we are asking a question that does not correspond to an experiment. What one does is send a flux of incident particles at each other (in the center of mass frame; in the laboratory, one particle is stationary, of course). If we want a flux of one particle per square centimeter per second, we must multiply the above by V divided by the volume of a cylinder with 1 cm² base, and the relative velocity of the particles in the center of mass frame in the initial state. The number of transitions for unit flux is just the cross section. We therefore have

$$d\sigma = \frac{1}{4\pi^2\hbar^4} \frac{1}{|v_{\text{rel}}|} d\Omega p_f m |\tilde{V}(\mathbf{\Delta})|^2 \qquad (24\text{-}80)$$

Since in the center of mass frame the two incident particles are moving toward each other with equal and opposite momenta of magnitude p_i, their relative velocity is

$$|v_{\text{rel}}| = \frac{p_i}{m_1} + \frac{p_i}{m_2} = p_i\left(\frac{1}{m_1} + \frac{1}{m_2}\right) = \frac{p_i}{m_{\text{red}}^{(i)}} \qquad (24\text{-}81)$$

if m_1 and m_2 are their masses. Thus, if the initial and final reduced masses and momenta are not the same, we have

$$\frac{d\sigma}{d\Omega} = \frac{1}{4\pi^2} \frac{p_f}{p_i} m_{\text{red}}^{(f)} m_{\text{red}}^{(i)} \left| \frac{1}{\hbar^2} \tilde{V}(\mathbf{\Delta}) \right|^2 \qquad (24\text{-}82)$$

When the initial and final particles are the same,

$$\frac{d\sigma}{d\Omega} = \frac{m_{\text{red}}^2}{4\pi^2} \left| \frac{1}{\hbar^2} \tilde{V}(\mathbf{\Delta}) \right|^2 \qquad (24\text{-}83)$$

When one particle is a great deal more massive than the other, $m_{\text{red}} \to m$, the mass of the lighter particle. When we compare the above with (24-15) we see that

$$f(\theta,\phi) = -\frac{m_{\text{red}}}{2\pi\hbar^2} \tilde{V}(\mathbf{\Delta}) \qquad (24\text{-}84)$$

Actually, to determine the sign, one must go through a more detailed comparison with the partial wave expansion. We will not bother to do this here.

As an illustration of the application of the Born approximation, we will calculate the cross section for the scattering of a particle of mass m and charge Z_1 by a Coulomb potential of charge Z_2. The source of the Coulomb field is taken to be infinitely massive, so that the mass in (24-83) is the mass of the incident particle. For generality (and, as we will see, for technical reasons) we take the Coulomb field to be screened, so that

$$V(\mathbf{r}) = Z_1 Z_2 e^2 \frac{e^{-r/a}}{r} \qquad (24\text{-}85)$$

where a is the screening radius. We thus need to evaluate

$$\tilde{V}(\mathbf{\Delta}) = Z_1 Z_2 e^2 \int d^3\mathbf{r} \, e^{-i\mathbf{\Delta}\cdot\mathbf{r}} \frac{e^{-r/a}}{r} \qquad (24\text{-}86)$$

We choose the direction of $\boldsymbol{\Delta}$ as z-axis, and then get

$$
\int d^3\mathbf{r}\ e^{-i\boldsymbol{\Delta}\cdot\mathbf{r}}\frac{e^{-r/a}}{r} = \int_0^{2\pi} d\phi \int_0^\pi \sin\theta d\theta \int_0^\infty r^2 dr\ e^{-i\Delta r\cos\theta}\frac{e^{-r/a}}{r}
$$

$$
= 2\pi \int_0^\infty r dr\ e^{-r/a} \int_{-1}^1 d(\cos\theta)\ e^{-i\Delta r\cos\theta}
$$

$$
= \frac{2\pi}{i\Delta} \int_0^\infty dr\ e^{-r/a} \left(e^{i\Delta r} - e^{-i\Delta r} \right)
$$

$$
= \frac{2\pi}{i\Delta} \left(\frac{1}{(1/a) - i\Delta} - \frac{1}{(1/a) + i\Delta} \right) = \frac{4\pi}{(1/a^2) + \Delta^2}
$$

$$(24\text{-}87)$$

Now

$$
\Delta^2 = \frac{1}{\hbar^2} (\mathbf{p}_f - \mathbf{p}_i)^2 = \frac{1}{\hbar^2} (2p^2 - 2\mathbf{p}_f\cdot\mathbf{p}_i) = \frac{2p^2}{\hbar^2} (1 - \cos\theta) \quad (24\text{-}88)
$$

so that the cross section becomes

$$
\frac{d\sigma}{d\Omega} = \frac{m^2}{4\pi^2} \frac{1}{\hbar^4} (Z_1 Z_2 e^2)^2 \frac{16\pi^2}{[(2p^2/\hbar^2)(1 - \cos\theta) + (1/a^2)]^2}
$$

$$
= \left(\frac{2m Z_1 Z_2 e^2}{4p^2 \sin^2(\theta/2) + (\hbar^2/a^2)} \right)^2
$$

$$
= \left(\frac{Z_1 Z_2 e^2}{4E \sin^2(\theta/2) + (\hbar^2/2ma^2)} \right)^2
$$

$$(24\text{-}89)$$

In the last line we replaced $p^2/2m$ by E, and we used $\frac{1}{2}(1 - \cos\theta) = \sin^2(\theta/2)$. The angle θ defined in (24-88) is the center of mass scattering angle. In the absence of screening ($a \to \infty$) this reduces to the well-known Rutherford formula. There is no $\hbar$ in it, and it is the same as the classical formula. Had we left out the screening factor in (24-86) we would have had an ill-defined integral. One often evaluates ambiguous integrals with the aid of such convergence factors.

The Born approximation has its limitations. For example, we found that $\tilde{V}(\boldsymbol{\Delta})$ was purely real so that $f(\theta)$ is also real in this approximation. This implies, by the optical theorem, that the cross section is zero. In fact, the Born Approximation is only good when either (a) the potential is weak, so that the cross section is of second order in a small parameter; this would make the use of it consistent with the optical theorem, or (b) at high energies for potentials such that the cross section goes to zero. This is true for most smooth potentials. It is not true for real particles; there it seems that the cross sections stay constant at very high energies, and one cannot expect the Born approximation to serve as more than a guide of the behavior of the scattering amplitude.

As a final comment, we observe that if the potential V has a spin dependence, then (24-77) is trivially modified by the requirement that the initial and final states be described by their spin wave functions, in addition to the spatial wave functions. Thus, for example, if the neutron-proton potential has the form

$$V(r) = V_1(r) + \mathbf{\sigma}_P \cdot \mathbf{\sigma}_N V_2(r)$$

the Born Approximation reads

$$M_{fi} = \frac{1}{V} \int d^3r \, e^{-i\Delta \cdot \mathbf{r}} \, \xi_f^+ \, V(r) \, \xi_i$$

where ξ_i and ξ_f represent the initial and final spin states of the neutron-proton system.

D. Scattering of Identical Particles

When two identical particles scatter, there is no way of distinguishing a deflection of a particle through an angle θ and a deflection of $\pi - \theta$ in the center of mass frame, since momentum conservation demands that if one of the particles

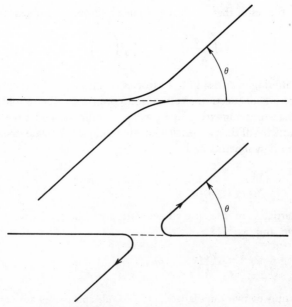

Fig. 24-7. Asymptotic directions in the scattering of two identical particles through a center of mass angle θ.

scatters through θ, the other goes in the direction $\pi - \theta$ (Fig. 24-7). Classically, too, the cross section for scattering is affected by the identity of particles, since the number of counts at a certain counter will be the sum of the counts due to the two particles. Thus

$$\sigma_{cl}(\theta) = \sigma(\theta) + \sigma(\pi - \theta) \tag{24-90}$$

In quantum mechanics there is no way of distinguishing the two final states, so that the two *amplitudes* $f(\theta)$ and $f(\pi - \theta)$ can interfere. Thus the cross section for the scattering of two identical spin zero (boson) particles, for example, α-particles, is

$$\frac{d\sigma}{d\Omega} = |f(\theta) + f(\pi - \theta)|^2 \tag{24-91}$$

This differs from the classical result by the interference term

$$\frac{d\sigma}{d\Omega} = |f(\theta)|^2 + |f(\pi - \theta)|^2 + [f^*(\theta) f(\pi - \theta) + f(\theta) f^*(\pi - \theta)] \tag{24-92}$$

and it leads to an enhancement at $\pi/2$, for example,

$$\left(\frac{d\sigma}{d\Omega}\right)_{\pi/2} = 4 \left| f\left(\frac{\pi}{2}\right) \right|^2 \tag{24-93}$$

compared to the result that would be obtained without interference:

$$\left(\frac{d\sigma}{d\Omega}\right)_{\pi/2} = 2 \left| f\left(\frac{\pi}{2}\right) \right|^2 \tag{24-94}$$

When the scattering of two spin 1/2 particles is considered, for example, proton-proton scattering or electron-electron scattering, then the amplitude should reflect the basic antisymmetry of the total wave function under the interchange of the two particles. If the two particles are in a spin singlet state, then the spatial wave function is symmetric, and

$$\frac{d\sigma_s}{d\Omega} = |f(\theta) + f(\pi - \theta)|^2 \tag{24-95}$$

If the two particles are in a spin triplet state, then the spatial wave function is antisymmetric, and

$$\frac{d\sigma_t}{d\Omega} = |f(\theta) - f(\pi - \theta)|^2 \tag{24-96}$$

In the scattering of two unpolarized protons, all spin states are equally likely, and thus the probability of finding the two protons in a triplet state is three times as large as finding them in a singlet state, so that

$$
\begin{aligned}
\frac{d\sigma}{d\Omega} &= \frac{3}{4}\frac{d\sigma_t}{d\Omega} + \frac{1}{4}\frac{d\sigma_s}{d\Omega} \\
&= \tfrac{3}{4}|f(\theta) - f(\pi - \theta)|^2 + \tfrac{1}{4}|f(\theta) + f(\pi - \theta)|^2 \\
&= |f(\theta)|^2 + |f(\pi - \theta)|^2 - \tfrac{1}{2}[f(\theta)f^*(\pi - \theta) + f^*(\theta)f(\pi - \theta)]
\end{aligned}
\tag{24-97}
$$

For proton-proton scattering as well as for $\alpha - \alpha$ scattering, the basic amplitude $f(\theta)$ is the sum of a nuclear term (if the energies are not too low) and a Coulomb term. Whether the identical particles be bosons or fermions, there is symmetry under the interchange $\theta \to \pi - \theta$.

Symmetry considerations also play a role in the scattering of particles by a crystal lattice. If we ignore spin, so that we do not have to worry whether the electron does or does not flip its spin ("up" $\to$ "down" or vice versa), then at low energies, the scattering amplitude $f(\theta)$ is independent of angle (S-wave scattering), and the solution of the Schrödinger equation by a single atom located at the lattice point $\mathbf{a}_i$ has the asymptotic form

$$
\psi(\mathbf{r}) \sim e^{i\mathbf{k}\cdot(\mathbf{r}-\mathbf{a}_i)} + f\frac{e^{ik|\mathbf{r}-\mathbf{a}_i|}}{|\mathbf{r}-\mathbf{a}_i|}
\tag{24-98}
$$

Now

$$
\begin{aligned}
k|\mathbf{r}-\mathbf{a}_i| &= k(\mathbf{r}^2 - 2\mathbf{r}\cdot\mathbf{a}_i + \mathbf{a}_i{}^2)^{1/2} \\
&\cong kr\left(1 - \frac{2\mathbf{r}\cdot\mathbf{a}_i}{r^2}\right)^{1/2} \\
&\cong kr - k\hat{\imath}_r\cdot\mathbf{a}_i
\end{aligned}
\tag{24-99}
$$

and since $k\hat{\imath}_r$ is a vector of magnitude k and it points in the direction $\mathbf{r}$, the point of observation, it is the final momentum $\mathbf{k}'$. If we divide out the phase factor $e^{-i\mathbf{k}\cdot\mathbf{a}_i}$, the wave function has the asymptotic form

$$
\psi \sim e^{i\mathbf{k}\cdot\mathbf{r}} + f e^{-i\mathbf{k}'\cdot\mathbf{a}_i} e^{i\mathbf{k}\cdot\mathbf{a}_i}\frac{e^{ikr}}{r} + 0\left(\frac{1}{r^2}\right)
\tag{24-100}
$$

so that the scattering amplitude is

$$
f(\theta) = f e^{-i\boldsymbol{\Delta}\cdot\mathbf{a}_i} \qquad \boldsymbol{\Delta} = \mathbf{k}' - \mathbf{k}
\tag{24-101}
$$

The total amplitude is the sum of all individual scattering amplitudes when we have a situation in which we cannot tell which atom in the crystal did the scattering. This is indeed the case for elastic low-energy scattering when recoil is not observed and spins are not measured. Thus, for the *coherent* process we have

$$
\frac{d\sigma}{d\Omega} = \left| f\sum_{\text{atoms}} e^{-i\boldsymbol{\Delta}\cdot\mathbf{a}_i}\right|^2
\tag{24-102}
$$

If we have a simple cubic array of latice points, such that

$$\mathbf{a}_i = a(n_x \hat{\imath}_x + n_y \hat{\imath}_y + n_z \hat{\imath}_z) \qquad -N \le n_x, n_y, n_z \le N \qquad (24\text{-}103)$$

(spacings are integral multiples of a in all directions) then

$$\sum e^{-i\mathbf{\Delta}\cdot\mathbf{a}_i} = \sum_{n_x=-N}^{N} \sum_{n_y=-N}^{N} \sum_{n_z=-N}^{N} e^{-ia\Delta_x n_x} e^{-ia\Delta_y n_y} e^{-ia\Delta_z n_z}$$

We use

$$\sum_{n=-N}^{N} e^{i\alpha n} = e^{-i\alpha N}(1 + e^{i\alpha} + e^{2i\alpha} + \ldots e^{2i\alpha N})$$

$$= e^{-i\alpha N} \frac{e^{i\alpha(2N+1)} - 1}{e^{i\alpha} - 1} = \frac{e^{i\alpha(N+1)} - e^{-i\alpha N}}{e^{i\alpha} - 1}$$

$$= \frac{e^{i\alpha(N+1/2)} - e^{-i\alpha(N+1/2)}}{e^{i\alpha/2} - e^{-i\alpha/2}} = \frac{\sin \alpha (N + \frac{1}{2})}{\sin \alpha/2} \qquad (24\text{-}104)$$

to obtain the result

$$\frac{d\sigma}{d\Omega} = |f|^2 \frac{\sin^2 \alpha_x(N+\frac{1}{2})}{\sin^2 \alpha_x/2} \cdot \frac{\sin^2 \alpha_y(N+\frac{1}{2})}{\sin^2 \alpha_y/2} \cdot \frac{\sin^2 \alpha_z(N+\frac{1}{2})}{\sin^2 \alpha_z/2} \qquad (24\text{-}105)$$

where

$$\alpha_x = a\,\Delta_x - 2\pi\nu_x \qquad (\nu_x = \text{integer}), \text{ etc.} \qquad (24\text{-}106)$$

We can make the generalization exhibited above, since a change $\alpha \to \alpha - 2\pi\nu$, with ν an integer, does not change (24-105). The expression (24-105) is not very transparent. However, when N is large, each of the factors becomes very strongly peaked when $\alpha_x, \ldots$ are near zero. In fact, using

$$\frac{\sin^2 Nu}{u^2/4} \to 4\pi N\,\delta(u) \qquad (24\text{-}107)$$

a formula easily derived from (22-36) by a simple change of variables, we get

$$\frac{d\sigma}{d\Omega} = |f|^2 (2\pi)^3 (2N)^3 \delta(a\mathbf{\Delta} - 2\pi\mathbf{\nu}) \qquad (24\text{-}108)$$

Now the total number of atoms is $(2N)^3$ and hence the cross section per atom is

$$\frac{d\sigma}{d\Omega} = |f|^2 \frac{(2\pi)^3}{a^3} \delta\left(\mathbf{k}' - \mathbf{k} - \frac{2\pi\mathbf{\nu}}{a}\right) \qquad (24\text{-}109)$$

Thus the differential cross section is very small, except in the directions given by

$$\mathbf{k}' - \mathbf{k} = \frac{2\pi}{a} \mathbf{\nu} \qquad (24\text{-}110)$$

where it is strongly peaked. The conditions above are called the *Bragg conditions*, and the integers ν_x, ν_y, ν_z are called the *Miller indices of the Bragg planes*.

The relations just derived can be generalized to more complicated crystals. They are used to study crystal structure, using neutrons or X-rays as incident particles, or using a known crystal to study X-rays that are emitted in atomic transitions involving energetic photons.

Problems

1. Show that for a central potential $V(\mathbf{r}) = V(r)$, the matrix element M_{fi} in (24-77) may be written in the form

$$M_{fi} = \frac{1}{V} \frac{4\pi\hbar}{\Delta} \int_0^\infty r\, dr V(r) \sin r\Delta$$

Note that this is an even function of Δ, that is, a function of

$$\Delta^2 = (\mathbf{p}_f - \mathbf{p}_i)^2/\hbar^2$$

2. Consider a potential of the form

$$V(r) = V_G\, e^{-r^2/a^2}$$

Calculate, using the Born Approximation, the differential cross section $d\sigma/d\Omega$ as a function of the center-of-mass scattering angle θ. Compare your result with the differential cross section for a Yukawa potential

$$V(r) = V_0 b\, \frac{e^{-r/b}}{r}$$

[Already done in (24-85)–(24-89)]. To make the comparison, adjust the parameters in the two cases so that the two differential cross sections and their slopes are the same in the forward direction at $\Delta = 0$. It might be convenient to pick some definite numerical values for V_G, V_0, a, and b to depict this graphically. Can you give a qualitative argument explaining the large difference between the predictions for large momentum transfers?

3. Consider the potential

$$V(r) = V_0 a\, \frac{e^{-r/a}}{r}$$

If the range parameter is $a = 1.2$ fm $= 1.2 \times 10^{-13}$ cm and $V_0 = 100$ MeV in magnitude, what is the total cross section for proton-proton scattering at 100 MeV center-of-mass energy, calculated in Born approximation?

(*Note.* This calculation involves a numerical integration. It is useful to use the relation

$$\hbar^2 \Delta^2 = (\mathbf{p}_f - \mathbf{p}_i)^2 = 2p^2(1 - \cos \theta)$$

to write

$$d\Omega = 2\pi \, d \, (\cos \theta) = \frac{\hbar^2 \pi}{p^2} \, d(\Delta^2)$$

and do the integration between $\Delta^2 = 0$ and $\Delta^2 = 4p^2/\hbar^2$.) Give your answer in millibarns (1 mb $= 10^{-27}$ cm^2).

4. Suppose the scattering amplitude for neutron-proton scattering is given by the form

$$f(\theta) = \xi_f^\dagger \, (A + B\mathbf{\sigma}_P \cdot \mathbf{\sigma}_N) \, \xi_i$$

where ξ_i and ξ_f are the initial and final spin states of the neutron-proton system. The possible states are

$$\xi_i = \chi_\uparrow^{(P)} \chi_\uparrow^{(N)} \qquad\qquad \xi_f = \chi_\uparrow^{(P)} \chi_\uparrow^{(N)}$$
$$\chi_\uparrow^{(P)} \chi_\downarrow^{(N)} \qquad\qquad \chi_\uparrow^{(P)} \chi_\downarrow^{(N)}$$
$$\chi_\downarrow^{(P)} \chi_\uparrow^{(N)} \qquad\qquad \chi_\downarrow^{(P)} \chi_\uparrow^{(N)}$$
$$\chi_\downarrow^{(P)} \chi_\downarrow^{(N)} \qquad\qquad \chi_\downarrow^{(P)} \chi_\downarrow^{(N)}$$

Use

$$\mathbf{\sigma}_P \cdot \mathbf{\sigma}_N = \sigma_z^{(P)}\sigma_z^{(N)} + 2(\sigma_+^{(P)}\sigma_-^{(N)} + \sigma_-^{(P)}\sigma_+^{(N)})$$

where

$$\sigma_+ = \frac{\sigma_x + i\sigma_y}{2} = \begin{pmatrix} 0 & 1 \\ 0 & 0 \end{pmatrix} \qquad \sigma_- = \frac{\sigma_x - i\sigma_y}{2} = \begin{pmatrix} 0 & 0 \\ 1 & 0 \end{pmatrix}$$

in the representation in which $\sigma_z = \begin{pmatrix} 1 & 0 \\ 0 & -1 \end{pmatrix}$ and $\chi_\uparrow = \begin{pmatrix} 1 \\ 0 \end{pmatrix}$, $\chi_\downarrow = \begin{pmatrix} 0 \\ 1 \end{pmatrix}$

to calculate all 16 scattering amplitudes. Make a table of your results and also tabulate the cross sections.

5. If any one of the spin states (e.g., initial proton, or initial neutron, etc.) is not measured, the cross section is the sum over the unmeasured spin states. Suppose both the initial and final proton spins are not measured. Write down expressions for the cross sections for the final neutron "up" and the final neutron "down," given that the initial neutron state is "up." What is the polarization P, defined by

$$P = \frac{\sigma \uparrow - \sigma \downarrow}{\sigma \uparrow + \sigma \downarrow}$$

where $\sigma_\uparrow$ is the cross section with the final neutron up and so on.

6. Use the table computed in Problem 4 to calculate the cross sections for triplet $\rightarrow$ triplet and singlet $\rightarrow$ singlet scattering, respectively. Show that triplet $\rightarrow$

singlet scattering vanishes. Check your results by observing that since (in units of $\hbar$)

$$\tfrac{1}{2}\mathbf{\sigma}_P + \tfrac{1}{2}\mathbf{\sigma}_N = \mathbf{S}$$

one has

$$\mathbf{\sigma}_P \cdot \mathbf{\sigma}_N = 2\mathbf{S}^2 - 3$$

$$= 1 \qquad \text{when acting on triplet state}$$
$$-3 \qquad \text{when acting on singlet state}$$

Note that the amplitude is independent of m_S so that m_S must be the same in the initial and final spin states. There are three states in the triplet, all contributing an equal amount to the cross section, and only one to the singlet cross section. (*Caution.* In calculating amplitudes such as

$$\frac{1}{\sqrt{2}}\,(\chi_\uparrow^{(P)}\chi_\downarrow^{(N)} - \chi_\downarrow^{(P)}\chi_\uparrow^{(N)})(A + B\mathbf{\sigma}_P\cdot\mathbf{\sigma}_N)\frac{1}{\sqrt{2}}\,(\chi_\uparrow^{(P)}\chi_\downarrow^{(N)} - \chi_\downarrow^{(P)}\chi_\uparrow^{(N)})$$

the amplitudes are added for the four terms before squaring. Can you explain why?)

9. It can be shown that the solution of the $l = 0$ Schrödinger equation for the potential

$$V(r) = -2\beta\lambda^2\,\frac{e^{-\lambda r}}{(\beta\,e^{-\lambda r} + 1)^2}$$

which behaves asymptotically like e^{-ikr} is

$$F(k,r) = e^{-ikr}\cdot\frac{2k(\beta\,e^{-\lambda r} + 1) + i\lambda(\beta\,e^{-\lambda r} - 1)}{(\beta\,e^{-\lambda r} + 1)(2k - i\lambda)}$$

The solution that behaves asymptotically like e^{+ikr} is $F(-k,r)$. Thus the regular solution, which vanishes at $r = 0$, is

$$u(r) = [F(k,0)\,F(-k,r) - F(-k,0)\,F(k,r)]$$

Use this information to obtain the scattering amplitude $F(k) = [S(k) - 1]/2ik$. Discuss the solution for various limiting cases.

References

Scattering theory is discussed in all of the textbooks listed at the end of this volume. In addition there exist a number of advanced treatises that are devoted to this subject alone. Most accessible to students at this level is

N. F. Mott and H. S. W. Massey, *The Theory of Atomic Collisions* (Third Edition) Oxford, The Clarendon Press (1965). More formal is

L. S. Rodberg and R. M. Thaler, *Introduction to the Quantum Theory of Scattering*, Academic Press, New York, 1967. More advanced are

M. L. Goldberger and K. M. Watson, *Collison Theory*, John Wiley & Sons, 1965.

R. Newton, *Scattering Theory of Waves and Particles*, McGraw-Hill Co., 1966.

chapter 25

The Absorption of Radiation
in Matter

The process that is the inverse of radiative decay of atoms, namely the capture of photons accompanied by the excitation of atoms, can also take place. For photon energies exceeding the ionization energy of the atom, the electron is excited to the continuum. This is called the *photoelectric effect*, and is an important mechanism in the absorption of radiation in matter.

According to the Golden Rule (22-55), the transition rate for the process

$$\gamma + (\text{atom}) \rightarrow (\text{atom})' + e \qquad (25\text{-}1)$$

is given by

$$
\begin{aligned}
R &= \frac{2\pi}{\hbar} \int \frac{V \, d^3\mathbf{p}_e}{(2\pi\hbar)^3} \, |M_{fi}|^2 \, \delta\left(\hbar\omega - E_B - \frac{p_e^2}{2m}\right) \\
&= \frac{2\pi}{\hbar} \int \frac{d\Omega V}{(2\pi\hbar)^3} \int m p_e \, d\left(\frac{p_e^2}{2m}\right) |M_{fi}|^2 \, \delta\left(\hbar\omega - E_B - \frac{p_e^2}{2m}\right) \\
&= \frac{2\pi V}{\hbar} \int d\Omega \, \frac{\cdot m p_e}{(2\pi\hbar)^3} \, |M_{fi}|^2 \qquad (25\text{-}2)
\end{aligned}
$$

In the above expression, m is the electron mass, the delta function represents energy conservation, E_B is the magnitude of the binding energy of the electron in the atom, and in the last line, p_e is evaluated at the vanishing of the argument of the delta function.

The matrix element is given by

$$\frac{e}{mc} \left(\frac{2\pi\hbar c^2}{\omega V}\right)^{1/2} \int d^3\mathbf{r} \psi_f^*(\mathbf{r}) \, \boldsymbol{\varepsilon} \cdot \mathbf{p} \, e^{i\mathbf{k}\cdot\mathbf{r}} \, \psi_i(\mathbf{r}) \qquad (25\text{-}3)$$

The vector potential is normalized, as in Chapter 22, to one photon in the volume V, and $\psi_i(\mathbf{r})$, $\psi_f(\mathbf{r})$ are the wave functions for the electron in the initial and final states. If we consider a hydrogenlike atom, and assume that the electron is in the ground state, we have

$$\psi_i(\mathbf{r}) = \frac{1}{\sqrt{\pi}} \left(\frac{Z}{a_0}\right)^{3/2} e^{-Zr/a_0} \tag{25-4}$$

The final-state wave function should be taken to be a solution of the Schrödinger equation with a Coulomb potential with $E > 0$. We did not discuss these solutions when we studied the hydrogen atom. They can be written in closed form but they are quite complicated, as is the integral in (25-3). If the photon energy is much larger than the ionization energy, then the residual interaction of the outgoing electron with the ion that it leaves behind becomes less important, and we may approximate $\psi_f(\mathbf{r})$ by a plane wave. Since we assume that we only have one atom in our volume, we will have only one electron in the volume, and hence the normalization is such that

$$\psi_f(\mathbf{r}) = \frac{1}{\sqrt{V}} e^{i\mathbf{p}_e \cdot \mathbf{r}/\hbar} \tag{25-5}$$

The factor V that appears in the phase space $[V\, d^3\mathbf{p}/(2\pi\hbar)^3]$ corresponds to the same normalization, that is, the two factors are not independent. The square of the matrix element is somewhat simplified since the final state is an eigenstate of momentum, so that

$$\langle f| \boldsymbol{\varepsilon} \cdot \mathbf{p}_{op}\, e^{i\mathbf{k}\cdot\mathbf{r}}|i\rangle = \boldsymbol{\varepsilon} \cdot \mathbf{p}_e \langle f| e^{i\mathbf{k}\cdot\mathbf{r}}|i\rangle \tag{25-6}$$

Hence the square of the matrix element is

$$|M_{fi}|^2 \simeq \left(\frac{e}{mc}\right)^2 \frac{2\pi\hbar c^2}{\omega V} \cdot \frac{1}{V}\frac{1}{\pi}\left(\frac{Z}{a_0}\right)^3 (\boldsymbol{\varepsilon}\cdot\mathbf{p}_e)^2$$
$$\times \left| \int d^3\mathbf{r}\, e^{i(\mathbf{k}-\mathbf{p}_e/\hbar)\cdot\mathbf{r}}\, e^{-Zr/a_0}\right|^2 \tag{25-7}$$

We will evaluate the integral later. At this point we note that the rate again has a $1/V$ behavior due to the fact that we are dealing with a single photon in the volume V. Instead, we will consider the cross section for the photoelectric effect. To have a flux of one photon per square centimeter we must have a density of photons $1/c$ per cubic centimeter (so that a cylinder of unit area base and length c corresponding to a time interval of 1 sec contain one photon), that is, we must multiply the rate by V/c. We get, combining (25-2) and (25-7) the differential cross section

$$\frac{d\sigma}{d\Omega} = \frac{2\pi}{\hbar}\frac{mp_e}{(2\pi\hbar)^3}\left(\frac{e}{mc}\right)^2 \frac{2\pi\hbar c^2}{\omega}\frac{1}{\pi}\left(\frac{Z}{a_0}\right)^3 (\boldsymbol{\varepsilon}\cdot\mathbf{p}_e)^2$$
$$\times \frac{1}{c}\left| \int d^3\mathbf{r}\, e^{i(\mathbf{k}-\mathbf{p}_e/\hbar)\cdot\mathbf{r}}\, e^{-Zr/a_0}\right|^2 \tag{25-8}$$

In this expression $d\Omega$ is the solid angle into which $\mathbf{p}_e$ points. The integral over all electron directions yields the total cross section σ for the photoelectric effect. If the target atoms are distributed with a density of N atoms per cubic centimeter, then in a slab of target material of area A and thickness dx, there are $NA\,dx$ target atoms. Each atom has a cross section σ for the reaction under consideration, so that the total effective area presented to the beam is $NA\,\sigma\,dx$. If there are n incident particles in the bombarding beam, then the number of particles that interact in the thickness dx of the target is given by

$$\frac{\text{interacting particles}}{\text{incident particles}} = \frac{\text{cross section}}{\text{total area}}$$

that is,

$$\frac{dn}{n} = -\frac{NA\sigma\,dx}{A} = -N\sigma\,dx \qquad (25\text{-}9)$$

The minus sign indicates that particles are removed from the beam. Integration gives

$$n(x) = n_0\,e^{-N\sigma x} \qquad (25\text{-}10)$$

where n_0 is the number of incident particles and $n(x)$ is the number of particles left in the beam after traversing a thickness x of the target. The quantity $\lambda = 1/N\sigma$ has the dimensions of a length, and is called the *mean free path*. One sometimes speaks of the mean free path for the photoelectric effect, for pair production, and so on, even though what is measured is the cross section.

To get an idea of the magnitudes of mean free paths, note that $N = N_0\rho/A$ where $N_0 = 6.02 \times 10^{23}$ is Avogadro's Number, ρ is the density in grams per cubic centimeter and A is the atomic weight. Cross sections for molecular collisions can be estimated from the properties of gases, and they turn out to have magnitudes of the order of 10^{-16} cm^2, consistent with the fact that atomic dimensions are of the order of 10^{-8} cm.[1] Is this a reasonable guess for the photoelectric cross section? We shall soon examine the reasons for why it is not. In the meantime we write the mean free path in centimeters, in a material of density ρ and atomic weight A, with the cross section expressed in units of 10^{-24} cm^2, called *barns* thus

$$\lambda = \frac{1}{N\sigma} = \frac{A}{\rho}\frac{1}{6.02 \times 10^{23}\,\sigma}$$

$$= \frac{A}{\rho}\frac{1.67}{\sigma\,(\text{barns})} \qquad (25\text{-}11)$$

---

[1] By the same token, nuclear cross sections tend to be of the order of 10^{-24} cm^2 (barns), particle physics cross sections are of the order of 10^{-27} cm^2 (millibarns), going down to microbarns for rarer reactions, and even down to 10^{-44} cm^2 for the extremely rare neutrino reactions at low energies.

To evaluate the cross section in (25-8) we need to work out the integral

$$\int d^3\mathbf{r}\; e^{i(\hbar\mathbf{k}-\mathbf{p}e)\cdot\mathbf{r}/\hbar}\; e^{-Zr/a_0} \tag{25-12}$$

If we use the integral evaluated in (24-87), which, with a slight change of notation reads

$$\int d^3\mathbf{r}\; e^{-i\Delta\cdot\mathbf{r}}\,\frac{e^{-\mu r}}{r} = \frac{4\pi}{\mu^2 + \Delta^2} \tag{25-13}$$

we can, by differentiating with respect to μ obtain

$$\int d^3 r\; e^{-i\Delta\cdot\mathbf{r}}\, e^{-\mu r} = \frac{8\pi\mu}{(\mu^2 + \Delta^2)^2} \tag{25-14}$$

so that we can finally calculate the cross section. After some judicious combining of factors, we end up with

$$\frac{d\sigma}{d\Omega} = 32 Z^5 a_0{}^2 \left(\frac{p_e c}{\hbar\omega}\right)\left(\frac{\boldsymbol{\varepsilon}\cdot\mathbf{p}_e}{mc}\right)^2 \frac{1}{(Z^2 + a_0{}^2\Delta^2)^4} \tag{25-15}$$

where

$$\Delta = \frac{(\hbar\mathbf{k} - \mathbf{p}_e)}{\hbar} = \frac{(\mathbf{p}_\gamma - \mathbf{p}_e)}{\hbar}$$

Since the electron and photon energies are related by

$$\hbar\omega = E_B + \frac{p_e{}^2}{2m} \tag{25-16}$$

we see that for energies quite a bit above the binding energies, $\hbar\omega \cong p_e{}^2/2m$. Hence

$$\frac{p_e c}{\hbar\omega}\left(\frac{\boldsymbol{\varepsilon}\cdot\mathbf{p}_e}{mc}\right)^2 \cong \frac{2 p_e}{mc}\,(\boldsymbol{\varepsilon}\cdot\hat{\mathbf{p}}_e)^2$$

$$\Delta^2 = \frac{1}{\hbar^2}\,(\mathbf{p}_\gamma - \mathbf{p}_e)^2 = \frac{1}{\hbar^2}\left[\left(\frac{\hbar\omega}{c}\right)^2 - 2\,\frac{\hbar\omega}{c}\,p_e\hat{\mathbf{p}}_\gamma\cdot\hat{\mathbf{p}}_e + p_e{}^2\right]$$

$$\cong \frac{1}{\hbar^2}\left[p_e{}^2 - \left(\frac{p_e{}^3}{mc}\right)\hat{\mathbf{p}}_\gamma\cdot\hat{\mathbf{p}}_e\right]$$

$$\cong \frac{p_e{}^2}{\hbar^2}\left(1 - \frac{v_e}{c}\,\hat{\mathbf{p}}_\gamma\cdot\hat{\mathbf{p}}_e\right) \tag{25-17}$$

for nonrelativistic electrons, $p_e \ll mc$.[2] We have used the notation $\mathbf{p}_\gamma = \hbar\mathbf{k}$ for

---

[2] For relativistic electrons one should really use the Dirac equation to describe the process. Effects other than the photoelectric effect are more important when $E_e \simeq 1$ MeV.

the photon momentum, and as usual the $\hat{}$ denotes the unit vector. Thus

$$
\frac{d\sigma}{d\Omega} = 64Z^5 a_0{}^2 \left(\frac{p_e}{mc}\right) \frac{(\boldsymbol{\varepsilon} \cdot \hat{\mathbf{p}}_e)^2}{\left[Z^2 + \dfrac{p_e{}^2}{\alpha^2 m^2 c^2}\left(1 - \dfrac{v_e}{c}\,\hat{\mathbf{p}}_e \cdot \hat{\mathbf{p}}_\gamma\right)\right]^4}
$$

$$
= \frac{64Z^5 \alpha^8 a_0{}^2 \left(\dfrac{p_e}{mc}\right)(\boldsymbol{\varepsilon} \cdot \hat{\mathbf{p}}_e)^2}{\left[(\alpha Z)^2 + \dfrac{p_e{}^2}{m^2 c^2}\left(1 - \dfrac{v_e}{c}\,\hat{\mathbf{p}}_e \cdot \hat{\mathbf{p}}_\gamma\right)\right]^4}
\tag{25-18}
$$

If we choose the photon direction to define the z-axis, and the two photon polarization directions $\boldsymbol{\varepsilon}^{(1)}$, $\boldsymbol{\varepsilon}^{(2)}$ to point in the x- and y-directions, respectively, then, writing

$$
\hat{\mathbf{p}}_e = (\sin\theta\cos\phi,\ \sin\theta\sin\phi,\ \cos\theta)
\tag{25-19}
$$

we have $(\mathbf{p}_e \cdot \boldsymbol{\varepsilon}^{(1)})^2 = \sin^2\theta\cos^2\phi$ and $(\mathbf{p}_e \cdot \boldsymbol{\varepsilon}^{(2)})^2 = \sin^2\theta\sin^2\phi$ so that the *average* of the numerator over the two polarization directions (we are calculating the photoeffect cross section with unpolarized photons) is

$$
\overline{(\hat{\mathbf{p}}_e \cdot \boldsymbol{\varepsilon})^2} = \tfrac{1}{2}(\sin^2\theta\sin^2\phi + \sin^2\theta\cos^2\phi) = \tfrac{1}{2}\sin^2\theta
\tag{25-20}
$$

also

$$
\hat{\mathbf{p}}_e \cdot \hat{\mathbf{p}}_\gamma = \cos\theta
\tag{25-21}
$$

so that, writing $p_e{}^2/2m = E$, we get

$$
\frac{d\sigma}{d\Omega} = \frac{32\sqrt{2}\ Z^5 \alpha^8 a_0{}^2 (E/mc^2)^{1/2} \sin^2\theta}{\left[(\alpha Z)^2 + \dfrac{2E}{mc^2}\left(1 - \dfrac{v_e}{c}\cos\theta\right)\right]^4}
\tag{25-22}
$$

For light elements, the condition that we imposed earlier, $\hbar\omega \gg E_B$, which is equivalent to

$$
E \gg \tfrac{1}{2}\,mc^2(Z\alpha)^2
\tag{25-23}
$$

is satisfied over a reasonably wide range of energies. If we insert (25-23) into the cross section, we find that the denominator simplifies, and we get

$$
\frac{d\sigma}{d\Omega} = 2\sqrt{2}\ Z^5 \alpha^8 a_0{}^2 \left(\frac{E}{mc^2}\right)^{-7/2} \frac{\sin^2\theta}{\left(1 - \dfrac{v_e}{c}\cos\theta\right)^4}
\tag{25-24}
$$

Let us discuss various aspects of this formula.

(1) First, the vague guess that since atomic sizes tend to be of the order of 10^{-8} cm, the cross sections should be of order 10^{-16} cm^2 is wrong! It is true that the factor $a_0{}^2$ is of that magnitude, but it is multiplied by $(1/137)^8$, which is

dimensionless, but hardly negligible! We should try to understand how one could be so wrong in order to have some guidance what one must be careful about in making estimates. If we ignore the last angular factor, which we will discuss later, we see that we may, with the help of

$$E = \tfrac{1}{2} m v_e^2$$

write the factor in front as

$$2\sqrt{2}\, a_0^2 Z^5 \alpha^8 \left(\frac{mc^2}{E}\right)^{7/2} = 32 a_0^2 Z^5 \alpha^8 \left(\frac{c}{v_e}\right)^7$$

$$= 32 \left(\frac{a_0}{Z}\right)^2 \alpha \left(\frac{\alpha Z c}{v_e}\right)^7 \qquad (25\text{-}25)$$

This is a more useful form. It shows, first of all, the presence of a single factor α, which should always be present when a single photon is emitted or absorbed. The coupling of the vector potential to a charge is proportional to the charge e, and the square of this will lead to the α. The factor $(a_0/Z)^2$ is a better measure of the area of the atom than a_0^2, since we are considering a hydrogenlike atom of charge Z. What remains is a rather high power of the ratio of the "orbital" velocity of the electron in the atom to the velocity of the outgoing free electron.

It is the ratio $(\alpha Z c/v_e)$ [rather than just (c/v_e), which is also dimensionless] that appears, because the matrix element involves the overlap between the free electron wave function and the bound electron wave function, that is, the square matrix element is related to the probability that a measurement of the momentum of the bound electron yields p_e. The functional dependence $f(\alpha Z c/v_e)$, in this case the eighth power,[3] cannot be guessed at on general qualitative grounds. For example, if the electron wave function were Gaussian [$\psi_i(\mathbf{r}) \propto e^{-r^2/a^2}$], the falloff with increasing velocity would be much faster than the eighth power. The reason why a guess is hard to make is that the momentum distribution of the electron is localized in a region of spread

$$\Delta p \sim \frac{\hbar}{a_0/Z} \sim \frac{\hbar Z}{\hbar/mc\alpha} \sim Z \alpha mc \qquad (25\text{-}26)$$

and for $p_e \gg Z\alpha mc$ one is far out in the tail of the momentum distribution. This, again by the uncertainty relation, depends on the small r-distribution of the wave function, and depends sensitively on the state, in particular on the angular momentum. This does make photo-disintegration in nuclear physics a very useful tool.

[3] There is a factor p_e in the phase space so that the matrix element squared gives an eighth power of $(\alpha c Z/v_e)$.

(2) The angular distribution of $d\sigma/d\Omega$ is given by

$$F(\theta) = \frac{\sin^2 \theta}{[1 - (v_e/c) \cos \theta]^4} \qquad (25\text{-}27)$$

We note, first of all, that the cross section vanishes in the forward direction. This is a consequence of the fact that photons are transversely polarized. The matrix element is proportional to $\mathbf{p}_e \cdot \boldsymbol{\varepsilon}$, and when $\mathbf{p}_e$ is parallel to the photon momentum, this factor vanishes. The factor in the denominator has, because of the fourth power, a strong influence on the angular distribution. When v_e/c approaches unity, this becomes very dramatic, but even for moderate v_e/c there is significant peaking in the near forward direction, where the denominator is at its smallest. This corresponds to the minimum value of the momentum transfer between photon and electron $(\mathbf{p}_\gamma - \mathbf{p}_e)^2$.

More detailed calculations need to be done to cover the relativistic region. The formula derived above works well in the region of its validity.[4] At very low energies, a more accurate wave function for the outgoing electron must be used. Such a wave function will reflect the Coulomb interaction between the nucleus and the electron. There will, of course, be no photoeffect below the threshold for ionizing the least-bound electron from an outer shell. As the energy increases above the threshold, electrons from deeper shells will be photoproduced. If one plots the integrated cross section, or preferably the *mass absorption coefficient*[5] $N\sigma/\rho$, as a function of the photon wave length, one finds the data shown on Fig. 25-1. The so-called K-edge corresponds to the ejection of the $n = 1$ electrons; the L-edges correspond to the various electrons in the $n = 2$ states. The edges occur at the binding energies of the various electrons. Moseley's empirical law states that they are located at

$$E = 13.6 \frac{(Z - \sigma_n)^2}{n^2} \text{ eV} \qquad (25\text{-}28)$$

where σ_n, the "screening constants," are approximately given by $\sigma_n = 2n + 1$. This formula is just what we expect for the ns orbitals, and the screening is the effect of all the other s electrons.

At relativistic energies the cross section drops less precipitously, with an $(E/m)^{-1}$ behavior instead of $(E/m)^{-7/2}$ but by the time energies of 0.5 MeV are reached, the photoelectric effect ceases to be of any importance as far as the absorption of radiation is concerned. In the energy region of 0.5–5 MeV, say, it is the *Compton Effect* that is the dominant absorptive effect.

Here free electrons scatter photons. At low frequencies the effect can be

---

[4] In calculating absorption of radiation, the result that we derived must be multiplied by 2, since there are two electrons in the ground state, except in hydrogen.

[5] This is equal to $N_0 \sigma/A$ where N_0 is Avogadro's number and A is the atomic weight.

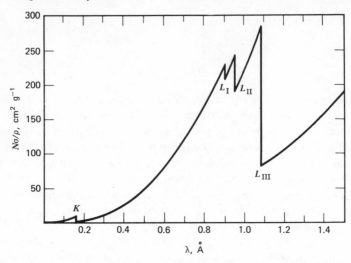

Fig. 25-1. Mass absorption coefficient $N\sigma/\rho$ for platinum as a function of photon wavelength.

understood classically; electromagnetic radiation impinging on the electron accelerates it, and the radiation emitted by the accelerated charge is the scattered radiation. The classically calculated *Thomson* cross section is

$$\sigma_T = \frac{8\pi}{3}\left(\frac{e^2}{mc^2}\right)^2 \tag{25-29}$$

In quantum mechanics, the scattering amplitude (matrix element) must be proportional to e^2, since two photons are involved. Since the perturbation in the Hamiltonian is

$$\frac{e}{mc}\,\mathbf{p}\cdot\mathbf{A}(\mathbf{r},t) + \frac{e^2}{2mc^2}\,\mathbf{A}^2(\mathbf{r},t) \tag{25-30}$$

when both terms in the expansion of (22-11) are kept, we see than an e^2 contribution to the scattering amplitude can come from two sources.

(i) the first source is a first-order contribution from the term $e^2\mathbf{A}^2(\mathbf{r},t)/2mc^2$.

(ii) the second source is a second-order perturbation term from the coupling $e\mathbf{p}\cdot\mathbf{A}(\mathbf{r},t)/mc$. Since we have not developed the second order perturbation formalism, we will restrict ourselves to stating the results.

(a) At threshold, with the gauge that we have been using, $\mathbf{\nabla}\cdot\mathbf{A}(\mathbf{r},t) = 0$, the whole amplitude comes from the term involving $e^2\mathbf{A}^2(\mathbf{r},t)/2mc^2$.

(b) The matrix element in second order has the form

$$-\sum_n \frac{\langle f|e\mathbf{p}\cdot\mathbf{A}/mc|n\rangle\langle n|e\mathbf{p}\cdot\mathbf{A}/mc|i\rangle}{E_n - E_i} \tag{25-31}$$

where the "sum" over intermediate states "n" also implies integration over all the momenta, when "n" includes continuum states.[6] It is not enough to include intermediate one-electron states corresponding to the sequence

$$\gamma_i + e_i \rightarrow e' \rightarrow \gamma_f + e_f$$

and the intermediate states containing an electron and two photons, corresponding to the process

$$e_i + \gamma_i \rightarrow \gamma_i + \gamma_f + e' \rightarrow \gamma_f + e_f$$

It turns out that it is necessary to include the possibility of the "virtual" creation of an electron-positron pair by the incident photon, followed by the annihilation of the positron by the incident electron, with the emission of the final photon, as in

$$e_i + \gamma_i \rightarrow e_i + e_f + e^{+\prime} \rightarrow \gamma_f + e_f$$

and the process

$$e_i + \gamma_i \rightarrow e_i + \gamma_i + \gamma_f + e_f + e^{+\prime} \rightarrow \gamma_f + e_f$$

The calculation leads to the *Klein-Nishina formula*

$$\sigma = 2\pi \left(\frac{e^2}{mc^2}\right)^2 \left\{ \frac{1+x}{x^2} \left[\frac{2(1+x)}{1+2x} - \frac{1}{x} \log\,(1+2x) \right] \right.$$
$$\left. + \frac{1}{2x} \log\,(1+2x) - \frac{1+3x}{(1+2x)^2} \right\}$$
$$x = \frac{\hbar\omega}{mc^2} \tag{25-32}$$

which is in excellent agreement with experiment. At low frequencies this becomes

$$\sigma = \frac{8\pi}{3} \left(\frac{e^2}{mc^2}\right)^2 (1 - 2x) \tag{25-33}$$

and at high frequencies ($x \gg 1$) this reads

$$\sigma = \pi \left(\frac{e^2}{mc^2}\right)^2 \frac{1}{x} (\log 2x + \tfrac{1}{2}) \tag{25-34}$$

Thus the Compton cross section, too, drops off at high energies. At energies above a few MeV, the dominant absorptive process is *pair production*.

It is a remarkable fact that a photon at high enough energies, $\hbar\omega > 2mc^2$ can "materialize" into an electron and a positron (Fig. 25-2). The latter can be

---

[6] The fact that (25-31) looks like an off-diagonal version of the second-order energy shift is, of course, no accident.

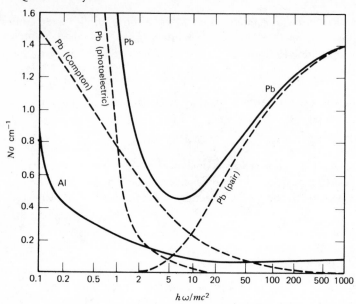

Fig. 25-2. Total absorption coefficient for lead and aluminum as a function of energy, in units of the electron rest energy (0.51 MeV). The photoelectric cross section for Al is negligible on the scale depicted here.

properly called an "antielectron"; it has the same mass as the electron and the same spin, but its charge and magnetic moment have the same value with *opposite sign* as those for the electron, and the nonrelativistic coupling with the electromagnetic field is obtained by the replacement of $\mathbf{p}$ by $\mathbf{p} - e\mathbf{A}(\mathbf{r},t)/c$. Such a materialization can only occur in the presence of a third particle, a nucleus, for example, since energy and momentum conservation cannot hold for the process

$$\gamma \rightarrow e + e^+$$

To see this without going through a long kinematical calculation, consider the inverse process $e + e^+ \rightarrow \gamma$ in the center of mass frame. The electron and positron have equal and opposite momenta, so that the final state has energy $2(m^2c^4 + p^2c^2)^{1/2}$ and momentum 0. A photon of energy E must carry momentum E/c. If there is a nucleus present, it can absorb momentum and energy (for a massive nucleus this will be very small, $p^2/2M$), so that it becomes possible to balance energy and momentum.

The calculation of

$$\gamma + \text{nucleus} \rightarrow e + e^+ + \text{nucleus}$$

is beyond the scope of this book. The theory of quantum electrodynamics that is used in these calculations also shows that we can transfer particles from one side of the equation to the other, provided we change the transferred particles to their antiparticles. Thus one predicts that

$$\text{nucleus} + e^{\pm} \rightarrow \text{nucleus} + e^{\pm} + \gamma$$

should also occur, with a matrix element very closely related to that of pair production. This is in agreement with experiment, and the last process is responsible for *cosmic ray showers.*

An incident γ ray of very high energy (it may come from the decay $\pi^{\circ} \rightarrow 2\gamma$, with the π° produced when a primary cosmic ray proton hits a nucleus at the top of the atmosphere) will make a pair, with each member carrying roughly half the original energy. Each member can produce a photon, as indicated above,[7] and the end products can make further photons and pairs. Showers coming from extremely high energy events occurring at the top of the atmosphere can cover areas of several square miles! Less spectacular showers in counters are used to identify photons or electrons. An incident particle that is charged, but much heavier, will be deflected less, and will therefore radiate less.

Detailed calculations show that energy lost in material through these processes follows the law

$$E(x) = E_{\text{inc}}\, e^{-x/L} \tag{25-35}$$

where the "radiation length" is given by

$$L = \frac{(m^2c^2/\hbar^2)\, A}{4Z^2\alpha^3 N_0 \rho \, \log\,(183/Z^{1/3})} \tag{25-36}$$

where $N_0 = 6.02 \times 10^{23}$ is Avogadro's number, m is the electron mass, A is the atomic weight, Z is the charge of the nucleus, and ρ is the density of the material in grams per cubic centimeter. The "pair production length" is given by

$$L_{\text{pair}} = \frac{9}{7}\, L \tag{25-37}$$

The formula is not good for very low Z. Typical values of L are

Air	330 m
Al	9.7 cm
Pb	0.53 cm

Bremsstrahlung is the dominant energy loss mechanism for electrons at high energies. At low energies ionization dominates. Lack of space keeps us from discussing this essentially classical effect.

---

[7] This process is called *Bremsstrahlung*, and can be understood classically; a charge deflected in the Coulomb field of the nucleus is accelerated, and hence radiates.

Problems

1. Calculate the cross section for the process

$$\gamma + \text{deuteron} \rightarrow N + P$$

The procedure is the same as that for the photoelectric effect. In the calculation of the matrix element, the final state wave function is again

$$\psi_f(\mathbf{r}) = \frac{1}{\sqrt{V}} e^{i\mathbf{p}\cdot\mathbf{r}/\hbar}$$

where $\mathbf{p}$ is the proton momentum. At low energies the wavelength of the radiation is much larger than the "size" of the deuteron, so that $e^{i\mathbf{k}\cdot\mathbf{r}} \approx 1$. To calculate

$$\int d^3r \, e^{-i\mathbf{p}\cdot\mathbf{r}/\hbar} \, \psi_i(\mathbf{r})$$

observe that

$$\nabla^2\psi_i + \frac{M_P E_B}{\hbar^2}\psi_i - \frac{M_P}{\hbar^2}V(r)\psi_i = 0$$

where the reduced mass $M_P/2$ was used. Now use integration by parts to show that

$$0 = \int d^3r \, e^{-i\mathbf{p}\cdot\mathbf{r}/\hbar}\left[\nabla^2\psi_i + \frac{M_P E_B}{\hbar^2}\psi_i - \frac{M_P V(r)}{\hbar^2}\psi_i\right]$$

$$= \int d^3r \, e^{-i\mathbf{p}\cdot\mathbf{r}/\hbar}\left(-\frac{p^2}{\hbar^2} + \frac{M_P E_B}{\hbar^2}\right)\psi_i - \frac{M_P}{\hbar^2}\int d^3r \, e^{-i\mathbf{p}\cdot\mathbf{r}/\hbar} V(r)\psi_i$$

leading to

$$(E - E_B)\int d^3r \, e^{-i\mathbf{p}\cdot\mathbf{r}/\hbar}\psi_i = -\int d^3r \, e^{-i\mathbf{p}\cdot\mathbf{r}/\hbar} V(r)\psi_i$$

Since the integral is only over the range of the potential, which is very short, we can replace $e^{-i\mathbf{p}\cdot\mathbf{r}/\hbar}$ on the right side by 1. Another use of the Schrödinger equation leads to

$$-\int d^3r \, e^{-i\mathbf{p}\cdot\mathbf{r}/\hbar} V(r)\psi_i = E_B\int d^3r \, \psi_i(\mathbf{r})$$

with E_B the deuteron binding energy $= -2.23$ MeV. For the calculation of this integral take

$$\psi_i(\mathbf{r}) = \frac{N}{\sqrt{4\pi}} e^{-\alpha(r-r_0)} \qquad r > r_0$$

$$= 0 \qquad r < r_0$$

properly normalized. For what energies would you expect the photon wavelength to be much larger than the range of the potential $r_0 \cong 1.2$ fm? (The justification for ignoring the integral inside the potential range ($r < r_0$) lies in that $r_0 \ll 1/\alpha$ where $\alpha^2 = M_P E_B/\hbar^2$).

2. The principle of detailed balance relates the matrix elements for the reactions

$$A + a \rightarrow B + b \qquad \text{I}$$

and

$$B + b \rightarrow A + a \qquad \text{II}$$

thus

$$\sum |M_{\text{I}}|^2 = \sum |M_{\text{II}}|^2$$

where the sum is over both initial and final spin states. Taking into account that in the calculation of a rate or cross section one averages over the initial spin states and sums over the final spin states, show that for the rates

$$\frac{(2J_A + 1)(2J_a + 1)}{p_b{}^2(dp_b/dE_b)} \frac{dR_{\text{I}}}{d\Omega_b} = \frac{(2J_B + 1)(2J_b + 1)}{p_a{}^2(dp_a/dE_a)} \frac{dR_{\text{II}}}{d\Omega_a}$$

where J_a, J_A, J_b, J_B are the spins of the particles, p_b and p_a are the center of mass momenta of particles b and a (I and II must take place at the same total energy), E_b and E_a are the corresponding energies of the particles, and $d\Omega_b$, $d\Omega_a$ are the solid angles in which b and a are observed. Use this result to express the cross section for the radiative capture process

$$N + P \rightarrow D + \gamma$$

in terms of the cross section calculated in problem 1. Note that the factor $(2J + 1)$ for photons is 2 since there are only two polarization states, and also the spin of the deuteron is 1.

3. The cross section for the reaction

$$\pi^+ + D \rightarrow P + P$$

has been measured for incident π^+ laboratory kinetic energy of 24 MeV, and found to be equal to 3.0×10^{-27} cm^2.

(a) At what laboratory energy would one be able to carry out a test of detailed balance by measuring the cross section for

$$P + P \rightarrow \pi^+ + D$$

(The pion mass is $m_\pi c^2 = 140$ MeV; $M_p c^2 = 940$ MeV; $M_D \cong 2M_p$).

(b) Given that the spin of the π^+ is 0, what is the predicted cross section for this reaction?

4. What is the radiation length in liquid Xenon, for which $Z = 54$, $A = 131$, and $\rho = 3.09$ gm cm^{-3}?

5. Suppose the electron were bound to the nucleus by a square well potential. Calculate the energy dependence of the cross section for the photoelectric

effect. Assume that the photon energy is much larger than the binding energy of the electron, and that the potential has a short range. (*Hint.* See problem 1.)

References

The mechanisms responsible for the absorption of radiation in matter are discussed in most of the modern physics textbooks listed at the end of Chapter 1. For a very complete discussion of the experimental techniques used in the measurement of the various effects, see also

E. Segre, *Nuclei and Particles*, W. A. Benjamin, Inc., New York, 1964.

chapter 26

Elementary Particles and Their Symmetries

In this chapter we discuss a number of topics related to the fundamental interactions of elementary particles. Although this chapter is necessarily more qualitative than the others, because whatever theory exists involves concepts too advanced to be discussed in a quantitative way, we will see that quantum concepts are essential in the analysis of a variety of complex phenomena.

A. Electrons and Positrons

In Chapter 25 we mentioned that in the presence of a second body (to conserve momentum) a photon can materialize into an electron and a positron. The positron can very appropriately be called the *antiparticle* corresponding to the electron; it is identical to it, except for the sign of the electric charge and other electromagnetic properties (e.g., opposite magnetic dipole moment). When an electron-positron pair is produced in a bubble chamber, say, a very characteristic pattern caused by the fact that the two particles bend oppositely in a magnetic field is apparent (Fig. 26-1). Antiparticles are a necessary consequence of relativistic quantum mechanics, and the positron was predicted in 1928 by Dirac, two years before Anderson discovered it experimentally.

We live in a world built of protons, neutrons, and electrons. Positrons are rare, since they have to be produced with the expenditure of energies of at least $2m_e c^2$ (1 MeV) (corresponding to the production of a pair at rest), and thus they have been studied only under special experimental circumstances. Nevertheless, all experience and the fact that there exists a remarkably successful theory of *quantum electrodynamics* symmetric under $e^+ \leftrightarrow e^-$ interchange indicate that electromagnetic interactions do not distinguish, except for the sign of the electric charge, between electrons and positrons. It is a bold extrapolation from this to

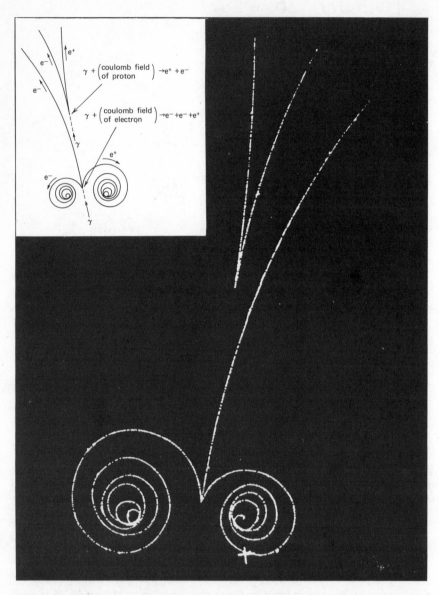

Fig. 26-1. Bubble chamber picture of the conversion of γ-rays into electron-positron pairs in the presence of matter (hydrogen). In one case the pair is produced in the coulomb field of an electron, in the other case it is produced in the coulomb field of a proton. The tracks curve in the magnetic field of the bubble chamber, with the faster particles bending less. (Courtesy of the Lawrence Berkeley Laboratory, University of California).

the general notion of the invariance of the laws of physics under particle-antiparticle conjugation, in which not only electrons are replaced by positrons (and vice versa) but protons by antiprotons, neutrons by antineutrons, and so forth. With this generalization, we can conceive of antimatter, consisting of anti-nuclei, with positrons bound to them by attractive Coulomb forces. A consequence of the invariance principle is that all the physical observables are the same, so that an observer could not tell whether he and his environment were made of matter or antimatter. Whether the laws of physics obey this conjectured invariance law must be settled by experiment. It now appears that the laws of the strong (nuclear) and electromagnetic interactions do, but that the laws of the weak interactions do not. The *existence* of antiparticles follows from quite general laws of relativistic quantum mechanics.

Let us concentrate on positrons for the time being and ask what happens when they are produced in the vicinity of electrons. Sooner or later they will collide with an electron, and in the inverse of the production process, they will annihilate each other, with the liberation of the total energy consisting of the kinetic energies and the sum of the rest masses $2m_ec^2$. This energy will be liberated in the form of radiation. To conserve energy and momentum, a third body (e.g., a nucleus) must be present, or at least two photons must be produced, as was pointed out earlier. A fascinating possibility is that the positron loses energy by ionization, that is, long-range collisions with electrons, so that slowing down rather than annihilation in flight occurs. The positron will generally be captured into an orbit about an electron. The positron and electron attract each other and form an atom that we call *positronium*. The atom is described, in first approximation, by the same equation as was used for the hydrogen atom, except that the reduced mass is $\mu = m_e/2$, so that the ionization energy is 6.8 eV, the "radius" is 1 Å instead of 0.5 Å and so on. Positronium does exist and all of its properties are in accord with expectations.[1] The ground state of positronium is an $l = 0$ state, and of the two possible states, $^1S_0$ and $^3S_1$, the former lies lower. This is what the hyperfine splitting (Chapter 17) would indicate, but the situation here is complicated by the fact that the possibility of annihilation really makes this problem different from that of the hydrogen atom, and the hyperfine interaction is partly cancelled by purely relativistic effects.

Once positronium is in an S-state, the wave functions of the two particles overlap significantly, and annihilation becomes likely. The $^1S_0$ state can decay into two photons, and the transition rate can be estimated as follows. For each photon emitted, the matrix element will have an e in it, so that the square of the matrix element will involve e^4, or, equivalently, α^2. The annihilation rate must also be proportional to the probability that the two particles overlap. A reason-

[1] This includes such esoteric effects as the spin-orbit coupling, hyperfine structure, and an effective imaginary part in its potential due to the possibility of annihilation into two photons.

able measure of this is $|\psi(0)|^2$, where $\psi(0)$ is the hydrogenlike wave function of the ground state, evaluated at zero separation between the particles. The value of this is $(1/\pi)(1/a_0)^3$, where $a_0 = \hbar/\mu c\alpha$. To make the dimensions come out right, we must multiply by a (length)^3 (time)$^{-1}$ made up of $\hbar$, c, and μ, but without any factors of e. The estimate is thus

$$
\begin{aligned}
R &= \eta\, \frac{1}{\pi}\, \alpha^2 \left(\frac{\mu c\alpha}{\hbar}\right)^3 \left(\frac{\hbar}{\mu c}\right)^3 \frac{\mu c^2}{\hbar} \\[2mm]
&= \left(\frac{\eta}{2\pi}\right) \alpha^5 \left(\frac{mc^2}{\hbar}\right)
\end{aligned}
\tag{26-1}
$$

The constant η is not determined by this dimensional argument. The rate thus is

$$
R = 0.26\,\eta \times 10^{10}\ \text{sec}^{-1} \tag{26-2}
$$

Comparison with the experimental rate of $0.8 \times 10^{10}\ \text{sec}^{-1}$ shows that the constant η is approximately 3.0. A proper evaluation of this constant really requires relativistic quantum mechanics, since we require the matrix element for the annihilation of a particle and an antiparticle, and this concept does not enter into the nonrelativistic Schrödinger equation.

Positronium in the $^3S_1$ state can also annihilate, but must do so with the emission of three photons. To understand this, we must consider the properties of positronium under charge conjugation, that is, under the interchange $e^+ \leftrightarrow e^-$. We observe (cf. Fig. 26-2) that charge conjugation can be accomplished by (1) an inversion, accompanied by (2) a spin exchange. A singlet state, with $S = 0$, is odd, and a triplet state with $S = 1$ is even under the latter exchange, so that the effect of spin exchange is $(-1)^{S+1}$. The effect of space reflection is usually $(-1)^l$, that is, even angular momentum states are even under reflection, and so on, but there is an additional factor of (-1) that arises from the fact that antiparticles of spin 1/2 (also 3/2, 5/2, ...) have parity opposite to that of particles.[2] This implies that we can write for the effect of charge conjugation

$$
C = (-1)^{S+l} \tag{26-3}
$$

Thus the ground state $^1S_0$ is even under charge conjugation, and the state just above it, $^3S_1$, is odd. Similarly, $^1P_1$ is odd under charge conjugation, while $^3P_{2,1,0}$ is even, and so forth. Now the electromagnetic field is *odd* under conjugation. For example, the equation

$$
\nabla \cdot \mathbf{E} = 4\pi\rho \tag{26-4}
$$

---

[2] This follows from quite general properties of relativistic quantum mechanics, and has experimental consequences aside from the one mentioned above. For example, the polarization vectors of the two photons produced in positronium annihilation are predicted to be preferentially perpendicular to each other. This is borne out by experiment.

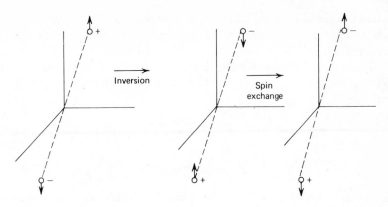

Fig. 26-2. Equivalence of charge conjugation and (inversion) × (spin exchange) for electron-positron system.

can only be invariant under charge conjugation if **E** changes sign when the charge density does. In effect, under charge conjugation,[3] $\varepsilon \to -\varepsilon$. This leads to a series of selection rules, which state that positronium in a state with a given S and l can only decay into an even number of photons if $S + l$ is even, and into an odd number of photons if $S + l$ is odd. Thus the $^3S_1$ state can only decay into an odd number of photons. Since each additional photon reduces the rate of decay by a factor of α, at least, the smallest allowed number of photons is favored.

B. Baryons, Antibaryons, and Mesons

In the last section we extrapolated the notion of charge conjugation to particles other than electrons and on that basis made a conjecture about the existence of antiprotons and antineutrons. These particles have actually been found to exist, although it is difficult to make them, since it takes a center of mass energy of at least $2M_p c^2 \cong 1880$ MeV to make a pair at rest. The antiproton has been found to have charge -1, mass equal to that of the proton, magnetic moment equal and opposite to that of the proton, and other properties expected from theory. Since the neutron is neutral, one might ask what distinguishes the antineutron from the neutron. The answer is that there appears to exist a

---

[3] Under charge conjugation $\mathbf{p} - (e/c)\,\mathbf{A} \to \mathbf{p} + (e/c)\,\mathbf{A}$. We do not change the sign of the *number* e/c but instead change the sign of $\mathbf{A}$, or, equivalently, $\mathbf{E}$ and $\mathbf{B}$, when we carry out charge conjugation. This means that the polarization vector ε is transformed into its negative. Since transition rates involve the squares of quantities involving ε, we see that these observable quantities are invariant under the transformation.

quantum number like charge, called *Baryon Number*, N_B, which is conserved, and which has value $+1$ (by definition) for nucleons (neutrons and protons) and -1 for antinucleons.

The notion of a conserved baryon number arises from the empirical observation that the reaction in which e^- and P annihilate does not seem to occur. If the reaction

$$e^- + P \rightarrow \text{"stuff"}$$

or, equivalently,

$$P \rightarrow e^+ + \text{"stuff"}$$

could occur, then matter would not be stable. We can trivially set a limit on the lifetime of the proton; the existence of very old rocks shows that it is longer than 4.5×10^9 years. Actually one can do better. Either of the above reactions occurring in a scintillator would give rise to radiation that could be detected. By properly shielding a large scintillator to eliminate pulses caused by external sources such as cosmic rays, and looking for "spontaneous" pulses, one can set a limit on how often the above reactions take place in, say 10^{25} atoms. The absence of such spontaneous pulses over a certain period of observation has been translated into a lower limit on the proton lifetime equal to 2×10^{28} years! It is fair to say that the proton is stable.

Actually the reaction

$$e^- + P \rightarrow N + \nu$$

does occur in nuclei, so that it is not quite right to say that the number of protons is conserved. The correct statement here is that the number of protons plus the number of neutrons is conserved, just like charge. This number is what we call the Baryon Number. We postulate that under charge conjugation this quantum number, just like charge, changes sign. Thus the antiproton and the antineutron have $N_B = -1$. Electrons and positrons have $N_B = 0$, and this is the formal explanation for the absence of the reactions $e^- + P \rightarrow \text{"stuff,"}$ and so on. Baryons and antibaryons *can* annihilate into "stuff" that must have $N_B = 0$, but may have charge 1, 0, -1 in the processes $P + \overline{N}$, $P + \overline{P}$ (or $N + \overline{N}$), and $\overline{P} + N$. Whatever the annihilation products are, they must have a total energy of at least 1880 MeV and this is a very distinctive feature of nucleon-antinucleon annihilation.

Although it is consistent with the symmetry laws to expect

$$P + \overline{P} \rightarrow (e^+e^- \text{ pairs}) + \text{photons}$$

to occur, it turns out that most of the time the annihilation products are π-*mesons* also called *pions*. These are particles first predicted by Yukawa in 1935 to explain the short-range nuclear forces in a manner analogous to the long-range electromagnetic forces. The pions were viewed as counterparts of photons.[4] Pions are not part of our everyday experience, since they have short lifetimes, 10^{-8} sec for the $\pi^{\pm}$ and 10^{-16} sec for the π°. They all have rest mass of the order $m_\pi c^2 \cong$ 140 MeV and $N_B = 0$. The π^- is the antiparticle of the π^+; their masses and decay patterns are the same. Pions are responsible for at least part of the nuclear forces, and as such must be rather strongly coupled to nucleons. Thus the analog of $e^2/\hbar c$ ($\cong 1/137$) is a number $g^2/\hbar c \cong 15$. Hence the rule that the smallest number of photons in a reaction is most probable does not hold for pions. In the annihilation

$$P + \bar{P} \rightarrow \text{pions}$$

the number can vary quite a lot, consistent with the constraint provided by energy conservation. The average number appears to be around 5, when the nucleon and antinucleon are at rest. Pions have been found to have spin 0 and negative parity, and thus are called *pseudoscalar* particles. Since they are spinless, they cannot have any electrical moments, for example, magnetic dipole moment. Thus the only difference between a π^+ and a π^- is its charge, and the π° is its own antiparticle, like the photon.

C. Isotopic Spin Conservation

The neutron and the proton are really very much alike. They differ in (a) their charge, (b) their magnetic moments, and (c) their masses, the last to about one part in a thousand. The nuclear forces, on the other hand, do not appear to distinguish between neutrons and protons. Thus the binding energies of mirror nuclei, that is, nuclei that transform into each other when neutrons are replaced by protons and vice versa, are almost equal, with the discrepancy explainable in terms of the difference in the Coulomb energies. If the neutron-proton mass difference also were an electromagnetic effect (and its si   consistent with this possibility), then one could blame all the differences between neutrons and protons on electromagnetism.

Heisenberg and Condon made the bold proposal that the nature of the nucleons must be such that if it were possible to "turn off" their coupling to the electromagnetic field, that is, the electric charges, then there would be no way of distinguishing between proton and neutron, and that these two particles should really be viewed as two substates of a single entity, the nucleon. This notion grew out of the realization that an electron with spin "up" and an electron with spin "down" in a magnetic field are still the same electron, even though the energy is different. In this case the magnetic field can be turned off, and the "symmetry breaking" be made to disappear. For nucleons, the symmetry-

---

[4] See Special Topics section 5 on "The Yukawa Theory."

breaking electromagnetic interaction cannot really be turned off experimentally, but this is no barrier to imagining a world without electromagnetism. The proposal was that the analogy with spin should be quite exact, that is, that *the proton and the neutron should be "up" and "down" states of an "isotopic" spin* $1/2$ *entity called the nucleon.* The nucleon is an $I = 1/2$ state and the proton and neutron are eigenstates of I_z, with eigenvalue $+1/2$ and $-1/2$, respectively. The antiproton and the antineutron also form a doublet, but now it is the antiproton that has $I_z = -1/2$, and the antineutron $I_z = +1/2$. Nuclear forces are now "nucleon-nucleon forces," and the equality of *P-P, N-P,* and *N-N* forces can readily be understood if one assumes that the nucleon-nucleon potential conserves the total angular momentum in isotopic spin space, that is, that *isotopic spin is conserved.*

Two nucleons, each having $I = 1/2$ can form an $I = 1$ triplet and an $I = 0$ singlet. The states, in complete analogy to the spin triplet and singlet, have the form

$$
\text{triplet} \begin{cases} PP \\ \dfrac{1}{\sqrt{2}} (PN + NP) \\ NN \end{cases} \qquad \text{singlet} \; \dfrac{1}{\sqrt{2}} (PN - NP) \qquad (26\text{-}5)
$$

The notation P and N here is the isotopic spin analog of spin "up" and spin "down" spinors χ_+, χ_-.

By the Pauli Exclusion Principle, the P-P states are limited to totally antisymmetric states $^1S_0$, $^3P_{2,1,0}$, $^1D_2$, ... *I*-spin conservation demands that the *N*-*P* system in the $I = 1$ state also obeys the same symmetry. Thus the *P*-*P* and *N*-*P* forces in these states will be equal (and equal to the *N*-*N* force), but the forces in the $^3S_1$, $^1P_1$, $^3D_{3,2,1}$, ... states of the *N*-*P* system can be different, since these correspond to $I = 0$. That the forces are different is shown by the fact a deuteron without a corresponding *P*-*P* and *N*-*N* bound state exists. This situation is analogous to the existence of spin-dependent potentials that give a different force for the triplet and singlet spin states, even though the total angular momentum is conserved. If we introduce operators $\mathbf{I}$ $(= I_x, I_y, I_z)$ obeying the "angular momentum" commutation relations

$$
[I_x, I_y] = iI_z \qquad \text{(cycl.)} \qquad (26\text{-}6)
$$

we can construct the whole isotopic angular momentum formalism in exact analogy with the ordinary angular momentum formalism, except that there is no analog of orbital angular momentum connected with motion in space. *I*-spin conservation implies that in the absence of electromagnetic interactions, the Hamiltonian has the property that

$$
[H, \mathbf{I}] = 0 \qquad (26\text{-}7)
$$

Furthermore, eigenstates of $\mathbf{I}^2$ and I_z, with eigenvalues $I(I + 1)$ and $-I \leq I_z \leq I$ can be used as basis states. The existence of three pions, π^+, π°, and π^-, all with spin 0 and parity -1 and all of almost the same mass, fits in very well with the existence of the symmetry; we can say that *pions form an $I = 1$ triplet*, with the $I_z = 1, 0, -1$ states represented by π^+, π°, and π^-.

The mere existence of these i-spin multiplets may be viewed as evidence for the underlying symmetry. We can, however, point to several other more direct manifestations of the symmetry of the *strong interactions* that are seen in the nuclear forces.

(a) Nuclei consisting of Z protons and $A–Z$ neutrons will have

$$I_z = \frac{Z - (A - Z)}{2}$$

$$= Z - A/2 \qquad (26\text{-}8)$$

so that any one of its states belongs to a multiplet of total I at least as large as $|Z - A/2|$. One might hope to find evidence for other members of these multiplets in neighboring nuclei, and such evidence has indeed been found. The i-spin partners of a given set of levels are called analog states and have been the object of intensive study by nuclear physicists. Figure 26-3 shows a particularly clean example of multiplets. ^{18}O, ^{18}F, and ^{18}Ne have proton/neutron numbers (8,10), (9,9), and (10,8), respectively. For the first and third, the ground states could belong to the same i-spin triplet; the spectrum shows that ^{18}F has a ground state that has spin-parity 1^+, so that it cannot belong with the 0^+ ground states of ^{18}Ne and ^{18}O, but that there is an excited 0^+ state that could be the $I_z = 0$ member of the triplet. Furthermore, there is a remarkable correspondence between a whole sequence of energy levels in the three nuclei, indicating that they are all parts of an $I = 1$ multiplet. The figure shows that they are not degenerate in mass, but that they differ by several MeV. This is to be expected because the Coulomb repulsion between the protons does not respect i-spin symmetry. The energy differences can be accounted for quantitatively in this way.

(b) I-spin multiplets also appear in excited states of nucleons. If one examines pions and nucleons emerging from a high-energy collision of a pion or proton with a target proton, one can determine their momenta and energies. If a particular nucleon and pion were to be decay products of a single entity, then that entity would have to have $I = 3/2$ or $1/2$, since the addition of $I = 1/2$ and $I = 1$ can only lead to such states.[5] Furthermore, if the decaying state were at rest and of mass M, then the pion and nucleon would have equal and opposite momenta, and the sum of their energies would equal Mc^2. More generally, if $(E,\mathbf{p})$ denote the nucleon energy and momentum and $(\epsilon,\mathbf{q})$ denote

[5] It is, of course, assumed that i-spin is conserved in such a "decay" of the excited state. More about this later!

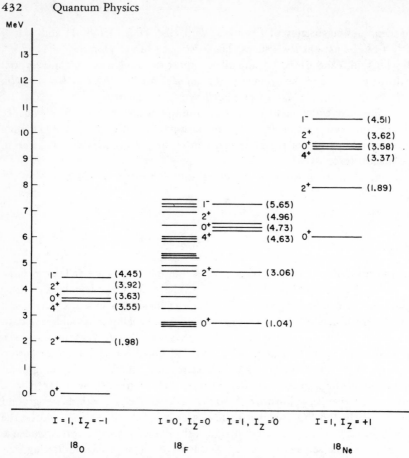

Fig. 26-3. The level schemes for nuclei with $A = 18$. The $I = 1$ levels for ^{18}O $(I_z = -1)$, ^{18}F $(I_z = 0)$, and ^{18}Ne $(I_z = 1)$ show a remarkable correspondence. The $I = 0$ levels for ^{18}F are also sketched in. (Data taken from F. Ajzenberg-Selove, *Nuclear Physics* A190, 1 (1972).)

the pion energy and momentum, then relativity tells us that the more general relation is

$$(E + \epsilon)^2 - (\mathbf{p} + \mathbf{q})^2 c^2 = M^2 c^4 \tag{26-9}$$

Thus by measuring energies and momenta, and studying such combinations of energies and momenta for pion-nucleon pairs, it is possible to look for such decaying states. Figure 26-4 shows some examples of such mass spectra. One of the most common states found is an $I = 3/2$ state with $Mc^2 = 1236$ MeV. One knows that it is $I = 3/2$, since the mass peak occurs in the $P\pi^+$ system, which has $I_z = 3/2$. It has, of course, been checked that the peak occurs in the other I_z states. The multiplet is usually denoted by $\Delta(1236)$. It is a particle, in the sense that a detailed study of the angular correlations between the pion and the nucleon indicate that the spin and parity are $3/2^+$. The mass distribution has a width of about 120 MeV. This is the "natural line width," and it indicates that the lifetime of the $I = 3/2$ state is

$$\tau \simeq \frac{\hbar}{\Delta M c^2} \simeq \frac{10^{-27}}{120 \times 1.6 \times 10^{-6}} \simeq 0.5 \times 10^{-23} \text{ sec} \qquad (26\text{-}10)$$

No wonder the Δ cannot be detected as a particle leaving a track in a bubble chamber!

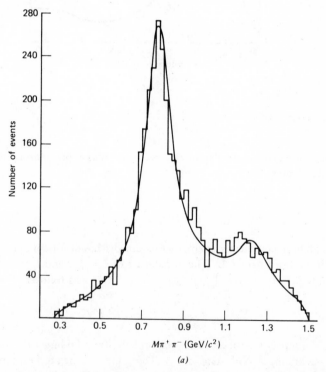

$M\pi^+\pi^-$ (GeV/c^2)

(a)

Fig. 26-4. Evidence for resonances. (a) Plot of $M_{\pi^+\pi^-} = (p_\mu{}^+ + p_\mu{}^-)^2$ for $\pi^+\pi^-$ pairs from reaction $\pi^- + P \rightarrow \pi^+\pi^- N$. (b) Plot of events $(\pi^+ P)$ as a function of $M_{\pi^+ P}$. Since the initial state involved $\pi^+ P$ and only two particle final states were studied, this is equivalent to a measurement of the $\pi^+ P$ cross section. The $\pi^+ P$ state is a pure $I = 3/2$, $I_z = 3/2$ state.

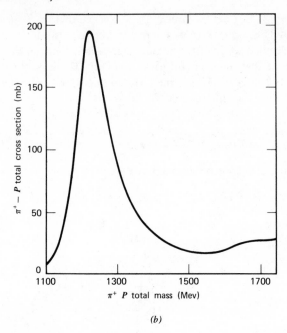

(b)

Fig. 26.4   continued

The conservation of i-spin allows us to make predictions about the relative decay rates of

$$\Delta^+ \begin{array}{c} \nearrow P\pi^\circ \\ \searrow N\pi^+ \end{array}$$

The procedure is completely analogous to our discussion of spin and intensity rules at the end of Chapter 22. The initial state of $I = 3/2$ and $I_z = 1/2$ may be written in terms of $I = 1$ and $I = 1/2$ wave functions, as follows

$$Y_{3/2,1/2} = \sqrt{\frac{2}{3}}\,\pi^\circ P + \sqrt{\frac{1}{3}}\,\pi^+ N \qquad (26\text{-}11)$$

where $Y_{3/2,1/2}$ represents the Δ^+. Thus the probability of finding a $P\pi^\circ$ state is $2/3$ and that of finding a $N\pi^+$ state is $1/3$. These predictions are borne out by an analysis of the data.

I-spin conservation has been tested in many reactions, and there is no question as to its correctness, subject to small electromagnetic corrections. The fact that these corrections are so small, for example, that the neutron-proton mass difference is so small, made it possible to identify the symmetry. When symmetry breaking is large, this becomes much more difficult.

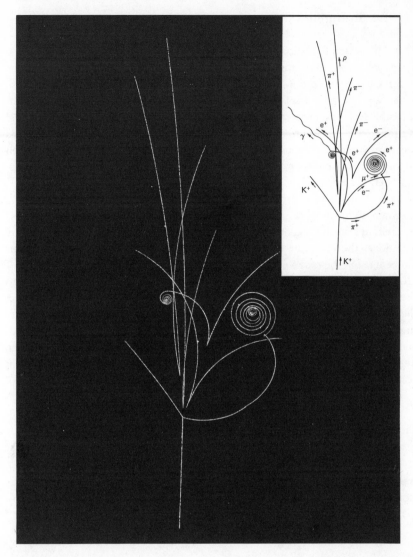

Fig. 26-5. The reaction $K^+ + P \rightarrow K^+ + K^\circ + \Lambda^\circ + \pi^+ + \pi^\circ$ seen in a hydrogen bubble chamber. The strong reaction conserves strangeness. The picture shows a large number of interesting secondary reactions: the π^+ scatters off a proton and subsequently decays according to $\pi^+ \rightarrow \mu^+ + \bar{\nu}$, with the μ^+ decaying $\mu^+ \rightarrow e^+ + \nu + \bar{\nu}$; the π° decays $\pi^\circ \rightarrow \gamma_1 + \gamma^2$, with $\gamma_1 \rightarrow e^+ + e^-$ and $\gamma_2 \rightarrow e^+ + e^-$, the e^+ undergoing Bremsstrahlung $e^+ \rightarrow e^+ + \gamma$; the Λ° decays according to $\Lambda^\circ \rightarrow P + \pi^-$; the K° undergoes the decay $K^\circ \rightarrow \pi^+ + \pi^-$. (Courtesy of Lawrence Berkeley Laboratory, University of California.)

435

D. Strangeness

In the late 1940s it appeared as if the basic ingredients of a theory of the *strong interactions*, those responsible for the nuclear forces, were established. The forces acted between the nucleon doublets, and the "glue" giving rise to these forces was the pion triplet, as predicted by Yukawa. There remained the insuperable technical problem of making reliable calculations (because the potential is strong, perturbation theory cannot be used), but the expectation was that this problem would, sooner or later, be solved. It was therefore very exciting when around 1950 cosmic ray experiments, and later, the first high-energy accelerators, indicated the existence of a new set of particles. The first particle so discovered was the Λ°: it was found that when a cloud chamber was exposed to cosmic rays, a certain number of V-shaped tracks were seen. When the experiment was repeated in a strong magnetic field, it was found that the tracks bent in opposite directions; the curvature of the tracks in the magnetic field determined their momenta, and the range determined their energies. From this, the application of (26-9) allowed the determination of the mass of the particle that decayed into the P and π^-: it was given by $M_\Lambda c^2 = 1115$ MeV. The apex of the V, marking the point of decay, appeared a certain distance from the point where the production interaction occurred. From this the lifetime of the Λ° could be determined, and was found to be 2.5×10^{-10} sec. From the number of Λ's seen in a given number of photographs it was determined that they were produced with a cross section of the order of 10^{-27} cm^2. In addition to the decay mode (see Fig. 26-5).

$$\Lambda^\circ \rightarrow P + \pi^-$$

the decay mode

$$\Lambda^\circ \rightarrow N + \pi^\circ$$

was later established. Another pair of particles were also found; these were the sigmas, $\Sigma^\pm$, with $M_\Sigma c^2 = 1190$ MeV, lifetimes $\tau_{\Sigma^+} = 0.8 \times 10^{-10}$ sec, $\tau_{\Sigma^-} = 1.5 \times 10^{-10}$ sec, and dominant decay modes

$$\Sigma^+ \begin{array}{l} \nearrow P + \pi^\circ \\ \searrow N + \pi^+ \end{array}$$

$$\Sigma^- \longrightarrow N + \pi^-$$

At about the same time other tracks appeared that were finally interpreted as caused by new particles with $N_B = 0$. Their decay modes were (see, for example, Fig. 26-5).

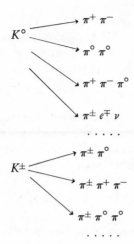

$$K° \longrightarrow \pi^+ \ \pi^-$$
$$\longrightarrow \pi° \ \pi°$$
$$\longrightarrow \pi^+ \ \pi^- \ \pi°$$
$$\longrightarrow \pi^{\pm} \ e^{\mp} \ \nu$$

. . . . .

$$K^{\pm} \longrightarrow \pi^{\pm} \ \pi°$$
$$\longrightarrow \pi^{\pm} \ \pi^+ \ \pi^-$$
$$\longrightarrow \pi^{\pm} \ \pi° \ \pi°$$

. . . . .

In spite of the different final states, these were attributed to single particles because in each case the masses came out $m_K c^2 = 494$ MeV and the lifetimes also clustered around the values $\tau_{K\pm} = 1.2 \times 10^{-8}$ sec and $\tau_{K°} = 0.8 \times 10^{-10}$ sec. These K mesons were also produced with cross sections of the order of 10^{-27} cm^2.

The discovery of these particles and their properties caused a crisis; the data, taken at face value, were not compatible and if taken seriously implied that quantum mechanics could not describe the behavior of these particles. To see this, let us describe the matrix element for the decay $\Lambda° \rightarrow P\pi^-$ by the number G. According to the Golden Rule, the decay rate is given by

$$R = \frac{2\pi}{\hbar} G^2 \frac{d^3p/dE}{(2\pi\hbar)^3} = \frac{2\pi}{\hbar} G^2 \frac{4\pi p^2 \ dp/dE}{(2\pi\hbar)^3} \qquad (26\text{-}12)$$

With relativistic kinematics (p is the center of mass momentum)

$$E = [(M_p c^2)^2 + p^2 c^2]^{1/2} + [(m_\pi c^2)^2 + p^2 c^2]^{1/2}$$

dp/dE evaluated at $E = M_\Lambda c^2$ can be calculated, and we get, writing G^2 in dimensionless form as

$$G^2 = \beta \frac{\hbar^3 c}{m_\pi} \qquad (26\text{-}13)$$

the rate

$$R = \frac{\beta}{\pi} \left(\frac{m_\pi c^2}{\hbar} \right) \left(\frac{pc}{m_\pi c^2} \right) \left(\frac{E_\pi}{m_\pi c^2} \right) \left(\frac{E_p}{M_p c^2} \right) \qquad (26\text{-}14)$$

The dimensionless number β is the analog of the fine structure constant α. Putting in numbers we get for the decay rate

$$R \cong 0.6 \ \beta \times 10^{23} \ \text{sec}^{-1} \qquad (26\text{-}15)$$

from which we find that $\beta \cong 0.7 \times 10^{-13}$. This number is very much smaller than the fine structure constant and suggests that the interaction responsible for the decay is much weaker than the electromagnetic interaction. On the other hand, if the $\Lambda°$ is produced in a conjectured reaction such as

$$\pi^- + P \to \Lambda° + \pi°$$

one can use the $\Lambda°P\pi^-$ coupling to estimate the cross section. Estimates of the type we used in our discussion of the photoelectric effect suggest that the cross section must be proportional to β. The largest "area" involving the masses of the particles involved thus gives, with the above β

$$\sigma \simeq \pi\beta \left(\frac{\hbar}{m_\pi c}\right)^2 = 4.5 \times 10^{-39} \text{ cm}^2 \tag{26-16}$$

This differs from the experimental value by a factor of 10^{12} and no minor changes in the estimates can save us. It thus appears that the $\Lambda°$ is produced by a strong interaction and decays through a weak interaction.

The way out of the dilemma was suggested by Pais, who proposed that the production process necessarily involves another one of the new particles, so that reactions involving pairs of the new particles could proceed strongly, whereas reactions involving only one of them would have to go slowly. Thus the conjectured reaction

$$\pi^- + P \to \Lambda° + \pi°$$

should not take place, but that

$$\pi^- + P \to \Lambda° + K°$$

for example, could. The suggestion of *Associated Production* turned out to be correct, and it was soon determined that the production of a $\Lambda°$ was always accompanied by the production of a K. The limitations of the Pais proposal were seen when still another particle, named the Cascade (Ξ^-), was discovered. Preliminary evidence showed it to decay, with a lifetime of 1.7×10^{-10} sec, according to

$$\Xi^- \to \Lambda° + \pi^-$$

but it *did not* decay according to the mode

$$\Xi^- \to N + \pi^-$$

Its mass was found to be $M_\Xi c^2 = 1321$ MeV. If the Ξ "belonged with" the $\Lambda°$, then the first decay mode has a pair of the "new" particles, and should go rapidly; if the Ξ^- is "normal" then the decay

$$\Xi^- \to N + \pi^-$$

should go rapidly, instead of not at all.

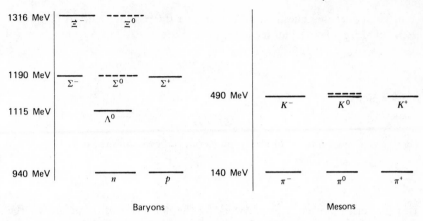

Fig. 26-6. Hadron spectrum as known in 1953 before the discovery of other particles predicted by strangeness theory. The energies are only approximate, and the new particles predicted by the strangeness theory are given by the dotted levels.

Order was brought into the jumbled situation by Gell-Mann and by Nishijima, who independently proposed the extension of the notion of i-spin to the new particles, and introduced a new quantum number, the *strangeness* S. The spectrum of baryons (the name for nucleons and the new particles that ultimately end up as nucleons) and mesons was, by 1953, believed to have the form shown in Fig. 26-6. It is clear that the $\Sigma^{\pm} - \Lambda^{\circ}$ mass difference is too large to put them into an i-spin triplet. Thus, in the absence of equal mass partners, the Λ° had to be an $I = 0$ state. Since the production cross section in the reaction

$$\pi^- + P \to \Lambda^{\circ} + K^{\circ}$$

was large enough to be viewed as a strong (rather than electromagnetic) process, i-spin should be conserved. The left side has $I = 3/2$ or $1/2$, and thus the K° must belong to one of these multiplets. The absence of any observed K^{++} or K^{--}, needed to make up a quartet, showed that the K° would have to be part of an $I = 1/2$ doublet, with $I_z = -1/2$ equal to the I_z of the $P\pi^-$ system. The K^+ was undoubtedly the $I_z = +1/2$ partner of the K°. The lifetimes of these two particles could be very different, since the weak interactions need not conserve i-spin any more than the electromagnetic ones do. The K^- is undoubtedly the antiparticle of the K^+, and its $I_z = 1/2$ partner is denoted by $\overline{K^{\circ}}$. Note that the $\overline{K^{\circ}}$ cannot be identical with the K° since they have different values of I_z. How is it possible for a neutral system not to go into itself under charge conjugation? We saw that the antineutron differed from the neutron, because they differed in the value of baryon number N_B. What is the quantum number that distinguishes between the $\overline{K^{\circ}}$ and the K°?

To answer this question, let us consider the relation between I_z and the electrical charge. For the nucleon system we have

$$Q = I_z + \frac{1}{2}$$

and for pions

$$Q = I_z$$

These two cases, and that of the antinucleon, can be combined in

$$Q = I_z + \frac{N_B}{2}$$

This formula, however, does not work for the new particles. If we modify the formula by the introduction of a new quantum number S called the strangeness, so that

$$Q = I_z + \frac{N_B}{2} + \frac{S}{2} \qquad (26\text{-}17)$$

we find that for nucleons and pions $S = 0$, but for the Λ° we must have $S = -1$. Thus the K°, and its partner, the K^+, must have $S = +1$, and hence the K^- and the $\overline{K^\circ}$ must have $S = -1$, if we assume that charge conjugation also changes the sign of S, as suggested by (26-17). This is the reason that the K° and the $\overline{K^\circ}$ are different. How this manifests itself we will see later.

Continuing with our examination of the baryon spectrum, we see that in the absence of Σ^{++} and/or Σ^{--}, it seems natural to assign $I = 1$ to the Σ's. This however *predicts* the existence of a Σ°, of mass close to 1190 MeV. Why was the decay

$$\Sigma^\circ \rightarrow P + \pi^-$$

(analogous to the Λ° decay) never seen? Gell-Mann pointed out that the electromagnetic decay

$$\Sigma^\circ \rightarrow \Lambda^\circ + \gamma$$

did not involve a change in strangeness, in contrast to the former decay mode. With the postulate that a decay in which strangeness changed by one unit,

$$|\Delta S| = 1 \qquad (26\text{-}18)$$

should go "weakly" with typical lifetime of the order of 10^{-10} sec, while strangeness conserving reactions should go strongly, or electromagnetically, it was possible to predict that the $\Sigma^\circ \rightarrow \Lambda^\circ + \gamma$ decay should be very rapid ($\tau \sim 10^{-20}$ sec), leaving no opportunity for the weak decay to take place. A careful study of Λ° decays showed that they frequently originated in a reaction in which some momentum and energy was missing. An examination of the missing momentum and energy showed that the mass of the missing particle was consistent with 0, and that the photon and Λ° were decay products of something with mass 1192 MeV, confirming the existence of the Σ°. How do these notions, that is, conservation of S in the strong and electromagnetic interactions, and $|\Delta S| = 1$ in the weak ones, work for the Ξ^-? The weak decay

$$\Xi^- \to \Lambda^\circ + \pi^-$$

suggests that the strangeness of the Ξ^- must be 0 or -2. The former assignment is incompatible with the absence of

$$\Xi^- \to N + \pi^-$$

and hence we must take $S = -2$. This implies that the Ξ^- has $I_z = -1/2$, and in the absence of multiply charged Ξ's we assign to it $I = 1/2$. This predicts a partner, the Ξ°, which should have mass around 1320 MeV and decay according to

$$\Xi^\circ \to \Lambda^\circ + \pi^\circ$$

(the decays $\Xi \to \Sigma + \pi$ are not possible because of the masses of the particles involved). The Ξ° was looked for and found! The explanation for the absence of

$$\Xi^- \to N + \pi^-$$

lies in that it is characterized by $|\Delta S| = 2$, which is presumably doubly weak, characterized perhaps by a β of magnitude 10^{-26}.

The notion of strangeness conservation in the strong and electromagnetic interactions, and $|\Delta S| = 1$ in the weak ones, has passed every test. The classification of particles by i-spin and strangeness, or equivalently by *hypercharge Y* defined by

$$Y = N_B + S \tag{26-19}$$

is given in the following table.

Y	I	Baryons $N_B = 1$	Mesons $N_B = 0$	Antibaryons $N_B = -1$
1	$\dfrac{1}{2}$	P, N	K^+, K^0	$\overline{\Xi^-} \ \overline{\Xi^0}$
0	1	$\Sigma^+, \Sigma^0, \Sigma^-$	π^+, π^0, π^-	$\overline{\Sigma^-}, \overline{\Sigma^0}, \overline{\Sigma^+}$
0	0	Λ^0	?	$\overline{\Lambda^0}$
-1	$\dfrac{1}{2}$	Ξ^0, Ξ^-	$\overline{K^0}, K^-$	$\overline{N}, \overline{P}$

Subsequent determination of the spins and parities of these particles showed that all the particles in the first column were $1/2^+$ and those in the second column were 0^-. If we are looking for patterns, where is the missing $I = 0$, $Y = 0$ pseudoscalar meson?

E. Unitary Symmetry

The search for a familial relation among the baryons and among the mesons was actively pursued in the late 1950s. Finally in 1961 Gell-Mann, and independently Ne'eman, discovered a generalization of i-spin, bearing the

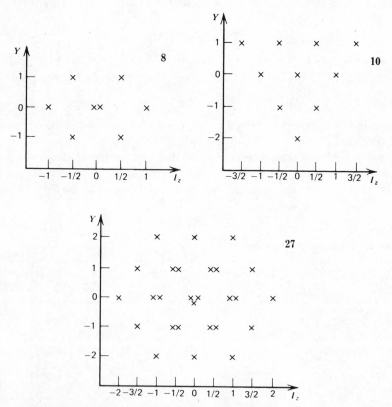

Fig. 26-7. Some SU(3) representations. The number of crosses at each site represents the multiplicity. Thus the 27 consists of a $Y = 2$, $I = 0$ state, $Y = 1$, $I = 3/2$ and $1/2$ states, $Y = 0$, $I = 2$, 1, 0 states, and so on.

technical name of SU(3). Since group theory, the tool most widely used in the search, is beyond the scope of this book, we can only give a very qualitative idea of how the familial relation looks. In SU(3), states come in supermultiplets. Each supermultiplet will consist of a number of states that can be labeled by i-spin as well as hypercharge. Figure 26-7 shows the several supermultiplets, the octet, **8**, the decuplet, **10**, and the **27**. Figures 26-8 and 26-9 show how the $1/2^+$ baryons and the 0^- mesons fit into the octet pattern. The missing $I = 0$, $Y = 0$ pseudoscalar meson was found in the examination of $\pi^+\pi^-\pi^°$ masses in bubble-chamber pictures. A study of

$$(m_\eta c^2)^2 = (E_+ + E_0 + E_-)^2 - c^2(\mathbf{p}_+ + \mathbf{p}_- + \mathbf{p}_0)^2$$

showed a strong peaking at $m_\eta c^2 = 550$ MeV, and an analysis of the distribution of the energies and momenta among the three particles showed that the decay pattern implied the quantum numbers 0^- for that particle. The absence of similar correlations in $\pi^\pm\pi^+\pi^-$, say, showed that the i-spin had to be zero. Since the decay was clearly not weak, the strangeness had to be zero.

If the SU(3) symmetry was indeed a badly broken (badly, compared to the only slightly broken i-spin symmetry) symmetry, then it should also be relevant to the excited states of the nucleon. The $I = 3/2$, $Y = 1$ $\Delta(1236)$ resonances needed partners. In 1960 a set of $\Lambda^°\pi$ resonances were discovered. The i-spin was clearly 1 and the hypercharge was 0, and detailed tests showed that the spin and parity were $3/2^+$, so that these so-called $\Sigma^*(1385)$ were most likely partners of the $\Delta(1236)$ in a supermultiplet. The simplest possible assignments were the **10** and the **27**. Gell-Mann and others conjectured that the **10** was the appropriate choice. The discovery in 1962 of a $\Xi\pi$ resonance of mass 1531 MeV, with $I = 1/2$

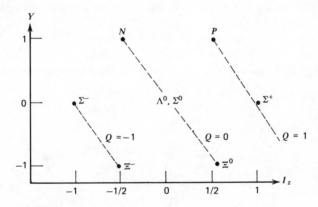

Fig. 26-8. Octet pattern for spin $1/2^+$ baryons. The diagonal lines are lines of constant charge Q.

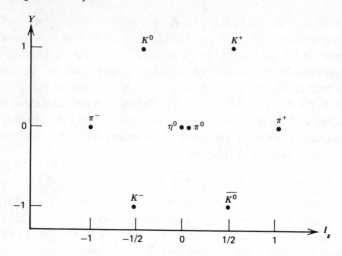

Fig. 26-9. Octet pattern for mesons. Included is the $T = 0$, $Y = 0$ particle $\eta°$ discovered in 1961.

(no $\Xi^-\pi^-$ resonance, for example) strongly supported this assignment. What was still missing was the last member of the decuplet, an $I = 0$, $Y = -2$ negatively charged particle. What would it look like? Here the pattern of masses of the other members of the decuplet, $\Delta(1236)$, $\Sigma^*(1385)$, and $\Xi^*(1531)$ suggested equal spacing increasing linearly with $|Y|$. If this were to be maintained, the missing particle, called the Ω^-, would have to have mass in the vicinity of $M_\Omega c^2$ $= 1675$ MeV. This would give the Ω^- a unique signature; its mass is too low to decay strongly into $\Xi + \overline{K}$, and hence it must undergo a $|\Delta S| = 1$ decay to $\Lambda°\overline{K}$ or $\Xi\pi$. Its production would also have a special signature, since the lowest possible Y-value for an initial state is $Y = 0$ for $K^- + P$ collisions. Thus, to make a $Y = -2$ Ω^-, two K's would have to be produced.

A massive search for such a particle was undertaken, and in 1964 the first Ω^- picture was published (Fig. 26-10). The production process was

$$K^- + P \to \Omega^- + K^+ + K°$$

and the decay

$$\Omega^- \longrightarrow \pi^- + \Xi°$$
$$ \longrightarrow \Lambda° + \pi°$$
$$ \longrightarrow \gamma + \gamma$$
$$ \longrightarrow e^+e^-$$
$$ \longrightarrow e^+e^-$$
$$ \longrightarrow P + \pi^-$$

The fact that in the π° decay

$$\pi^\circ \rightarrow \gamma + \gamma$$

both γ-rays produced pairs was very fortuitous and makes the picture a textbook example of what theoreticians dream of, but experimentalists seldom see. The mass of the Ω^- was found to lie remarkably close to the predicted value, at 1672 MeV. The spin and parity have not yet been measured, since to date there exist only 28 Ω^- pictures, but there is no doubt in anybody's mind of what the outcome will be.

The SU(3) symmetry, just like i-spin conservation, makes predictions about decay rates among particles in the same supermultiplet. Thus there are predictions relating $\Delta^{++} \rightarrow P + \pi^+$, $\Sigma^{+*} \rightarrow \Lambda^\circ + \pi^+$, and $\Xi^{\circ*} \rightarrow \Xi^- + \pi^+$ that are in good agreement with experiment, given the fact that the symmetry is broken. The generally accepted view is that SU(3) is indeed an underlying symmetry of the strong interactions. The mechanism by which it is broken is not yet well understood, although some patterns (such as the equal spacing rule in

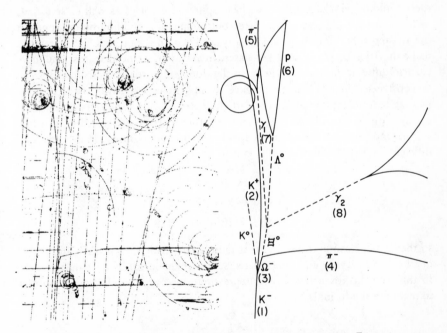

Fig. 26-10. The first Ω^-. The mass of the Ω^- is below the $\Xi + \bar{K}$ mass, so that a strangeness-violating (weak) decay is involved. What is seen in the bubble-chamber picture is the sequence $\Omega^- \rightarrow \Xi^\circ + \pi^-$; $\Xi^\circ \rightarrow \Lambda^\circ + \pi^\circ(\rightarrow 2\gamma)$; $\Lambda^\circ \rightarrow P + \pi^+$. [From Barnes *et al.*, *Phys. Rev. Letters*, *12*, 204 (1964), courtesy of Brookhaven National Laboratory and Dr. N. P. Samios.]

the decuplet) follow from some simple postulates about the nature of the symmetry breaking. All of the many resonances that have been found can be fitted into supermultiplets, although frequently there are still undiscovered partners. It is a remarkable fact that all of the mesonic resonances (and there are now 1^-, 1^+, 2^+ and (probably) 0^+ octets) and all of the baryonic resonances fit into octets for mesons, and octets or decuplets for baryons. This absence of higher supermultiplets, for example, the 27, cannot be understood on the basis of SU(3) alone. It does follow from a simple composite model of elementary particles, called the *quark model* first proposed by Gell-Mann and by Zweig.

F. The Quark Model

The question of which particles are elementary has been a pressing question in this fundamental field, and with the recognition that the "elementary" proton and neutron had six partners, it became clear that *if* there were some elementary building blocks, there would most probably be fewer than eight. What might these building blocks be like? We have spin 1/2 particles as well as spin 0, 1, 3/2, . . . particles. These can be made out of spin 1/2 building blocks, but not out of spin 0 building blocks; similarly we need an i-spin doublet, at least, to make up i-spin 0, 1/2, 1, and 3/2 states. In addition, we need at least one more particle differing in hypercharge from the doublet, so that various Y states can be constructed. SU(3) happens to have, as its simplest nontrivial states, a three-particle representation and its antiparticle representation. These representations named "quarks" can be used to build up other SU(3) representations, just like angular momentum 1/2 can be used to build up angular momentum states of J different from 1/2. The rules turn out to be

$$3 \otimes \bar{3} = \ 8 + 1$$
$$3 \otimes 3 = \ 6 + \bar{3}$$
$$3 \otimes 3 \otimes 3 = 10 + 8 + 8 + 1$$

so that it is plausible to assume that mesons are made of "quarks" and "anti-quarks," and baryons and their excited states are made out of three quarks each. In this way the absence of higher representations can be understood. In order to maintain the formula

$$Q = I_z + \frac{Y}{2}$$

the following quantum numbers are assigned to the quarks, which are labeled with lowercase letters related to their i-spin content, p, n, and λ.

Particle	N_B	T	I_z	Y	Q
p	$\frac{1}{3}$	$\frac{1}{2}$	$\frac{1}{2}$	$\frac{1}{3}$	$\frac{2}{3}$
n	$\frac{1}{3}$	$\frac{1}{2}$	$-\frac{1}{2}$	$\frac{1}{3}$	$-\frac{1}{3}$
λ	$\frac{1}{3}$	0	0	$-\frac{2}{3}$	$-\frac{1}{3}$
$\bar{\lambda}$	$-\frac{1}{3}$	0	0	$\frac{2}{3}$	$\frac{1}{3}$
$\bar{n}$	$-\frac{1}{3}$	$\frac{1}{2}$	$\frac{1}{2}$	$-\frac{1}{3}$	$\frac{2}{3}$
$\bar{p}$	$-\frac{1}{3}$	$\frac{1}{2}$	$-\frac{1}{2}$	$-\frac{1}{3}$	$-\frac{2}{3}$

To construct the composite wave functions for a quark-antiquark system, we start with the highest Y and highest I_z states; lowering I_z can be done by successively converting $p \to n$ or $\bar{n} \to \bar{p}$. To get the states of one lower unit of Y, convert a p to a λ or an n to a λ. For example,

$$K^+ = (p\bar{\lambda})$$

Hence

$$K^\circ = (n\bar{\lambda})$$

To get the π^+ state, convert $\bar{\lambda} \to \bar{n}$ in the K^+. This yields

$$\pi^+ = (p\bar{n})$$

and then, successively[6]

$$\pi^\circ = \frac{1}{\sqrt{2}}(-p\bar{p} + n\bar{n})$$

$$\pi^- = (n\bar{p})$$

The K^- and $\overline{K^\circ}$ are just the antiparticles of the K doublet, so that

$$\overline{K^\circ} = (\lambda\bar{n})$$

$$K^- = (\lambda\bar{p})$$

There remains the $I = 0$, $Y = 0$, η° state. It can be obtained by converting $p \to \lambda$ in the K^+. What one gets is part η° and part π°, but since we already know what part π° is, the η° can be found by insisting that it be orthogonal to the π°. The choice turns out to be

$$\eta_0 = \frac{1}{\sqrt{6}}(2\lambda\bar{\lambda} - p\bar{p} - n\bar{n})$$

---

[6] The minus sign in the π° is a technical subtlety. In effect, the antispinor to (p, n) is $(\bar{n}, -\bar{p})$.

There remains a final possibility:

$$X_0 = \frac{1}{\sqrt{3}}(p\bar{p} + n\bar{n} + \lambda\bar{\lambda})$$

orthogonal to both π° and η°. This is, in fact, the $\mathbf{1}$ in the decomposition $\mathbf{3} \times \bar{\mathbf{3}} = \mathbf{8} + \mathbf{1}$, and it does correspond to a particle that has been identified, the $\eta'(958)$, which also has spin and parity 0^-.

The bound states with $L = 0$, that is, the $^1S_0$ states, are the pseudoscalar mesons (recall that the parity of an antiparticle has an additional minus sign). One can imagine $L = 1$ states, for example, $^1P_1$ (i.e., 1^+) bound states, and also $^3P_{2,1,0}$ states (2^+, 1^+, 0^+), and so on. Many of these have been found. The quark model, being more specific than SU(3) yields more predictions, correlating decays of particles that have different spins.

The three-quark wave functions can be worked out just like the quark-antiquark wave functions. The highest Y, I_z state is the (ppp) state, which can be identified with the Δ^{++}. Its partners are again obtained by successive conversion of $p \rightarrow n$:

$$\Delta^{++} = (ppp)$$

$$\Delta^+ = \frac{1}{\sqrt{3}}(ppn + pnp + npp)$$

$$\Delta^\circ = \frac{1}{\sqrt{3}}(pnn + npn + nnp)$$

$$\Delta^- = (nnn)$$

The Σ^{*+} is obtained from (ppp) by changing p to λ. Thus

$$\Sigma^{*+} = \frac{1}{\sqrt{3}}(pp\lambda + p\lambda p + \lambda pp)$$

$$\Sigma^{*\circ} = \frac{1}{\sqrt{6}}(pn\lambda + np\lambda + p\lambda n + n\lambda p + \lambda pn + \lambda np)$$

$$\Sigma^{*-} = \frac{1}{\sqrt{3}}(nn\lambda + n\lambda n + \lambda nn)$$

Successively

$$\Xi^{*\circ} = \frac{1}{\sqrt{3}}(p\lambda\lambda + \lambda p\lambda + \lambda\lambda p)$$

$$\Xi^{*-} = \frac{1}{\sqrt{3}}(n\lambda\lambda + \lambda n\lambda + \lambda\lambda n)$$

and

$$\Omega^- = \lambda\lambda\lambda$$

The four multiplets differ in the number of λ quarks. If it is assumed that the symmetry breaking completely resides in the fact that the λ quark is some 150 MeV more massive than the (p,n) doublet, the mass pattern of the decuplet can be understood. When this argument is applied to the meson octet, it does not work as well with a 150 MeV mass difference. However, the relations

$$m_\pi = 2\, m_p + \text{binding}$$

$$m_K = m_p + m_\lambda + \text{binding}$$

$$m_\eta = \frac{2}{3}\,(2m_\lambda) + \frac{1}{3}\,(2m_p) + \text{binding}$$

(the factor 2/3 coming from the probability of finding the η° in the $\lambda\lambda$ state) lead to

$$m_K = \frac{1}{4}\,(3m_\eta + m_\pi) \tag{26-20}$$

which is known as the Gell-Mann Okubo mass formula. It works to about a 10 percent accuracy, but is very good if the relation is written for the squares of the masses. Incidentally, the same formula will work for other octets, and for the baryons it will have the form

$$\frac{1}{2}\,(m_P + m_\Xi) = \frac{1}{4}\,(3m_\Lambda + m_\Sigma) \tag{26-21}$$

It is quite accurate, and is sometimes used to estimate where partners of incomplete octets might be located.

The quark model leads to many other predictions. For example, if it is assumed that at high energies all quarks and antiquarks interact identically, leading to equal cross sections, then it follows that in PP collisions there are nine possible interactions and in πP collisions there are six, so that

$$\frac{\sigma(PP)}{\sigma(\pi P)} = \frac{3}{2} \tag{26-22}$$

Surprisingly, this simpleminded counting works very well, both in this instance and in many more. It is an urgent and as yet unsolved problem in particle physics to understand why quarks, which must be very massive if they exist at all, since otherwise they would have beeen seen, act in such a simple additive manner. Other questions remain. Why do three quarks bind, but not two? More detailed considerations show that quarks, even though assumed to have spin 1/2, act as if they did not obey Fermi-Dirac statistics. Why is this so? We do not know.

G. Parity Nonconservation

In addition to the strong interactions and the electromagnetic interactions, there exist, in nature weak interactions. They were first discovered in beta decay, that is the reaction

$$N \rightarrow P + e^- + \bar{\nu}$$

and related reactions such as positron decay

$$P \rightarrow N + e^+ + \nu$$

and the capture reaction

$$e^- + P \rightarrow N + \nu$$

with the last two occurring only in nuclei. What was observed was a nuclear decay of the form

$$(A,Z) \rightarrow (A,Z + 1) + e^-$$

The electrons did not come out with a fixed energy, as they would have to if this was a two-body decay, although the maximum electron energy matched that available for a two-body decay. Faced with the choice of giving up energy conservation or proposing a new particle, Pauli in 1931 postulated that there exist a neutral particle emitted in the reaction with the electron. The properties of the new particle, named the *neutrino*, were the following:

1. Charge conservation required that it be electrically neutral.
2. The equality of the maximum electron energy to the available energy required that the neutrino mass be very tiny; it is now believed to be zero.
3. Studies of the spins of the initial and final nuclei required the neutrino to be a fermion. It is now known to have spin $1/2$.
4. The neutrino was not found when it was first postulated. The reason is that it interacts very weakly with matter. The cross section for neutrino absorption could be calculated with a detailed theory proposed by Fermi in 1932, and it was shown to be 10^{-44} cm^2 at low energies. Thus, in spite of its esoteric nature, the existence of the neutrino was accepted by most physicists, and it was finally identified in 1954. Nowadays neutrinos coming from the decay of high energy pions are used to study high-energy neutrino-nucleus collisions.

The Fermi theory of beta decay explained a class of weak decays, including those involving a new particle, the muon (μ), which is for all practical purposes an electron of mass $m_\mu c^2 = 105$ MeV, which was discovered in the 1940s. It could also explain in principle, if not in detail, decays such as

$$\pi^+ \rightarrow \begin{pmatrix} e^+ \\ \mu^+ \end{pmatrix} + \nu$$

since these could occur through the steps

$$\pi^+ \to P + \overline{N}$$

$$P \to N + \begin{pmatrix} e^+ \\ \mu^+ \end{pmatrix} + \nu$$

with the N–$\overline{N}$ annihilating. With the discovery of the strange particles, weak interactions not involving electrons, muons, and neutrinos appeared on the scene, as in

$$\Lambda^\circ \to P + \pi^-$$

and

$$K^+ \to \pi^+ + \pi^\circ$$

$$K^+ \to \pi^+ + \pi^+ + \pi^-$$

The latter decays attracted much attention. As the experimental data began to point to the fact that the K meson had spin 0, a paradox arose. The decay

$$K \to 2\pi$$

into two spinless particles implied that the orbital angular momentum also had to vanish. Since both pions were of negative parity, the implication was that the K had positive parity. On the other hand, a detailed study of the energy distributions in the decay

$$K \to 3\pi$$

suggested very strongly that all three pions were in S states relative to each other, as might have been guessed from the small amount of kinetic energy available to the three pions in the decay. This meant that the parity had to be $(-1)^3$, that is, odd, since there were three negative parity particles in the final state. These conclusions were inconsistent with the well-established principle of the invariance of the laws of physics under space reflection.

In 1956, Lee and Yang, in a very important paper, raised the question, *How do we really know that parity is conserved in the weak interactions?* There was no doubt about the validity of parity conservation in the electromagnetic interactions. Parity conservation implies some selection rules, and these are satisfied to a high degree of accuracy. This degree of accuracy is not high enough, however, to say anything about the conservation of parity at the weak interaction level. In the direct study of weak interactions, there are also some selection rules; for example, the K should not be able to decay into 2π's *and* 3π's! In general, what is needed to check parity nonconservation is to examine a physical observable that allows us to distinguish between our world and a "world reflected in a mirror." The question immediately arises, Would not a state of an electron, moving with momentum **p**, be reflected into a state with momentum

−**p**, and would not this distinguish between the two worlds? The answer is that this would not, provided that both states are equally probable, so that if we see an electron with momentum −**p** we are not forced to the conclusion that we live in the "mirror" world. The existence of elliptical orbits in planetary physics is not evidence against the invariance of the laws of gravitation under rotations, unless some elliptical orbits are preferred to others. Thus to distinguish between our world and the "mirror" world more subtlety is needed.

Suppose we had a one-dimensional potential that violated parity conservation, that is, suppose it was of the form

$$V(x) = V_{even}(x) + V_{odd}(x) \tag{26-23}$$

and suppose that $V_{odd}(x)$ is very much smaller than $V_{even}(x)$. If $u_n(x)$ are the eigenvalues of

$$H_0 = \frac{p^2}{2m} + V_{even}(x) \tag{26-24}$$

then the lowest order energy change due to the presence of $V_{odd}(x)$ is

$$\Delta E_n = \int_{-\infty}^{\infty} dx \, u^*(x) \, V_{odd}(x) \, u_n(x) \tag{26-25}$$

Now the Hamiltonian H_0 is even in x, and hence, as discussed in Chapter 4, the eigenfunctions $u_n(x)$ can be chosen as eigenstates of the parity operator, that is, they are either even or odd. Consequently, ΔE_n vanishes, that is, a measurement of the energy cannot be used to distinguish between a world and a "mirror" world. The second-order energy shift will not vanish, but since it is quadratic in $V_{odd}(x)$, its contribution will be the same in the two worlds. The argument can be generalized.

We can similarly argue that a determination of a decay rate cannot distinguish between the two worlds. If parity is not conserved, it is possible for the matrix element for some transition to have the form

$$M = M_{even} + M_{odd} \tag{26-26}$$

By the Golden Rule, the decay rate has the form

$$R = \frac{2\pi}{\hbar} \sum \left| M_{even} + M_{odd} \right|^2 \rho(E) \tag{26-27}$$

In the "mirror" world this takes the form

$$R' = \frac{2\pi}{\hbar} \sum \left| M_{even} - M_{odd} \right|^2 \rho(E) \tag{26-28}$$

On the face of it, it looks as if the decay rates are different. We must, however, be careful in describing the difference between the two terms. First of all, M_{even} must be a scalar; it cannot be a vector, since this would single out a direction in space and ultimately imply lack of angular momentum conservation. Thus if there are momenta $\mathbf{p}_i$ in the process, then M_{even} can be a function of various products $\mathbf{p}_i \cdot \mathbf{p}_j$. If there are spin vectors present, then it can also depend on $\mathbf{S}_i \cdot \mathbf{S}_j$, and on $(\mathbf{p}_i \cdot \mathbf{S}_j)^2$, but not on $\mathbf{p}_i \cdot \mathbf{S}_j$, since the last is a pseudoscalar quantity; whereas momentum changes sign under an inversion, the angular momentum does not (e.g., $\mathbf{r} \times \mathbf{p}$ does not, and hence the spins cannot). On the other hand, M_{odd} must be linear in a pseudoscalar quantity. Thus if there are more than three independent momenta in the final state of a decay, a possible pseudoscalar is $\mathbf{p}_1 \cdot \mathbf{p}_2 \times \mathbf{p}_3$. In two or three body decays, the pseudoscalars must be of the form $\mathbf{S} \cdot \mathbf{p}$ where $\mathbf{S}$ is one of the spin vectors and $\mathbf{p}$ is one of the momenta. Hence, if for definiteness we assume that

$$M = A + B\mathbf{S} \cdot \mathbf{p} \tag{26-29}$$

then

$$\sum |M_{\text{even}} \pm M_{\text{odd}}|^2 = \sum |A|^2 + \sum |B|^2 (\mathbf{S} \cdot \mathbf{p})^2$$
$$\pm \sum (AB^* + A^*B)\,\mathbf{S} \cdot \mathbf{p} \tag{26-30}$$

In a decay rate the spin states are usually summed over, that is, no measurements involving the correlation of the spin and momentum are made. In that case the last term vanishes, and the rates are the same for the world and the "mirror" world. It is only if the presence of a correlation such as $\mathbf{S} \cdot \mathbf{p}$ is measured, that is, *if a pseudoscalar quantity is measured*, that parity nonconservation can be detected.

Lee and Yang then suggested several experiments involving the weak interactions in which such correlations could be measured. Within a few months of the appearance of their paper a number of experiments showed that *parity was indeed violated in the weak interactions*. What does this do to the cherished notion that the laws of nature should be invariant under inversion, that, so to speak, it should not be possible to instruct an extragalactic being on how to make a right-handed screw? It now appears that the weak interactions not only violate parity conservation, but are also not invariant under charge conjugation, *They are invariant under combined charge conjugation and inversion, CP*. Thus, from a routine attempt to determine the parity of the K meson from its decay grew (a) the experimental verification that parity is not conserved in the weak interactions, (b) subsequent clarification of many aspects of the weak interactions, not possible before parity nonconservation had been observed, and (c) the discovery that nature shows more imagination than physicists, as in supplanting C and P by CP invariance.

H. The $K^\circ - \overline{K}^\circ$ System

As our final topic we discuss the implications of the consequences of the strangeness theory that the K^0 is not identical to its antiparticle, the $\overline{K}^\circ$, since these implications make use of simple quantum theory and are quite startling. As noted before, what distinguishes the K° from the $\overline{K}^\circ$ is the strangeness, and in a production process it is clear which of these is produced. Thus in the reaction

$$\pi^- + P \rightarrow \Lambda^\circ + K^\circ$$

we know that a K° is produced; in the reaction

$$K^- + P \rightarrow N + \overline{K}^\circ$$

we also know that it is the $\overline{K}^\circ$ that is produced. Given that the particles are pseudoscalar, we find that

$$CP|K^\circ\rangle = -|\overline{K}^\circ\rangle$$

$$CP|\overline{K}^\circ\rangle = -|K^\circ\rangle$$

Thus both K° and $\overline{K}^\circ$ may be viewed as linear superpositions of CP eigenstates; if we write

$$K^\circ = \frac{1}{\sqrt{2}}(K_1 + K_2) \qquad \overline{K}^\circ = \frac{1}{\sqrt{2}}(-K_1 + K_2) \qquad (26\text{-}31)$$

it follows that

$$CP|K_1\rangle = |K_1\rangle \qquad CP|K_2\rangle = -|K_2\rangle \qquad (26\text{-}32)$$

Since CP is conserved in the weak interactions, and the $\pi^+\pi^-$ system with zero angular momentum is even under CP, in the decays

$$K^\circ \rightarrow \pi^+\pi^-$$

$$\overline{K}^\circ \rightarrow \pi^+\pi^-$$

it is really only K_1 that is decaying. Both K_1 and K_2 can decay into some of the other modes, for example,

$$\pi^+\pi^-\pi^\circ$$

$$\pi^\pm e^\mp \nu$$

$$\cdots \cdots$$

In general, both K° and $\overline{K}^\circ$ or K_1 and K_2 are equivalent basis states in a two-state space. The strong interaction production process acts as a polarizer, producing a particular particle K°, say, or equivalently, a particular coherent mixture of K_1 and K_2 (coherent, in the sense that the phase relationship is fixed). After 10^{-10}

sec strangeness no longer means anything. The weak decay into $\pi^+\pi^-$ acts as an analyzer, and it picks out K_1. Gell-Mann and Pais in 1955 pointed out that the remaining K_2 component persists, and, since it cannot decay into the 2π channel available to the K_1, presumably has a longer lifetime and should be looked for, with one of the alternative decay modes. The K_2 was looked for and found. It had a lifetime of about 5×10^{-8} sec compared with 0.8×10^{-10} sec for the K_1. Other interesting effects emerge. Pais and Piccioni noted that if one starts with a K° beam, then after 10^{-10} sec one is left with $1/\sqrt{2}\ K_2$, that is, a beam of the form $\frac{1}{2}(K^\circ + \overline{K}^\circ)$. If, before the K_2 decay can take place, matter is interposed, then, because of the different strong interactions of the K° and the $\overline{K}^\circ$ components, for example,

$$K^\circ + P \rightarrow K^\circ + P \qquad\qquad \overline{K}^\circ + P \rightarrow \overline{K}^\circ + P$$

$$K^\circ + P \rightarrow K^+ + N \qquad\qquad \overline{K}^\circ + P \rightarrow \Sigma^+ + \pi^\circ$$

$$K^\circ + N \rightarrow K^\circ + N \qquad\qquad\qquad \rightarrow \Sigma^\circ + \pi^+$$

$$\cdot \quad \cdot \quad \cdot \quad \cdot \quad \cdot \quad \cdot \quad \cdot$$

the particular phase relation is destroyed, and one no longer has a pure K_2 beam. Hence, 2π pairs will again be seen, since the mixture now will involve some K_1. Thus under the idealized conditions that all the $\overline{K}^\circ$ are absorbed, and the K° merely scattered in the forward direction, what emerges from the interposed slab of material is $\frac{1}{2}K^\circ$. The phenomenon, known as *regeneration*, has been observed and studied in detail. (See Fig. 26-11.) The verification of the prediction of Gell-Mann and Pais adds strong support to our belief in the validity of quantum mechanics as the proper framework for the description of subatomic phenomena.

This is a good note on which to end. The reader, having mastered the material that we have presented, is ready to go deeper into the study of quantum mechanics, in which more sophisticated mathematical tools are necessary. Such a study will bring him or her to the frontiers of knowledge, be it in the investigation of the structure of elementary particles at energies of billions of volts, the properties of matter at 10^{-3} K$^\circ$, or the nature of nuclear matter on the surface and inside a neutron star. Wherever he or she chooses to go, there will be excitement and surprises.

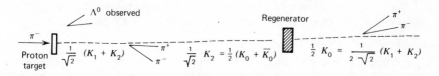

Fig. 26-11. Schematic drawing of regeneration experiment.

Problems

1. A 10 MeV positron collides with a hydrogen atom at rest. Write down the energy and momentum conservation relations. Taking into account the fact that the proton mass is $M_p c^2 = 940$ MeV, what will be the energy of the emitted photon from the reaction $e^+ + H \rightarrow P + \gamma$?

2. What is the threshold energy for the production of an antiproton in the reaction

$$P + P \rightarrow P + \bar{P} + P + P$$

One of the initial protons is at rest.

3. He³ (PPN) and H³ (PNN) are likely candidates for an i-spin doublet (there is no trineutron nucleus). Consider the reactions

$$P + D \left\langle \begin{array}{l} \rightarrow \pi^\circ + \text{He}^3 \\ \rightarrow \pi^+ + \text{H}^3 \end{array} \right.$$

Show that i-spin conservation predicts that

$$\frac{\sigma(\text{He}^3)}{\sigma(\text{H}^3)} = 2$$

Hint. Write out the initial state (why is it an i-spin eigenstate?) in terms of the π's (analog of $Y_{1,m}$) and He³, H³ (analog of χ_+, χ_- in spin 1/2 wave functions), as was done in (26-11).

4. Which of the following reactions can proceed strongly, which weakly, and which not at all, and why?

$$P + P \rightarrow P + \Lambda^\circ + K^+$$
$$P + P \rightarrow P + \Lambda^\circ + \pi^+$$
$$\pi^- + P \rightarrow \Lambda^\circ + \overline{K^\circ}$$
$$\pi^+ + P \rightarrow \Lambda^\circ + K^+$$
$$\bar{P} + P \rightarrow K^+ + \overline{K^\circ} + \pi^-$$
$$\bar{P} + P \rightarrow \Lambda^\circ + \Lambda^\circ + \overline{N}$$
$$K^+ \rightarrow \pi^+ + e^+ + e^-$$
$$K^\circ \rightarrow \pi^+ + e^-$$

5. Consider a beam of pions impinging on a proton target. What is the threshold for K° production? What is the threshold for K^- production? (*Hint.* Start in the center of mass frame.)

6. Calculate the parameter β introduced in Section D of this chapter that characterizes the decays (a) $\Sigma^+ \rightarrow N + \pi^+$ and (b) $K^+ \rightarrow \pi^+ + \pi^\circ$. The rates are 0.6×10^{10} sec^{-1} and 1.7×10^7 sec^{-1} respectively.

7. A particle is seen to decay weakly (lifetime of the order of 10^{-10} sec) as follows:

$$X \to \pi^+ + \pi^+$$

What can you say about the particle on the basis of this information? Consider (a) limits on its mass, (b) i-spin, and (c) spin, parity. Into which SU(3) super-multiplet, of the ones discussed in this chapter, could this particle fit? What would be the significance of the observation of such a particle?

References

Elementary particle physics is discussed at the advanced undergraduate and first-year graduate level in

D. H. Perkins, *Introduction to High Energy Physics*, Addison-Wesley Publishing Co., 1972.

See also

E. Segre, *Nuclei and Particles*, W. A. Benjamin, Inc., New York, 1964.

An elementary, nonmathematical discussion may be found in the fascinating little volume

K. W. Ford, *The World of Elementary Particles*, Blaisdell Publishing Company, New York, 1963.

Recent developments are usually written up in a readable form in *Scientific American*.

special topics

Relativistic Kinematics

In this section we summarize some formulas that are useful in simplifying the effects of relativistic transformations from one reference frame to another. A typical application arises in scattering: theory deals with the center of mass frame, experiment with the laboratory frame, and the results of the two must be compared. The simplifying technique to be used is based on two results from the theory of special relativity:

(a) The scalar product of two four-vectors, $A_\mu = (A_0, \mathbf{A})$ and $B_\mu = (B_0, \mathbf{B})$, defined by

$$A \cdot B = A_\mu B_\mu \equiv (A_0 B_0 - \mathbf{A} \cdot \mathbf{B}) \qquad \text{(ST 1-1)}$$

is invariant under Lorentz transformations.

(b) The energy and momentum of a particle transform as a four vector

$$p_\mu = \left(\frac{E}{c}, \mathbf{p} \right) \qquad \text{(ST 1-2)}$$

whose square of "length" is given in terms of the rest mass of the particle

$$p^2 = p_\mu p_\mu = \frac{E^2}{c^2} - \mathbf{p}^2 = m^2 c^2 \qquad \text{(ST 1-3)}$$

In general, a collision between two particles, leading to two particles in the final state, for example,

$$A(p_A) + B(p_B) \rightarrow C(p_C) + d(p_D)$$

will be characterized by just two numbers. The reason is that there are $4 \times 4 = 16$ different components of four momenta; these are restricted by four mass conditions (ST 1-3), and four energy and momentum conservation conditions; furthermore invariance under translation and under rotation implies that six more coodinates, the center of mass momentum, the orientation of the scattering plane in space and the choice of axes in that plane are irrelevant.

For our two invariants, we take

$$s = (p_A + p_B)^2 = (p_C + p_D)^2 \qquad \text{(ST 1-4)}$$

with the second term following from four-momentum conservation, and

$$t = (p_C - p_A)^2 = (p_D - p_B)^2 \qquad \text{(ST 1-5)}$$

Another possible choice is

$$u = (p_D - p_A)^2 = (p_C - p_B)^2 \qquad \text{(ST 1-6)}$$

These three are not independent, since the reader can easily convince himself that $p_{A\mu} + p_{B\mu} = p_{C\mu} + p_{D\mu}$, the energy-momentum conservation law implies

$$s + t + u = m_A{}^2c^2 + m_B{}^2c^2 + m_C{}^2c^2 + m_D{}^2c^2 \qquad \text{(ST 1-7)}$$

The invariants have the following significance.

s: Consider the center of mass frame, in which

$$\mathbf{p}_A^* + \mathbf{p}_B^* = 0 \qquad \text{(ST 1-8)}$$

There

$$
\begin{aligned}
s &= (p_{0A}^* + p_{0B}^*)^2 - (\mathbf{p}_A^* + \mathbf{p}_B^*)^2 \\
&= \left(\frac{E_A^*}{c} + \frac{E_B^*}{c}\right)^2 \\
&= \frac{1}{c^2}\,(E_A^* + E_B^*)^2 \qquad \text{(ST 1-9)}
\end{aligned}
$$

that is, it is, within the factor of c^2, the square of the total center of mass energy. We follow custom in labeling the center of mass coordinates with an asterisk.

t: The significance of t is somewhat clearer in the special (but very common) case that particles A and C, and B and D are the same, as, for example, in the reactions

$$\pi + P \to \pi + P$$

and

$$\gamma + e \to \gamma + e$$

In that case, in the center of mass frame,

$$
\begin{aligned}
\mathbf{p}_B^* &= -\mathbf{p}_A^* \qquad \mathbf{p}_D^* = -\mathbf{p}_C^* \\
E_A^* + E_B^* &= E_C^* + E_D^* \qquad \text{(ST 1-10)}
\end{aligned}
$$

and

$$m_A = m_C \qquad m_B = m_D \qquad \text{(ST 1-11)}$$

imply that

$$
\begin{aligned}
(\mathbf{p}_A^{*2}\,c^2 + m_A^2\,c^4)^{1/2} + (\mathbf{p}_A^{*2}\,c^2 + m_B^2\,c^4)^{1/2} \\
= (\mathbf{p}_C^{*2}\,c^2 + m_A^2\,c^4)^{1/2} + (\mathbf{p}_C^{*2}\,c^2 + m_B^2\,c^4)^{1/2}
\end{aligned}
$$

that is,

$$E_A^* = E_C^* \qquad E_B^* = E_D^* \tag{ST 1-12}$$

Then

$$t = (p_A - p_C)^2 = \left(\frac{E_A^*}{c} - \frac{E_C^*}{c}\right)^2 - (\mathbf{p}_A^* - \mathbf{p}_C^*)^2$$

$$= -(\mathbf{p}_A^* - \mathbf{p}_C^*)^2 \tag{ST 1-13}$$

that is, it is minus the square of the momentum transfer in the center of mass frame.

Note that t is related to the center of mass scattering angle. The above yields

$$t = -\mathbf{p}_A^{*2} - \mathbf{p}_C^{*2} + 2\mathbf{p}_A^* \cdot \mathbf{p}_C^*$$

$$= -\mathbf{p}_A^{*2} - \mathbf{p}_C^{*2} + 2|\mathbf{p}_A^*||\mathbf{p}_C^*| \cos \theta^* \tag{ST 1-14}$$

The laboratory frame is characterized by $\mathbf{p}_B{}^L = 0$

$$p_{B\mu} = (m_B c, \mathbf{0}) \tag{ST 1-15}$$

Thus

$$s = (p_A + p_B)^2 = p_A{}^2 + p_B{}^2 + 2p_A \cdot p_B$$

$$= m_A{}^2 c^2 + m_B{}^2 c^2 + 2m_B E_A{}^L \tag{ST 1-16}$$

and

$$t = (p_D - p_B)^2$$

$$= m_D{}^2 c^2 + m_B{}^2 c^2 - 2m_B E_D{}^L$$

$$= (p_A - p_C)^2$$

$$= m_A{}^2 c^2 + m_C{}^2 c^2 - 2E_A{}^L E_C{}^L / c^2 + 2\mathbf{p}_A{}^L \cdot \mathbf{p}_C{}^L$$

$$= m_A{}^2 c^2 + m_C{}^2 c^2 - 2E_A{}^L E_C{}^L / c^2 + 2|\mathbf{p}_A{}^L||\mathbf{p}_C{}^L| \cos \theta^L \tag{ST 1-17}$$

This, with the help of

$$E_A{}^L + m_B c^2 = E_C{}^L + E_D{}^L \tag{ST 1-18}$$

and *the invariance of s and t*, that is, the fact that s and t have the same values in the center of mass and the laboratory frames (or any other frames) allows us to compute the relation between center of mass scattering angle and laboratory scattering angle, and between the energies in the two frames.

The transformation properties of differential cross sections, $d\sigma/d (\cos \theta)$ are obtained from the statement, which can be established when one formulates scattering theory relativistically, that $d\sigma$ is an invariant. Hence

$$\frac{d\sigma}{dt} \tag{ST 1-19}$$

is invariant and, to exhibit the cross section in a form in which the transformations from one frame to another are most easily done, it is best to write it as a function s and t. We will not do this here. As a final peripheral comment we note that the expression for the

$$\int \frac{d^3\mathbf{p}}{(2\pi\hbar)^3}$$

is not relativistically invariant. However the manifest invariance of

$$\int .. \int d^4p \, \delta(p^2 - m^2c^2)$$

$$= \int d^3\mathbf{p} \int_{p_0>0} dp_0 \delta(p_0{}^2 - \mathbf{p}^2 - m^2c^2)$$

$$= \int d^3\mathbf{p} \frac{1}{2\sqrt{\mathbf{p}^2 + m^2c^2}} = \frac{c}{2} \int \frac{d^3\mathbf{p}}{(\mathbf{p}^2c^2 + m^2c^4)^{1/2}}$$

$$\text{(ST 1-20)}$$

shows that

$$\int \frac{d^3\mathbf{p}}{E} \frac{1}{(2\pi\hbar)^3} \qquad \text{(ST 1-21)}$$

is invariant. Matrix elements in relativistic theories always have the particles normalized not according to

$$\frac{1}{\sqrt{V}} e^{i\mathbf{p}\cdot\mathbf{r}/\hbar}$$

but according to

$$\frac{1}{\sqrt{V}} \frac{1}{\sqrt{E}} e^{i\mathbf{p}\cdot\mathbf{r}/\hbar}$$

so that the necessary factors emerge from the square of the matrix element.

The Equivalence Principle

According to the relativity principle, the laws of physics must be such as to make it impossible to distinguish between two inertial frames that differ from each other only in that one is moving with a constant velocity relative to the other. Common experience suggests another equivalence; it is not possible to distinguish by simple mechanical observations whether a system is in a uniform gravitational field or whether it is in a gravity-free region, but subject to a constant acceleration of the appropriate magnitude and direction. In real gravitational fields, the equivalence only holds on a scale small enough so that the difference in gravitational potentials between two points $\mathbf{r}_1$ and $\mathbf{r}_2$ is linear in $|\mathbf{r}_1 - \mathbf{r}_2|$. Einstein proposed that this equivalence be a fundamental principle of nature, and that *all* laws of physics be in accord with it.

The acceptance of it has some far-reaching consequences. First of all we note that the uniform acceleration $\mathbf{a}$ must be produced by a force $\mathbf{F}$, and these are related by

$$\mathbf{F} = m\mathbf{a} \qquad\qquad (\text{ST 2-1})$$

The equivalent gravitational field may be produced by a mass M a distance R away, provided the gravitational force

$$\mathbf{F} = G\frac{mM}{R^3}\mathbf{R} \qquad\qquad (\text{ST 2-2})$$

is made equal to $m\mathbf{a}$ by an appropriate choice of M and (large) R. There is, however, no *a priori* reason why the inertial mass of the object which appears in the relation (ST 2-1) should equal the gravitational mass that enters into the gravitational potential energy

$$V(\mathbf{r}) = -G\frac{Mm}{r} \qquad\qquad (\text{ST 2-3})$$

The gravitational mass enters into the above as a "coupling constant" just as the charges do into the Coulomb potential, and one could imagine, for example, that

465

the "masses" that enter into (ST2-3) are different from the inertial masses that appear in (ST2-1). This is not possible if the equivalence principle really holds. Suppose, for example, that a rather extreme situation were to hold: electrons are not subject to gravitational forces. In that case the inertial mass of an atom is (approximately) $M(A,Z) = AM + Zm$ where M and m are the nucleon and electron masses, respectively. On the other hand, the gravitational mass that enters into (ST2-3) is just AM, to the same approximation. Thus, inside a satellite a mass of lead might be floating, whereas a mass of material for which the Z/A ratio is different would fall. This suggests an experimental test of the equivalence principle; according to the principle, all materials should behave in the same way under a combination of gravitational and centrifugal forces. This test, showing the equivalence between inertial and gravitational mass, has been carried out, and established to an accuracy of one part in 10^{11} in recent experiments by Roll, Krotkov, and Dicke,[1] but was known to be true to an accuracy of a few parts in 10^9 from early experiments of Eötvös (1890, 1922). The principle of the experiment involves suspending two equal masses of different material (gold and aluminum) from a torsion balance. Any difference in the acceleration of the two masses toward the sun as the earth moves in its orbit would result in a deflection, which in fact, was not observed.

Another consequence was mentioned in Chapter 2 and in Chapter 22 (Section B). A photon of energy E has gravitational mass E/c^2: to establish this, we consider an atom in an excited state, with mass M^* at a height x above some reference level. The work done to lift it to that height is M^*gx. When the atom decays to the ground state, of mass M, a photon of energy $M^*c^2 - Mc^2$ is emitted.

Suppose that the photon is absorbed at the reference level. The energy absorbed is $(E + E_g)$ where E_g is the gravitational energy that it acquired in the fall. If the atom in the ground state also drops to the reference level, the total work done by the gravitational field is Mgx on the atom. If the absorbed energy is used to excite the atom back to the state of mass M^*, we are back to the original situation, with M^* at the reference level, provided

$$M^*gx = E_g + Mgx \qquad (ST\ 2\text{-}4)$$

Thus

$$E_g = (M^* - M)\,gx = \frac{E}{c^2}\,gx \qquad (ST\ 2\text{-}5)$$

Consider now a photon at a height x. Let its frequency there be ν, so that

$$E = h\nu \qquad (ST\ 2\text{-}6)$$

[1] For a detailed description of the experiments, see R. H. Dicke, *The Theoretical Significance of Experimental Relativity*, Gordon and Breach, Science Publishers, New York (1964).

The photon energy at the reference level is

$$E + E_g = E\left(1 + \frac{gx}{c^2}\right)$$

$$= h\nu\left(1 + \frac{gx}{c^2}\right) \tag{ST 2-7}$$

This must equal $h\nu'$, where ν' is the frequency of the photon measured at the reference level. Thus the relation

$$\nu' = \nu\left(1 + \frac{gx}{c^2}\right) \tag{ST 2-8}$$

implies that the frequency of the photon is raised and

$$\frac{\Delta\nu}{\nu} = \frac{gx}{c^2} \tag{ST 2-9}$$

The period, which is reciprocal to ν, is thus changed according to

$$\frac{\Delta T}{T} = -\frac{gx}{c^2} \tag{ST 2-10}$$

This prediction was confirmed in a terrestrial experiment done with the Mössbauer effect (Chapter 22, section B). The shift is more dramatic if we compare the frequency of an atomic emission line on the surface of a massive star with the frequency of the line on earth. There

$$\frac{\Delta\nu}{\nu} = \frac{GM}{Rc^2} \tag{ST 2-11}$$

For the sun, whose mass is 1.99×10^{33} gm, and whose radius is 6.96×10^{10} cm, the shift is small, $\Delta\nu/\nu = 2.12 \times 10^{-6}$. This has been measured for the sodium line in sunlight (Dicke, loc. cit.) and observation agrees with theory to 5%. The shift is called the *gravitational red shift* since the earth is "up" relative to the sun.

Another prediction follows by analogy with electrostatics. Just like a charge is deflected by a Coulomb field, with a deflection angle given by

$$\tan\frac{\theta}{2} = \frac{Ze^2}{mv^2}\frac{1}{b} \tag{ST 2-12}$$

where Ze and e are the two charges, m the mass of the moving charge, v is its asymptotic velocity, and b is the impact parameter, so will the photon be deflected by a large mass, for example, the sun. Since the force law is the same, we just make the substitution $Ze^2 \rightarrow GMm$ and $v \rightarrow c$. The impact parameter is roughly the radius of the sun, so that for two stars,

$$\delta\theta \simeq \frac{GM}{Rc^2} \qquad \text{(ST 2-13)}$$

The numerical value of this is 0.83″. The actual measured value of this, observed by looking at stars near the rim of the sun during a total eclipse, is just twice this value. The explanation of this discrepancy lies in the Einstein Theory of Gravitation, which is beyond the scope of this book.

It should be stressed that Planck's constant does not enter (ST 2-8). The result is a purely classical one, and can be derived without the use of the connection $E = h\nu$, which is a convenience, but not necessary.[2]

---

[2] The original paper of Einstein is very readable, and is reprinted in translation in *The Principle of Relativity*, a collection of original papers, published by Dover Publications, Inc.

The Wentzel-Kramers-Brillouin Approximation

This approximation method is particularly useful when one is dealing with slowly varying potentials. Exactly what this means will become clear later. One wants to solve the equation

$$\frac{d^2\psi(x)}{dx^2} + \frac{2m}{\hbar^2} [E - V(x)] \, \psi(x) = 0 \qquad \text{(ST 3-1)}$$

and to do so, it is useful to write

$$\psi(x) = R(x) \, e^{iS(x)/\hbar} \qquad \text{(ST 3-2)}$$

Then

$$\frac{d^2\psi}{dx^2} = \left[\frac{d^2R}{dx^2} + \frac{2i}{\hbar} \frac{dR}{dx} \frac{dS}{dx} + \frac{i}{\hbar} R \frac{d^2S}{dx^2} - \frac{1}{\hbar^2} R \left(\frac{dS}{dx} \right)^2 \right] e^{iS(x)/\hbar} \quad \text{(ST 3-3)}$$

so that the differential equation splits into two, by taking the real and imaginary part of (ST 3-1) after (ST 3-3) has been substituted. The imaginary part gives

$$R \frac{d^2S}{dx^2} + 2 \frac{dR}{dx} \frac{dS}{dx} = 0 \qquad \text{(ST 3-4)}$$

that is,

$$\frac{d}{dx} \left(\log \frac{dS}{dx} + 2 \log R \right) = 0$$

whose solution is

$$\frac{dS}{dx} = \frac{C}{R^2} \qquad \text{(ST 3-5)}$$

469

The real part reads

$$\frac{d^2R}{dx^2} - \frac{1}{\hbar^2} R \left(\frac{dS}{dx} \right)^2 + \frac{2m[E - V(x)]}{\hbar^2} R = 0$$

which, when (ST 3-5) is substituted, becomes

$$\frac{d^2R}{dx^2} - \frac{C^2}{\hbar^2} \frac{1}{R^3} + \frac{2m[E - V(x)]}{\hbar^2} R = 0 \qquad \text{(ST 3-6)}$$

At this point we make the approximation that

$$\frac{1}{R} \frac{d^2R}{dx^2} \ll \frac{C^2}{\hbar^2} \frac{1}{R^4} = \frac{1}{\hbar^2} \left(\frac{dS}{dx} \right)^2 \qquad \text{(ST 3-7)}$$

so that the equation becomes

$$\frac{C^2}{R^4} = 2m[E - V(x)] \qquad \text{(ST 3-8)}$$

Thus

$$\frac{C}{R^2} = \frac{dS}{dx} = \sqrt{2m[E - V(x)]} \qquad \text{(ST 3-9)}$$

and hence

$$S(x) = \int_{x_1}^{x} dy \sqrt{2m[E - V(x)]} \qquad \text{(ST 3-10)}$$

The condition for the validity can be translated into a statement about the variation of $V(x)$. It will be satisfied if $V(x)$ varies slowly in a wavelength, which varies from point to point, but which for slowly varying $V(x)$ is defined by

$$\lambda(x) = \frac{\hbar}{p(x)} = \frac{\hbar}{\{2m[E - V(x)]\}^{1/2}} \qquad \text{(ST 3-11)}$$

At the points where

$$E - V(x) = 0 \qquad \text{(ST 3-12)}$$

special treatment is required, because in the approximate Eq. ST 3-8 $R(x)$ appears to be singular. This cannot be, and this means that the approximation (ST 3-7) must be poor there. The special points are called *turning points* because it is there that a classical particle would turn around; it can only move where $E - V(x) \geq 0$. The way of handling solutions near turning points is a little too technical to be presented here. The basic idea is that we have a solution to the left of the turning point [where $E > V(x)$, say], of the form

$$\psi(x) = R \, e^{i \int_{x_1}^{x} dy \sqrt{(2m/\hbar^2)[E-V(y)]}} \qquad \text{(ST 3-13)}$$

and a solution to the right of the turning point [where $E < V(x)$], and what we need is a formula that interpolates between them. In the vicinity of the turning point one can approximate $\sqrt{(2m/\hbar^2)} [E - V(x)]$ by a straight line over a small interval, and solve the Schrödinger equation exactly. Since it is a second-order equation, there are two adjustable constants, one of which is fixed by fitting the solution to (ST 3-13) and the other by fittting it to

$$\psi(x) = R \, e^{-\int_{x_1}^{x} dy \, \sqrt{(2m/\hbar^2)[V(y)-E]}} \qquad \text{(ST 3-14)}$$

the solution to the right of the turning point.[1] The above solution decreases in amplitude as x increases. The total attenuation at the next turning point, when $E \geq V(x)$ again, is

$$\frac{\psi(x_{II})}{\psi(x_I)} \simeq \exp \left\{ - \int_{x_I}^{x_{II}} dy \, \sqrt{(2m/\hbar^2)[V(y)-E]} \right\} \qquad \text{(ST 3-15)}$$

which is just the square root of the transmission probability that we found in Chapter 5.

---

[1] For more details, see almost any of the more advanced books on quantum mechanics, for example, J. L. Powell and B. Crasemann, *Quantum Mechanics*, Addison-Wesley Publishing Co. (1961); L. I. Schiff, *Quantum Mechanics*, McGraw-Hill Book Co. (1968).

Lifetimes, Line Widths, and Resonances

In this section we will discuss a slightly improved treatment of transition rates, which will indicate how the exponential decay behavior comes about.[1] The limited mathematical sophistication assumed of the reader will make the treatment somewhat less elegant than is possible.

To simplify the problem as much as possible, we consider an atom with just two levels, the ground state, with energy 0, and a single excited state, with energy E. The two states are coupled to the electromagnetic field, which we will take to be scalar, so that no polarization vectors appear. We will only consider the subset of eigenstates of H_0 consisting of the excited state ϕ_1, for which

$$H_0\phi_1 = E\phi_1 \qquad \text{(ST 4-1)}$$

and of the ground state + one photon, $\phi(\mathbf{k})$, for which

$$H_0\phi(\mathbf{k}) = \epsilon(\mathbf{k})\,\phi(\mathbf{k}) \qquad \text{(ST 4-2)}$$

and limit ourselves to these in an expansion of an arbitrary function. This is certainly justified when the coupling between the two states, ϕ_1 and $\phi(\mathbf{k})$ through the potential V, is small, as in electromagnetic coupling, since then the influence of two-, three-, ... photon states is negligible. Note that

$$\langle \phi_1 | \phi(\mathbf{k}) \rangle = 0 \qquad \text{(ST 4-3)}$$

even when the $\mathbf{k}$ is such that the energies $\epsilon(\mathbf{k})$ and E are the same. The states are orthogonal because one has a photon in it and the other does not, and because for one of them the atom is in an excited state, and for the other it is not.

The solution of the equation

$$i\hbar\,\frac{d\psi(t)}{dt} = (H_0 + V)\,\psi(t) \qquad \text{(ST 4-4)}$$

---

[1] This was first derived by Weisskopf and Wigner (1930).

may be written in terms of the complete set

$$\psi(t) = a(t)\,\phi_1\,e^{-iEt/\hbar} + \int d^3k\, b(\mathbf{k},t)\,\phi(\mathbf{k})\,e^{-i\epsilon(\mathbf{k})t/\hbar} \qquad \text{(ST 4-5)}$$

When this is inserted into (ST 4-4),

$$i\hbar\,\frac{da}{dt}\,e^{-iEt/\hbar}\,\phi_1 + Ea\,e^{-iEt/\hbar}\,\phi_1$$

$$+\; i\hbar \int d^3k\,\frac{db(\mathbf{k},t)}{dt}\,e^{-i\epsilon(\mathbf{k})t/\hbar}\,\phi(\mathbf{k}) + \int d^3k\,\epsilon(\mathbf{k})\,b(\mathbf{k},t)\,e^{-i\epsilon(\mathbf{k})t/\hbar}\,\phi(\mathbf{k})$$

$$=\; Ea(t)\,e^{-iEt/\hbar}\,\phi_1 + \int d^3k\,\epsilon(\mathbf{k})\,b(\mathbf{k},t)\,e^{-i\epsilon(\mathbf{k})t/\hbar}\,\phi(\mathbf{k})$$

$$+\; a(t)\,e^{-iEt/\hbar}\,V\phi_1 + \int d^3k\, b(\mathbf{k},t)\,e^{-i\epsilon(\mathbf{k})t/\hbar}\,V\phi(\mathbf{k})$$

results. If we take the scalar product with ϕ_1, we get

$$i\hbar\,\frac{da}{dt} = a(t)\langle\phi_1|V|\phi_1\rangle + \int d^3k\, b(\mathbf{k},t)\,e^{-i[\epsilon(\mathbf{k})-E]t/\hbar}\,\langle\phi_1|V|\phi(\mathbf{k})\rangle$$

Since V, acting on a state, is supposed to change the photon number by one, $\langle\phi_1|V|\phi_1\rangle = 0$. With the notation

$$\epsilon(\mathbf{k}) - E = \hbar\omega(\mathbf{k})$$

$$\langle\phi_1|V|\phi(\mathbf{k})\rangle = M(\mathbf{k}) \qquad \text{(ST 4-6)}$$

the equation becomes

$$i\hbar\,\frac{da(t)}{dt} = \int d^3k\, b(\mathbf{k},t)\,e^{-i\omega(\mathbf{k})t}\,M(\mathbf{k}) \qquad \text{(ST 4-7)}$$

If we take the scalar product with $\phi(\mathbf{q})$, and again use photon counting to set $\langle\phi(\mathbf{q})|V|\phi(\mathbf{k})\rangle = 0$, we get, after a little manipulation, using a normalization that

$$\langle\phi(\mathbf{q})|\phi(\mathbf{k})\rangle = \delta(\mathbf{k} - \mathbf{q}) \qquad \text{(ST 4-8)}$$

the equation

$$i\hbar\,\frac{db(\mathbf{q},t)}{dt} = a(t)\,e^{i\omega(\mathbf{q})t}\,M^*(\mathbf{q}) \qquad \text{(ST 4-9)}$$

Since $b(\mathbf{k},0) = 0$ if the excited state is occupied at $t = 0$, a solution of this equation is

$$b(\mathbf{k},t) = \frac{1}{i\hbar}\,M^*(\mathbf{k}) \int_0^t dt'\, e^{i\omega(\mathbf{k})t'}\,a(t') \qquad \text{(ST 4-10)}$$

We now insert this into (ST 4-7) to get

$$\frac{da(t)}{dt} = -\frac{1}{\hbar^2} \int d^3\mathbf{k} \, |M(\mathbf{k})|^2 \, e^{-i\omega(\mathbf{k})t} \int_0^t dt' \, a(t') \, e^{i\omega(\mathbf{k})t'} \quad \text{(ST 4-11)}$$

Next multiply both sides of the equation by $e^{-\lambda t}$ and integrate over time from 0 to ∞. On the left hand side

$$\int_0^\infty dt \, e^{-\lambda t} \frac{da(t)}{dt} = \int_0^\infty dt \, \frac{d}{dt}[e^{-\lambda t} a(t)] + \lambda \int_\infty^\infty dt \, e^{-\lambda t} a(t)$$

$$= \lambda \int_0^\infty dt \, e^{-\lambda t} a(t) - 1 \quad \text{(ST 4-12)}$$

where we have used

$$a(0) = 1 \quad \text{(ST 4-13)}$$

On the right side we have

$$-\frac{1}{\hbar^2} \int d^3\mathbf{k} \, |M(\mathbf{k})|^2 \int_0^\infty dt \, e^{-\lambda t} \, e^{-i\omega(\mathbf{k})t} \int_0^t dt' a(t') \, e^{i\omega(\mathbf{k})t'}$$

Now, as can be seen from Figure ST 4-1, the integral

$$\int_0^\infty dt \, e^{-[\lambda + i\omega(\mathbf{k})]t} \int_0^t dt' a(t') \, e^{i\omega(\mathbf{k})t'}$$

over the first octant in the t–t' plane, can also be written as

$$\int_0^\infty dt' a(t') \, e^{i\omega(\mathbf{k})t'} \int_{t'}^\infty dt \, e^{-[\lambda + i\omega(\mathbf{k})]t}$$

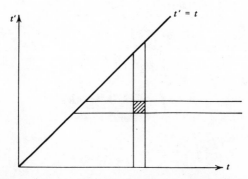

Fig. ST4-1.   The integration over the first octant can be done either by holding t fixed, integrating along the vertical strip, and then summing the strips from $t = 0$ to $t = \infty$, or by first taking the integral along the horizontal strip from $t = t'$ to $t = \infty$ and then summing over all the horizontal strips from $t' = 0$ to $t' = \infty$.

and here, the last integral can be done, so that

$$\lambda \int_0^\infty dt\, e^{-\lambda t}\, a(t) - 1 = -\frac{1}{\hbar^2} \int d^3\mathbf{k}\, |M(\mathbf{k})|^2 \int_0^\infty dt'\, a(t') e^{i\omega(\mathbf{k})t'} \frac{e^{-[\lambda+i\omega(\mathbf{k})]t'}}{\lambda + i\omega(\mathbf{k})}$$

$$= -\frac{1}{\hbar^2} \int d^3\mathbf{k}\, \frac{|M(\mathbf{k})|^2}{\lambda + i\omega(\mathbf{k})} \int_0^\infty dt'\, a(t')\, e^{-\lambda t'} \quad \text{(ST 4-14)}$$

We can solve for $\int_0^\infty dt\, a(t)\, e^{-\lambda t}$ to get

$$\int_0^\infty dt\, a(t)\, e^{-\lambda t} = \frac{1}{\lambda + \dfrac{1}{\hbar^2} \displaystyle\int d^3\mathbf{k}\, \dfrac{|M(\mathbf{k})|^2}{\lambda + i\omega(\mathbf{k})}} \quad \text{(ST 4-15)}$$

The reader familiar with the theory of Laplace transformations will recognize the above as such. The inversion of the Laplace transformation of the above form needs some discussion, which can be found in the more advanced literature. We will argue as follows. Although we do not know how to extract $a(t)$ from the above, we can examine the relation in the limit that $\lambda \to 0$. If we make the *Ansatz*

$$a(t) = e^{-zt} \quad \text{(ST 4-16)}$$

we get,

$$\frac{1}{z + \lambda} = \frac{1}{\lambda - \dfrac{i}{\hbar^2} \displaystyle\int d^3\mathbf{k}\, \dfrac{|M(\mathbf{k})|^2}{\omega(\mathbf{k}) - i\lambda}}$$

which implies in the limit $\lambda \to 0$ that

$$z = \operatorname*{Lim}_{\lambda \to 0+} -\frac{i}{\hbar^2} \int d^3\mathbf{k}\, \frac{|M(\mathbf{k})|^2}{\omega(\mathbf{k}) - i\lambda} \quad \text{(ST 4-17)}$$

We can write this as

$$z = \operatorname*{Lim}_{\lambda \to 0+} -\frac{i}{\hbar^2} \int d^3\mathbf{k}\, \frac{|M(\mathbf{k})|^2}{\omega^2(\mathbf{k}) + \lambda^2} (\omega(\mathbf{k}) + i\lambda)$$

$$= \operatorname*{Lim}_{\lambda \to 0+} -\frac{i}{\hbar^2} \int d^3\mathbf{k}\, \frac{|M(\mathbf{k})|^2\, \omega(\mathbf{k})}{\omega^2(\mathbf{k}) + \lambda^2}$$

$$+ \frac{1}{\hbar^2} \int d^3\mathbf{k}\, |M(\mathbf{k})|^2 \frac{\lambda}{\omega^2(\mathbf{k}) + \lambda^2} \quad \text{(ST 4-18)}$$

In evaluating the real part of z we make use of the relation

$$\operatorname*{Lim}_{\lambda \to 0+} \frac{\lambda}{\omega^2(\mathbf{k}) + \lambda^2} = \pi\, \delta[\omega(\mathbf{k})] \quad \text{(ST 4-19)}$$

so that finally, we get

$$z = - \frac{i}{\hbar^2} \int d^3\mathbf{k}\, \frac{|M(\mathbf{k})|^2}{\omega(\mathbf{k})} + \frac{\pi}{\hbar} \int d^3\mathbf{k}\, |M(\mathbf{k})|^2\, \delta[\hbar\omega(\mathbf{k})]$$

(ST 4-20)

When this is exponentiated, we find that the coefficient of ϕ_1 in $\psi(t)$ is

$$e^{-\gamma t/2}\, e^{-iE't/\hbar}$$

(ST 4-21)

where

$$\gamma = \frac{2\pi}{\hbar} \int d^3\mathbf{k}\, |\langle \phi_1 | V | \phi(\mathbf{k}) \rangle|^2\, \delta[\epsilon(\mathbf{k}) - E]$$

$$E' = E + \int d^3\mathbf{k}\, |\langle \phi_1 | V | \phi(\mathbf{k}) \rangle|^2\, \frac{1}{E - \epsilon(\mathbf{k})}$$

(ST 4-22)

Thus the probability of finding $\psi(t)$ in the state ϕ_1 after a time t is, to the extent that our solution is approximately correct,

$$|a(t)|^2 = e^{-\gamma t}$$

(ST 4-23)

where γ is the decay rate calculated in perturbation theory. Furthermore, the oscillatory behavior of $a(t)$ is characterized by the energy E of the state ϕ_1, *shifted by the second order perturbation energy shift*, as comparison with (16-16) shows. The only difference is that the intermediate states summed over here form a continuum, and thus the limiting process shown in (ST 4-18) must be used to define the integral when the energy denominator can vanish.

Another quantity of interest is the probability that the state $\psi(t)$ ends up in the state $\phi(\mathbf{k})$ at $t = \infty$. This is given by $|b(\mathbf{k}, \infty)|^2$, where, according to (ST 4-10) and (ST 4-16), we have

$$b(\mathbf{k}, \infty) = \frac{1}{i\hbar}\, M^*(\mathbf{k}) \int_0^\infty dt'\, e^{-[z - i\omega(\mathbf{k})]t'}$$

$$= \frac{M^*(\mathbf{k})}{i\hbar}\, \frac{1}{\frac{\pi}{\hbar} \int d^3\mathbf{k}\, |M(\mathbf{k})|^2\, \delta[\hbar\omega(\mathbf{k})] - i\,\Delta}$$

where

$$\Delta = \frac{[\epsilon(\mathbf{k}) - E]}{\hbar} - \frac{1}{\hbar} \int d^3\mathbf{k}'\, \frac{|M(\mathbf{k}')|^2}{E - \epsilon(\mathbf{k}')}$$

Thus

$$b(\mathbf{k}, \infty) = \frac{M^*(\mathbf{k})}{\epsilon(\mathbf{k}) - E - \int d^3\mathbf{k}'\, \frac{|M(\mathbf{k}')|^2}{E - \epsilon(\mathbf{k}')} + i\hbar\gamma/2}$$

(ST 4-24)

and the absolute square of this,

$$|b(\mathbf{k}, \infty)|^2 = \frac{|M(\mathbf{k})|^2}{[\epsilon(\mathbf{k}) - E - \Delta E]^2 + (\hbar\gamma/2)^2} \qquad \text{(ST 4-25)}$$

yields the Lorentzian shape for the line width, that is, the photon energy is centered about the (shifted) energy of the excited level, with the width described by $\hbar\gamma/2$. The energy shift is small, and usually ignored.

The same form appears in the scattering problem. Consider the scattering of a "photon" of momentum $\mathbf{k}_i$ by the atom in the ground state. The state of the system is again described by equations (ST 4-1, ST 4-2, ST 4-7, and ST 4-9) except that initially, which here means at $t = -\infty$, the state is specifically given as $\phi(\mathbf{k}_i)$, so that

$$b(\mathbf{q},t) = \delta(\mathbf{q} - \mathbf{k}_i) \qquad \text{at } t = -\infty \qquad \text{(ST 4-26)}$$

Hence the integration of (ST 4-9) gives

$$b(\mathbf{q},t) = \delta(\mathbf{q} - \mathbf{k}_i) + \frac{1}{i\hbar} M^*(\mathbf{q}) \int_{-\infty}^{t} dt' a(t') e^{i\omega(\mathbf{q})t'} \qquad \text{(ST 4-27)}$$

The quantity of interest is the amplitude for a transition into a final state in which the photon has momentum $\mathbf{k}_f$ at $t = +\infty$, that is, it is

$$\langle \phi(\mathbf{k}_f) | \psi(+\infty) \rangle = b(\mathbf{k}_f, +\infty)$$

$$= \delta(\mathbf{k}_f - \mathbf{k}_i) - \frac{i}{\hbar} M^*(\mathbf{k}_f) \int_{-\infty}^{\infty} dt' a(t') e^{i\omega_f t'}$$

$$(\omega_f \equiv \omega(\mathbf{k}_f)) \qquad \text{(ST 4-28)}$$

using the previous equation.

Substituting (ST 4-27) into (ST 4-7) yields the equation

$$\frac{da(t)}{dt} = \frac{1}{i\hbar} e^{-i\omega_i t} M(\mathbf{k}_i) - \frac{1}{\hbar^2} \int d^3k \, |M(\mathbf{k})|^2 e^{-i\omega(\mathbf{k})t} \int_{-\infty}^{t} dt' a(t') e^{i\omega(\mathbf{k})t'} \qquad \text{(ST 4-29)}$$

This may be integrated, taking into account that $a(-\infty) = 0$, to give

$$a(t) = \frac{M(\mathbf{k}_i)}{i\hbar} \int_{-\infty}^{t} dt' \, e^{-i\omega_i t'}$$

$$- \frac{1}{\hbar^2} \int d^3k \, |M(\mathbf{k})|^2 \int_{-\infty}^{t} dt' \, e^{-i\omega(\mathbf{k})t'} \int_{-\infty}^{t'} dt'' a(t'') e^{i\omega(\mathbf{k})t''} \qquad \text{(ST 4-30)}$$

Now the integral $\int_{-\infty}^{t} dt' \, e^{-i\omega_i t'}$ is not well defined. The standard procedure is to write it in the form

$$\text{Lim}_{\epsilon \to 0} \int_{-\infty}^{t} dt' \, e^{-i(\omega_i + i\epsilon)t'} = \text{Lim}_{\epsilon \to 0} \, i \, \frac{e^{-i(\omega_i + i\epsilon)t}}{\omega_i + i\epsilon} \tag{ST 4-31}$$

The use of a convergence factor, which is then allowed to vanish in a well-defined way, is somewhat similar to the treatment of the Coulomb potential as the limiting case of a screened Coulomb potential in Chapter 24. Next, as can be seen from Figure ST 4-2,

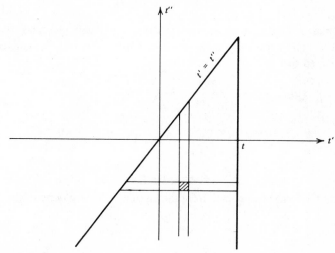

Fig. ST4-2. The integral in (ST4-30) can either be written as a "sum" of the vertical strips, as in the equation, as or, a sum of the horizontal strips, as in (ST4-32). The same interchange of orders is used in Eq. (ST4-33) except .that the vertical line at $t' = t$ is shifted to $+\infty$.

$$\int_{-\infty}^{t} dt' \, e^{-i\omega(\mathbf{k})t'} \int_{-\infty}^{t'} dt'' a(t'') \, e^{i\omega(\mathbf{k})t''} = \int_{-\infty}^{t} dt'' a(t'') \, e^{i\omega(\mathbf{k})t''} \int_{t''}^{t} dt' \, e^{-i\omega(\mathbf{k})t'}$$

$$= \frac{i}{\omega(\mathbf{k})} \int_{-\infty}^{t} dt'' a(t'') [e^{-i\omega(\mathbf{k})(t-t'')} - 1]$$

so that

$$a(t) = \frac{M(\mathbf{k}_i) \, e^{-i\omega_i t}}{h(\omega_i + i\epsilon)} - \frac{i}{\hbar^2} \int d^3\mathbf{k} \, \frac{|M(\mathbf{k})|^2}{\omega(\mathbf{k})} \int_{-\infty}^{t} dt'' a(t'') [e^{-i\omega(\mathbf{k})(t-t'')} - 1]$$

$$\tag{ST 4-32}$$

According to (ST 4-28) the quantity of interest for nonforward scattering (so that the first term can be ignored) is

$$\int_{-\infty}^{\infty} dt a(t) \, e^{i\omega_f t} = \frac{M(\mathbf{k}_i)}{\hbar(\omega_i + i\epsilon)} \int_{-\infty}^{\infty} dt \, e^{i(\omega_f - \omega_i)t}$$

$$- \frac{i}{\hbar^2} \int d^3\mathbf{k} \, \frac{|M(\mathbf{k})|^2}{\omega(\mathbf{k})} \int_{-\infty}^{\infty} dt \, e^{i\omega_f t} \int_{-\infty}^{t} dt'' a(t'') \, [e^{-i\omega(\mathbf{k})(t-t'')} - 1]$$

$$= \frac{2\pi M(\mathbf{k}_i)}{\hbar(\omega_i + i\epsilon)} \, \delta(\omega_f - \omega_i)$$

$$- \frac{i}{\hbar^2} \int d^3\mathbf{k} \, \frac{|M(\mathbf{k})|^2}{\omega(\mathbf{k})} \int_{-\infty}^{\infty} dt'' a(t'') \int_{t''}^{\infty} dt \, \{e^{i\omega(\mathbf{k})t''} \, e^{i[\omega_f - \omega(\mathbf{k})]t} - e^{i\omega_f t}\}$$

(ST 4-33)

where in the last term we again rewrote the integral with the help of Figure ST 4-2. The t-integration can again be done with the help of the convergence factor trick, so that (ST 4-33) now reads

$$\int_{-\infty}^{\infty} dt a(t) \, e^{i\omega_f t} = \frac{2\pi M(\mathbf{k}_i) \, \delta(\omega_f - \omega_i)}{\hbar(\omega_i + i\epsilon)}$$

$$+ \frac{1}{\hbar^2} \int d^3\mathbf{k} \, \frac{|M(\mathbf{k})|^2}{\omega(\mathbf{k})} \int_{-\infty}^{\infty} dt'' a(t'') \, e^{i\omega_f t''}$$

$$\times \left[\frac{1}{\omega_f - \omega(\mathbf{k}) + i\epsilon} - \frac{1}{\omega_f + i\epsilon} \right]$$

which is in the form of an equation for the unknown. This may be solved to give

$$\int_{-\infty}^{\infty} dt a(t) \, e^{i\omega_f t} = \frac{2\pi M(\mathbf{k}_i) \, \delta(\omega_f - \omega_i)}{\hbar(\omega_i + i\epsilon)}$$

$$\times \frac{1}{1 - \frac{1}{\hbar^2 \omega_f} \int d^3\mathbf{k} \, \frac{|M(\mathbf{k})|^2}{\omega_f - \omega(\mathbf{k}) + i\epsilon}} \qquad \text{(ST 4-34)}$$

Hence, in the nonforward direction

$$b(\mathbf{k}_f, \infty) = - \frac{i}{\hbar} M^*(\mathbf{k}_f) \cdot 2\pi M(\mathbf{k}_i) \, \delta(\omega_f - \omega_i)$$

$$\times \frac{1}{\hbar\omega_i + i\epsilon - \int d^3\mathbf{k} \, \frac{|M(\mathbf{k})|^2}{\hbar\omega_f - \hbar\omega(\mathbf{k}) + i\epsilon}}$$

$$= \frac{-2\pi i \, \delta(\hbar\omega_f - \hbar\omega_i) \, M(\mathbf{k}_i) \, M^*(\mathbf{k}_f)}{\epsilon(\mathbf{k}_i) - E - \int d^3\mathbf{k} \, \frac{|M(\mathbf{k})|^2}{\epsilon(\mathbf{k}_i) - \epsilon(\mathbf{k})} + i\pi \int d^3\mathbf{k} \, |M(\mathbf{k})|^2 \, \delta[\epsilon(\mathbf{k}_i) - \epsilon(\mathbf{k})]}$$

(ST 4-35)

The amplitude peaks strongly when the incident (and final) energy $\epsilon_i \, (= \epsilon_f)$ is near the energy of the excited state of the atom, shifted to $E + \Delta E$, as in (ST 4-25). This justifies the comments made at the end of Chapter 18.

The Yukawa Theory

The earliest nuclear physics experiments such as the scattering of α particles by light nuclei, and the study of α decay lifetimes,[1] showed that the nuclear forces were such that the radius of the nucleus (with A nucleons) was

$$R = r_0 A^{1/3} \qquad \text{(ST 5-1)}$$

with $r_0 \simeq 1.1 \times 10^{-13}$ cm. Thus the nuclear density (nucleons/unit volume) was a constant independent of A. With long-range forces, such as the electrostatic forces, there would be $A(A - 1)/2$ "bonds" and one would expect the density to increase with A, since the kinetic energy only increases linearly with A. The constancy of the nuclear density thus strongly suggested short-range forces. These forces, leading to binding energies measured in MeV, rather than electron volts, had to be significantly stronger than electromagnetic forces.

In search for a mechanism that would give rise to such forces, Yukawa in 1935 drew on the insights gained from the successes of quantum electrodynamics and proposed his *meson theory of nuclear forces*. At the level of this book, only a qualitative description of that theory can be given. The interaction between two charged particles at rest (or very low velocities) can be described in terms of an "action-at-a-distance" Coulomb interaction. A more accurate description involves the field concept; the charged particles are sources of, and interact with electromagnetic fields $(\mathbf{E}, \mathbf{B})$ and thus they interact with each other through the intermediary of the field. In a description that is accurate on the quantum level, the electromagnetic field is quantized, and the quanta, the photons, are the carriers of the field.[2] Two charges can interact by the following mechanism. Charge "1" emits a photon. We know from energy-momentum conservation that this cannot be a real photon, or equivalently, a real process; either the photon has an energy that does not correspond to its momentum

---

[1] See the discussion of tunnelling in Chapter 5.

[2] This is somewhat fuzzy language, and should not be taken literally. In particular, it would be wrong to think of a photon somehow carrying a cloud of field with it.

(i.e., $E = pc$), or, the photon is real, but energy is not conserved in the reaction[3]

$$e_1 \rightarrow e_1 + \gamma$$

When that photon is absorbed by charge "2" the imbalance in energy can be corrected, since that process

$$\gamma + e_2 \rightarrow e_2$$

also cannot conserve energy. The question is, why should such an "exchange" of photons give rise to an attraction or repulsion between charges? The answer really is that our visual description of the exchange is just an interpretation of the second-order perturbation energy shift due to a perturbing potential H_1. The formula (16-16), describing the energy shift from the interaction-free energy of two charges e_1 and e_2, reads

$$E_{e_1 e_2}^{(2)} = - \sum_{n \neq e_1 e_2} \frac{\langle e_1 e_2 | H_1 | n \rangle \langle n | H_1 | e_1 e_2 \rangle}{E_n - E_{e_1 e_2}^{(0)}} \qquad \text{(ST 5-2)}$$

The sum is over all intermediate states that can be obtained from H_1 acting on the state $| e_1 e_2 \rangle$. Our verbal description corresponds to the intermediate state in which e_2 is in its initial state, and e_1 has, through the action of H_1, emitted a photon, so that $| n \rangle = | e_1' \gamma, e_2 \rangle$ here. The energy of the intermediate state is the recoil energy of the e_1 $\sqrt{(pc)^2 + (m_e c^2)^2}$, plus the photon energy pc. The sum over intermediate states corresponds to an integration over all possible photon momenta consistent with momentum conservation, that is, an integration over all directions. Since the charge "2" could be the one that emits the photon, one must also calculate the contribution from the emission of the photon by e_2 and its absorption by e_1. We will not do the calculation showing that the Coulomb potential emerges, because, in fact, electrodynamics is rather subtle. The calculation will be done for mesons.

What Yukawa suggested in his extremely significant paper is that there exist a meson field, that is, a field that is different from the electromagnetic field, whose quanta, the *mesons* have a finite mass, which is coupled to protons and neutrons in a manner analogous to the coupling of photons to charged particles. The exchange of these quanta will then give rise to an interaction between nucleons. In this way, interactions as different as electromagnetism and the nuclear forces would at least share a common universal mechanism.[4] Let us go through a

---

[3] We assume momentum conservation in both points of view. They turn out to be completely equivalent.

[4] The idea that all interactions proceed through an "exchange" of quanta has gained such wide acceptance that even the weak interactions are believed by many people to be mediated by a "weak intermediate vector meson" whose properties are deduced from what is known about the weak interactions. Such a particle has not been discovered, but it is quite consistent to assume that it is very massive, which would explain why it has not been seen at existing accelerators.

very simple calculation assuming that the meson field is scalar, and that we can write the simple Hamiltonian for the nucleon in interaction

$$H = \frac{p^2}{2m} - g\phi(\mathbf{r},t)$$ (ST 5-3)

The meson field will be written in analogy to the vector potential in (22-27) as

$$\phi(\mathbf{r},t) = \left(\frac{2\pi c^2 \hbar}{\omega V}\right)^{1/2} e^{i\mathbf{k}\cdot\mathbf{r}} e^{-i\omega t} \quad \text{(absorption)}$$

$$+ \left(\frac{2\pi c^2 \hbar}{\omega V}\right)^{1/2} e^{-i\mathbf{k}\cdot\mathbf{r}} e^{i\omega t} \quad \text{(emission)}$$ (ST 5-4)

We are not in a position to justify all the factors. The appearance of ω in the normalization factor has the same source as in the photon problem; it comes from the fact that the energy of a meson quantum is $\hbar\omega$. The momentum of the meson is $\hbar\mathbf{k}$, and since the meson has a mass, μ, we now have the energy-momentum relation

$$(\hbar\omega)^2 = (\hbar\mathbf{k}c)^2 + (\mu c^2)^2$$ (ST 5-5)

Let us now calculate the various terms in the second-order energy shift (ST 5-2). When nucleon "1" emits the meson, we have

$$\langle N_1 + \text{meson} \,|g\phi|\, N_1\rangle = g\left(\frac{2\pi c^2 \hbar}{\omega V}\right)^{1/2} e^{-i\mathbf{k}\cdot\mathbf{r}_1} e^{i\omega t}$$ (ST 5-6)

No reference is made to nucleon "2", since it is unaffected by H_1 during the emission by "1." The absorption by "2" leaves "1" unaltered, and what enters is

$$\langle N_2|g\phi| N_2 + \text{meson}\rangle = g\left(\frac{2\pi c^2 \hbar}{\omega V}\right)^{1/2} e^{i\mathbf{k}\cdot\mathbf{r}_2} e^{-i\omega t}$$ (ST 5-7)

The energy denominator is

$$E_{\text{interm}} - E_{\text{initial}} = (E'_{N1} + E_{N2} + \hbar\omega) - (E_{N1} + E_{N2}) \simeq \hbar\omega$$ (ST 5-8)

E'_{N1} differs from E_{N1} since nucleon "1" recoils upon the emission of the meson. However, the recoil energy is $(\hbar k)^2/2M_N$ and this is generally small in the non-relativistic approximation, since the nucleon mass is so large. Hence the energy denominator is just $\hbar\omega$. Thus the energy shift is given by

$$\Delta E = -\sum \frac{2\pi c^2 \hbar}{\omega V} g^2 e^{i\mathbf{k}\cdot\mathbf{r}_2} e^{-i\mathbf{k}\cdot\mathbf{r}_1} \frac{1}{\hbar\omega}$$

The sum is over all meson momentum states, and as always, this means an integration over the phase space

$$\sum = \int \frac{V\, d^3\mathbf{p}}{(2\pi\hbar)^3} = \int \frac{V\, d^3\mathbf{k}}{(2\pi)^3}$$ (ST 5-10)

Hence

$$\Delta E = -\frac{2\pi c^2}{V}\frac{g^2}{(2\pi)^3}\,V\int d^3\mathbf{k}\,\frac{e^{i\mathbf{k}\cdot(\mathbf{r}_2-\mathbf{r}_1)}}{\omega^2}$$

$$= -\frac{g^2c^2}{4\pi^2}\int d^3\mathbf{k}\,\frac{e^{i\mathbf{k}\cdot(\mathbf{r}_2-\mathbf{r}_1)}}{\mathbf{k}^2c^2+(\mu c^2/\hbar)^2}$$

$$= -\frac{g^2}{4\pi^2}\int d^3\mathbf{k}\,\frac{e^{i\mathbf{k}\cdot(\mathbf{r}_2-\mathbf{r}_1)}}{\mathbf{k}^2+(\mu c/\hbar)^2} \qquad \text{(ST 5-11)}$$

Note that we did not take into account momentum conservation, but integrated over all momenta for the meson. The reason is that, in effect, we treat the nucleons as infinitely massive (they do not recoil and are always at $\mathbf{r}_1$ and $\mathbf{r}_2$, respectively) and thus any meson momentum is allowed in the intermediate state. Remember that this is a very crude calculation!

The integral can be done [it is in fact the three-dimensional Fourier transform of (24-87)], and it yields for

$$\Delta E = -\frac{g^2}{4\pi^2}\frac{(2\pi)^3}{4\pi}\frac{e^{-\mu c|\mathbf{r}_1-\mathbf{r}_2|/\hbar}}{|\mathbf{r}_1-\mathbf{r}_2|}$$

This should be doubled, because the nucleon "2" could be doing the emitting. Thus the energy change due to the meson field is

$$\Delta E = -g^2\frac{e^{-\mu c|\mathbf{r}_1-\mathbf{r}_2|/\hbar}}{|\mathbf{r}_1-\mathbf{r}_2|} \qquad \text{(ST 5-12)}$$

The energy depends on the separation, $\mathbf{r}$, and drops off fast for $r > \hbar/\mu c$. The range is therefore

$$a = \frac{\hbar}{\mu c} \qquad \text{(ST 5-13)}$$

Given that $a \simeq 1.4 \times 10^{-13}$ cm, we obtain

$$\mu c^2 = \frac{\hbar c}{a} \simeq \frac{10^{-27} \times 3 \times 10^{10}}{1.4 \times 10^{-13}}\ \text{ergs}$$

$$\simeq \frac{3 \times 10^{-17}}{1.4 \times 10^{-13}}\frac{1}{1.6 \times 10^{-6}}\ \text{MeV}$$

$$\simeq 130\ \text{MeV} \qquad \text{(ST 5-14)}$$

If the mesons are not scalar, but pseudoscalar, then a coupling like that shown in (ST 5-3) does not conserve parity, since the kinetic energy is even and the potential is odd under inversion. One must therefore make a scalar out of psuedoscalar meson field, *or its derivatives*. There are, on reflection, two alternatives; one is to have mesons always emitted in pairs, with the coupling

$$g_1 \phi^2(\mathbf{r},t) \tag{ST 5-15}$$

as for the pairwise emission of photons due to the term $(e^2/2mc^2)\mathbf{A}^2$; the other is to construct an axial vector (an axial vector is like a magnetic field), and dot it into the nucleon spin operator, so that the coupling would be[5]

$$f \frac{\hbar}{\mu c} \; \boldsymbol{\sigma} \cdot \boldsymbol{\nabla} \phi(\mathbf{r},t) \tag{ST 5-16}$$

This would allow single emission of mesons. Both couplings could, of course, be present. If the mesons are vector mesons, that is, essentially "heavy photons," then the coupling could be

$$-\frac{g}{Mc} \mathbf{p} \cdot \boldsymbol{\phi}(\mathbf{r},t) \tag{ST 5-17}$$

In all cases, however, the range is still $\hbar/\mu c$.

The mesons predicted by Yukawa were finally found in 1947. The long range part of the nucleon-nucleon force is due to pi-mesons, whose mass was found to be 140 MeV! They were found to be pseudoscalar, and, like photons, they can be emitted in collisions or transitions. The coupling (ST 5-16) explains a great deal about pion-nucleon scattering (the analog of the Compton effect) in the low energy region, and Yukawa's idea is fundamental to all the understanding we have about the strong interactions. In detail, much more has happened. There are also vector mesons (spin-parity 1^-) and spin 2 mesons, and many others. They can all be exchanged, and emitted, and, since the coupling to the nucleons is strong, they can be exchanged not just once, but many times. The calculations are beyond present-day mathematics, and it is a curious fact that the nuclear force problem, which started all this, is still less well understood than, for example, high-energy scattering. From our point of view, it is very important to note that even in this new realm of short distances and strong forces, there is no reason to suppose that quantum mechanics is not the correct way to describe nature.

---

[5] The factor $\hbar/\mu c$ is just there to make f dimensionless.

appendices

appendix A

The Fourier Integral and Delta Functions

Consider a function $f(x)$ that is periodic, with period $2L$, so that

$$f(x) = f(x + 2L) \qquad \text{(A-1)}$$

Such a function may be expanded in a Fourier Series in the interval $(-L, L)$, and the series has the form

$$f(x) = \sum_{n=0}^{\infty} A_n \cos \frac{n\pi x}{L} + \sum_{n=1}^{\infty} B_n \sin \frac{n\pi x}{L} \qquad \text{(A-2)}$$

We may rewrite the series in the form

$$f(x) = \sum_{n=-\infty}^{\infty} a_n \, e^{in\pi x/L} \qquad \text{(A-3)}$$

which is certainly possible, since

$$\cos \frac{n\pi x}{L} = \frac{1}{2} \left(e^{in\pi x/L} + e^{-in\pi x/L} \right)$$

$$\sin \frac{n\pi x}{L} = \frac{1}{2i} \left(e^{in\pi x/L} - e^{-in\pi x/L} \right)$$

The coefficients may be determined with the help of the orthonormality relation

$$\frac{1}{2L} \int_{-L}^{L} dx \, e^{in\pi x/L} \, e^{-im\pi x/L} = \delta_{mn} = \begin{cases} 1 & m = n \\ 0 & m \neq n \end{cases} \qquad \text{(A-4)}$$

Thus

$$a_n = \frac{1}{2L} \int_{-L}^{L} dx f(x) \, e^{-in\pi x/L} \qquad \text{(A-5)}$$

489

Let us now rewrite (A-3) by introducing Δn, the difference between two successive integers. Since this is unity, we have

$$f(x) = \sum_n a_n e^{in\pi x/L} \Delta n$$

$$= \frac{L}{\pi} \sum_n a_n e^{in\pi x/L} \frac{\pi \Delta n}{L} \tag{A-6}$$

Let us change the notation by writing

$$\frac{\pi n}{L} = k \tag{A-7}$$

and

$$\frac{\pi \Delta n}{L} = \Delta k \tag{A-8}$$

We also write

$$\frac{L a_n}{\pi} = \frac{A(k)}{\sqrt{2\pi}} \tag{A-9}$$

Hence (A-6) becomes

$$f(x) = \sum \frac{A(k)}{\sqrt{2\pi}} e^{ikx} \Delta k \tag{A-10}$$

If we now let $L \rightarrow \infty$, then k approaches a continuous variable, since Δk becomes infinitesimally small. If we recall the Riemann definition of an integral, we see that in the limit (A-10) may be written in the form

$$f(x) = \frac{1}{\sqrt{2\pi}} \int_{-\infty}^{\infty} A(k) e^{ikx} dk \tag{A-11}$$

The coefficient $A(k)$ is given by

$$A(k) = \sqrt{2\pi} \frac{L}{\pi} \cdot \frac{1}{2L} \int_{-L}^{L} dx f(x) e^{-in\pi x/L}$$

$$\rightarrow \frac{1}{\sqrt{2\pi}} \int_{-\infty}^{\infty} dx f(x) e^{-ikx} \tag{A-12}$$

Equations A-11 and A-12 define the Fourier integral transformations. If we insert the second equation into the first we get

$$f(x) = \frac{1}{2\pi} \int_{-\infty}^{\infty} dk \, e^{ikx} \int_{-\infty}^{\infty} dy f(y) e^{-iky} \tag{A-13}$$

Suppose now that we interchange, without question, the order of integrations. We then get

$$f(x) = \int_{-\infty}^{\infty} dy f(y) \left[\frac{1}{2\pi} \int_{-\infty}^{\infty} dk \, e^{ik(x-y)} \right] \qquad \text{(A-14)}$$

For this to be true, the quantity $\delta(x - y)$ defined by

$$\delta(x - y) = \frac{1}{2\pi} \int_{-\infty}^{\infty} dk \, e^{ik(x-y)} \qquad \text{(A-15)}$$

and called the *Dirac Delta function* must be a very peculiar kind of function; it must vanish when $x \neq y$, and it must tend to infinity in an appropriate way when $x - y = 0$, since the range of integration is infinitesimally small. It is therefore not a function in the usual mathematical sense, but it is rather a "generalized function" or a "distribution."[1] It does not have any meaning by itself, but it can be defined provided it always appears in the form

$$\int dx f(x) \, \delta(x - a)$$

with the function $f(x)$ sufficiently smooth in the range of values that the argument of the delta function takes. We will take that for granted and manipulate the delta function by itself, with the understanding that at the end all the relations that we write down only occur under the integral sign.

The following properties of the delta function can be demonstrated:

(i)

$$\delta(ax) = \frac{1}{|a|} \, \delta(x) \qquad \text{(A-16)}$$

This can be seen to follow from

$$f(x) = \int dy f(y) \, \delta(x - y) \qquad \text{(A-17)}$$

If we write $x = a\xi$ and $y = a\eta$, then this reads

$$f(a\xi) = |a| \int d\eta f(a\eta) \, \delta[a(\xi - \eta)]$$

On the other hand,

$$f(a\xi) = \int d\eta f(a\eta) \, \delta(\xi - \eta)$$

which implies our result.

---

[1] The theory of distributions was developed by the mathematician Laurent Schwartz. An introductory treatment may be found in M. J. Lighthill, *Introduction to Fourier Analysis and Generalized Functions*, Cambridge University Press (1958).

(ii) A relation that follows from (A-16) is

$$\delta(x^2 - a^2) = \frac{1}{2|a|} [\delta(x - a) + \delta(x + a)] \qquad \text{(A-18)}$$

This follows from the fact that the argument of the delta function vanishes at $x = a$ and $x = -a$. Thus there are two contributions:

$$\delta(x^2 - a^2) = \delta[(x - a)(x + a)]$$

$$= \frac{1}{|x + a|} \delta(x - a) + \frac{1}{|x - a|} \delta(x + a)$$

$$= \frac{1}{2|a|} [\delta(x - a) + \delta(x + a)]$$

More generally, one can show that

$$\delta[f(x)] = \sum_i \frac{\delta(x - x_i)}{|df/dx|_{x=x_i}} \qquad \text{(A-19)}$$

where the x_i are the roots of $f(x)$ in the interval of integration.

In addition to the representation (A-15) of the delta function, there are other representations that may prove useful. We discuss several of them.

(a) Consider the form (A-15), which we write in the form

$$\delta(x) = \frac{1}{2\pi} \lim_{L \to \infty} \int_{-L}^{L} dk\, e^{ikx} \qquad \text{(A-20)}$$

The integral can be done, and we get

$$\delta(x) = \lim_{L \to \infty} \frac{1}{2\pi} \frac{e^{iLx} - e^{-iLx}}{ix}$$

$$= \lim_{L \to \infty} \frac{\sin Lx}{\pi x} \qquad \text{(A-21)}$$

(b) Consider the function $\Delta(x,a)$ defined by

$$\Delta(x,a) = 0 \qquad x < -a$$

$$= \frac{1}{2a} \qquad -a < x < a \qquad \text{(A-22)}$$

$$= 0 \qquad a < x$$

Then

$$\delta(x) = \lim_{a \to 0} \Delta(x,a) \qquad \text{(A-23)}$$

It is clear that an integral of a product of $\Delta(x,a)$ and a function $f(x)$ that is smooth near the origin will pick out the value at the origin

$$\operatorname*{Lim}_{a\to 0} \int dx f(x)\Delta(x,a) = f(0) \operatorname*{Lim}_{a\to 0} \int dx\Delta(x,a)$$

$$= f(0)$$

(c) By the same token, any peaked function, normalized to unit area under it, will approach a delta function in the limit that the width of the peak goes to zero. We will leave it to the reader to show that the following are representations of the delta function

$$\delta(x) = \operatorname*{Lim}_{a\to 0} \frac{1}{\pi} \frac{a}{x^2 + a^2} \tag{A-24}$$

and

$$\delta(x) = \operatorname*{Lim}_{\alpha\to\infty} \frac{\alpha}{\sqrt{\pi}} e^{-\alpha^2 x^2} \tag{A-25}$$

(d) We will have occasion to deal with *orthonormal polynomials*, which we denote by the general symbol $P_n(x)$. These have the property that

$$\int dx P_m(x)\, P_n(x)\, w(x) = \delta_{mn} \tag{A-26}$$

where $w(x)$ may be unity or some simple function, called the weight function. For functions that may be expanded in a series of these orthogonal polynomials, we can write

$$f(x) = \sum_n a_n P_n(x) \tag{A-27}$$

If we multiply both sides by $w(x)P_m(x)$ and integrate over x, we find that

$$a_m = \int dy w(y)\, f(y)\, P_m(y) \tag{A-28}$$

If we insert this into (A-27) and prepared to deal with "generalized functions," we freely interchange sum and integral, we get

$$f(x) = \sum_n P_n(x) \int dy w(y)\, f(y)\, P_n(y)$$

$$= \int dy f(y) \left(\sum_n P_n(x)\, w(y)\, P_n(y) \right) \tag{A-29}$$

Thus we get still another representation of the delta function. Examples of the $P_n(x)$ are Legendre polynomials, Hermite polynomials, and Laguerre polynomials, all of which make their appearance in quantum mechanical problems.

Since the delta function always appears multiplied by a smooth function under an integral sign, we can give meaning to its derivatives. For example

$$\int_{-\epsilon}^{\epsilon} dx f(x) \frac{d}{dx} \delta(x) = \int_{-\epsilon}^{\epsilon} dx \frac{d}{dx} [f(x) \delta(x)] - \int_{-\epsilon}^{\epsilon} dx \frac{df(x)}{dx} \delta(x)$$

$$= - \int_{-\epsilon}^{\epsilon} dx \frac{df(x)}{dx} \delta(x)$$

$$= - \left(\frac{df}{dx} \right)_{x=0} \tag{A-30}$$

and so on. The delta function is an extremely useful tool, and the student will encounter it in every part of mathematical physics.

The integral of a delta function is

$$\int_{-\infty}^{x} dy\, \delta(y - a) = 0 \qquad x < a$$

$$= 1 \qquad x > a \tag{A-31}$$

$$\equiv \theta(x - a)$$

which is the standard notation for this discontinuous function. Conversely, the derivative of the so-called *step function* is the Dirac delta function:

$$\frac{d}{dx} \theta(x - a) = \delta(x - a) \tag{A-32}$$

appendix B

Operators

In this appendix we discuss some topics related to linear operators. The set of admissible wave packets are square integrable functions. Since

$$\psi(x) = \alpha\psi_1(x) + \beta\psi_2(x) \tag{B-1}$$

is square integrable, if $\psi_1(x)$ and $\psi_2(x)$ are square integrable and α,β are arbitrary complex numbers, we say that the ψ's form a *linear space*. An operator A on this space is a mapping:

$$A\psi(x) = \phi(x) \tag{B-2}$$

where $\phi(x)$ is also square integrable. Among all the operators there is a subset called *linear operators*, which have the property that

$$A\alpha\psi(x) = \alpha A\psi(x) \tag{B-3}$$

where α is an arbitrary complex constant, and

$$A[\alpha\psi_1(x) + \beta\psi_2(x)] = \alpha A\psi_1(x) + \beta A\psi_2(x). \tag{B-4}$$

with α,β being complex numbers. A further subset is the *hermitian operators* for which the expectation value for all admissible $\psi(x)$,

$$\langle A \rangle_\psi = \int dx\psi^*(x)\, A\psi(x) \tag{B-5}$$

is real. First we prove that for all admissible ψ_1 and ψ_2

$$\int \psi_2^*(x)\, A\psi_1(x)\, dx = \int [A\psi_2(x)]^*\, \psi_1(x)\, dx \tag{B-6}$$

holds.

The reality of $\langle A \rangle$ implies that

$$\int dx\psi^*(x)\, A\psi(x) = \int dx\, [A\psi(x)]^*\, \psi(x) \tag{B-7}$$

495

Now substitute for $\psi(x)$

$$\psi(x) = \psi_1(x) + \lambda\psi_2(x) \tag{B-8}$$

This implies that

$$\int dx(\psi_1^* + \lambda^*\psi_2)\, A(\psi_1 + \lambda\psi_2) = \int dx(\psi_1 + \lambda\psi_2)\,(A\psi_1 + \lambda A\psi_2)^* \tag{B-9}$$

Using hermiticity, that is,

$$\int dx\psi_i^* A\psi_i = \int dx\psi_i(A\psi_i)^* \qquad i = 1.2 \tag{B-10}$$

we obtain

$$\lambda^* \int \psi_2^* A\psi_1 + \lambda \int \psi_1^* A\psi_2 = \lambda \int \psi_2(A\psi_1)^* + \lambda^* \int \psi_1(A\psi_2)^* \tag{B-11}$$

Since λ is an arbitrary complex number, the relations for the coefficient of λ and for the coefficient of λ^* must separately hold. Thus

$$\int dx\psi_2^* A\psi_1 = \int dx(A\psi_2)^*\,\psi_1 \tag{B-12}$$

The next result that we wish to prove is that *eigenfunctions of a hermitian operator corresponding to different eigenvalues are orthogonal.* Consider the two equations

$$A\psi_1(x) = a_1\psi_1(x)$$

and

$$[A\psi_2(x)]^* = a_2\psi_2^*(x) \tag{B-13}$$

Note that a_2 is real since the eigenvalues of a hermitian operator are real. Take the scalar product of the first equation with ψ_2^* and the second equation with ψ_1. Thus

$$\int dx\psi_2^* A\psi_1(x) = a_1 \int \psi_2^*(x)\,\psi_1(x)\,dx$$

$$\int dx(A\psi_2)^*\,\psi_1(x) = a_2 \int \psi_2^*(x)\,\psi_1(x)\,dx \tag{B-14}$$

Subtracting, we get

$$(a_1 - a_2) \int \psi_2^*(x)\,\psi_1(x)\,dx = \int dx\psi_2^* A\psi_1 - \int dx(A\psi_2)^*\,\psi_1$$

$$= 0 \tag{B-15}$$

Thus, if $a_1 \neq a_2$, we have

$$\int \psi_2^*(x)\,\psi_1(x)\,dx = 0 \tag{B-16}$$

If we define the hermitian conjugate of the operator A by $A^\dagger$, so that

$$\int dx (A\psi_2)^* \, \psi_1 \equiv \int dx \psi_2^* A^\dagger \psi_1 \tag{B-17}$$

then for a hermitian operator

$$A = A^\dagger \tag{B-18}$$

We can prove that

$$(AB)^\dagger = B^\dagger A^\dagger \tag{B-19}$$

To do so, we note that

$$\int \psi_2^* (AB)^\dagger \, \psi_1 = \int (AB\psi_2)^* \, \psi_1$$

$$\int (B\psi_2)^* \, (A^\dagger \psi_1)$$

$$\int \psi_2^* B^\dagger (A^\dagger \psi_1)$$

$$\int \psi_2^* B^\dagger A^\dagger \psi_1 \tag{B-20}$$

A generalization of this is

$$(ABC \ldots Z)^\dagger = Z^\dagger \ldots C^\dagger B^\dagger A^\dagger \tag{B-21}$$

Thus, a product of two hermitian operators is only hermitian if the two operators commute:

$$(AB)^\dagger = B^\dagger A^\dagger = BA = AB + [B,A] \tag{B-22}$$

Another result is that for any operator A, the following

$$A + A^\dagger$$
$$i(A - A^\dagger) \tag{B-23}$$
$$AA^\dagger$$

will be hermitian.

Next we prove the "uncertainty relations." We define

$$(\Delta A)^2 = \langle A^2 \rangle - \langle A \rangle^2 = \langle (A - \langle A \rangle)^2 \rangle \tag{B-24}$$

Let

$$U = A - \langle A \rangle$$
$$V = B - \langle B \rangle \tag{B-25}$$

and consider

$$\phi = U\psi + i\lambda V\psi \tag{B-26}$$

Then

$$I(\lambda) = \int dx \phi^* \phi \geq 0 \tag{B-27}$$

With A and B hermitian, so are U and V. We may thus rewrite:

$$\begin{aligned}
I(\lambda) &= \int dx (U\psi + i\lambda V\psi)^* (U\psi + i\lambda V\psi) \\
&= \int dx (U\psi)^* (U\psi) + \lambda^2 \int dx (V\psi)^* (V\psi) \\
&\quad + i\lambda \int dx [(U\psi)^* (V\psi) - (V\psi)^* (U\psi)] \\
&= \int dx \psi^* (U^2 + \lambda^2 V^2 + i\lambda [U,V]) \psi \\
&= (\Delta A)^2 + \lambda^2 (\Delta B)^2 + i\lambda \int dx \psi^* [U,V] \psi \geq 0 \\
&= (\Delta A)^2 + \lambda^2 (\Delta B)^2 + i\lambda \langle [A,B] \rangle \tag{B-28}
\end{aligned}$$

The minimum will occur when

$$2\lambda (\Delta B)^2 + i \langle [A,B] \rangle = 0 \tag{B-29}$$

Substituting the solution

$$\lambda = -i \frac{\langle [A,B] \rangle}{2 (\Delta B)^2} \tag{B-30}$$

into $I(\lambda)$, we get

$$(\Delta A)^2 - \frac{\langle [A,B] \rangle^2}{4(\Delta B)^2} + \frac{\langle [A,B] \rangle^2}{2(\Delta B)^2} \geq 0$$

that is,

$$(\Delta A)^2 (\Delta B)^2 \geq \tfrac{1}{4} \langle i[A,B] \rangle^2 \tag{B-31}$$

Incidentally, the minimum value occurs when ψ is such that $U\psi$ and $V\psi$ are proportional to each other. For the case of the operators x and p, this means that

$$\frac{\hbar}{i} \frac{d\psi(x)}{dx} + i\beta x \psi(x) = 0 \tag{B-32}$$

whose solution is

$$\psi(x) = C e^{-\beta(x^2/2\hbar)} \tag{B-33}$$

a ground state eigenfunction of the harmonic oscillator. It is important to note that the uncertainty relation

$$(\Delta A)^2 (\Delta B)^2 \geq \tfrac{1}{4} (\langle i[A,B] \rangle)^2 \qquad \text{(B-34)}$$

was derived without any use of wave concepts or the reciprocity between a wave form and its fourier transform. The results depends entirely on the operator properties of the observables A amd B.

We conclude the appendix by listing some properties of commutators.
(i)

$$[A,B] = -[B,A] \qquad \text{(B-35)}$$

(ii)

$$\begin{aligned} [A,B]^\dagger &= (AB)^\dagger - (BA)^\dagger \\ &= B^\dagger A^\dagger - A^\dagger B^\dagger \\ &= [B^\dagger, A^\dagger] \end{aligned} \qquad \text{(B-36)}$$

(iii) If A and B are hermitian, so is $i[A,B]$. This follows directly from the preceding properties.
(iv)

$$\begin{aligned} [AB,C] &= ABC - CAB \\ &= ABC - ACB + ACB - CAB \\ &= A[B,C] + [A,C]B \end{aligned} \qquad \text{(B-37)}$$

(v) It may be shown term by term that

$$e^A B e^{-A} = B + [A,B] + \frac{1}{2!}[A,[A,B]] + \frac{1}{3!}\Big[A,[A,[A,B]]\Big] + \dots \qquad \text{(B-38)}$$

This is known as the Baker-Hausdorff lemma, and is of some utility in manipulations of operators.

(vi) It is easily established that

$$[A,[B,C]] + [B,[C,A]] + [C,[A,B]] = 0 \qquad \text{(B-39)}$$

This is called the Jacobi identity.

A more extensive discussion of operators and the linear spaces that they are defined on may be found in J. D. Jackson, *Mathematics for Quantum Mechanics*, W. A. Benjamin, Inc., New York (1962).

references[1]

G. Baym, *Lectures on Quantum Mechanics*, W. A. Benjamin, Inc., New York, 1969.

This is a very appealing book, with just the right mixture of formalism, intuitive arguments, and applications. It should be considered an advanced book, accessible to the student who has covered the material in this book.

D. B. Beard and G. B. Beard, *Quantum Mechanics with Applications*, Allyn and Bacon, Inc., Boston, 1970.

A number of quantum mechanics topics are developed around particular applications. By itself, the book is probably not very useful for self-study, but the variety of applications make it a useful reference book. The demands made by the book are probably comparable to those of this book.

R. Becker, *Electromagnetic Fields and Interactions*, Blaisdell Publishing Co., New York, 1964.

This is not really a textbook on quantum mechanics, but about half of Volume 2 is devoted to the description of properties of matter with the help of quantum mechanics. Certain applications are discussed, and the student may find this a useful collateral reference, on the level of the present book.

H. A. Bethe and R. W. Jackiw, *Intermediate Quantum Mechanics* (second edition), W. A. Benjamin, Inc., New York, 1968.

This book contains detailed discussions of calculational methods applicable to the theory of atomic structure, multiplet splittings, the photoelectric effect, and atomic collisions. Much of the material is not to be found in any other textbook. The book is thus an advanced text, as well as an exhaustive reference book.

H. A. Bethe and E. E. Salpeter, *Quantum Mechanics of One- and Two-Electron Atoms*, Springer Verlag, 1957.

This reprint of the authors' article in the *Handbuch der Physik* is an elaborate, detailed, definitive treatment of the problem at hand. It is a book about atoms and not about quantum mechanics, and the level is high. It is an excellent reference book.

D. Bohm, *Quantum Theory*, Prentice-Hall, Inc., 1951.

The book is discursively written, on a level comparable to the present book. The author pays much attenton to the principles of quantum theory, and gives an excellent discussion of the quantum theory of the measurement process. There are few applications and not many problems.

[1] Many books have been written about quantum mechanics. I have studied from some of them, read others, glanced at a few, and probably missed some others. The list given here is not meant to be exhaustive, and no book is criticized by its omission. In particular, no books on quantum chemistry are listed.

S. Borowitz, *Fundamentals of Quantum Mechanics*, W. A. Benjamin, Inc., New York, 1967.

This is a well-written book, about half of which is devoted to the theory of waves and to classical mechanics. The level is comparable to that of the present book.

E. U. Condon and G. H. Shortley, *The Theory of Atomic Spectra*, Cambridge University Press, Cambridge, 1959.

This is a very detailed reference book on all aspects of atomic spectra, although it does not make use of the recent techniques that depend on group theory. The book is very advanced, and thus its shortcomings in the technical developments are not important for anybody but the specialist. It is very useful for the student.

A. S. Davydov, *Quantum Mechanics*, Addison-Wesley Publishing Co., Inc., 1965.

This is an advanced, comprehensive textbook. The book is a little weak on fundamentals, but treats many physical systems. There are excellent discussions of relativistic equations, group theory, second quantization, and some aspects of solid-state theory.

R. H. Dicke and J. P. Wittke, *Introduction to Quantum Mechanics*, Addison-Wesley Publishing Co., Inc., 1960.

I enjoyed this book very much. It is on a level comparable to the present book, and discusses a few topics, notably quantum statistics, that are not treated here. The problems are excellent.

P. A. M. Dirac, *The Principles of Quantum Mechanics* (fourth edition), Oxford, Clarendon Press, 1958.

This is a superb book by one of the major creators of quantum mechanics. The student who has studied the material in this book will have no trouble with Dirac; if he is at all serious about mastering quantum mechanics, he should sooner or later go through Dirac's book.

R. P. Feynman and A. R. Hibbs, *Quantum Mechanics and Path Integrals*, McGraw-Hill Book Co., 1965.

In 1948, R. P. Feynman proposed a different formulation of quantum mechanics. In this book the equivalence of this formulation to the standard theory is demonstrated, and the "path integral" expression for the general amplitude is exploited in a number of calculations. The selection of material is very interesting, and the point of view is different from the one developed by the author. Thus this somewhat more advanced book presents an excellent complement to this book.

R. P. Feynman, R. B. Leighton, and M. Sands, *The Feynman Lectures on Physics, Vol. 3, Quantum Mechanics*, Addison-Wesley Publishing Co., Inc., 1965.

In this introduction to quantum mechanics, Feynman abandons the path integral and approaches the subject from the point of view of state vectors. A large number of fascinating examples are discussed with the minimum of formal apparatus. A superb complementary book, whose only short-coming is the absence of problems.

K. Gottfried, *Quantum Mechanics, Vol. 1, Fundamentals*, W. A. Benjmain, Inc., 1966.

This is a very advanced book, distinguished by the care with which the various topics are discussed. The treatment of the measurement process and of invariance principles is excellent. The student who has mastered the material in this book should be able to read Gottfried's book, provided he has acquired the necessary mathematical equipment.

W. Heisenberg, *The Physical Principles of the Quantum Theory*, Dover Publications, Inc., 1930.

This reprint of some 1930 lectures given by Heisenberg on the physical sig-nificance of the quantum theory still makes good reading. The discussion of the uncertainty relations is particularly useful.

F. A. Kaempffer, *Concepts in Quantum Mechanics*, Academic Press, 1965.

This is not a textbook in any sense. A variety of topics are discussed. The selection of topics is imaginative, and the discussion informative. The level is somewhat above that of the present book.

H. A. Kramers, *Quantum Mechanics*, Interscience Publishers, Inc., 1957.

This book by one of the founders of the subject is at its best in the discussion of spin and the introduction to relativistic quantum theory, both rather advanced subjects. The student who is comfortable with quantum me-chanics will find browsing through this book enjoyable and rewarding.

L. D. Landau and E. M. Lifshitz, *Quantum Mechanics (Nonrelativistic Theory)* (second edition), Addison-Wesley Publishing Co., Inc., 1965.

The book by Landau and Lifshitz is one of a series of superb books covering all of theoretical physics. It is hard to think of this as a textbook for any but the most sophisticated students. Any student, however, once he reaches the advanced level, will find much that is useful in this book. There is an assumed mathematical facility on the part of the student.

F. Mandl, *Quantum Mechanics*, Butterworths Scientific Publications, London, 1957.

This book contains a good discussion of the foundation of wave mechanics at a level that is comparable with the present book.

A. Messiah, *Quantum Mechanics* (in 2 volumes), John Wiley and Sons, Inc., 1968.

This book gives a complete coverage of quantum theory from the treatment of one-dimensional potentials through the quantization of the electro-magnetic field and the relativistic wave equation of Dirac. It is an advanced book, and it assumes a mathematical sophistication that few first year graduate students possess. It is an extremely worthwhile book.

E. Merzbacher, *Quantum Mechanics* (second edition) John Wiley and Sons, Inc., 1970.

Together with the book by Schiff, this is the standard first year graduate textbook, and deservedly so. The complete range of concepts and phenomena is treated with economy and taste. The book should be available to the student who has gone through the material in this book.

D. Park, *Introduction to the Quantum Theory*, McGraw-Hill Co., 1964.

This attractive book is written on the same level as the present book. Among the topics discussed by Park, and absent in this volume, is the subject of quantum statistics, which is treated with clarity.

W. Pauli, *Die Allgemeinen Prinzipien der Wellenmechanik*, *Handbuch der Physik*, Vol. 5/1, Springer Verlag, 1958.

The advanced student who reads German will find in this reprint of a 1930 article by Pauli a concise definitive discussion of quantum mechanics. There are no applications, but all of the important matters are there.

J. L. Powell and B. Crasemann, *Quantum Mechanics*, Addison-Wesley Publishing Co., Inc., 1961.

The strength of this book is in the painstaking working out of all of the mathematical details of wave mechanics and matrix mechanics. Probably all of the mathematical aspects of these subjects that have been bypassed in our book can be found here. There is a good discussion of the WKB approximation and of the general properties of second order differential equations. There are relatively few applications, and there are more exercises than problems.

M. E. Rose, *Elementary Theory of Angular Momentum*, John Wiley & Sons, Inc., 1937.

An advanced treatment of angular momentum and the many applications to atomic and nuclear physics.

J. J. Sakurai, *Advanced Quantum Mechanics*, Addison-Wesley, Inc., 1967.

An excellent book at a level just above Merzbacher and Schiff. For the advanced student.

D. S. Saxon, *Elementary Quantum Mechanics*, Holden-Day, San Francisco, 1968.

This book is on the same level as the present one, and it is a useful reference, since the selection of topics is just a little different, as is the emphasis and the choice of applications. The book contains an excellent set of problems.

L. I. Schiff, *Quantum Mechanics* (third edition), McGraw-Hill Book Co., 1968.

This is one of the standard first year graduate textbooks. It is perhaps a little too compact, and thus most suitable for the well-prepared student. The level of mathematical sophistication assumed is above that of the reader of the present book.

D. ter Haar, *Selected Problems in Quantum Mechanics*, Academic Press, Inc., New York, 1964.

Both the student and the teacher will find this a very useful volume. Most of the problems are on the level of this book. There are a few other books worth listing as part of the general reference list. These are books that deal in detail with some of the topics that are touched on illustratively in this volume.

B. L. Cohen, *Concepts of Nuclear Physics*, McGraw-Hill Book Co., 1972 (for nuclear physics).

C. Kittel, *Introduction to Solid State Physics* (fourth edition), John Wiley & Sons, Inc., New York, 1971 (for solid-state physics).

E. Segre, *Nuclei and Particles*, W. A. Benjamin, Inc., New York, 1964. (for experimental aspects of nuclear and particle physics).

D. H. Perkins, *Introduction to High Energy Physics*, Addison-Welsley Publishing Co., 1972.

M. Abramowitz and I. A. Stegun (Eds.), *Handbook of Mathematical Functions*, National Bureau of Standards, 1964.

W. Magnus and F. Oberhettinger, *Formulas and Theorems for the Special Functions of Mathematical Physics*, Chelsea Publishing Co., New York, 1949.

A. Erdelyi, W. Magnus, F. Oberhettinger, and F. G. Tricomi (Eds.), *Higher Transcendental Functions* (The Bateman Project), Volumes 1, 2, and 3, McGraw-Hill Book Co., New York, 1953.

I. S. Gradshteyn and I. M. Ryzhik, *Tables of Integrals, Series and Products*, Academic Press, New York, 1965.

physical constants[1]

N_0 (Avogadro's number)	$= 6.022169(40) \times 10^{23}$ mole^{-1}
c (velocity of light)	$= 2.9979250(10) \times 10^{10}$ cm sec^{-1}
e (electron charge)	$= 4.803250(21) \times 10^{-10}$ esu
	$= 1.6021917(70) \times 10^{-19}$ coulomb
1 MeV	$= 1.6021917(70) \times 10^{-6}$ erg
$\hbar$ (Planck constant/2π)	$= 6.582183(22) \times 10^{-22}$ MeV sec
	$= 1.0545919(80) \times 10^{-27}$ erg sec
α (fine structure constant $e^2/\hbar c$)	$= 1/137.03602(21)$
k (Boltzmann constant)	$= 1.380622(59) \times 10^{-16}$ erg K^{-1}
m_e (electron mass)	$= 9.109558(54) \times 10^{-28}$ gm
	$= 0.5110041(16)$ MeV/c^2
m_p (proton mass)	$= 938.2592(52)$ MeV/c^2
m_p/m_e	$= 1836.109(11)$
1 amu $(1/12 \times m_{C^{12}})$	$= 931.4812(52)$ MeV/c^2
$a_0 = (\hbar/m_e c\alpha)$	$= 0.52917715(81) \times 10^{-8}$ cm
R_∞ $(= m_e c^2 \alpha^2/2)$	$= 13.605826(45)$ eV = 1 Rydberg
G (gravitational constant)	$= 6.6732(31) \times 10^{-8}$ cm^3 gm^{-1} sec^{-2}

$$\mu_{\text{Bohr}} \text{ (Bohr magneton)} = \frac{e\hbar}{2m_e c} = 0.5788381(18) \times 10^{-14} \text{ MeV gauss}^{-1}$$

[1] Compiled by Stanley J. Brodsky, as presented in *Reviews of Modern Physics*, 45,(2), Part II. The figures in parentheses correspond to the statistical uncertainty (one standard deviation) in the last digits of the main number.

index

509